VOLUME ONE HUNDRED AND FIFTY EIGHT

INTERNATIONAL REVIEW OF NEUROBIOLOGY

What does Medial Frontal Cortex Signal During Behavior? Insights from Behavioral Neurophysiology

INTERNATIONAL REVIEW OF NEUROBIOLOGY

VOLUME 158

SERIES EDITOR

PATRICIA JANAK

Janak Lab, Department of Neuroscience
Dunning Hall, John Hopkins University
Baltimore, MD, USA

PETER JENNER

Division of Pharmacology and Therapeutics
GKT School of Biomedical Sciences
King's College, London, UK

VOLUME ONE HUNDRED AND FIFTY EIGHT

INTERNATIONAL REVIEW OF NEUROBIOLOGY

What does Medial Frontal Cortex Signal During Behavior? Insights from Behavioral Neurophysiology

Edited by

ADAM T. BROCKETT
Department of Psychology;
Program in Neuroscience and Cognitive Science,
University of Maryland, College Park, MD, United States

LINDA M. AMARANTE
Department of Neuroscience, Johns Hopkins University,
Baltimore, MD, United States

MARK LAUBACH
Department of Neuroscience, American University,
Washington, DC, United States

MATTHEW R. ROESCH
Department of Psychology;
Program in Neuroscience and Cognitive Science,
University of Maryland, College Park, MD, United States

ELSEVIER

ACADEMIC PRESS
An imprint of Elsevier

Academic Press is an imprint of Elsevier
50 Hampshire Street, 5th Floor, Cambridge, MA 02139, United States
525 B Street, Suite 1650, San Diego, CA 92101, United States
The Boulevard, Langford Lane, Kidlington, Oxford OX5 1GB, United Kingdom
125 London Wall, London, EC2Y 5AS, United Kingdom

First edition 2021

ISBN: 978-0-323-85336-1
ISSN: 0074-7742

For information on all Academic Press publications
visit our website at https://www.elsevier.com/books-and-journals

Publisher: Zoe Kruze
Acquisitions Editor: Sam Mahfoudh
Developmental Editor: Shellie Bryant
Production Project Manager: Abdulla Sait
Cover Designer: Victoria Pearson

Typeset by SPi Global, India

Contents

Contributors

Linda M. Amarante
Department of Neuroscience, Johns Hopkins University, Baltimore, MD, United States

Bilal A. Bari
The Solomon H. Snyder Department of Neuroscience, Brain Science Institute, Kavli Neuroscience Discovery Institute, Johns Hopkins University, Baltimore, MD, United States

Adam T. Brockett
Department of Psychology; Program in Neuroscience and Cognitive Science, University of Maryland, College Park, MD, United States

Jessica P. Caballero
Neuroscience and Behavior Graduate Program, University of Massachusetts Amherst, Amherst, MA, United States

Marcelo S. Caetano
Center for Mathematics, Computing, and Cognition, Universidade Federal do ABC (UFABC), Santo André, Brazil

Marie Carlén
Department of Biosciences and Nutrition; Department of Neuroscience, Karolinska Institutet, Stockholm, Sweden

Steve W.C. Chang
Department of Psychology, Yale University; Department of Neuroscience; Kavli Institute for Neuroscience, Yale University School of Medicine, New Haven, CT, United States

Jeremiah Y. Cohen
The Solomon H. Snyder Department of Neuroscience, Brain Science Institute, Kavli Neuroscience Discovery Institute, Johns Hopkins University, Baltimore, MD, United States

Austin A. Coley
Salk Institute for Biological Studies, La Jolla, CA, United States

Bruno Delord
Institute of Intelligent Systems and Robotics (ISIR), Sorbonne Université, Centre National de la Recherche Scientifique, UMR 7222, Paris, France

Geoffrey W. Diehl
Department of Neuroscience, University of Minnesota, Minneapolis, MN, United States

Ilka Diester
Optophysiology, University of Freiburg, Faculty of Biology, BrainLinks-BrainTools, Bernstein Center, Intelligent Machine-Brain Interfacing Technology (IMBIT), Freiburg, Germany

Vincent Fontanier
Univ Lyon, Université Claude Bernard Lyon 1, Inserm, Stem Cell and Brain Research Institute U1208, Bron, France

Clément Goussi
Univ Lyon, Université Claude Bernard Lyon 1, Inserm, Stem Cell and Brain Research Institute U1208, Bron, France

Stefanie Hardung
Optophysiology, University of Freiburg, Faculty of Biology, Freiburg, Germany

Benjamin Y. Hayden
Department of Neuroscience, University of Minnesota, Minneapolis, MN, United States

Nicole K. Horst
Centre for Teaching and Learning and Postdoc Academy, University of Cambridge, Cambridge, England

Zoe Jäckel
Optophysiology, University of Freiburg, Faculty of Biology, Freiburg, Germany

David S. Jacobs
Department of Behavioral Neuroscience, Oregon Health & Science University, Portland, OR, United States

Felix Jung
Department of Biosciences and Nutrition, Karolinska Institutet, Stockholm, Sweden

Beata Kaminska
Neuroscience and Behavior Graduate Program, University of Massachusetts Amherst, Amherst, MA, United States

Rachel Knoebl
Department of Neuroscience, University of Minnesota, Minneapolis, MN, United States

Alex C. Kwan
Department of Psychiatry; Department of Neuroscience, Yale University School of Medicine, New Haven, CT, United States

Mark Laubach
Department of Neuroscience, American University, Washington, DC, United States

Amber E. McLaughlin
Department of Neuroscience, University of Minnesota, Minneapolis, MN, United States

Bita Moghaddam
Department of Behavioral Neuroscience, Oregon Health & Science University, Portland, OR, United States

David E. Moorman
Neuroscience and Behavior Graduate Program; Department of Psychological and Brain Sciences, University of Massachusetts Amherst, Amherst, MA, United States

Nandakumar S. Narayanan
Department of Neurology, University of Iowa, Iowa City, IA, United States

Nancy Padilla-Coreano
Salk Institute for Biological Studies, La Jolla, CA, United States

Reesha Patel
Salk Institute for Biological Studies, La Jolla, CA, United States

Emmanuel Procyk
Univ Lyon, Université Claude Bernard Lyon 1, Inserm, Stem Cell and Brain Research Institute U1208, Bron, France

Philip T. Putnam
Department of Psychology, Yale University, New Haven, CT, United States

A. David Redish
Department of Neuroscience, University of Minnesota, Minneapolis, MN, United States

Erin L. Rich
Nash Family Department of Neuroscience and Friedman Brain Institute, Icahn School of Medicine at Mount Sinai, New York, NY, United States

Matthew R. Roesch
Department of Psychology; Program in Neuroscience and Cognitive Science, University of Maryland, College Park, MD, United States

Peter H. Rudebeck
Nash Family Department of Neuroscience and Friedman Brain Institute, Icahn School of Medicine at Mount Sinai, New York, NY, United States

Matthieu Sarazin
Institute of Intelligent Systems and Robotics (ISIR), Sorbonne Université, Centre National de la Recherche Scientifique, UMR 7222, Paris, France

Jeremy K. Seamans
Department of Psychiatry, University of British Columbia, Vancouver, BC, Canada

Joseph Simon IV
Nash Family Department of Neuroscience and Friedman Brain Institute, Icahn School of Medicine at Mount Sinai, New York, NY, United States

Kay M. Tye
Salk Institute for Biological Studies, La Jolla, CA, United States

Benjamin Voloh
Department of Neuroscience, University of Minnesota, Minneapolis, MN, United States

Matthew A. Weber
Department of Neurology, University of Iowa, Iowa City, IA, United States

Charles R.E. Wilson
Univ Lyon, Université Claude Bernard Lyon 1, Inserm, Stem Cell and Brain Research Institute U1208, Bron, France

Jen-Hau Yang
Department of Psychiatry, Yale University School of Medicine, New Haven, CT, United States

Qiang Zhang
Department of Neurology, University of Iowa, Iowa City, IA, United States

Jan Zimmermann
Department of Neuroscience, University of Minnesota, Minneapolis, MN, United States

Preface

Before saying anything about what this book is or how it came to be, the editors feel it is important to acknowledge that each chapter of this book was written, revised, reviewed, and edited not within the context of the normal academic background most of us are accustomed to (i.e., doing experiments, teaching, advising, mentoring) but rather in the midst of a global pandemic that has profoundly disrupted science, university life, and, most critically, the lives of each author. Importantly, this acknowledgment is in no way meant to serve as a disclaimer about the quality of each respective chapter. While we may be biased, we think each chapter is excellent and full of truly brilliant insights into the field. Instead this acknowledgment is meant to serve as recognition of the tremendous dedication, professionalism, and commitment to science each author has displayed, as well as a thank you to each of them for committing to this process and seeing it through to the end. Again, as editors and fellow scientists, we cannot thank these authors enough for their time and commitment to this book during what has been an incredibly challenging period.

Much like the adaptability exhibited by the authors throughout this process, adaptability has been a defining characteristic of medial frontal cortex (MFC) research since its inception. From emotional regulation to social learning to cognitive control to motor planning most, if not all, behaviors you can think of either elicit some response from or have been in some way linked to the MFC. The unique ability of the MFC to adapt and jump between sensory and motor domains so fluidly has made the exact function of the region as a whole—and the function of the individual brain regions that make up the MFC—a near universal source of debate among the neuroscience community. Behaviorists, cognitive psychologists, physiologists, and computational neuroscientists alike have filled the neuroscience literature with reports of MFC function, many of which either directly conflict with one another or seem to align briefly, only to then diverge dramatically in terms of takeaway message and planned future directions. Navigating these waters is difficult and frequently scientists, who may ultimately be interested in similar behaviors and similar brain regions, find themselves adrift, steadfastly paddling in their own stream.

In an effort to paddle smarter and not harder, the editors, Adam T. Brockett, Linda M. Amarante, Mark Laubach, and Matthew R. Roesch, met in the summer of 2018 for a joint lab meeting. The goal of this meeting was to get to know one another and the work that each lab did. To facilitate this, each person was given 5 min to share a little bit about the types of questions they were interested in. Despite different behavioral approaches and nomenclatures, it became clear that both labs were centrally interested in understanding what the MFC does or does not do; and despite the relatively short nature of the talks, it was readily apparent how subtle changes in behavioral dynamics seemed to dramatically alter MFC responses. Moreover, regardless of the differences in results and interpretation, the interest in and the ideas about action selection and flexibility shared in the description of what led to each study were so similar that both labs were left with many more questions than answers. Independently of course, these questions permeated many follow up discussions, and when the editor of this series, Dr Patricia Janak, offered Dr Matthew Roesch the opportunity to edit a book on a topic of his choosing, we collectively felt it was time to ask these questions of a wider audience.

What does medial frontal cortex signal during behavior? Insights for behavioral neurophysiology is our attempt to begin to answer the questions posed in that joint lab meeting in 2018. Given our established difficulty in scheduling joint lab meetings, attempting to answer this question by planning a cross lab experiment(s) was deemed untenable (at least for now). Instead, we decided to survey and curate the insights of 16 leading animal behavioral neurophysiologists and their labs in MFC research. Much like our one joint lab meeting, we hoped this experience would help to both highlight what has been done in the field thus far, as well as document and bring together—in one place—the current ideas of leading MFC scientists.

Accordingly, authors were given rather vague instructions—to simply write a chapter trying to answer what they thought the MFC is encoding in the context of behavior. While readers may notice a wide array of anatomical terms used by the authors across chapters, regardless of the term or species used, all of the chapters reveal information about the overarching medial regions of the frontal cortex. We specifically instructed authors to focus more on MFC as opposed to more rostral and lateral prefrontal regions (e.g., orbitofrontal cortex, dorsal lateral prefrontal cortex) since there have been several excellent reviews and special journal issues covering those brain regions and there is more of consensus about homologies across species. Beyond that, the editors encouraged authors to highlight work or ideas that excited them, but few, if any, other restrictions were placed on them. What emerged was at first a very dense and complicated view of MFC function.

While at first glance, individual chapters seemed to emphasize only one aspect of MFC function or one region independent of others, having now read each chapter several times, the editors began seeing overarching themes that we felt we should lay out here to help guide you, the reader, in how to approach this book. First, central to all chapters is the idea that the MFC does many things. Despite its functional diversity, four general functions of the MFC appeared to emerge. First, the MFC is involved in the integration and collection of *valenced information*. This information is emotional in nature and can be based on either positive or negative experiences. Regardless of whether it is based on a positive or on a negative experience, this information is used to ensure decisions are made by considering the context and underlying state of the organism making the decision. Second, and often in combination with information about valence, knowledge about the environment influences *social interaction*. The MFC is essential for social cognition, and in turn, social interaction has profound effects on decision making. Third, the MFC is tasked with maintaining an organism's sense of *adaptability*. Whether faced with overwhelmingly positive or negative outcomes, assessing the outcomes of a partner, or mitigating conflict between competing internal goals, the MFC employs cognitive control and utilizes information about error commission to shift and adapt behavior in a relatively flexible manner. Finally, valence, social, and adaptability features of MFC function ultimately to serve the eventual generation and biasing of appropriate *motor responses*—the ultimate measure of successful cognitive processing.

With these four ideas of MFC function in mind, another guiding principle emerged regarding the organization of this information across the MFC. Cross species differences in homology aside, rather than having a system whereby each functional theme is localized to a single area, several chapters in this book envision a system where information is spread across the MFC like a gradient, in which the relative proportion of neural signals representing valence, social, adaptability, and motor information changes with respect to the rostral to caudal axis of the brain. As such, the table of contents of this book has been designed to reflect both the functional themes and their anatomical organization. In other words, this book is designed to mirror the emerging idea of a functional gradient in MFC, with more valence-focused chapters at the beginning giving rise to more social-focused chapters, giving rise to more adaptability-focused chapters, which finally give rise to more motor-focused chapters at the end. We hope this strategy, while rhetorical in nature, helps to guide your eventual thinking about MFC function.

Lastly, as you will see in this book and likely know from your own experience with the field, there is no single correct way to answer the question: what does medial frontal cortex signal during behavior? Behavioral neurophysiological approaches offer great control and potential to closely link behavioral outcomes with the underlying neural signals. Moreover, these tools, when used in combination with pharmacological, optogenetic, and chemogenetic/kinetic approaches, can only further these attempts to understand the neural processes involved. While we think this is a major advantage to behavioral neurophysiological/systems neuroscience approaches, one immediately obvious shortcoming of this book is in its exclusion of experts in human imaging and cognitive psychology/neuroscience fields. This is regrettable, and while the goal of this book was ultimately to offer a forum for behavioral neurophysiologists to share their ideas on MFC function, we recognize that the real value of these ideas can only be fully realized when considered in combination with the ideas and findings of these other disciplines. We encourage readers of this book to actively seek out this body of literature and suggest that future volumes be devoted to these perspectives, perhaps with a focus on merging more human and animal focused ideas of MFC function.

Ultimately, we hope this book helps you as you either begin to or continue to navigate the research waters of MFC function. Furthermore, we hope that the ideas contributed here by our incredibly talented (and adaptable!) authors, serve to further discussion in this very important and exciting field of Neuroscience.

Adam T. Brockett
Linda M. Amarante
Mark Laubach
Matthew R. Roesch

CHAPTER ONE

From affective to cognitive processing: Functional organization of the medial frontal cortex

Joseph Simon IV, Peter H. Rudebeck, and Erin L. Rich*
Nash Family Department of Neuroscience and Friedman Brain Institute, Icahn School of Medicine at Mount Sinai, New York, NY, United States
*Corresponding author: e-mail address: erin.rich@mssm.edu

Contents

Abstract

The medial wall of the primate frontal lobe encompasses multiple anatomical subregions. Based on distinct neurophysiological correlates and effects of lesions, individual areas are thought to play unique roles in behavior. Further, evidence suggests that dysfunction localized to specific subregions is commonly found in different neuropsychiatric disorders. The neurobiological underpinnings of these disorders, however, remain far from clear. Here, to better understand the functions of medial frontal cortex (MFC) and its role in psychiatric disease, we focus on its functional organization. We describe the emerging pattern in which more dorsal regions subserve temporally extended cognitive functions and more ventral regions predominantly subserve affective functions.

International Review of Neurobiology, Volume 158
ISSN 0074-7742
https://doi.org/10.1016/bs.irn.2020.11.011

We focus on two specific domains, decision-making and social cognition, that require integration across emotion and cognition. In each case, we discuss the current understanding of the functions believed to depend on subregions of MFC as a stepping-stone to speculate on how they might work in unison. We conclude with an overview of how symptoms of certain psychiatric disorders relate to our understanding of MFC functional organization and how further discovery could fuel advances in circuit-based therapies.

1. Introduction

Our most advanced cognitive abilities, including learning, planning and decision-making rely on an intact frontal cortex (Murray & Rudebeck, 2018). How the activity of frontal cortical neurons underlies such complex processes remains largely unknown but has been the subject of intense focus. The attention dedicated to uncovering its functions is in part motivated by the fact that nearly all neuropsychiatric disorders are associated with dysfunction in some region of frontal cortex (Phillips, Drevets, Rauch, & Lane, 2003). Here, we focus on a frontal region strongly linked to disorders of mood, emotion, and cognitive functioning, medial frontal cortex (MFC). As a whole, primate MFC has been implicated in a wide range of behaviors, from cognitive control and goal-directed behaviors to reward processing and socio-emotional regulation (Rushworth, Noonan, Boorman, Walton, & Behrens, 2011). Despite this understanding, the combined cognitive and affective nature of the processes ascribed to MFC has made characterizing its exact functionality difficult.

One way to begin to get a handle on MFC function is through its organization, both anatomical and neurophysiological. In this chapter, we first take a detailed look at the connectivity within MFC and with other parts of the brain. With this foundation, we map anatomy onto the patterns of neural activity during behavior, highlighting how this might shape the impact of damage or dysfunction within specific subregions. From this view, a pattern of functional organization emerges along the dorsal-ventral axis of the MFC, in which processing shifts from predominantly motor and cognitive to affective functions. We make a case that the dorsal MFC (dMFC) is more involved in aspects of cognition that require integration of information over extended time frames, while the ventral MFC (vMFC) is critical for reward and affect-related processing (Gillihan et al., 2011).

Orchestration of these different MFC functions is likely critical for selecting advantageous courses of action and facilitating voluntary behaviors. Indeed, MFC damage or dysfunction can lead to behavioral disturbances, and is implicated in a range of neuropsychiatric disorders, including mood and anxiety disorders and autism (Lord et al., 2020; Phillips et al., 2003). Because of this link to human psychiatric conditions, we concentrate on insights gained from work with human and non-human primates. This focus is not because work in other species is less useful for understanding MFC function. Instead, it is motivated by a desire to more readily translate findings to the human brain, in order to understand how the MFC contributes to psychiatric disorders. Using cross-species comparisons to map basic cognitive functions onto this heterogeneous brain region may improve our understanding of fundamental pathologies in psychiatric disease, and could be used to directly inform future therapeutic interventions (Heilbronner & Chafee, 2019; Rudebeck, Rich, & Mayberg, 2019).

2. Anatomical organization

2.1 Cross-species definition of MFC

When comparing MFC across monkeys and humans, a few similarities and differences are of note. The following is a summary derived from investigations of cytoarchitecture, and to some extent connectivity, which allow comparisons across species (for an in-depth discussion see Passingham & Wise, 2012 and Öngür & Price, 2000). Here we use the nomenclature adopted by Price and colleagues (Carmichael & Price, 1994; Öngür, Ferry, & Price, 2003) (Fig. 1), although we note that others exist (e.g., Barbas & Pandya, 1989; Picard & Strick, 1996).

Starting from frontal pole, medial area 10 (10m) in humans takes up a greater volume than in monkeys, extending further into the ventral portion of MFC. The expansion of area 10m means that the prelimbic area (32pl) is situated more caudal and ventral of the corpus callosum (Öngür et al., 2003). Two areas in humans have been designated 32: prelimbic (32pl) ventrally, and a portion of the anterior cingulate cortex (32ac) dorsally. Posterior to area 32ac in humans is area 24ac, which wraps around the genu of the corpus callosum, meeting area 25 to form the subgenual region of the anterior cingulate cortex (sgACC). Cytoarchitecture of subgenual 24a/c in humans shares anatomical features to anterior portions of 24b in monkeys (Carmichael & Price, 1994; Öngür et al., 2003).

The most dorsal areas of MFC lie on the superior frontal gyrus, which comprises the anterior component of area 6aβ, area 8Bm, and area 9 (9m) in humans, and appears comparable to the analogous regions in the monkey

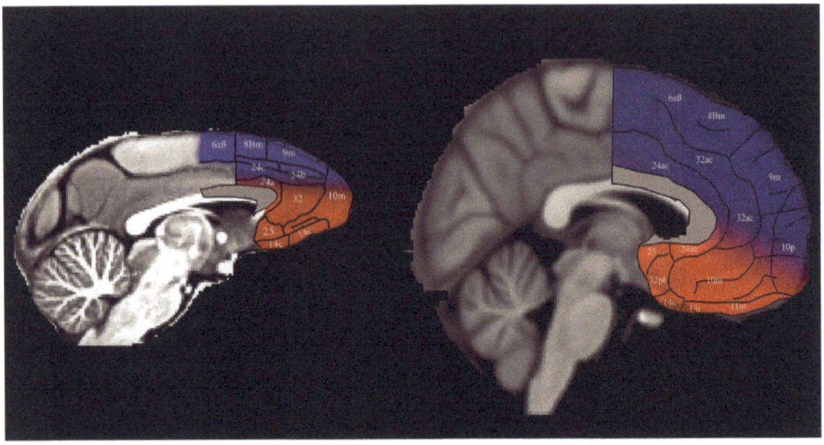

Fig. 1 Anatomical regions of the primate MFC. Midsagittal MR images of a macaque monkey (left) and human brain (right), overlaid with demarcations of anatomical subregions based on Carmichael and Price (1994) and Öngür et al. (2003). Shading denotes the dorsal-ventral axis described in the main text.

(Petrides, Tomaiuolo, Yeterian, & Pandya, 2012). The most caudal of these (6aβ) is referred to as the pre-supplementary motor area (pre-SMA), which, despite its name, is not typically classified as premotor. Although there is evidence for short-latency functional interactions with motor cortex (Mars et al., 2009), it lacks direct connections with primary motor cortex and is therefore seen as more transitional between motor and prefrontal function (Luppino, Matelli, Camarda, & Rizzolatti, 1993; Picard & Strick, 2001).

The most ventral reaches of MFC include the cortex between the gyrus rectus and medial orbital sulcus. This location, often referred to as medial orbitofrontal cortex (mOFC) or ventromedial prefrontal cortex (vmPFC), includes primarily area 14, as well as medial portions of areas 11 and 13 (Carmichael & Price, 1994; Öngür et al., 2003). Posterior and dorsal to area 14, in sgACC, lies area 25, which appears to have homology across species (Öngür & Price, 2000; Wise, 2008).

2.2 Anatomical connectivity

Beyond cytoarchitecture, the anatomical arrangement of MFC can also be considered in terms of its dominant connectivity with other brain structures. As a whole, MFC is reciprocally connected with the thalamus and wide ranging cortical and subcortical structures, among the most notable are the dorsal and ventral striatum, amygdala, and cortical association areas. Differences in anatomical connectivity across MFC subregions may relate to their overall contributions to behavior. For instance, dMFC, including pre-SMA and dorsal anterior cingulate cortex (dACC) are more strongly associated with motor function and receive inputs from premotor and dorsolateral prefrontal cortex (Luppino et al., 1993). In contrast, the most ventral MFC, encompassing areas 14 and 11m have stronger connectivity with limbic structures including the amygdala, entorhinal and perirhinal cortices (Carmichael & Price, 1995), consistent with its potential role in higher-order reward processing.

Beyond cortical connections, thalamic connectivity of the anterior cingulate portions of MFC differ from those of other frontal cortex areas, suggesting unique origins and functions (Vogt, Pandya, & Rosene, 1987). In particular, the dACC is interconnected with parts of the ventral anterior nucleus of the thalamus, which also sends efferents to premotor cortex, indicating a role in motor planning and movement initiation.

Also of note is different striatal connectivity between dMFC and vMFC (Haber, 2016). Studies in humans and monkeys show gradients within striatum that follow a dorsolateral to ventromedial organization, and these form distinct but overlapping connections with cortex spanning from dorsal to ventral MFC (Haber, 2016). The most dorsal regions of MFC, including areas on superior frontal gyrus, connect most strongly to dorsomedial sections of the striatum, and areas 24b and 24c project widely to most of rostral striatum (Calzavara, Mailly, & Haber, 2007). Corticostriatal interactions in these circuits are believed to facilitate learning and coordinate motor-functions. Regions situated more centrally and ventrally in MFC project to more central and ventromedial targets in the striatum, and these circuits are more strongly implicated in reward-related functions (Haber, Kim, Mailly, & Calzavara, 2006).

2.3 Functional connectivity

Studies of fMRI functional connectivity can corroborate proposed homologies between humans and monkeys and identify functional circuits, as imaging techniques can be readily applied to both species (Mars et al., 2016). The premise of functional connectivity is essentially that regions whose activity covaries are likely interconnected and together subserve the same functions. Thus, for a given region, there is a set of brain–wide interactions, and from these a connectivity "fingerprint" can be constructed and compared across brain areas and species (Mars et al., 2016). Taking this approach, Neubert and colleagues revealed that, compared to vMFC, the coupling strength of cortical areas 9m and the neighboring portion of 32ac in the dMFC was stronger with cingulate motor areas (CMA), pre-SMA, dorsolateral and ventrolateral PFC (dlPFC and vlPFC), and head of the caudate. In contrast, ventral areas 11m and 14 coupled more with orbitofrontal cortex (OFC), inferior temporal cortex, and ventral striatum (Neubert, Mars, Sallet, & Rushworth, 2015). Importantly, MFC connectivity fingerprints were highly similar in both humans and monkeys, indicating that homologies in functional connectivity follow with anatomical homologies reviewed above. Thus, in terms of anatomy and connectivity, MFC is organized in an overall dorsal to ventral trend, and this pattern is conserved in both humans and macaques. As we will describe in the ensuing sections, distinctions in function appear to map onto this anatomical organization.

3. Organization of function

The fundamental functions of primate MFC have been a subject of investigation for decades. Interestingly, these investigations and the ensuing debates have played out in two largely separate arenas, focused on dMFC, particularly dACC, and vMFC. Despite many unresolved questions, the theories and ideas put forth have generally supported the notion that dorsal regions are more strongly tied to cognitive processing, while ventral regions have a stronger role in affective processing. Indeed, studies have posited a role for dMFC in behaviors ranging from conflict monitoring (Botvinick, 2007), to action-selection (Kennerley, Walton, Behrens, Buckley, & Rushworth, 2006) and prediction error encoding (Kennerley, Behrens, & Wallis, 2011; Matsumoto, Matsumoto, Abe, & Tanaka, 2007). In part because such a range of cognitive and emotional processes have been linked with this region, there has been much speculation, and some contention about its function (Kolling et al., 2016; Shenhav, Cohen, & Botvinick, 2016). On the other hand, activity in vMFC, often referred to as vmPFC, frequently relates to reward processing, with neuroimaging studies showing that, given a variety of task designs, activation of vmPFC correlates with subjective evaluations (Rushworth et al., 2011). This contributed to the theory that vmPFC is specialized for processing positive valences, whereas vlPFC (primarily area 12/47) processes negative valences (Kringelbach & Rolls, 2004), however a number of more recent studies have discredited this view

(Plassmann, O'Doherty, & Rangel, 2010; Tom, Fox, Trepel, & Poldrack, 2007) and it has generally not held up at the level of neurophysiology in animal subjects (Rich & Wallis, 2014). Together, the rich histories of study in both regions support the notion that dMFC and vMFC have stronger ties to cognitive and affective processing respectively, though the specific functions of each region remain uncertain.

In the following sections, we discuss domains in which both dorsal and ventral MFC have been studied, with the aim of highlighting their unique contributions to behavior. Because subregions of MFC are heavily interconnected (Carmichael & Price, 1996) there is likely crosstalk between regions, and we also speculate on how putative subregional functions are ultimately integrated. While this likely plays out in many domains, we focus on decision-making and social cognition as complex operations that require the integration of cognitive and emotional processing.

3.1 Decision-making

A variety of evidence supports the notion that MFC is critical for optimal value-based decision-making. Much of this evidence has been drawn from choice tasks in which the subject directly compares two or more options before choosing one. MFC also appears to be involved in decisions to simply accept or reject an option in the context of foraging or search tasks. Below, we outline the current state of knowledge of MFC's contributions to these different types of value-based decisions, with an emphasis on functional heterogeneity within MFC.

3.1.1 Multi-alternative choice tasks

Every day we make simple value-based decisions. Would you like cereal or toast for breakfast? Would you prefer hiking or kayaking this weekend? Despite their apparent simplicity, these decisions involve a complex integration of information, starting from processing stimuli, concepts, relevant contexts, potential costs and internal states; to computing values, comparing options, planning and executing a behavioral choice; and finishing with perceiving and evaluating the ensuing outcome and updating relevant beliefs or estimates appropriately. Thus, a simple decision comprises many processes, and evidence from neuropsychology and neurophysiology suggests that different regions of MFC make unique contributions. Specifically, the more dorsal regions, including dACC, seem to play a role in conferring flexibility to choice behavior, while more ventral regions have been implicated in evaluating and comparing choice options. Below we review evidence supporting these dissociations.

There is broad consensus that neural activity in dorsal MFC relates to the outcomes of decisions. The activity of neurons in dMFC covaries with the subjective value of reward-predicting stimuli, but this largely reflects post-decision variables (Cai & Padoa-Schioppa, 2012; Ito, Stuphorn, Brown, & Schall, 2003). That is, the neurons encode the value of an item chosen and then received by the subject, but not values of each option considered. This is in contrast to neurons in OFC, which signal both pre- and post-decision variables, including the value of the available options (Padoa-Schioppa & Assad, 2006; Rich & Wallis, 2016; Thorpe, Rolls, & Maddison, 1983), consistent with the idea that OFC causally contributes to decision formation (Padoa-Schioppa & Conen, 2017). Meanwhile, value coding in dMFC lags that of OFC, relates actions to their associated outcomes, and encodes prediction error signals (Kennerley et al., 2011; Luk & Wallis, 2013). As such, dMFC has been linked to a variety of functions dependent on outcome evaluation, including weighing history of rewards, updating beliefs and models of the environment, and exploring or searching for alternatives (Heilbronner & Hayden, 2016; Ito et al., 2003; Kolling et al., 2016).

Evidence from neuropsychology studies has corroborated and clarified the importance of dMFC in choice behavior. While earlier theories of dMFC function emphasized an importance in detecting or responding to errors (Gehring, Goss, Coles, Meyer, & Donchin, 1993; Ito et al., 2003; Shima & Tanji, 1998), recent work has shown that animals with lesions centered on dACC can detect changes in action-outcome contingencies and change their responses appropriately, but they cannot continue making optimal responses over time, suggesting a deficit in maintaining contingency information (Kennerley et al., 2006). This pattern is consistent with neural recordings that show that neurons in dACC have some of the longest time constants of areas sampled in monkey cortex (Murray et al., 2014). This means that events in the past have long-lasting effects on dACC activity, making it particularly well suited to integrate information over time. Together, these findings support a role for dACC in temporally extended cognitive processes that inform simple decisions.

The more ventral aspects of MFC are also involved in multi-alternative decision-making, but appear to play a different role. In humans and monkeys, vMFC lesions impair decision-making that requires either integration of multiple sources of information or comparison among multiple dimensions (Noonan et al., 2010; Pelletier & Fellows, 2019). For instance, patients with vMFC lesions have difficulty assessing the overall value of an object from a combination of component parts, each with learned values (Pelletier & Fellows, 2019). An alternative perspective suggests that vMFC activity is critical for comparing the value of options during decision-making (Noonan, Walton, et al., 2010). Consistent with this, neuron recordings in monkeys (Strait, Blanchard, & Hayden, 2014) and fMRI in humans (Hunt et al., 2012) both reveal patterns of vMFC activity consistent with mutual inhibition, a proposed mechanism of value comparison. vMFC's role in integration or comparison of values could be reconciled from the perspective that both processes require multiple sources of value information to converge on a consensus signal.

Similar to dMFC, vMFC's role in integrating information about value may depend in part on maintenance of information over time. Recent evidence from rhesus monkeys found that lesions of sgACC lead to an inability to sustain arousal, as measured by autonomic responses, in anticipation of rewards (Rudebeck et al., 2014). Importantly, lesions did not cause general autonomic dysfunction, as responses to the rewards themselves remained intact. sgACC is interconnected with subcortical areas regulating autonomic function, in addition to being connected with multimodal cortical and limbic areas (Öngür, An, & Price, 1998), supporting a long-held view that this area plays a role in integrating autonomic and cognitive functions (Critchley, 2005). Indeed, early data found that electrical stimulation affects breathing and heart rate (Kaada, Pribram, & Epstein, 1949; Smith, 1945), though sgACC appears less related to low-level autonomic regulation and more closely tied to cognitive influences, in particular those that maintain autonomic responses when predictive cues are not present. Whether this occurs through sustained attention, network dynamics that hold value information online (Enel, Wallis, & Rich, 2020), or other direct effects on autonomic control centers remains to be determined. Regardless, maintaining expectations about upcoming rewards is critical for optimizing attention and motor preparation, and also for updating value expectations through learning when the anticipated outcome does not match expectations.

3.1.2 Single alternative search and foraging tasks

Value-based decisions may also take the form of deciding to accept or reject a single option. This is most commonly encountered in situations where accepting an offer comes at a cost of time, resources or opportunity. For instance, one classic problem involves choosing whether to hire a job applicant or wait to see whether a more qualified person applies later

(Costa & Averbeck, 2015). Similarly, an animal may have to decide whether to invest the time and energy to pursue a prey or forage in a particular location while forgoing other opportunities (Calhoun & Hayden, 2015). These scenarios differ from multi-option choices in that they require the subject to estimate not only the value of the option, but also statistics of the environment such as richness, or average returns expected, so that ultimately the information to be compared is the offer against an aggregate of these environmental statistics. Thus, a small offer is more likely to be accepted in an impoverished environment than one in which rewards are plentiful.

Studies that examine the role of MFC in single-alternative choices generally parallel interpretations of multi-alternative choice but extend them in important ways. The ability to estimate reward statistics in an environment necessarily involves integration over time, a factor that appears to be important in MFC's role in multi-alternative choices, and specifically depends on more dorsal regions (Shidara & Richmond, 2002). Similarly, fMRI BOLD activity in dMFC tracks the volatility of potential outcomes, another temporally extended reward statistic, when participants performed a decision-making task where the probability of reward fluctuated to different degrees across blocks (Behrens, Woolrich, Walton, & Rushworth, 2007). Finally, the same region of dMFC is activated when a subject updates models of the current task or environment, leading to the suggestion that dMFC encodes these statistics in order to update behavioral policies (Karlsson, Tervo, & Karpova, 2012; O'Reilly et al., 2013).

A similar perspective has found that MFC activity correlates with dynamic appraisals of the value of persevering (McGuire & Kable, 2015). In these tasks, subjects reassess the subjective value of an upcoming reward as time passes, which requires continuously updating their estimate of the remaining delay time (McGuire & Kable, 2013). The regions in MFC activated by this dynamic evaluation were compared to those that encode subjective value in other types of decision tasks, and the results showed clear evidence of functional organization. Dynamic evaluations activated a more dorsal region of MFC that only partially overlapped with a more ventral region that responded to simpler evaluations (McGuire & Kable, 2015) (Fig. 2). This has similarities with the conclusions drawn from multi-alternative decisions, in that dynamic evaluations must evolve over longer timescales than simple ones. Nonetheless, complex tasks, such as integrating temporally extended reward information or deciding whether to persevere, tap into both of these cognitive domains and therefore likely require effective orchestration of MFC subregions.

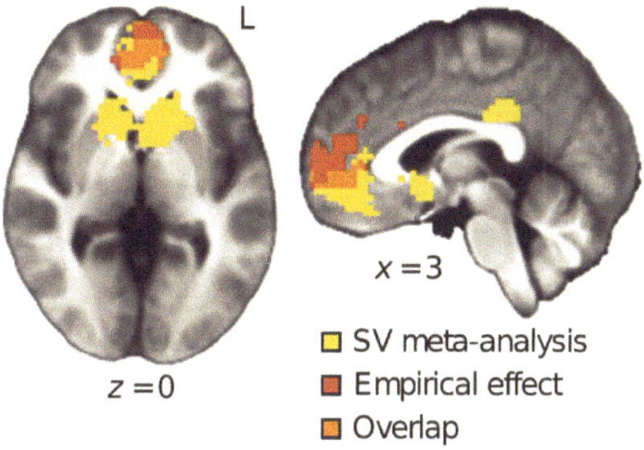

Fig. 2 Functional organization of simple and temporally-dependent subjective value. In Red is the region where activity significantly correlated with dynamically changing subjective values. In the task, participants decided how long to continue waiting for a monetary reward. On each trial, a token began with no value, but matured after a random delay. The token could be sold at any time, and a new trial started. The total task time was fixed, and participants aimed to maximize their earnings in that time by estimating the value of waiting. In Yellow is the MFC region activated by canonical evaluation tasks, as identified from a 206-study meta-analysis (Bartra, McGuire, & Kable, 2013). In Orange is the region of overlap. *Reprinted by permission from Springer Nature Customer Service Centre GmbH: Nature Publishing Group, Nature Neuroscience: McGuire, J. T., & Kable, J. W. (2015). Medial prefrontal cortical activity reflects dynamic re-evaluation during voluntary persistence.* Nature Neuroscience, 18*(5), 760–766. https://doi.org/10.1038/nn.3994.*

3.2 Social cognition and the MFC

Social cognition involves interpreting the perceptions and intentions of others to facilitate appropriate behavior (Adolphs, 2008) and is another domain that illustrates the distinctive and integrated functions of dorsal and ventral MFC. In humans, abnormal social behavior has classically been observed as a result of MFC lesions (Freemon, 1971; Harlow, 1869), and neuroimaging studies have shown that MFC activity correlates with processes critical for social cognition (Adolphs, 2008). While there is strong evidence that MFC is active during social tasks, it is debated whether these responses are specific to social contexts or reflect more general functions of the region, such as reward processing (Delgado et al., 2016). Below we discuss the potential contributions of the MFC to social cognition and behavior, with an emphasis on neurophysiological studies, which have the resolution to reveal network properties that may be unique to social information.

3.2.1 The dMFC and social interpretation

Some early accounts of monkeys with experimental lesions encompassing dMFC described an ill-defined syndrome characterized as a loss of "social consciousness" (Glees, Cole, Whitty, & Cairns, 1950; Ward, 1948). The lesioned animals behaved as if their cage-mate were inanimate, "leaping over it, knocking into it, sitting on it, and taking food from its hands" (Glees et al., 1950). More modern investigations have similarly found that large lesions involving nearly the full extent of MFC reduce macaque social interaction, while increasing interactions with inanimate objects (Hadland, Rushworth, Gaffan, & Passingham, 2003). More focal lesions of dMFC impair monkeys' valuation of social stimuli (Rudebeck, Buckley, Walton, & Rushworth, 2006), and ability to learn to associate specific options with prosocial behaviors (Basile, Schafroth, Karaskiewicz, Chang, & Murray, 2020). Interestingly, however, when presented with an inanimate object, dMFC lesioned animals tended to exhibit prosocial-like behaviors (Basile et al., 2020), paralleling the previous accounts of increased interest in inanimate objects. Together, these findings support the notion that MFC in the monkey, and particularly the dorsal region, plays a role in appropriate social behavior.

Mechanistically, orchestrating social behavior may relate to non-social decision-making by taxing the same functions that require dMFC, such as tracking and integrating contexts into ongoing behavior (Gehring et al., 1993). Supporting this, dMFC neurons recorded from monkeys tracked actions and errors committed by themselves as well as other monkeys to guide learning (Yoshida, Saito, Iriki, & Isoda, 2011, 2012). Importantly, rather than signaling choices generally, many of these neurons uniquely tracked either the subject's own selections or those of the partner. Thus, new information was uniquely attributed to actions performed by the self or a social partner, suggesting that different mechanisms in dMFC govern integration of information obtained by experience versus observation, both of which are relevant to social forms of decision-making.

3.2.2 The vMFC and social referencing

The classic case of Phineas Gage was one of the first reported links between vMFC lesions and social behavior (Harlow, 1869). More recent evaluations frequently note blunted affect and emotional lability following vMFC damage, in extreme cases describing the individuals as having acquired sociopathy (Saver & Damasio, 1991). Despite this, the exact role played by the vMFC in social behavior remains unclear (Delgado et al., 2016). Studies in monkeys with experimental lesions in a dorsal region of the vMFC (area 32) found changes in social behavior (Rudebeck et al., 2006), while lesions of the more ventral portions of vMFC (area 14) produced no notable social effects (Noonan, Sallet, Rudebeck, Buckley, & Rushworth, 2010). This contrast suggests heterogeneity within vMFC in which some subsections are more tuned to social information. Nevertheless, the combined functions of vMFC appear to be critical for interpreting complex value-based information (Pelletier & Fellows, 2019) which often includes social information (Vaidya & Fellows, 2019).

One challenge in assessing vMFC's role in complex social behavior is that the majority of studies come from human functional imaging (Adolphs, 2008), while most evidence from the activity of individual neurons derives from work in animal models. To translate more readily, recent investigations have assessed the neural underpinnings of complex social behavior in non-human primates, who afford the opportunity to study advanced cognitive processes related to conspecifics at single neuron resolution (Chang, 2017). For example, in a vicarious reinforcement task that allowed monkeys to choose between giving a juice reward to themselves, a partner, or neither, Chang and colleagues (Chang, Winecoff, & Platt, 2011) found that monkeys show prosocial preferences, frequently deciding to give the other monkeys juice over neither of them. In MFC areas 24a/b and 32, an equal proportion of neurons signaled rewards to self, partner or both, but in the neighboring OFC neurons were predominately active when the subject themselves received a reward (Chang, Gariépy, & Platt, 2013), indicating that tracking or engaging in prosocial behavior uniquely engages some parts of vMFC.

One important function of the vMFC may be taking into account internal variables when making decisions, and this could serve a critical role in self-referenced decisions (Critchley, 2005). Self-referencing refers to using one's own beliefs and values as a benchmark for expected beliefs or values of others, and some evidence suggests that vMFC is important for this process (Critchley, 2005; Tamir & Mitchell, 2010), though other recent work brings this into question (Koski, McHaney, Rigney, & Beer, 2020). In monkey studies, it is difficult to determine whether self-referencing is used, yet there are differences in the representation of internal (or self-referenced) versus external states in decision-making circuits. For instance, more neurons in area 14 compared to the OFC, preferentially encode internal versus external states (Bouret & Richmond, 2010). This supports the view that vMFC is critical for evaluating internal information, and may apply to self-referencing strategies that guide social decision-making.

3.2.3 *MFC and the "social brain" circuit*

So far, in social contexts dMFC has been linked to context monitoring and planning, while vMFC is implicated in self-referencing and evaluation, but little is known about how these regions may interact during behavior. Ultimately, understanding the orchestration of interactions within MFC and in broader circuits will improve our understanding of the large-scale neural networks that underlie social cognition. One step in this direction focused on interactions between dMFC and ventral tegmental area (VTA) when monkeys received reward (self-reward), or when an observed partner received reward (other-reward). Both outcomes were coded by dMFC neurons, and were translated to subjective values in VTA (Noritake, Ninomiya, & Isoda, 2018). Among simultaneously recorded local field potentials (LFPs), information flowed predominately from dMFC to the VTA (Noritake et al., 2018). A companion study also recorded activity in the lateral hypothalamus, where neurons tracked the probability of both self- and other-reward, and were also influenced by information from dMFC (Noritake, Ninomiya, & Isoda, 2020). Beyond LFP analyses, spike timing in relation to LFPs allows us to infer the coordination of information between recording sites. This approach revealed spike-LFP coherence between ACC gyrus (areas 24a/b and 32) and basolateral amygdala (BLA) when actor monkeys made decisions about a partner monkey's rewards (Dal Monte, Chu, Fagan, & Chang, 2020) (Fig. 3). Actor monkeys either chose between giving juice to a partner versus a bottle on some trials (Other-Bottle), or to themselves versus both monkeys on others (Self-Both), thus holding the Actor's own reward consistent within each decision type. Spike-field coherence between the Actor's ACCg and BLA differed when they made prosocial (Other, Both) versus antisocial (Bottle, Neither) choices, suggesting a critical role for inter-area directional coupling in social decisions. Future studies focused on how these oscillatory interactions differ with dorsal and ventral MFC, as well as interactions between these subregions, will be critical to understanding the functions of this region in social decision-making and overall behavior.

4. Disorders of function

A primary goal in understanding MFC function is to inform our diagnosis and treatment of mental illness. Disorders of personality, mood, motivation and cognition have often been associated with MFC dysfunction, in many cases with subregional localization (Phillips et al., 2003). We close this

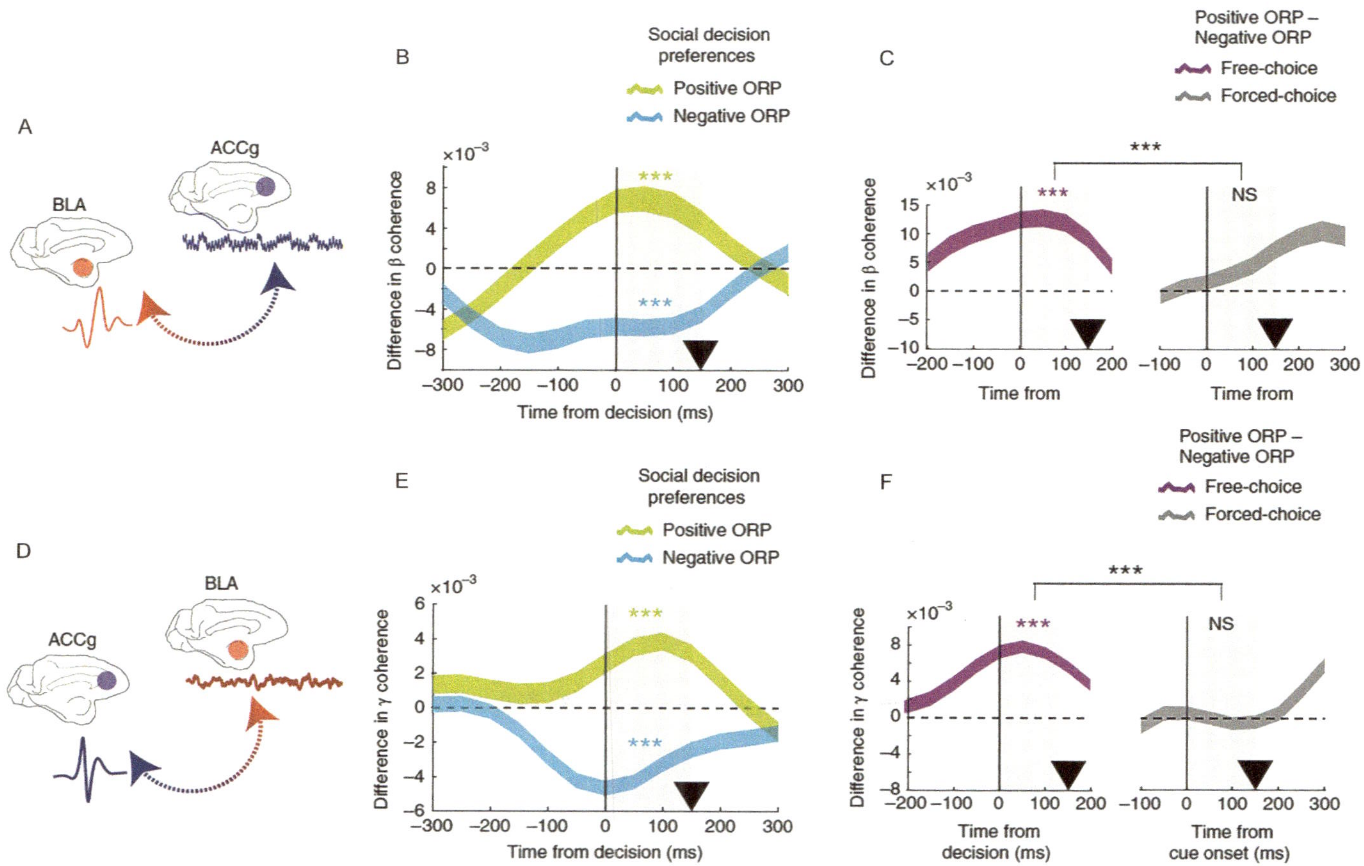

Fig. 3 See figure legend on next page.

chapter with a brief overview of three disorders linked to MFC function: autism, mood disorders, and obsessive-compulsive disorder (OCD). Our goal is not to provide an exhaustive review of all disorders with evidence of MFC involvement, or to fully elaborate on the pathology and pathophysiology of each disease below. Instead, we aim to highlight how our understanding of the functional organization of MFC is reflected in selected neuropsychiatric conditions.

4.1 Autism

Autism spectrum disorder (ASD) is a heterogeneous neurodevelopmental disorder that includes deficits in social functioning, particularly reciprocal social interactions, verbal and nonverbal communication impairments, and stereotyped behaviors. ASD is frequently accompanied by clinical anxiety, and severe cases may include intellectual disability or seizures (Lord et al., 2020). Neuroanatomical abnormalities have been reported in ASD in the cerebellum, amygdala, and frontal cortex, yet none of these abnormalities are seen as pathognomonic of the disease (Amaral, Schumann, & Nordahl, 2008; Varghese et al., 2017). Early neuropathological investigations of cortical involvement found consistent abnormalities in ACC, particularly area 24 of dMFC. Findings included cytoarchitecture that appeared poorly or irregularly laminated, with areas of cortical thickening and dysgenesis (Bailey et al., 1998; Kemper & Bauman, 1993), much of which is consistent with a developmental origin of the disorder (Blatt, 2012; Simms, Kemper, Timbie, Bauman, & Blatt, 2009). Neuroimaging studies have also shown abnormal activation patterns involving the dMFC in a variety of social, emotional, and reward-based tasks (reviewed in Minshew & Keller, 2010),

Fig. 3 Spike-Field Coherence between the ACCg and BLA in a social task. Monkeys chose between delivering juice to a partner or collection bottle (Other-Bottle), and themselves or both partners (Self-Both), and either exhibited positive (Other, Both) or negative (Bottle, Self) other-regarding preferences (ORP). (A) Schematic of coherence analysis between BLA spikes and ACCg field activity. (B) Differences in beta spike-field coherence in positive (green) and negative (blue) ORP choices. (C) Difference between positive and negative ORP in free-choice (purple) and forced-choice (gray) conditions. (D–F) The same analyses shown for ACCg spikes and BLA field activity. *Reprinted by permission from Springer Nature Customer Service Center GmbH: Nature Publishing Group, Nature Neuroscience: Dal Monte, O., Chu, C. C. J., Fagan, N. A., & Chang, S. W. C. (2020). Specialized medial prefrontal–amygdala coordination in other-regarding decision preference.* Nature Neuroscience, 23, *1–10. https://doi.org/10.1038/s41593-020-0593-y.*

as well as disrupted functional connectivity in widespread networks involving the dMFC (Rane et al., 2015; Rippon, Brock, Brown, & Boucher, 2007).

Thus, there is strong evidence that the pathophysiology of ASD, in part, involves dysfunction in MFC, often localized to the dACC region. The cognitive and emotional processes proposed to depend on dMFC have notable consistency with classic ASD symptoms and domains of dysfunction. For instance, social impairments may reflect dysregulation of dMFC's contributions to social cognition. Resistance to change characteristic of ASD, including repetitive behaviors and thoughts, rituals, routines, narrow interests (Lord et al., 2020) may also reflect disruption of dMFC contributions to updating behavioral policies in changing environments. Given MFC's role in regulating the autonomic nervous system, an additional feature of note is that abnormal autonomic activity, particularly in response to social or emotional stimuli, has also been reported (Kylliäinen & Hietanen, 2006; Vaughan Van Hecke et al., 2009). Physiological events such as changes in heart rate, respiration, pupil dilation, and sweating, are believed to be key features of perceiving and interpreting emotional experiences (Barrett, Mesquita, Ochsner, & Gross, 2007; Craig, 2002), and social situations may be one domain in which rapid interpretation of such emotional cues is particularly critical. Thus, it has been hypothesized that autonomic dysregulation could be related to the social-emotional deficits of ASD (Gu et al., 2015, 2018). For instance, a recent study demonstrated that patients had reduced spontaneous skin conductance responses at rest, as well as abnormal correlations between these responses and activation of the default mode network, including the ACC (Eilam-Stock et al., 2014). Given that the vMFC is more closely tied to autonomic regulation, however, there may be more widespread involvement of MFC regions in ASD.

4.2 Mood disorders

Mood disorders, including major depression and bipolar disorder, are among the most prevalent forms of psychiatric disease worldwide, and are frequently comorbid with anxiety disorders such as panic and posttraumatic stress disorders. These conditions are not associated with gross brain pathology, but close analyses of gray matter volumes and cell counts have found consistent reductions in MFC and limbic structures (Price & Drevets, 2010). Volumetric reductions are most prominent in the gray matter of the sgACC, primarily area 25 (Drevets et al., 1997). These abnormalities have been found in individuals with unipolar depression as well as bipolar disorder (Adler et al., 2007; Haznedar et al., 2005), suggesting sgACC dysfunction could play a role in mood disorders generally.

Functional imaging studies also support the idea that mood disorders arise from dysregulation of brain circuits involving sgACC and limbic regions. In particular, after correcting for reduced volumes, fMRI and PET imaging demonstrate sgACC hyperactivity in depressed patients (Mayberg et al., 1997), which is reduced with successful treatment (Mayberg et al., 2005). Whereas some studies have reported variable activity in sgACC in depressed individuals, the differences relate to responsiveness to antidepressant treatment (Mayberg et al., 1997; Pizzagalli et al., 2001), supporting a consistent link between sgACC and disorders of mood.

As we previously noted, there is a potential role for sgACC in value processing and autonomic regulation in healthy individuals, and from this view, there are different ways that sgACC dysfunction could contribute to mood disorders. First, impaired reward processing may be an underlying mechanism of anhedonia, the characteristic symptom of depression and a key symptom of bipolar disorder. Indeed, there is evidence that vMFC is also dysregulated in schizophrenia (Dowd & Barch, 2012; Schlagenhauf et al., 2009), and relates to anhedonic symptoms of the disorder (Dowd & Barch, 2010). These findings hint at a trans-diagnostic role for vMFC in mood. Second, dysregulation of autonomic responses found in depressed patients (Stange, Hamilton, Olino, Fresco, & Alloy, 2017) could arise from vMFC dysfunction. In particular, excessive and inappropriate autonomic activation related to sgACC hyperactivity could underlie the chronic sense of unease reported in depression. The consistent findings of dysregulated vMFC function have spurred the development of deep brain stimulation (DBS) approaches for treatment-resistant forms of depression, in which stimulating electrodes are targeted to sgACC and associated white matter tracks with the goal of introducing disruptive high-frequency stimulation (Holtzheimer et al., 2017; Mayberg et al., 2005; Riva-Posse et al., 2018). Advances gained by novel treatments for previously resistant individuals illustrate the importance of understanding how underlying function and dysfunction map onto anatomical regions of the MFC.

4.3 Obsessive compulsive disorder

OCD is a heterogeneous disorder characterized by repetitive and intrusive thoughts (obsessions) and/or behaviors (compulsions). Standard therapy combines pharmacotherapy and psychotherapy, but does not adequately treat a high proportion of patients (Pallanti & Quercioli, 2006). As with depression, novel DBS approaches have been investigated for treatment of severe, refractory cases of OCD, with the goal of modulating activity in circuits that appear dysfunctional (de Koning, Figee, van den Munckhof, Schuurman, & Denys, 2011).

Different theories have been proposed to describe how OCD symptoms could arise from dysfunction in frontal circuits. On one hand, OCD's characteristic obsessions have an anxiety-provoking quality, and it is believed that the rewarding effects of alleviating this anxiety can reinforce compulsive behaviors (Figee et al., 2011). Consequently, OCD has been conceptualized as a reward processing disorder, involving dysfunction in ventral frontal circuits centered on the OFC and vMFC. Alternatively, obsessions and compulsions have been interpreted as arising from dysfunction in dMFC circuits, resulting in cognitive and behavioral inflexibility, or impaired cognitive control (McGovern & Sheth, 2017) or action-monitoring systems (Pitman, 1987). From the latter perspective, deficits in signaling conflicts or misalignment of goals with actions may cause feelings of erroneous performance, creating an excessive concern for goals to be met. Corroborating this view, error-related negativities (ERNs) recorded on scalp EEG following commission of an error are believed to arise from dMFC, and are exaggerated in OCD patients (Gehring, Himle, & Nisenson, 2000). While some hypotheses involve separate functions of different MFC areas and their respective targets in the basal ganglia, recent views propose that disruption of more integrated circuit functions underlie OCD symptoms (McGovern & Sheth, 2017; Milad & Rauch, 2012). Indeed, anatomical studies have shown that projections from the OFC, dMFC and vMFC converge in adjacent regions of the ventral striatum (Haber & Knutson, 2010), supporting the notion of integrated circuit involvement. Further understanding of how different regions of MFC operate together and in broader brain networks will be crucial in advancing neuromodulatory approaches to OCD treatment such as DBS.

4.4 Application of functional organization to mental health disorders

As illustrated above, parcellating the MFC along the dorsal-ventral axis with respect to its cognitive and affective functions could help move toward the goal of circuit-based therapeutics in psychiatry. A key challenge in studying mental health disorders is the range of behavioral phenotypes that often present. We argue that understanding functional divisions and gradients in MFC, in the context of integrated circuits, has the potential to illuminate neuroanatomical sources of psychiatric symptoms and expand treatment options. For example, DBS approaches to treating severe OCD most commonly target to the anterior limb of the internal capsule, which connects the frontal cortex to thalamus, or the nucleus accumbens (NAcc). Interestingly, effective NAcc DBS was found to reduce excessive functional connectivity between the NAcc and dMFC (Figee et al., 2013), emphasizing the likely role of dMFC circuits in symptom generation. Thus, DBS has shown promise in treating both mood disorders and OCD, but the mechanisms of its therapeutic effects remain unclear. Further refinements of neuromodulatory approaches such as DBS require a better understanding of the functional organization of the circuits targeted for modulation and their relationship to behavior.

5. Summary

Whether it is adaptively searching for food or responding appropriately in social situations, MFC is involved in some of the most vital parts of our lives. When MFC becomes dysfunctional, maladaptive behaviors are evident and at the extremes, individuals are classified as having psychiatric disorders. As we detail above, there appear to be a core set of functions that MFC is involved in, for instance tracking situations where reality does not meet expectations in either reward-guided or social situations. Consequently, MFC plays an important role in both adaptive decision-making and social cognition. In animal subjects, neurophysiological studies are beginning to provide insight into how the underlying processes are encoded at the neuronal and circuit levels. Ultimately, such investigations can lead to better understanding of normal functions carried out by different subregions of MFC, and will be vital for shedding light on psychological disorders that result from MFC dysfunction.

Acknowledgments

The authors thank D.E. Croote, D. Schiller, and P. Croxson for providing structural brain images, and X. Gu for comments on a portion of the manuscript.

References

Adler, C. M., DelBello, M. P., Jarvis, K., Levine, A., Adams, J., & Strakowski, S. M. (2007). Voxel-based study of structural changes in first-episode patients with bipolar disorder. *Biological Psychiatry*, *61*(6), 776–781. https://doi.org/10.1016/j.biopsych.2006.05.042.

Adolphs, R. (2008). The social brain: Neural basis of social knowledge. *Annual Review of Psychology*, *60*(1), 693–716. https://doi.org/10.1146/annurev.psych.60.110707.163514.

Amaral, D. G., Schumann, C. M., & Nordahl, C. W. (2008). Neuroanatomy of autism. *Trends in Neurosciences*, *31*(3), 137–145. https://doi.org/10.1016/j.tins.2007.12.005.

Bailey, A., Luthert, P., Dean, A., Harding, B., Janota, I., Montgomery, M., et al. (1998). A clinicopathological study of autism. *Brain*, *121*(5), 889–905. https://doi.org/10.1093/brain/121.5.889.

Barbas, H., & Pandya, D. N. (1989). Architecture and intrinsic connections of the prefrontal cortex in the rhesus monkey. *Journal of Comparative Neurology*, *286*(3), 353–375. https://doi.org/10.1002/cne.902860306.

Barrett, L. F., Mesquita, B., Ochsner, K. N., & Gross, J. J. (2007). The experience of emotion. *Annual Review of Psychology*, *58*, 373–403. https://doi.org/10.1146/annurev.psych.58.110405.085709.

Bartra, O., McGuire, J. T., & Kable, J. W. (2013). The valuation system: A coordinate-based meta-analysis of BOLD fMRI experiments examining neural correlates of subjective value. *NeuroImage*, *76*, 412–427. https://doi.org/10.1016/j.neuroimage.2013.02.063.

Basile, B. M., Schafroth, J. L., Karaskiewicz, C. L., Chang, S. W. C., & Murray, E. A. (2020). The anterior cingulate cortex is necessary for forming prosocial preferences from vicarious reinforcement in monkeys. *PLoS Biology*, *18*(6), e3000677. https://doi.org/10.1371/journal.pbio.3000677.

Behrens, T. E. J., Woolrich, M. W., Walton, M. E., & Rushworth, M. F. S. (2007). Learning the value of information in an uncertain world. *Nature Neuroscience*, *10*(9), 1214–1221. https://doi.org/10.1038/nn1954.

Blatt, G. J. (2012). The neuropathology of autism. *Scientifica*, *2012*, 703675. https://doi.org/10.6064/2012/703675.

Botvinick, M. M. (2007). Conflict monitoring and decision making: Reconciling two perspectives on anterior cingulate function. *Cognitive, Affective, & Behavioral Neuroscience*, 7(4), 356–366. https://doi.org/10.3758/CABN.7.4.356.

Bouret, S., & Richmond, B. J. (2010). Ventromedial and orbital prefrontal neurons differentially encode internally and externally driven motivational values in monkeys. *Journal of Neuroscience*, *30*(25), 8591–8601. https://doi.org/10.1523/JNEUROSCI.0049-10.2010.

Cai, X., & Padoa-Schioppa, C. (2012). Neuronal encoding of subjective value in dorsal and ventral anterior cingulate cortex. *Journal of Neuroscience*, *32*(11), 3791–3808. https://doi.org/10.1523/JNEUROSCI.3864-11.2012.

Calhoun, A. J., & Hayden, B. Y. (2015). The foraging brain. *Current Opinion in Behavioral Sciences*, *5*, 24–31. https://doi.org/10.1016/j.cobeha.2015.07.003.

Calzavara, R., Mailly, P., & Haber, S. N. (2007). Relationship between the corticostriatal terminals from areas 9 and 46, and those from area 8A, dorsal and rostral premotor cortex and area 24c: An anatomical substrate for cognition to action. *European Journal of Neuroscience*, *26*(7), 2005–2024. https://doi.org/10.1111/j.1460-9568.2007.05825.x.

Carmichael, S. T., & Price, J. L. (1994). Architectonic subdivision of the orbital and medial prefrontal cortex in the macaque monkey. *Journal of Comparative Neurology*, *346*(3), 366–402. https://doi.org/10.1002/cne.903460305.

Carmichael, S. T., & Price, J. L. (1995). Limbic connections of the orbital and medial prefrontal cortex in macaque monkeys. *Journal of Comparative Neurology*, *363*(4), 615–641. https://doi.org/10.1002/cne.903630408.

Carmichael, S. T., & Price, J. L. (1996). Connectional networks within the orbital and medial prefrontal cortex of macaque monkeys. *Journal of Comparative Neurology, 371*(2), 179–207. https://doi.org/10.1002/(SICI)1096-9861(19960722)371:2<179::AID-CNE1>3.0.CO;2-#.

Chang, S. W. C. (2017). An emerging field of primate social neurophysiology: Current developments. *eNeuro, 4*(5), ENEURO.0295-17.2017. https://doi.org/10.1523/ENEURO.0295-17.2017.

Chang, S. W. C., Gariépy, J.-F., & Platt, M. L. (2013). Neuronal reference frames for social decisions in primate frontal cortex. *Nature Neuroscience, 16*(2), 243–250. https://doi.org/10.1038/nn.3287.

Chang, S. W. C., Winecoff, A. A., & Platt, M. L. (2011). Vicarious reinforcement in rhesus macaques (Macaca Mulatta). *Frontiers in Neuroscience, 5*, 27. https://doi.org/10.3389/fnins.2011.00027.

Costa, V. D., & Averbeck, B. B. (2015). Frontal–parietal and limbic-striatal activity underlies information sampling in the best choice problem. *Cerebral Cortex, 25*(4), 972–982. https://doi.org/10.1093/cercor/bht286.

Craig, A. D. (2002). How do you feel? Interoception: The sense of the physiological condition of the body. *Nature Reviews Neuroscience, 3*(8), 655–666. https://doi.org/10.1038/nrn894.

Critchley, H. D. (2005). Neural mechanisms of autonomic, affective, and cognitive integration. *Journal of Comparative Neurology, 493*(1), 154–166. https://doi.org/10.1002/cne.20749.

Dal Monte, O., Chu, C. C. J., Fagan, N. A., & Chang, S. W. C. (2020). Specialized medial prefrontal–amygdala coordination in other-regarding decision preference. *Nature Neuroscience, 23*(4), 565–574. https://doi.org/10.1038/s41593-020-0593-y.

de Koning, P. P., Figee, M., van den Munckhof, P., Schuurman, P. R., & Denys, D. (2011). Current status of deep brain stimulation for obsessive-compulsive disorder: A clinical review of different targets. *Current Psychiatry Reports, 13*(4), 274–282. https://doi.org/10.1007/s11920-011-0200-8.

Delgado, M. R., Beer, J. S., Fellows, L. K., Huettel, S. A., Platt, M. L., Quirk, G. J., et al. (2016). Viewpoints: Dialogues on the functional role of the ventromedial prefrontal cortex. *Nature Neuroscience, 19*(12), 1545–1552. https://doi.org/10.1038/nn.4438.

Dowd, E. C., & Barch, D. M. (2010). Anhedonia and emotional experience in schizophrenia: Neural and behavioral indicators. *Biological Psychiatry, 67*(10), 902–911. https://doi.org/10.1016/j.biopsych.2009.10.020.

Dowd, E. C., & Barch, D. M. (2012). Pavlovian reward prediction and receipt in schizophrenia: Relationship to anhedonia. *PLoS One*, 7(5), e35622. https://doi.org/10.1371/journal.pone.0035622.

Drevets, W. C., Price, J. L., Simpson, J. R., Todd, R. D., Reich, T., Vannier, M., et al. (1997). Subgenual prefrontal cortex abnormalities in mood disorders. *Nature, 386*(6627), 824–827. https://doi.org/10.1038/386824a0.

Eilam-Stock, T., Xu, P., Cao, M., Gu, X., Van Dam, N. T., Anagnostou, E., et al. (2014). Abnormal autonomic and associated brain activities during rest in autism spectrum disorder. *Brain, 137*(1), 153–171. https://doi.org/10.1093/brain/awt294.

Enel, P., Wallis, J. D., & Rich, E. L. (2020). Stable and dynamic representations of value in the prefrontal cortex. *eLife, 9*, e54313. https://doi.org/10.7554/eLife.54313.

Figee, M., Luigjes, J., Smolders, R., Valencia-Alfonso, C.-E., van Wingen, G., de Kwaasteniet, B., et al. (2013). Deep brain stimulation restores frontostriatal network activity in obsessive-compulsive disorder. *Nature Neuroscience, 16*(4), 386–387. https://doi.org/10.1038/nn.3344.

Figee, M., Vink, M., de Geus, F., Vulink, N., Veltman, D. J., Westenberg, H., et al. (2011). Dysfunctional reward circuitry in obsessive-compulsive disorder. *Biological Psychiatry, 69*(9), 867–874. https://doi.org/10.1016/j.biopsych.2010.12.003.

Freemon, F. R. (1971). Akinetic mutism and bilateral anterior cerebral artery occlusion. *Journal of Neurology, Neurosurgery & Psychiatry, 34*(6), 693–698. https://doi.org/10.1136/jnnp.34.6.693.

Gehring, W. J., Goss, B., Coles, M. G. H., Meyer, D. E., & Donchin, E. (1993). A neural system for error detection and compensation. *Psychological Science, 4*(6), 385–390. https://doi.org/10.1111/j.1467-9280.1993.tb00586.x.

Gehring, W. J., Himle, J., & Nisenson, L. G. (2000). Action-monitoring dysfunction in obsessive-compulsive disorder. *Psychological Science, 11*(1), 1–6. https://doi.org/10.1111/1467-9280.00206.

Gillihan, S. J., Xia, C., Padon, A. A., Heberlein, A. S., Farah, M. J., & Fellows, L. K. (2011). Contrasting roles for lateral and ventromedial prefrontal cortex in transient and dispositional affective experience. *Social Cognitive and Affective Neuroscience, 6*(1), 128–137. https://doi.org/10.1093/scan/nsq026.

Glees, P., Cole, J., Whitty, C. W. M., & Cairns, H. (1950). The effects of lesions in the cingular gyrus and adjacent areas in monkeys. *Journal of Neurology, Neurosurgery, and Psychiatry, 13*(3), 178–190.

Gu, X., Eilam-Stock, T., Zhou, T., Anagnostou, E., Kolevzon, A., Soorya, L., et al. (2015). Autonomic and brain responses associated with empathy deficits in autism spectrum disorder. *Human Brain Mapping, 36*(9), 3323–3338. https://doi.org/10.1002/hbm.22840.

Gu, X., Zhou, T. J., Anagnostou, E., Soorya, L., Kolevzon, A., Hof, P. R., et al. (2018). Heightened brain response to pain anticipation in high-functioning adults with autism spectrum disorder. *The European Journal of Neuroscience, 47*(6), 592–601. https://doi.org/10.1111/ejn.13598.

Haber, S. N. (2016). Corticostriatal circuitry. *Dialogues in Clinical Neuroscience, 18*(1), 7–21.

Haber, S. N., Kim, K.-S., Mailly, P., & Calzavara, R. (2006). Reward-related cortical inputs define a large striatal region in primates that Interface with associative cortical connections, providing a substrate for incentive-based learning. *Journal of Neuroscience, 26*(32), 8368–8376. https://doi.org/10.1523/JNEUROSCI.0271-06.2006.

Haber, S. N., & Knutson, B. (2010). The reward circuit: Linking primate anatomy and human imaging. *Neuropsychopharmacology, 35*(1), 4–26. https://doi.org/10.1038/npp.2009.129.

Hadland, K. A., Rushworth, M. F. S., Gaffan, D., & Passingham, R. E. (2003). The effect of cingulate lesions on social behaviour and emotion. *Neuropsychologia, 41*(8), 919–931. https://doi.org/10.1016/S0028-3932(02)00325-1.

Harlow, J. M. (1869). *Recovery from the passage of an iron bar through the head.* David Clapp & Son. https://journals.sagepub.com/doi/10.1177/0957154X9300401407.

Haznedar, M. M., Roversi, F., Pallanti, S., Baldini-Rossi, N., Schnur, D. B., LiCalzi, E. M., et al. (2005). Fronto-thalamo-striatal gray and white matter volumes and anisotropy of their connections in bipolar spectrum illnesses. *Biological Psychiatry, 57*(7), 733–742. https://doi.org/10.1016/j.biopsych.2005.01.002.

Heilbronner, S. R., & Chafee, M. V. (2019). Learning how neurons fail inside of networks: Nonhuman primates provide critical data for psychiatry. *Neuron, 102*(1), 21–26. https://doi.org/10.1016/j.neuron.2019.02.030.

Heilbronner, S. R., & Hayden, B. Y. (2016). Dorsal anterior cingulate cortex: A bottom-up view. *Annual Review of Neuroscience, 39*(1), 149–170. https://doi.org/10.1146/annurev-neuro-070815-013952.

Holtzheimer, P. E., Husain, M. M., Lisanby, S. H., Taylor, S. F., Whitworth, L. A., McClintock, S., et al. (2017). Subcallosal cingulate deep brain stimulation for treatment-resistant depression: A multisite, randomised, sham-controlled trial. *The Lancet. Psychiatry, 4*(11), 839–849. https://doi.org/10.1016/S2215-0366(17)30371-1.

Hunt, L. T., Kolling, N., Soltani, A., Woolrich, M. W., Rushworth, M. F. S., & Behrens, T. E. J. (2012). Mechanisms underlying cortical activity during value-guided choice. *Nature Neuroscience, 15*(3), 470–476. S1-3 https://doi.org/10.1038/nn.3017.

Ito, S., Stuphorn, V., Brown, J. W., & Schall, J. D. (2003). Performance monitoring by the anterior cingulate cortex during saccade countermanding. *Science (New York, N.Y.)*, *302*(5642), 120–122. https://doi.org/10.1126/science.1087847.

Kaada, B. R., Pribram, K. H., & Epstein, J. A. (1949). Respiratory and vascular responses in monkeys from temporal pole, insula, orbital surface and cingulate gyrus; a preliminary report. *Journal of Neurophysiology*, *12*(5), 347–356. https://doi.org/10.1152/jn.1949.12.5.347.

Karlsson, M. P., Tervo, D. G. R., & Karpova, A. Y. (2012). Network resets in medial prefrontal cortex mark the onset of behavioral uncertainty. *Science*, *338*(6103), 135–139. https://doi.org/10.1126/science.1226518.

Kemper, T. L., & Bauman, M. L. (1993). The contribution of neuropathologic studies to the understanding of autism. *Neurologic Clinics*, *11*(1), 175–187. https://doi.org/10.1016/S0733-8619(18)30176-2.

Kennerley, S. W., Behrens, T. E. J., & Wallis, J. D. (2011). Double dissociation of value computations in orbitofrontal and anterior cingulate neurons. *Nature Neuroscience*, *14*(12), 1581–1589. https://doi.org/10.1038/nn.2961.

Kennerley, S. W., Walton, M. E., Behrens, T. E. J., Buckley, M. J., & Rushworth, M. F. S. (2006). Optimal decision making and the anterior cingulate cortex. *Nature Neuroscience*, *9*(7), 940–947. https://doi.org/10.1038/nn1724.

Kolling, N., Wittmann, M. K., Behrens, T. E. J., Boorman, E. D., Mars, R. B., & Rushworth, M. F. S. (2016). Value, search, persistence and model updating in anterior cingulate cortex. *Nature Neuroscience*, *19*(10), 1280–1285. https://doi.org/10.1038/nn.4382.

Koski, J. E., McHaney, J. R., Rigney, A. E., & Beer, J. S. (2020). Reconsidering longstanding assumptions about the role of medial prefrontal cortex (MPFC) in social evaluation. *NeuroImage*, *214*, 116752. https://doi.org/10.1016/j.neuroimage.2020.116752.

Kringelbach, M. L., & Rolls, E. T. (2004). The functional neuroanatomy of the human orbitofrontal cortex: Evidence from neuroimaging and neuropsychology. *Progress in Neurobiology*, *72*(5), 341–372. https://doi.org/10.1016/j.pneurobio.2004.03.006.

Kylliäinen, A., & Hietanen, J. K. (2006). Skin conductance responses to another person's gaze in children with autism. *Journal of Autism and Developmental Disorders*, *36*(4), 517–525. https://doi.org/10.1007/s10803-006-0091-4.

Lord, C., Brugha, T. S., Charman, T., Cusack, J., Dumas, G., Frazier, T., et al. (2020). Autism spectrum disorder. *Nature Reviews. Disease Primers*, *6*(1), 5. https://doi.org/10.1038/s41572-019-0138-4.

Luk, C.-H., & Wallis, J. D. (2013). Choice coding in frontal cortex during stimulus-guided or action-guided decision-making. *The Journal of Neuroscience: The Official Journal of the Society for Neuroscience*, *33*(5), 1864–1871. https://doi.org/10.1523/JNEUROSCI.4920-12.2013.

Luppino, G., Matelli, M., Camarda, R., & Rizzolatti, G. (1993). Corticocortical connections of area F3 (SMA-proper) and area F6 (pre-SMA) in the macaque monkey. *The Journal of Comparative Neurology*, *338*(1), 114–140. https://doi.org/10.1002/cne.903380109.

Mars, R. B., Klein, M. C., Neubert, F.-X., Olivier, E., Buch, E. R., Boorman, E. D., et al. (2009). Short-latency influence of medial frontal cortex on primary motor cortex during action selection under conflict. *The Journal of Neuroscience: The Official Journal of the Society for Neuroscience*, *29*(21), 6926–6931. https://doi.org/10.1523/JNEUROSCI.1396-09.2009.

Mars, R. B., Verhagen, L., Gladwin, T. E., Neubert, F.-X., Sallet, J., & Rushworth, M. F. S. (2016). Comparing brains by matching connectivity profiles. *Neuroscience & Biobehavioral Reviews*, *60*, 90–97. https://doi.org/10.1016/j.neubiorev.2015.10.008.

Matsumoto, M., Matsumoto, K., Abe, H., & Tanaka, K. (2007). Medial prefrontal cell activity signaling prediction errors of action values. *Nature Neuroscience*, *10*(5), 647–656. https://doi.org/10.1038/nn1890.

Mayberg, H. S., Brannan, S. K., Mahurin, R. K., Jerabek, P. A., Brickman, J. S., Tekell, J. L., et al. (1997). Cingulate function in depression: A potential predictor of treatment response. *Neuroreport*, *8*(4), 1057–1061. https://doi.org/10.1097/00001756-199703030-00048.

Mayberg, H. S., Lozano, A. M., Voon, V., McNeely, H. E., Seminowicz, D., Hamani, C., et al. (2005). Deep brain stimulation for treatment-resistant depression. *Neuron, 45*(5), 651–660. https://doi.org/10.1016/j.neuron.2005.02.014.

McGovern, R. A., & Sheth, S. A. (2017). Role of the dorsal anterior cingulate cortex in obsessive-compulsive disorder: Converging evidence from cognitive neuroscience and psychiatric neurosurgery. *Journal of Neurosurgery, 126*(1), 132–147. https://doi.org/10.3171/2016.1.JNS15601.

McGuire, J. T., & Kable, J. W. (2013). Rational temporal predictions can underlie apparent failures to delay gratification. *Psychological Review, 120*(2), 395–410. https://doi.org/10.1037/a0031910.

McGuire, J. T., & Kable, J. W. (2015). Medial prefrontal cortical activity reflects dynamic re-evaluation during voluntary persistence. *Nature Neuroscience, 18*(5), 760–766. https://doi.org/10.1038/nn.3994.

Milad, M. R., & Rauch, S. L. (2012). Obsessive-compulsive disorder: Beyond segregated cortico-striatal pathways. *Trends in Cognitive Sciences, 16*(1), 43–51. https://doi.org/10.1016/j.tics.2011.11.003.

Minshew, N. J., & Keller, T. A. (2010). The nature of brain dysfunction in autism: Functional brain imaging studies. *Current Opinion in Neurology, 23*(2), 124–130. https://doi.org/10.1097/WCO.0b013e32833782d4.

Murray, J. D., Bernacchia, A., Freedman, D. J., Romo, R., Wallis, J. D., Cai, X., et al. (2014). A hierarchy of intrinsic timescales across primate cortex. *Nature Neuroscience, 17*(12), 1661–1663. https://doi.org/10.1038/nn.3862.

Murray, E. A., & Rudebeck, P. H. (2018). Specializations for reward-guided decision-making in the primate ventral prefrontal cortex. *Nature Reviews Neuroscience, 19*(7), 404–417. https://doi.org/10.1038/s41583-018-0013-4.

Neubert, F.-X., Mars, R. B., Sallet, J., & Rushworth, M. F. S. (2015). Connectivity reveals relationship of brain areas for reward-guided learning and decision making in human and monkey frontal cortex. *Proceedings of the National Academy of Sciences of the United States of America, 112*(20), E2695–E2704. https://doi.org/10.1073/pnas.1410767112.

Noonan, M. P., Sallet, J., Rudebeck, P. H., Buckley, M. J., & Rushworth, M. F. (2010). Does the medial orbitofrontal cortex have a role in social valuation? *European Journal of Neuroscience, 31*(12), 2341–2351. https://doi.org/10.1111/j.1460-9568.2010.07271.x.

Noonan, M. P., Walton, M. E., Behrens, T. E. J., Sallet, J., Buckley, M. J., & Rushworth, M. F. S. (2010). Separate value comparison and learning mechanisms in macaque medial and lateral orbitofrontal cortex. *Proceedings of the National Academy of Sciences of the United States of America, 107*(47), 20547–20552. https://doi.org/10.1073/pnas.1012246107.

Noritake, A., Ninomiya, T., & Isoda, M. (2018). Social reward monitoring and valuation in the macaque brain. *Nature Neuroscience, 21*(10), 1452–1462. https://doi.org/10.1038/s41593-018-0229-7.

Noritake, A., Ninomiya, T., & Isoda, M. (2020). Representation of distinct reward variables for self and other in primate lateral hypothalamus. *Proceedings of the National Academy of Sciences of the United States of America, 117*(10), 5516–5524. https://doi.org/10.1073/pnas.1917156117.

Öngür, D., An, X., & Price, J. L. (1998). Prefrontal cortical projections to the hypothalamus in Macaque monkeys. *Journal of Comparative Neurology, 401*(4), 480–505. https://doi.org/10.1002/(SICI)1096-9861(19981130)401:4<480::AID-CNE4>3.0.CO;2-F.

Öngür, D., Ferry, A. T., & Price, J. L. (2003). Architectonic subdivision of the human orbital and medial prefrontal cortex. *Journal of Comparative Neurology, 460*(3), 425–449. https://doi.org/10.1002/cne.10609.

Öngür, D., & Price, J. L. (2000). The organization of networks within the orbital and medial prefrontal cortex of rats, monkeys and humans. *Cerebral Cortex, 10*(3), 206–219. https://doi.org/10.1093/cercor/10.3.206.

O'Reilly, J. X., Schüffelgen, U., Cuell, S. F., Behrens, T. E. J., Mars, R. B., & Rushworth, M. F. S. (2013). Dissociable effects of surprise and model update in parietal and anterior cingulate cortex. *Proceedings of the National Academy of Sciences of the United States of America, 110*(38), E3660–E3669. https://doi.org/10.1073/pnas.1305373110.

Padoa-Schioppa, C., & Assad, J. A. (2006). Neurons in the orbitofrontal cortex encode economic value. *Nature, 441*(7090), 223–226. https://doi.org/10.1038/nature04676.

Padoa-Schioppa, C., & Conen, K. E. (2017). Orbitofrontal cortex: A neural circuit for economic decisions. *Neuron, 96*(4), 736–754. https://doi.org/10.1016/j.neuron.2017.09.031.

Pallanti, S., & Quercioli, L. (2006). Treatment-refractory obsessive-compulsive disorder: Methodological issues, operational definitions and therapeutic lines. *Progress in Neuro-Psychopharmacology & Biological Psychiatry, 30*(3), 400–412. https://doi.org/10.1016/j.pnpbp.2005.11.028.

Passingham, R. E., & Wise, S. P. (2012). *The neurobiology of the prefrontal cortex: Anatomy, evolution, and origin of insight.* Oxford University Press.

Pelletier, G., & Fellows, L. K. (2019). A critical role for human ventromedial frontal lobe in value comparison of complex objects based on attribute configuration. *Journal of Neuroscience, 39*(21), 4124–4132. https://doi.org/10.1523/JNEUROSCI.2969-18.2019.

Petrides, M., Tomaiuolo, F., Yeterian, E. H., & Pandya, D. N. (2012). The prefrontal cortex: Comparative architectonic organization in the human and the macaque monkey brains. *Cortex, 48*(1), 46–57. https://doi.org/10.1016/j.cortex.2011.07.002.

Phillips, M. L., Drevets, W. C., Rauch, S. L., & Lane, R. (2003). Neurobiology of emotion perception II: Implications for major psychiatric disorders. *Biological Psychiatry, 54*(5), 515–528. https://doi.org/10.1016/s0006-3223(03)00171-9.

Picard, N., & Strick, P. L. (1996). Motor areas of the medial wall: A review of their location and functional activation. *Cerebral Cortex, 6*(3), 342–353. https://doi.org/10.1093/cercor/6.3.342.

Picard, N., & Strick, P. L. (2001). Imaging the premotor areas. *Current Opinion in Neurobiology, 11*(6), 663–672. https://doi.org/10.1016/S0959-4388(01)00266-5.

Pitman, R. K. (1987). A cybernetic model of obsessive-compulsive psychopathology. *Comprehensive Psychiatry, 28*(4), 334–343. https://doi.org/10.1016/0010-440x(87)90070-8.

Pizzagalli, D., Pascual-Marqui, R. D., Nitschke, J. B., Oakes, T. R., Larson, C. L., Abercrombie, H. C., et al. (2001). Anterior cingulate activity as a predictor of degree of treatment response in major depression: Evidence from brain electrical tomography analysis. *American Journal of Psychiatry, 158*(3), 405–415. https://doi.org/10.1176/appi.ajp.158.3.405.

Plassmann, H., O'Doherty, J. P., & Rangel, A. (2010). Appetitive and aversive goal values are encoded in the medial orbitofrontal cortex at the time of decision making. *The Journal of Neuroscience: The Official Journal of the Society for Neuroscience, 30*(32), 10799–10808. https://doi.org/10.1523/JNEUROSCI.0788-10.2010.

Price, J. L., & Drevets, W. C. (2010). Neurocircuitry of mood disorders. *Neuropsychopharmacology, 35*(1), 192–216. https://doi.org/10.1038/npp.2009.104.

Rane, P., Cochran, D., Hodge, S. M., Haselgrove, C., Kennedy, D. N., & Frazier, J. A. (2015). Connectivity in autism: A review of MRI connectivity studies. *Harvard Review of Psychiatry, 23*(4), 223–244. https://doi.org/10.1097/HRP.0000000000000072.

Rich, E. L., & Wallis, J. D. (2014). Functional organization of the orbitofrontal cortex. *Journal of Cognitive Neuroscience, 26*(7), 1347–1362. https://doi.org/10.1162/jocn_a_00573.

Rich, E. L., & Wallis, J. D. (2016). Decoding subjective decisions from orbitofrontal cortex. *Nature Neuroscience, 19*(7), 973–980. https://doi.org/10.1038/nn.4320.

Rippon, G., Brock, J., Brown, C., & Boucher, J. (2007). Disordered connectivity in the autistic brain: Challenges for the 'new psychophysiology'. *International Journal of Psychophysiology, 63*(2), 164–172. https://doi.org/10.1016/j.ijpsycho.2006.03.012.

Riva-Posse, P., Choi, K. S., Holtzheimer, P. E., Crowell, A. L., Garlow, S. J., Rajendra, J. K., et al. (2018). A connectomic approach for subcallosal cingulate deep brain stimulation surgery: Prospective targeting in treatment-resistant depression. *Molecular Psychiatry, 23*(4), 843–849. https://doi.org/10.1038/mp.2017.59.

Rudebeck, P. H., Buckley, M. J., Walton, M. E., & Rushworth, M. F. S. (2006). A role for the macaque anterior cingulate gyrus in social valuation. *Science, 313*(5791), 1310–1312. https://doi.org/10.1126/science.1128197.

Rudebeck, P. H., Putnam, P. T., Daniels, T. E., Yang, T., Mitz, A. R., Rhodes, S. E. V., et al. (2014). A role for primate subgenual cingulate cortex in sustaining autonomic arousal. *Proceedings of the National Academy of Sciences of the United States of America, 111*(14), 5391–5396. https://doi.org/10.1073/pnas.1317695111.

Rudebeck, P. H., Rich, E. L., & Mayberg, H. S. (2019). From bed to bench side: Reverse translation to optimize neuromodulation for mood disorders. *Proceedings of the National Academy of Sciences of the United States of America, 116*(52), 26288–26296. https://doi.org/10.1073/pnas.1902287116.

Rushworth, M. F. S., Noonan, M. P., Boorman, E. D., Walton, M. E., & Behrens, T. E. (2011). Frontal cortex and reward-guided learning and decision-making. *Neuron, 70*(6), 1054–1069. https://doi.org/10.1016/j.neuron.2011.05.014.

Saver, J. L., & Damasio, A. R. (1991). Preserved access and processing of social knowledge in a patient with acquired sociopathy due to ventromedial frontal damage. *Neuropsychologia, 29*(12), 1241–1249. https://doi.org/10.1016/0028-3932(91)90037-9.

Schlagenhauf, F., Sterzer, P., Schmack, K., Ballmaier, M., Rapp, M., Wrase, J., et al. (2009). Reward feedback alterations in unmedicated schizophrenia patients: Relevance for delusions. *Biological Psychiatry, 65*(12), 1032–1039. https://doi.org/10.1016/j.biopsych.2008.12.016.

Shenhav, A., Cohen, J. D., & Botvinick, M. M. (2016). Dorsal anterior cingulate cortex and the value of control. *Nature Neuroscience, 19*(10), 1286–1291. https://doi.org/10.1038/nn.4384.

Shidara, M., & Richmond, B. J. (2002). Anterior cingulate: Single neuronal signals related to degree of reward expectancy. *Science (New York, N.Y.), 296*(5573), 1709–1711. https://doi.org/10.1126/science.1069504.

Shima, K., & Tanji, J. (1998). Role for cingulate motor area cells in voluntary movement selection based on reward. *Science, 282*(5392), 1335–1338. https://doi.org/10.1126/science.282.5392.1335.

Simms, M. L., Kemper, T. L., Timbie, C. M., Bauman, M. L., & Blatt, G. J. (2009). The anterior cingulate cortex in autism: Heterogeneity of qualitative and quantitative cytoarchitectonic features suggests possible subgroups. *Acta Neuropathologica, 118*(5), 673–684. https://doi.org/10.1007/s00401-009-0568-2.

Smith, W. K. (1945). The functional significance of the rostral cingular cortex as revealed by its responses to electrical excitation. *Journal of Neurophysiology, 8*(4), 241–255. https://doi.org/10.1152/jn.1945.8.4.241.

Stange, J. P., Hamilton, J. L., Olino, T. M., Fresco, D. M., & Alloy, L. B. (2017). Autonomic reactivity and vulnerability to depression: A multi-wave study. *Emotion (Washington, D.C.), 17*(4), 602–615. https://doi.org/10.1037/emo0000254.

Strait, C. E., Blanchard, T. C., & Hayden, B. Y. (2014). Reward value comparison via mutual inhibition in ventromedial prefrontal cortex. *Neuron, 82*(6), 1357–1366. https://doi.org/10.1016/j.neuron.2014.04.032.

Tamir, D. I., & Mitchell, J. P. (2010). Neural correlates of anchoring-and-adjustment during mentalizing. *Proceedings of the National Academy of Sciences of the United States of America, 107*(24), 10827–10832. https://doi.org/10.1073/pnas.1003242107.

Thorpe, S. J., Rolls, E. T., & Maddison, S. (1983). The orbitofrontal cortex: Neuronal activity in the behaving monkey. *Experimental Brain Research, 49*(1), 93–115. https://doi.org/10.1007/BF00235545.

Tom, S. M., Fox, C. R., Trepel, C., & Poldrack, R. A. (2007). The neural basis of loss aversion in decision-making under risk. *Science (New York, N.Y.)*, *315*(5811), 515–518. https://doi.org/10.1126/science.1134239.

Vaidya, A. R., & Fellows, L. K. (2019). Ventromedial frontal lobe damage affects interpretation, not exploration, of emotional facial expressions. *Cortex*, *113*, 312–328. https://doi.org/10.1016/j.cortex.2018.12.013.

Varghese, M., Keshav, N., Jacot-Descombes, S., Warda, T., Wicinski, B., Dickstein, D. L., et al. (2017). Autism spectrum disorder: Neuropathology and animal models. *Acta Neuropathologica*, *134*(4), 537–566. https://doi.org/10.1007/s00401-017-1736-4.

Vaughan Van Hecke, A., Lebow, J., Bal, E., Lamb, D., Harden, E., Kramer, A., et al. (2009). Electroencephalogram and heart rate regulation to familiar and unfamiliar people in children with autism spectrum disorders. *Child Development*, *80*(4), 1118–1133. https://doi.org/10.1111/j.1467-8624.2009.01320.x.

Vogt, B. A., Pandya, D. N., & Rosene, D. L. (1987). Cingulate cortex of the rhesus monkey: I. Cytoarchitecture and thalamic afferents. *The Journal of Comparative Neurology*, *262*(2), 256–270. https://doi.org/10.1002/cne.902620207.

Ward, A. A. (1948). The cingular gyrus: Area 24. *Journal of Neurophysiology*, *11*(1), 13–23. https://doi.org/10.1152/jn.1948.11.1.13.

Wise, S. P. (2008). Forward frontal fields: Phylogeny and fundamental function. *Trends in Neurosciences*, *31*(12), 599–608. https://doi.org/10.1016/j.tins.2008.08.008.

Yoshida, K., Saito, N., Iriki, A., & Isoda, M. (2011). Representation of others' action by neurons in monkey medial frontal cortex. *Current Biology*, *21*(3), 249–253. https://doi.org/10.1016/j.cub.2011.01.004.

Yoshida, K., Saito, N., Iriki, A., & Isoda, M. (2012). Social error monitoring in macaque frontal cortex. *Nature Neuroscience*, *15*(9), 1307–1312. https://doi.org/10.1038/nn.3180.

CHAPTER TWO

Medial prefrontal cortex encoding of stress and anxiety

David S. Jacobs and Bita Moghaddam*
Department of Behavioral Neuroscience, Oregon Health & Science University, Portland, OR, United States
*Corresponding author: e-mail address: bita@ohsu.edu

Contents

Abstract

The prefrontal cortex (PFC) is involved in adaptive control of behavior and optimizing action selection. When an organism is experiencing an aversive event, such as a sustained state of anxiety or an overt experience of fear or stress, the mechanisms that govern PFC regulation of action selection may be critical for survival. A large body of literature has shown that acute aversive states influence the activity of PFC neurons and the release of neurotransmitters in this region. These states also result in long-term neurobiological changes in the PFC and expression of PFC-dependent motivated behaviors. The mechanism for how these changes lead to modifying action selection is only recently beginning to emerge. Here, we review animal and human studies into the neural mechanisms which may mediate the adaptive changes in the PFC that emerge during negative affective states. We then highlight recent advances in approaches for understanding how anxiety influences action selection and related cortical processes. We conclude by proposing that PFC neurons selectively influence action encoding during conditions where actions toward obtaining a reward or avoiding harm are executed under a fog of fear and anxiety.

International Review of Neurobiology, Volume 158
ISSN 0074-7742
https://doi.org/10.1016/bs.irn.2020.11.014

1. Introduction

The prefrontal cortex (PFC) has been dubbed the "doer" cortex because of its extensive involvement in the planning and adaptive control of motivated actions (Fuster, 2015). The PFC integrates neural signals from multiple brain regions, including sensory, motor, and limbic systems, and encodes actions and outcomes toward goals. Ultimately these processes allow the PFC to flexibly plan and adapt behavioral processes in accordance with the demands of the environment. The executive functions of the PFC are far from simple, sitting within a hierarchy that takes multidimensional information into account (Fuster, 2001). One such piece of information is internal affective and motivational states, which can bias attention and plan actions in accordance with an organism's own goals such as acquiring resources or minimizing harm (Compton, 2003; Naqvi, Shiv, & Bechara, 2006). Affective states, in particular those associated with negative emotions such as fear and anxiety, are powerful influencers of goal-directed actions encoded by the PFC (Bagozzi & Pieters, 1998; Bogdan & Pizzagalli, 2006; Devilbiss, Jenison, & Berridge, 2012; Devilbiss, Spencer, & Berridge, 2017; Park & Moghaddam, 2017b).

How do PFC neurons respond and adapt to stress, fear, and anxiety? And how do these aversive states influence ongoing encoding of goal-directed actions and the adaptive control of behavior by the PFC? A large literature has been providing key data to begin to address these questions. Here, we will review some of this literature with an emphasis on the importance of the sensitivity of PFC neurons to anxiety, and how affective states can influence PFC encoding of motivated actions. Much of this work has been performed in rodents, which have a less developed PFC compared to primates. We, therefore, begin with a brief overview of functional and anatomical properties of rodent PFC before we delve into the literature related to the impact of aversive states on PFC.

2. Brief review of rodent medial prefrontal cortex subdivisions

Rodent models have been instrumental in increasing our understanding of the neuronal basis of behaviors modulated or mediated by the PFC. These models have been critical in informing us about the computational processes performed by the PFC and have typically focused on the medial wall of prefrontal cortical (mPFC) regions of the rodent brain. While broad outlines of the contributions of mPFC subregions to behavior have emerged, considerable debate still remains about the exact functions of each of these cortical domains and, in particular, whether there are strict functional-anatomical analogues between primate and rodent PFC (Laubach, Amarante, Swanson, & White, 2018; Preuss, 1995; Rushworth & Owen, 1998). The rodent mPFC has two primary subregions: the dorsal prelimbic (PL) region (which can also include the anterior cingulate cortex (AC)) and the ventral infralimbic (IL) region. A rich literature has addressed how these subregions differ in the encoding of stress and anxiety (see below). Furthermore, there has been a recent focus on including subregion distinctions given their possible differences in functionality and homologies to the primate cortex (Laubach et al., 2018). The exact homolog of the mPFC and its subregions in regards to the brain of the primates is also a matter of debate (Laubach et al., 2018; Preuss, 1995; Uylings, Groenewegen, & Kolb, 2003), with early studies suggesting that the rodent mPFC represents the primate dorsolateral PFC, and more recent accounts positing that it may represent the anterior midcingulate cortex, pregenual AC, and subgenual AC for rodent AC, PL, and IL regions, respectively (see Laubach et al., 2018). Regardless, there is general agreement that the mPFC in rodents is involved in several analogous functions of the dorsolateral, medial, and cingulate cortical regions in humans and monkeys such as working memory, response control, updating, and affective regulation including sensitivity to stressors (Arnsten, 2015; Floresco, Magyar, Ghods-Sharifi, Vexelman, & Maric, 2006; Granon & Poucet, 1995; Kesner, Hunt, Williams, & Long, 1996; Miller, 2000; Wager, Davidson, Hughes, Lindquist, & Ochsner, 2008). This makes rodents a valuable organism for assessing neural underpinnings of these PFC-dependent constructs.

3. Stress and medial prefrontal cortex function

Stress and isolation of a "stress response" can be hard to quantify. In this context, we will consider a stress response as any response to experimental manipulation involving a noxious physical or emotional stimulus. A stress response encapsulates a variety of physiological and psychological changes such as triggering the hypothalamic-pituitary-adrenal (HPA) axis and increasing vigilance, attention, and autonomic functions (heart rate, respiration, etc.). Several brain regions respond acutely and adaptively to stressful manipulations and are often grouped as "stress circuitry." These include the amygdala (particularly the basal lateral nucleus), bed nucleus of the stria terminalis (BNST), hippocampus, and hypothalamus (McEwen et al., 2015; McEwen, Nasca, & Gray, 2016; Walker, Toufexis, & Davis, 2003). The PFC is reciprocally and in a topographically organized matter connected to a number of these regions. For example, PL and IL regions show afferent and efferent connectivity to the brainstem, hypothalamus, and thalamus (Hoover & Vertes, 2007; Vertes, 2004). Subregion specificity is also present; while the PL-mPFC has extensive efferent connections to the basolateral nucleus of the amygdala, the IL-mPFC has greater connectivity to the BNST and central nucleus of the amygdala (Vertes, 2004). It is therefore not surprising that mPFC is exquisitely sensitive to stressful manipulations (Arnsten, 2015). Stress increases the expression of neural activity markers such as c-Fos, and glutamate and monoamine release in the mPFC (Abercrombie, Keefe, DiFrischia, & Zigmond, 1989; Moghaddam, 1993; Morrow, Elsworth, Lee, & Roth, 2000; Ostrander, Richtand, & Herman, 2003; see Holmes & Wellman, 2009 for review). Furthermore, PL/AC-mPFC principal neurons increase their firing rate and local field potential theta power in the PL/AC mPFC after exposure to acute stressors (Del Arco, Park, & Moghaddam, 2020) and adapt their response to repeated stress (Jackson & Moghaddam, 2006). In humans, it is similarly found that elevations in blood oxygen level dependent (BOLD) activity are seen in the dorsolateral and ventrolateral regions of the PFC following cold stress using functional magnetic resonance imaging (fMRI; Porcelli et al., 2008). Collectively, these data suggest that acute stress has a transient influence on neuronal activity in the mPFC.

While the acute stress response is a well-conserved adaptive response, chronic stress can have detrimental effects on health and result in brain-wide neurobiological changes (McEwen & Morrison, 2013). Critically, the overlapping neural circuitry between stress and affective states suggest that chronic stress may contribute to the etiology of affective disorders such as anxiety and panic disorders (Calhoon & Tye, 2015; Daviu, Bruchas, Moghaddam, Sandi, & Beyeler, 2019; Shin & Liberzon, 2010). Chronic stress has been found to decrease pyramidal neuron dendritic branching, decrease excitatory neural NMDA and AMPA currents, and modify the activity of γ-aminobutyric acid (GABA) interneurons which inhibit pyramidal neurons. Overall, chronic stress appears to generally decrease pyramidal neuron tone and increase GABAergic inhibition, resulting in decreased mPFC output (Cook & Wellman, 2004; Radley et al., 2006; Yuen et al., 2012; McKlveen et al., 2016, but see McKlveen, Moloney, Scheimann,

Myers, & Herman, 2019 for review). This state of PFC hypoactivity may allow for pathological and uncontrollable negative affective states to develop. Excitatory mPFC outputs control HPA axis activity via negative feedback and inhibit activation of the paraventricular nucleus of the hypothalamus (Radley, Arias, & Sawchenko, 2006; Radley, Williams, & Sawchenko, 2008). Thus, a decrease in mPFC output tone may attenuate inhibitory control of the stress response. There is some evidence for divergent roles of PL-mPFC and IL-mPFC regions, whereby each produces HPA axis inhibition and excitation, respectively. This is particularly intriguing because some studies have also shown IL-mPFC neurons display dendritic retraction in response to acute stressors (Radley, Arias, & Sawchenko, 2006), which would indicate a region-specific dissociation in stress responsivity. However, the precise roles for these subregions in the regulation of stress response remain unknown and an ongoing focus of research (see Holmes & Wellman, 2009 for review).

Collectively, studies so far support the idea that stress exposure, particularly when it is chronic, remodels the mPFC and may have important consequences for mPFC-mediated functions. This is consistent with clinical literature showing, for example, that stress and trauma are associated with changes in BOLD neural activity in the medial, subgenual cingulate, and dorsolateral regions of the human PFC (Han et al., 2016; Olson, Kaiser, Pizzagalli, Rauch, & Rosso, 2019; Pizzagalli et al., 2004), which are functionally analogous to the rodent mPFC and serve as executive network hubs for behavior. There have been a number of excellent reviews on preclinical work into the mechanisms of chronic stress on mPFC-dependent behaviors (see Arnsten, 2015; Sandi, 2013). An extensive discussion of this literature is outside the scope of this chapter but we will briefly summarize some of the key findings regarding stress- and mPFC-dependent behavior below.

Impact of stress on working memory has been extensively studied in the mPFC, which recruits the PL and IL subregions in rodents and dorsolateral regions in non-human primates (Cassaday, Nelson, & Pezze, 2014; Goldman-Rakic, 1995; Murphy, Arnsten, Jentsch, & Roth, 1996; Yang, Shi, Wang, Peng, & Li, 2014 for review). In rats, increases in firing rate and synchrony of neurons is observed during working memory tasks with one study reporting that 93% of recorded neurons showed time locked changes in activity during a Y-maze delayed alternation task (Yang et al., 2014). Complementing the impact of acute stress on working memory (Murphy et al., 1996), chronic stress paradigms have been shown to impair working memory. For example, using the morris water maze, a task where animals must remember and utilize external clues to find a platform hidden in a pool of water, chronic stress was found to impair working memory in association with glutamate neuron dendritic atrophy in the AC, PL, and IL-mPFC (Cerqueira, Mailliet, Almeida, Jay, & Sousa, 2007). Other studies have shown that chronic stress may be causing working memory deficits by attenuating dopamine neuromodulation (Mizoguchi et al., 2000) or disrupting place cell activity in the hippocampus (Kim et al., 2007).

Another critical function of mPFC is behavioral flexibility, which is often quantified using set-shifting tasks (Birrell & Brown, 2000; Floresco, Block, & Maric, 2008). Here, subjects are required to learn and switch between distinct rules (or sets) to guide actions which are motivated by a reward. Successful performance of these tasks requires the mPFC to encode actions and outcomes. Previous studies have shown that PL-mPFC neurons show time-locked firing rate changes at the time of action execution and the corresponding outcome period (Del Arco, Park, Wood, Kim, & Moghaddam, 2017; Park, Wood, Bondi, Del Arco, & Moghaddam, 2016). These neural activity patterns likely reflect real-time encoding of actions because (1) most of these neurons display a peak correlation with current trial actions and outcomes and (2) neural population signals during encoding periods (i.e., the post-action period) can discriminate between rules and be used to predict the current rule (Del Arco et al., 2017; Park et al., 2016). Finally, in a version of a set shifting task where actions are randomly reinforced (i.e., no rule is "correct" and there is no cognitive demand), significant levels of neural encoding in the post action period are not observed, while outcome responsivity of mPFC neurons is preserved (Del Arco et al., 2017). Taken together, these studies provide evidence that mPFC neurons generally encode outcomes regardless of cognitive demand but monitor and encode actions when cognitive engagement is required/beneficial.

In addition to acute stress, several studies have investigated the impact of chronic stress on set-shifting behavior. Bondi and colleagues investigated how 2 weeks of unpredictable stressors effected set shifting behavior. Chronic stress produced deficits in set shifting by increasing the number of trials needed to reach a learning criteria (Bondi, Rodriguez, Gould, Frazer, & Morilak, 2008). Additional studies have implicated the mPFC in chronic stress-induced deficits by demonstrating a positive correlation between impaired set shifting abilities and stress-related dendritic retraction in the AC-mPFC (Liston et al., 2006). Thus, these studies indicate that the mPFC is involved in encoding motivated actions and that chronic stress likely disrupts this function and compromises behavioral flexibility. The specific effects of chronic stress on single unit activity during set-shifting behavior remain to be determined. It is likely this will involve suppressed activity at peri-choice epochs, as this was recently demonstrated in chronically stressed mice which showed abnormal behavioral adaptation strategies in a cost-benefit choice task (Friedman et al., 2017).

These results highlight the sensitivity of mPFC neurons to stress and the associated neurobiological changes, which in the case of chronic stress reflects reduced excitatory neuron output and disrupted mPFC-dependent behavioral function. The mPFC disrupted activity may reduce top-down control of brain regions involved in the stress response, permitting enhanced subcortical control over behavior. This may result in actions being preferentially guided by stimuli present in the immediate environment and by habit, rather than by future goals (Schwabe & Wolf, 2011). Regardless of the exact cause of disruption from chronic stress, these findings indicate that stress can have powerful effects on mPFC neural activity and behavior.

4. The medial prefrontal cortex encoding of fear states and fear learning

Fear is generally defined as distressing emotions caused by impending and certain danger. MPFC encoding of fear is commonly assessed using Pavlovian fear conditioning. This procedure applies an aversive unconditioned stimulus (US), such as a footshock, paired with a neutral conditioned stimulus (CS), such as a tone, that elicits a conditioned behavioral fear response (CR) such as freezing or darting (Gruene, Flick, Stefano, Shea, & Shansky, 2015; Johansen, Cain, Ostroff, & LeDoux, 2011). Thus, the main measure in these and similar paradigms is the strength of the association between the CS and US through the CR.

It is well documented that expression of freezing behavior in Pavlovian conditioning requires the PL-mPFC. Inactivation of the PL-mPFC after fear learning prevents conditioned freezing (Corcoran and Quirk, 2007). However, when this region is inactivated during the acquisition of fear learning and the subject is tested for fear expression later, the fear expression remains intact (Corcoran & Quirk, 2007). Furthermore, fear responses to predator odor and in a novel open field, where learned associations are not required, remain intact following inactivation of the PL-mPFC. Thus, mPFC circuitry is critical for expression of learned fear responses but not necessary for the acquisition of fear learning itself nor for innate fear expression. At the neural level, single unit recording from the PL-mPFC neurons of putative excitatory and inhibitory subtypes shows heterogeneous but time-locked changes in firing rate when fear associated stimuli are presented and during freezing (Baeg et al., 2001; Courtin et al., 2014). In addition, the excitatory PL-mPFC response to a CS is associated with higher levels of fear expression across all phases of fear learning (i.e., habituation, conditioning, and extinction; Burgos-Robles, Vidal-Gonzalez, & Quirk, 2009). IL-mPFC neuron inactivation, however, does not influence fear learning (Sierra-Mercado, Padilla-Coreano, & Quirk, 2011) and IL-mPFC neurons fail to show time-locked responses to CS during the acquisition phase of fear learning (Milad & Quirk, 2002). These findings suggest that for the development of associations between CS and US, the PL-mPFC shows high involvement, while the IL-mPFC does not.

Fear conditioning paradigms can also assess behavioral flexibility by a process known as fear extinction. The extinction procedure involves extinguishing the CS-US association, after freezing in response to the CS is learned, by presenting the CS while omitting the US. Animals' ability to learn the change in contingencies is measured by a reduction (or extinction) in fear responses after the CS. While extinction of freezing is unchanged by lesions in the PL-mPFC, it is impaired by lesions of IL-mPFC (Quirk, Russo, Barron, & Lebron, 2000). Consistent with lesion studies, single unit recording from IL-mPFC and PL-mPFC neurons showed that IL-mPFC responses correlate with the extinguishing of fear behavior (Chang, Berke, & Maren, 2010). Moreover, animals that displayed a large IL-mPFC neural response to the CS showed deficits in an extinction retention test (Chang et al., 2010). This is intriguing because earlier studies showed that larger phasic responses of IL-mPFC neurons in response to the CS in extinction was associated with improved learning (Milad & Quirk, 2002). To resolve this, a recent study used optogenetics to selectively stimulate or inhibit IL-mPFC neurons at the time of CS delivery during extinction training. Optogenetic stimulation produced a dramatic effect nearly obliterating freezing during training and enhancing extinction learning in a later recall test in the absence of optogenetic stimulation. Surprisingly, inhibition of these neurons during extinction training had no effects on extinction *per se*, but produced deficits in extinction memory because freezing was elevated in the retrieval test the following day (Do-Monte, Manzano-Nieves, Quiñones-Laracuente, Ramos-Medina, & Quirk, 2015).

While fear conditioning paradigms lack explicit goal-directed actions, other procedures such as active avoidance can measure how the mPFC is involved in planning and encoding actions that are executed after cued footshock threat. In active avoidance paradigms, subjects learn to move to a safe location or execute an action (such as pressing a lever) with the goal of avoiding an imminent aversive outcome signaled by a cue. Thus the latency to escape the aversive stimulus is the main measure of how well subject learns this action-outcome association. A recent comprehensive study recording from PL-mPFC found that these neurons encode cues that signal avoidance, and that inhibiting the phasic response of mPFC neurons (through excitation) delays the latency to actively avoid the footshock (Diehl et al., 2018). Thus, PL-mPFC was interpreted to be necessary for facilitating actions toward the goal of avoiding an aversive outcome. This is similar to previous studies where IL-mPFC lesions before avoidance resulted in impaired active avoidance (Moscarello & LeDoux, 2013). While the results of the IL-mPFC study may be due to enhancement of freezing behavior seen after the lesion, facilitatory roles for PL and IL-mPFC in avoidance have been shown in more recent studies which required animals to discriminate active and inhibitory avoidance cues (Capuzzo & Floresco, 2020). However, when the active avoidance task was simple, i.e., required no discrimination, this study failed to show an effect of PL-mPFC lesion even though IL-mPFC lesions continued to perturb avoidance responses. This is interesting because the avoidance study outlined earlier (Diehl et al., 2018) also had a food reinforcement contingency in the task to coax animals away from the safety platform. Taken together these findings suggest that PL-mPFC may be involved in avoidance actions when multiple contingences are in play.

Collectively fear learning literature suggests that PL-mPFC activity subserves the expression of fear whereas the IL-mPFC activity exhibits greater involvement in promoting the extinction of fear. These regions, however, are also important for execution of actions motivated by the goal of avoiding stressors in active avoidance procedures. Thus while the computational and functional processes of neural activity in the mPFC in response to fear learning and avoidance remain to be fully determined, mPFC subregions are convincingly involved in encoding fear related information and behavior.

5. Medial prefrontal cortex encoding of anxiety

Anxiety may be defined as a feeling of dread when there is a probability of harm occurring in distant or near future (Grillon, 2008; also see NIMH Research Domain Criteria). Similar to stress or fear, anxiety can be an adaptive response. Anxiety may help orient organisms to harm producing contingencies by directing attention away from a particular goal in order to focus on cues or outcomes that have the potential of being harmful in the future. Anxiety, however, becomes maladaptive when there is an exaggerated attentional emphasis on potential risk of future harm to the point that it disrupts ongoing behaviors necessary for survival.

While fear and anxiety are related, they are distinct constructs. Anxiety results from potential and uncertain future occurrence of a threatening event whereas fear is a response to imminent threat. Consistent with the literature establishing that PFC neurons are known to encode internalized information at the service of distant/future actions (Fuster, 2000; Kesner & Churchwell, 2011), the major body of preclinical and clinical work has implicated PFC function in animal models of anxiety and in most forms of anxiety disorders (Balderston, Liu, Roberson-Nay, Ernst, & Grillon, 2017; Balderston et al., 2017; Lammel, Tye, & Warden, 2014; Roberts, 2020). In addition, some of the core symptoms of anxiety disorders are impairments in cognitive or attentional processes that are associated with PFC (Basten, Stelzel, & Fiebach, 2011, 2012). For example, in the *N*-back working memory task generalized anxiety disorder (GAD) patients show impairments in the speed of responding, and high trait anxiety is associated with problems in the ability to shift strategies and disregarding irrelevant information (Balderston, Vytal, et al., 2017; Basten et al., 2011, 2012).

One of the prominent theories related to anxiety's effects on PFC is attention control theory (Eysenck, Derakshan, Santos, & Calvo, 2007). This theory posits that anxiety taxes the cognitive load, taking away resources for maintaining selective attention and processing demands for goal-directed responding. More specifically anxiety may attenuate executive cortical control and permit enhanced drive from subcortical "emotional" regions such as the amygdala to bias attention toward threats and promote more stimulus driven responding (Bishop, Duncan, & Lawrence, 2004; Eysenck et al., 2007). This impact of anxiety on PFC activity is thought to manifest as a general decrease in PFC activity and efficiency, and may also result in disengagement of subcortical circuitry from top-down control.

Evidence for changes in PFC engagement under anxiety has been evaluated by assessing whether dorsolateral-PFC activity is associated with anxiety levels in the No Shock, Predictable Shock, and Unpredictable Shock (NPU) task (Schmitz & Grillon, 2012). In this task, subjects are given neutral stimuli as well as stimuli that represent predictable (fear) and unpredictable (anxiety) shock. Unpredictable shock is sufficient to increase subjective levels of state anxiety in both healthy controls and patients with GAD, and potentiates startle responses, a marker of hyperarousal and vigilance associated with anxiety (Grillon, 2008). Through measuring BOLD activity with fMRI after inducing anxiety, persons with less activity in the dorsolateral PFC during anxiety periods exhibited more experiment-induced anxiety (Balderston, Liu, et al., 2017). These findings are consistent with the idea that PFC activity is attenuated in anxiety states.

To assess PFC dependent activity and behavioral changes in long lasting forms of anxiety such as in trait anxiety or in GAD, neuroimaging studies have found that PFC BOLD activity during working memory performance is perturbed in patients with GAD compared to controls (Balderston, Vytal, et al., 2017). Specifically, GAD patients show decreased working memory load related activation of the PFC compared to controls, particularly in the dorsolateral and cingulate regions. Similarly, high trait anxiety is associated with reduced dorsolateral PFC activity during attentional control and conflict processing (Bishop, 2009). However, anxiety from threat of shock elicits robust changes in the PFC activation during working memory task performance in both GAD patients and controls and did not interact with diagnosis. Taken together these studies demonstrate that longstanding anxiety phenotypes or a transient state of anxiety can decrease PFC activity, while anxiety phenotypes appear to have altered PFC engagement during cognitive performance.

In rodents, monkeys, and humans, anxiety can also be induced through systemic administration of ligands which decrease $GABA_a$ receptor function such as the $GABA_a$ receptor inverse agonist FG-7142 (Crawley et al., 1985; Dorow, 1987; Pellow & File, 1986). In rats, FG-7142 produces a robust and consistent reduction in the spontaneous firing rate of a population of putative excitatory neurons in the PL-mPFC (Park et al., 2016). These findings corroborate the population level BOLD signal decreases in human fMRI studies, and further support the idea of disengagement of the PFC during anxiety states.

To behaviorally induce anxiety in rodents, canonical models include the open field (OF) and the elevated plus maze (EPM), whereby animals are placed in a novel context that has "safe" areas like enclosed arms and "threatening" areas like open arms (Lezak, Missig, & Carlezon, 2017). Regardless of arm, potential threats are temporally and probabilistically uncertain to the animals and it is worth noting that, in reality, threatening outcomes for actions are non-existent. These models leverage what is known as approach/avoidance conflict whereby rodents balance innate motivations to explore but also to avoid risky open areas. Thus these models of anxiety are also referred to as "innate" models of anxiety. These models have face level validity for assessing anxiety, but they can be difficult to interpret because the impetus for behavior in these tasks is ambiguous. For example, is the environment invoking a fear state because it is novel and unfamiliar? Do animals who avoid open areas have anxiety or less motivation to explore? These limitations are possibly reflected in the variability that has been seen in studies investigating the role of the mPFC using these models (see Roberts, 2020 for review). For example, lesions and pharmacological manipulations of the PL-mPFC subregion, and even AC-mPFC, have seemed to suggest the PL-mPFC activity is necessary for anxiety-like behavior in the OF and EPM, but other studies have shown either contradictory or null effects from lesions of these regions (Bissiere, McAllister, Olpe, & Cryan, 2006; Lacroix, Broersen, Weiner, & Feldon, 1998; Pati, Sood, Mukhopadhyay, & Vaidya, 2018; Stern, Do Monte, Gazarini, Carobrez, & Bertoglio, 2010). Inconsistencies have also been reported after manipulation of the IL-mPFC, though some studies which compared PL-mPFC vs. IL-mPFC excitation demonstrated anxiogenic effects of PL-mPFC stimulation with no effect of IL-mPFC stimulation (Bi et al., 2013; Shah & Treit, 2003; Sierra-Mercado et al., 2011; Suzuki et al., 2016).

PFC neurons monitor location of rodents in these innate anxiety paradigms. One study which recorded single unit activity in the PL-mPFC in the EPM demonstrated increases in mPFC neural firing rate when animals entered closed or open arms and these changes typically preceded transitions from safe to risky areas (Adhikari, Topiwala, & Gordon, 2011). This pattern of activity was replicated in recent work which recorded population neural calcium activity in the PL-mPFC in the elevated zero maze (Loewke, Minerva, Nelson, Kreitzer, & Gunaydin, 2020), suggesting that the mPFC tracks the entry and exit from anxiogenic contexts. At the behavioral level, firing patterns appear to have some relation to anxiety-like behavior. Animals with higher anxiety on the EPM also show lesser mPFC neuron discrimination of closed and open arms (Adhikari et al., 2011). This would argue that mPFC neurons have direct involvement in behavioral strategies under innate anxiety. However, optogenetic stimulation of the PL-PFC and PL-PFC → amygdala projection neurons have not supported this idea as stimulation had no effect on open field or EPM behavior (Adhikari et al., 2015; Kumar et al., 2013). Instead only IL-mPFC → amygdala projection neuron stimulation produced anxiolytic effects in the EPM (Adhikari et al., 2015). Thus, the role of the mPFC in innate models of anxiety has been far from consistent, though evidence for mPFC encoding of behavior in the task seems apparent. It is possible this is a reflection of diverse strategies and levels of anxiety states in these tasks or that such effects can only be seen in certain contexts as suggested by Kumar et al. (2013) showing that PL-mPFC stimulation had anxiolytic effects when subjects were exposed to chronic stress earlier in life.

6. Medial prefrontal cortex manipulation of goal-directed action under anxiety states

The innate anxiety models, while contributing a great deal to our knowledge of the nature of PFC involvement, do not inform us about the nature of cortical encoding of ongoing behavior under a fog of anxiety. In a real-life situation, the impact of anxiety extends beyond its general aversive state and influences, and often disrupts, ongoing goal-directed actions that utilize cognitive and affective processing. We have been interested in understanding how the PFC's encoding of motivated actions changes with anxiety. Addressing this has not been trivial as we have needed to design experimental manipulations where the animal performs the same goal-directed behavior with and without a background of anxiety (Park & Moghaddam, 2017a).

We have so far used two operationally different approaches to study the impact of anxiety on PFC encoding of goal-directed actions. The first has been to record from PFC neurons during the performance of a reward-guided set-shifting task in the presence and absence of anxiogenic GABA inverse agonist FG-7142 (Park et al., 2016). The extradimensional set-shifting task used required animals to flexibly change their choice based on feedback (Stefani & Moghaddam, 2005). This task is dependent on the functional integrity of the mPFC, and once animals learn it, the performance is stable allowing for within (in addition to between) subject comparison of the impact of the anxiogenic treatment on performance and encoding of behavioral events. Behaviorally, FG-7142-induced anxiety produces modest effects on set-shifting performance. Animals treated with FG-7142 complete the task, but take longer than vehicle treated animals. Moreover, FG-7142 treatment increases the number of errors following rule switches, suggesting disrupted cognitive flexibility. The population response of neurons across all trials did not reveal an effect of anxiety on action encoding or neural activity after correct or incorrect outcomes, suggesting outcome encoding is also not influenced by FG-7142-induced anxiety. More detailed analysis focusing on neurons that showed significant responses to rules themselves, however, revealed several intriguing results. Response of neurons that encoded the rules, particularly in the peri-action period, were attenuated by anxiogenic treatment. These "rule-sensitive" neurons under control conditions developed rule selective response in later periods of rule learning (i.e., after several trials of exposure to the rule change). FG-7142 attenuated this rule-sensitive response. This suggests that anxiety itself disrupts the phasic responses of PFC neurons which encode specific task rules, resulting in impairments in flexibly changing behavior based on new rules. Finally, some trials in the set shifting task were further classified as "conflict" trials. These were instances where the incorrect rule conflicted with the correct rule, requiring increased attention and discrimination. Treatment with FG-7142 had a stronger disruptive effect on behavioral performance in these conflict trials compared to non-conflict trials and decreased the percentage of rule encoding neurons in conflict trials. This suggested that anxiety, at least using this model, disrupts cognitive function by decreasing PFC responsivity to discriminate between rules when stimuli for a past rule is present and must be ignored. Such disruptions ultimately decrease animals' ability to make the correct choice because the presence of conflicting stimuli are either not properly processed under anxiety, or such rules and stimuli are not properly integrated to optimize action selection. Furthermore, the finding that anxiety affected rule encoding primarily around the time of action execution suggested that detrimental effects of anxiety on mPFC function involve disrupted encoding of action selection and execution.

Our approach to use a pharmacological induction of anxiety clearly had limitations. While it provided the advantage of studying ongoing behavior under an extended state of "anxiety," the drug could be producing brain-wide effects that are unrelated to anxiety. Thus, we set out to design behavioral models of anxiety that, as much as possible, mimic a naturalistic setting of human anxiety. In these settings, anxiety is often "learned" because it develops after the organism learns that certain actions or contexts have the potential of causing harm in the immediate or far future. For example, seeing a gun may elicit anxiety because the possibility that the gun can be used to harm people has been learned previously. Another example is learning that driving can be dangerous under certain conditions such as bad weather and thus experiencing anxiety if one has to drive in a snow storm. While these contexts produce "normal" levels of anxiety, in cases of anxiety disorders the risk of harm may be exaggerated (Hartley & Phelps, 2012) and severely disrupt ongoing behavior.

Learned anxiety states often emerge when there is conflict between the desire to acquire resources and the learned harmful possibilities of seeking those resources. To model this, early work by Vogel and colleagues utilized a paradigm where animals were trained to execute an action to acquire a reward, such as food or water, and after this, seeking such reward was accompanied by a small electric shock punishment (Vogel, Beer, & Clody, 1971). This is also similar the Geller-Seifter task, which utilized the same overall design, but split the experimental session into shock and no shock blocks to allow assessment of behavior under anxiety and safety in the same session (Geller & Seifter, 1960). When shock contingencies are active, animals will show lesser and slower rates of reward seeking. Thus animals learned that their actions could be accompanied by harmful outcomes and suppress behavioral responses accordingly. The idea that conflict-related behavioral suppression is anxiety related was supported by demonstrating that anxiolytic drugs, typically benzodiazepines, resulted in anti-conflict effects (Dalterio, Wayner, Geller, & Hartmann, 1988; Liljequist & Engel, 1984; Shimizu, Kumasaka, Tanaka, Hirose, & Nakamura, 1992). For example, doses of diazepam which reduce fear-potentiated startle also result in selective increases in response rate during conflict periods (Brodkin, Busse, Sukoff, & Varney, 2002) indicating that increases in response rate and action execution during conflict tasks are related to reductions in anxiety.

These behavioral approaches collectively suggest that conflict inducing paradigms provide a feasible approach to assess how anxiety states develop and influence goal-directed actions. A role for the mPFC in these models was subsequently demonstrated by local pharmacological manipulation of neuronal activity. Inactivating the PL and IL-mPFC with lidocaine resulted in anti-conflict effects in rats, which indicated that the mPFC activity promotes anxiety and/or sensitivity to anxiogenic conditions (Resstel, Souza, & Guimaraes, 2008). Blocking dopaminergic activity in the PL-mPFC also reduces anxiety states in conflict procedures by promoting action under conflict. This finding is intriguing because increases in monoamines, such as dopamine, in the mPFC are a common observation in response to stressful and aversive stimuli and are generally attenuated by anxiolytic treatment (Broersen et al., 1995). Taken together, these findings highlighted that a complex interplay of neuromodulators in addition to the predominant excitatory and inhibitory cell types in the mPFC are likely utilized in learned anxiety states seen in conflict procedures.

Guided by these approaches, Park and colleagues developed a task where the probability of harm or conflict changed during reward-guided action execution in the same behavioral series (Park & Moghaddam, 2017b). The probabilistic harm contingency started at 0% and increased to 10%, changing from safe to low, within the same recording session to track neural responses with and without anxiety. This model is relevant to human anxiety models such as the NPU task, which assesses action performance under both safe and unpredictable threat of harm. Under these conditions, PL-mPFC neurons showed distinct changes in firing patterns in safe vs. risky (i.e., anxiety) blocks. To assess if anxiety had unique time-specific effects on these neurons, the authors determined if anxiety-related changes in PL-mPFC neuron spiking activity occurred during the trial initiation cue, the action encoding period, and reward delivery. This analysis revealed that the largest anxiety-related neural activity change occured during the action encoding period rather than reward delivery or the trial initiation cue. In addition, these anxiety-related changes in encoding were not associated with time on the task, as a control group who performed the task but with no risk of shock for the entire session did not show such robust changes. Thus PL-mPFC neurons are exquisitely sensitive to this model of anxiety particularly during the encoding of motivated actions.

Population level activity in the PL-mPFC was also influenced by anxiety. The population level state derived from single units was predictive of anxiety-like action suppression on a trial by trial basis. Furthermore, assessing oscillatory population activity through local field potentials revealed anxiety-related changes in theta activity. Theta band activity in the mPFC is thought to synchronize mPFC neurons and aid in learning processes and behavioral responses (Buschman, Denovellis, Diogo, Bullock, & Miller, 2012; Paz, Bauer, & Paré, 2008). In anxiety blocks, oscillatory mPFC theta activity was found to decrease in the action encoding period, and synchrony between the mPFC and the ventral tegmental area (VTA), the dopamine containing region implicated in reward processing, was also attenuated under anxiety. Thus, not only does anxiety appear to change how the mPFC encodes actions toward goals, it also results in large scale changes in regional and network coordination that could disrupt reward processing.

In a recent study, we adapted this procedure by designing a probabilistic punishment task in which two sequential actions were performed to acquire a reward (Jacobs & Moghaddam, 2020; see Fig. 1). The first "seeking" action was risky because it could probabilistically (increased progressively from 0% to 18% over the session) result in a mild footshock, but deterministically resulted in the ability to perform the "taking" response to acquire a reward. Similar to other conflict studies, these contingencies resulted in behavioral suppression with increasing risk of shock (see Fig. 2).

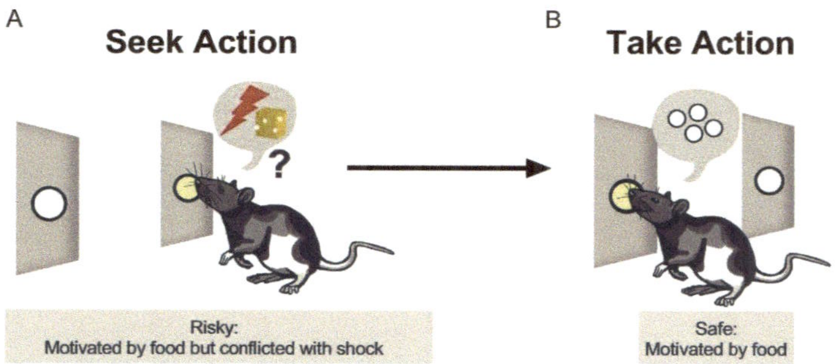

Fig. 1 Demonstration of trial structure in the probabilistic punishment task to allow assessment of motivated action performed under anxiety from shock risk and when safe from risk. Animals perform two sequential actions which are motivated by food. (A) The first "seek" action contains a risk of footshock. (B) The second "take" action is safe and results in food pellet delivery. Long Evans rat graphic drawn by Alina Bogachuk.

A

Probabilistic Punishment Task

Block	Maximum Trials	Seek Action Shock Risk (%)
1	20	0
2	20	6
3	20	10
4	20	18

B **Behavioral Suppression Due to Shock Risk After Task Learning (Session 5-8)**

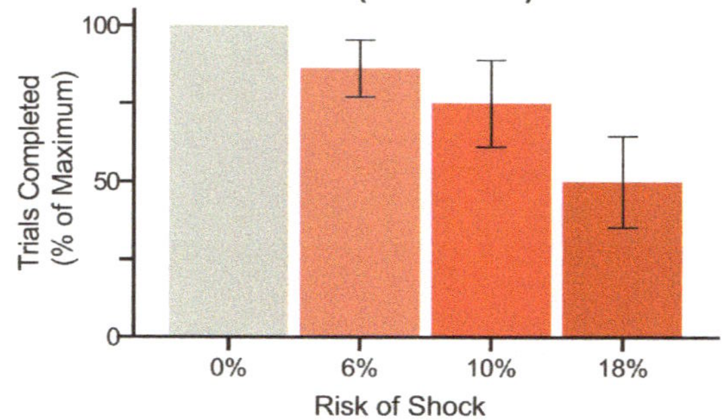

Fig. 2 Probabilistic punishment task structure to assess motivated action with varying risk of shock. (A) Task structure is comprised of successive 14-min blocks where up to 20 food pellets can be earned. Shock risk increases in an ascending order for each block. (B) After the task is learned, which was achieved between Session 5–8, behavioral suppression is selectively seen for trial completion when risk is present (pink and red) but not when seek actions are safe (0% risk) in block 1 (gray). Data represent the mean ± SEM for $n = 8$ rats. *Figure adapted from Jacobs, D. S., & Moghaddam, B. (2020). Prefrontal cortex representation of learning of punishment probability during reward-motivated actions. The Journal of Neuroscience, 40(26), 5063–5077. doi:10.1523/jneurosci.0310-20.2020.*

Importantly, having one action be safe while another action (which was operationally similar) be risky in the process of acquiring the goal, allowed us to control for whether changes in encoding under anxiety are specific to risky actions or are generalized to any motivated action. We utilized *in vivo* calcium imaging to measure neural activity in the PL-mPFC both before and after animals had learned the anxiogenic task. Measuring calcium activity provided the advantage of investigating if mPFC action encoding changed as learning of the anxiogenic context progressed over multiple sessions and as the animal received the stressor (footshock) itself. Thus we could assess whether specific actions (safe vs. risky—anxiogenic) or outcomes (footshock vs. reward) were encoded differently with the learning of risk of harm. Up to this point, studies that have assessed anxiety using similar procedures only measured neural activity after the task was learned and thus did not address the learning process.

It was found that the phasic neural response during encoding of the risky action in the PL-mPFC changed with learning of anxiety (see Fig. 3).

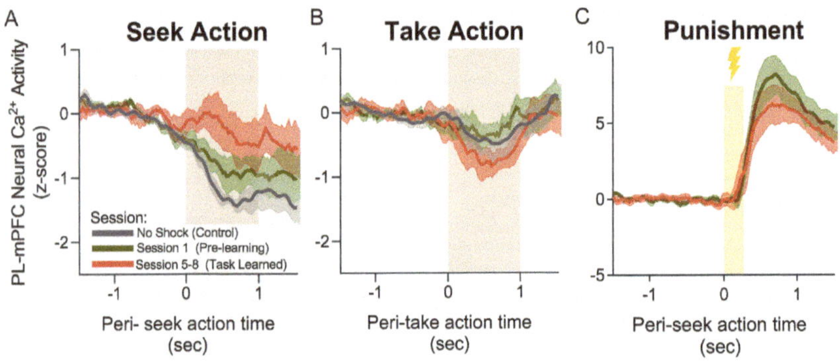

Fig. 3 Recording PL-mPFC neural calcium activity during action and punishment encoding before, during, an after learning of the probabilistic punishment task. (A) The seek action was accompanied by a phasic decrease in activity during the action encoding period (shaded area) when no shock risk was present for the entire session (No-Shock-gray) and before the task was learned (Session 1-green). After task learning (Session 5–8-red), this phasic response was attenuated. (B) The phasic response during safe take action encoding was more modest and did not change after learning. (C) Punishment (yellow shaded area) following the seek action produced a phasic increase in activity that did not change over task learning. Data represent the mean ± SEM for seek actions that were risky but did not result in shock in panel A, take actions that were not preceded by shock in panel B, or for seek actions that resulted in shock in panel C across all risk blocks (i.e., blocks 2–4) for $n = 5$–7 rats. *Figure adapted from results in Jacobs, D. S., & Moghaddam, B. (2020). Prefrontal cortex representation of learning of punishment probability during reward-motivated actions. The Journal of Neuroscience, 40(26), 5063–5077. doi:10.1523/jneurosci.0310-20.2020.*

Initially, the risky action encoding periods displayed phasic decreases in neural activity both in an anxiety-free "No-Shock" session and in the first session where footshock was introduced (i.e., before learning). Behavior was monitored until anxiogenic contingencies were "learned" based on observation of selective suppression of behavior for the risky action and in risk blocks. After learning, risky action encoding in the PL-mPFC changed such that the previously phasic decrease was attenuated to a more neutral state. The activity state for safe action encoding, however, did not change with learning. These findings suggested that plasticity during the learning of anxiety is tuned to those actions and contingencies that carry risk of harm (and are, therefore, anxiogenic) further highlighting a role for the PL-mPFC in tuning its action encoding in response to conflict. Finally, the footshock itself was represented by large increases in neural activity, and these responses did not change with learning. This indicates that the PL-mPFC does not desensitize or stop tracking stressful stimuli, at least over the initial process of learning (Fig. 3). Collectively these findings highlight a novel role for the PL-PFC in the learning of anxiety, by selectively changing neural activity patterns for risky actions. While these learning-related neural changes were correlated with the overall impacts of anxiety on motivated behavior, much like those seen in human cognition studies, the causal role of these changes is likely to be complex and presents a significant opportunity for future studies.

Evidence for causality for mPFC neural activity during conflict are beginning to emerge. One such study which reported on responsive cells in dorsal and ventral mPFC regions during ethanol seeking under conflict, found that optogenetic inhibition of IL-mPFC neurons but not PL-mPFC neurons was able to attenuate some, but not all, aspects of anxiety-like suppression of motivated behavior. This suggests that signaling in IL-mPFC supports action suppression under conflict (Halladay et al., 2020). In another study, PL-mPFC projections to the nucleus accumbens (NAc) or the ventral tegmental area (VTA) displayed discriminate encoding between rewarded and punished (shocked) outcomes for the same action, and greater changes in PL-mPFC → NAc projections just before the time of action execution were correlated with increased suppression of reward seeking. In this study, optogenetic stimulation of the PL-mPFC → NAc projection neurons had little effect on behavior, *but* selective stimulation of PL-mPFC → NAc neurons activated by footshock were able to suppress reward motived behavior, suggesting that increased PL-mPFC output in specific subpopulations of neurons can control anxiety-like reward suppression under conflict (Kim et al., 2017). Effects have also been seen for the PL-mPFC in compulsive cocaine seeking, where PL-mPFC excitatory neuron stimulation or inhibition results in suppression or enhancement of reward seeking under footshock risk, respectively (Chen et al., 2013). Thus both PL and IL regions of mPFC appear to have conflict-related roles by increasing anxiety-like suppression of motivated behavior, though there is much work to be done regarding mPFC neural subtypes and projections which underlie conflict-related anxiety.

In summay, findings in learned anxiety through conflict models indicate that at the single neuron and population levels, action encoding and execution are selectively influenced by anxiety. These studies also raise a number of questions that challenge conventional views of the mPFC in action execution. For example, the aforementioned optogenetic studies demonstrated that both subdivisions having pro-suppressive (anxiety-like) effects on motivated behavior, which is different from the go/no-go dichotomy outlined for action execution in the PL/IL mPFC, respectively (for review on this see Gourley & Taylor, 2016). Thus some canonical views of mPFC function may have to be modified when anxiety states are involved in motivated processes.

7. Conclusion

This review expands on original work demonstrating that subregions of the mPFC are critical in action encoding during reinforced behavior and in response to stress and other negative affective states (Balleine & Dickinson, 1998; Del Arco et al., 2017; Killcross & Coutureau, 2003; Park & Moghaddam, 2017b). The roles for the mPFC in action selection and encoding are complex and encapsulate behavior that is driven by both positive and negative stimuli and outcomes. This complexity highlights the need for future work to better understand the neural representations of motivated behavior in the mPFC under fear and anxiety states.

Acknowledgments

This work was supported by the NIMH R01MH115027 (BM), NIDA T32DA007262 (DSJ), and a scholar award from the ARCS foundation (DSJ).

References

Abercrombie, E. D., Keefe, K. A., DiFrischia, D. S., & Zigmond, M. J. (1989). Differential effect of stress on in vivo dopamine release in striatum, nucleus accumbens, and medial frontal cortex. *Journal of Neurochemistry*, *52*(5), 1655–1658.

Adhikari, A., Lerner, T. N., Finkelstein, J., Pak, S., Jennings, J. H., Davidson, T. J., et al. (2015). Basomedial amygdala mediates top-down control of anxiety and fear. *Nature*, *527*(7577), 179–185. https://doi.org/10.1038/nature15698.

Adhikari, A., Topiwala, M. A., & Gordon, J. A. (2011). Single units in the medial prefrontal cortex with anxiety-related firing patterns are preferentially influenced by ventral hippocampal activity. *Neuron, 71*(5), 898–910. https://doi.org/10.1016/j.neuron.2011.07.027.

Arnsten, A. F. (2015). Stress weakens prefrontal networks: Molecular insults to higher cognition. *Nature Neuroscience, 18*(10), 1376.

Baeg, E. H., Kim, Y. B., Jang, J., Kim, H. T., Mook-Jung, I., & Jung, M. W. (2001). Fast spiking and regular spiking neural correlates of fear conditioning in the medial prefrontal cortex of the rat. *Cerebral Cortex, 11*(5), 441–451. https://doi.org/10.1093/cercor/11.5.441.

Bagozzi, R. P., & Pieters, R. (1998). Goal-directed emotions. *Cognition & Emotion, 12*(1), 1–26.

Balderston, N. L., Liu, J., Roberson-Nay, R., Ernst, M., & Grillon, C. (2017). The relationship between dlPFC activity during unpredictable threat and CO(2)-induced panic symptoms. *Translational Psychiatry*, 7(12), 1266. https://doi.org/10.1038/s41398-017-0006-5.

Balderston, N. L., Vytal, K. E., O'Connell, K., Torrisi, S., Letkiewicz, A., Ernst, M., et al. (2017). Anxiety patients show reduced working memory related dlPFC activation during safety and threat. *Depression and Anxiety, 34*(1), 25–36. https://doi.org/10.1002/da.22518.

Balleine, B. W., & Dickinson, A. (1998). Goal-directed instrumental action: Contingency and incentive learning and their cortical substrates. *Neuropharmacology, 37*(4–5), 407–419. https://doi.org/10.1016/s0028-3908(98)00033-1.

Basten, U., Stelzel, C., & Fiebach, C. J. (2011). Trait anxiety modulates the neural efficiency of inhibitory control. *Journal of Cognitive Neuroscience, 23*(10), 3132–3145. https://doi.org/10.1162/jocn_a_00003.

Basten, U., Stelzel, C., & Fiebach, C. J. (2012). Trait anxiety and the neural efficiency of manipulation in working memory. *Cognitive, Affective, & Behavioral Neuroscience, 12*(3), 571–588. https://doi.org/10.3758/s13415-012-0100-3.

Bi, L.-L., Wang, J., Luo, Z.-Y., Chen, S.-P., Geng, F., & Chen, Y.-h., … Gao, T.-M. (2013). Enhanced excitability in the infralimbic cortex produces anxiety-like behaviors. *Neuropharmacology, 72*, 148–156.

Birrell, J. M., & Brown, V. J. (2000). Medial frontal cortex mediates perceptual attentional set shifting in the rat. *The Journal of Neuroscience, 20*(11), 4320–4324. https://doi.org/10.1523/jneurosci.20-11-04320.2000.

Bishop, S. J. (2009). Trait anxiety and impoverished prefrontal control of attention. *Nature Neuroscience, 12*(1), 92–98.

Bishop, S. J., Duncan, J., & Lawrence, A. D. (2004). State anxiety modulation of the amygdala response to unattended threat-related stimuli. *Journal of Neuroscience, 24*(46), 10364–10368.

Bissiere, S., McAllister, K. H., Olpe, H.-R., & Cryan, J. F. (2006). The rostral anterior cingulate cortex modulates depression but not anxiety-related behaviour in the rat. *Behavioural Brain Research, 175*(1), 195–199.

Bogdan, R., & Pizzagalli, D. A. (2006). Acute stress reduces reward responsiveness: Implications for depression. *Biological Psychiatry, 60*(10), 1147–1154. https://doi.org/10.1016/j.biopsych.2006.03.037.

Bondi, C. O., Rodriguez, G., Gould, G. G., Frazer, A., & Morilak, D. A. (2008). Chronic unpredictable stress induces a cognitive deficit and anxiety-like behavior in rats that is prevented by chronic antidepressant drug treatment. *Neuropsychopharmacology, 33*(2), 320–331.

Brodkin, J., Busse, C., Sukoff, S. J., & Varney, M. A. (2002). Anxiolytic-like activity of the mGluR5 antagonist MPEP: A comparison with diazepam and buspirone. *Pharmacology Biochemistry and Behavior, 73*(2), 359–366.

Broersen, L., Heinsbroek, R., De Bruin, J., Laan, J.-B., Joosten, R., & Olivier, B. (1995). Local pharmacological manipulations of prefrontal dopamine affect conflict behaviour in rats. *Behavioural Pharmacology*, *6*(4), 395–404.

Burgos-Robles, A., Vidal-Gonzalez, I., & Quirk, G. J. (2009). Sustained conditioned responses in prelimbic prefrontal neurons are correlated with fear expression and extinction failure. *The Journal of Neuroscience*, *29*(26), 8474–8482. https://doi.org/10.1523/jneurosci.0378-09.2009.

Buschman, T. J., Denovellis, E. L., Diogo, C., Bullock, D., & Miller, E. K. (2012). Synchronous oscillatory neural ensembles for rules in the prefrontal cortex. *Neuron*, *76*(4), 838–846. https://doi.org/10.1016/j.neuron.2012.09.029.

Calhoon, G. G., & Tye, K. M. (2015). Resolving the neural circuits of anxiety. *Nature Neuroscience*, *18*(10), 1394–1404.

Capuzzo, G., & Floresco, S. B. (2020). Prelimbic and Infralimbic prefrontal regulation of active and inhibitory avoidance and reward-seeking. *The Journal of Neuroscience*, *40*(24), 4773–4787. https://doi.org/10.1523/jneurosci.0414-20.2020.

Cassaday, H. J., Nelson, A. J., & Pezze, M. A. (2014). From attention to memory along the dorsal-ventral axis of the medial prefrontal cortex: Some methodological considerations. *Frontiers in Systems Neuroscience*, *8*, 160. https://doi.org/10.3389/fnsys.2014.00160.

Cerqueira, J. J., Mailliet, F., Almeida, O. F., Jay, T. M., & Sousa, N. (2007). The prefrontal cortex as a key target of the maladaptive response to stress. *The Journal of Neuroscience*, *27*(11), 2781–2787. https://doi.org/10.1523/jneurosci.4372-06.2007.

Chang, C. H., Berke, J. D., & Maren, S. (2010). Single-unit activity in the medial prefrontal cortex during immediate and delayed extinction of fear in rats. *PLoS One*, *5*(8), e11971. https://doi.org/10.1371/journal.pone.0011971.

Chen, B. T., Yau, H. J., Hatch, C., Kusumoto-Yoshida, I., Cho, S. L., Hopf, F. W., et al. (2013). Rescuing cocaine-induced prefrontal cortex hypoactivity prevents compulsive cocaine seeking. *Nature*, *496*(7445), 359–362. https://doi.org/10.1038/nature12024.

Compton, R. J. (2003). The interface between emotion and attention: A review of evidence from psychology and neuroscience. *Behavioral and Cognitive Neuroscience Reviews*, *2*(2), 115–129.

Cook, S. C., & Wellman, C. L. (2004). Chronic stress alters dendritic morphology in rat medial prefrontal cortex. *Journal of Neurobiology*, *60*(2), 236–248.

Corcoran, K. A., & Quirk, G. J. (2007). Activity in prelimbic cortex is necessary for the expression of learned, but not innate, fears. *The Journal of Neuroscience*, *27*(4), 840–844. https://doi.org/10.1523/jneurosci.5327-06.2007.

Courtin, J., Chaudun, F., Rozeske, R. R., Karalis, N., Gonzalez-Campo, C., Wurtz, H., et al. (2014). Prefrontal parvalbumin interneurons shape neuronal activity to drive fear expression. *Nature*, *505*(7481), 92–96. https://doi.org/10.1038/nature12755.

Crawley, J. N., Ninan, P. T., Pickar, D., Chrousos, G. P., Linnoila, M., Skolnick, P., et al. (1985). Neuropharmacological antagonism of the beta-carboline-induced "anxiety" response in rhesus monkeys. *The Journal of Neuroscience*, *5*(2), 477–485. https://doi.org/10.1523/jneurosci.05-02-00477.1985.

Dalterio, S., Wayner, M., Geller, I., & Hartmann, R. (1988). Ethanol and diazepam interactions on conflict behavior in rats. *Alcohol*, *5*(6), 471–476.

Daviu, N., Bruchas, M. R., Moghaddam, B., Sandi, C., & Beyeler, A. (2019). Neurobiological links between stress and anxiety. *Neurobiology of Stress*, *11*, 100191.

Del Arco, A., Park, J., & Moghaddam, B. (2020). Unanticipated stressful and rewarding experiences engage the same prefrontal cortex and ventral tegmental area neuronal populations. *eNeuro*, 7(3), 1–13. https://doi.org/10.1523/eneuro.0029-20.2020.

Del Arco, A., Park, J., Wood, J., Kim, Y., & Moghaddam, B. (2017). Adaptive encoding of outcome prediction by prefrontal cortex ensembles supports behavioral flexibility. *The Journal of Neuroscience*, *37*(35), 8363–8373. https://doi.org/10.1523/jneurosci.0450-17.2017.

Devilbiss, D. M., Jenison, R. L., & Berridge, C. W. (2012). Stress-induced impairment of a working memory task: Role of spiking rate and spiking history predicted discharge. *PLoS Computational Biology*, *8*(9), e1002681. https://doi.org/10.1371/journal.pcbi.1002681.

Devilbiss, D. M., Spencer, R. C., & Berridge, C. W. (2017). Stress degrades prefrontal cortex neuronal coding of goal-directed behavior. *Cerebral Cortex*, *27*(5), 2970–2983. https://doi.org/10.1093/cercor/bhw140.

Diehl, M. M., Bravo-Rivera, C., Rodriguez-Romaguera, J., Pagan-Rivera, P. A., Burgos-Robles, A., Roman-Ortiz, C., et al. (2018). Active avoidance requires inhibitory signaling in the rodent prelimbic prefrontal cortex. *eLife*, *7*, 1–19. https://doi.org/10.7554/eLife.34657.

Do-Monte, F. H., Manzano-Nieves, G., Quiñones-Laracuente, K., Ramos-Medina, L., & Quirk, G. J. (2015). Revisiting the role of infralimbic cortex in fear extinction with optogenetics. *The Journal of Neuroscience*, *35*(8), 3607–3615. https://doi.org/10.1523/jneurosci.3137-14.2015.

Dorow, R. (1987). FG 7142 and its anxiety-inducing effects in humans. *British Journal of Clinical Pharmacology*, *23*(6), 781–782.

Eysenck, M. W., Derakshan, N., Santos, R., & Calvo, M. G. (2007). Anxiety and cognitive performance: Attentional control theory. *Emotion*, *7*(2), 336–353. https://doi.org/10.1037/1528-3542.7.2.336.

Floresco, S. B., Block, A. E., & Maric, T. (2008). Inactivation of the medial prefrontal cortex of the rat impairs strategy set-shifting, but not reversal learning, using a novel, automated procedure. *Behavioural Brain Research*, *190*(1), 85–96.

Floresco, S. B., Magyar, O., Ghods-Sharifi, S., Vexelman, C., & Maric, T. (2006). Multiple dopamine receptor subtypes in the medial prefrontal cortex of the rat regulate set-shifting. *Neuropsychopharmacology*, *31*(2), 297–309.

Friedman, A., Homma, D., Bloem, B., Gibb, L. G., Amemori, K. I., Hu, D., et al. (2017). Chronic stress alters Striosome-circuit dynamics, leading to aberrant decision-making. *Cell*, *171*(5), 1191–1205.e1128. https://doi.org/10.1016/j.cell.2017.10.017.

Fuster, J. M. (2000). Prefrontal neurons in networks of executive memory. *Brain Research Bulletin*, *52*(5), 331–336.

Fuster, J. M. (2001). The prefrontal cortex- -an update: Time is of the essence. *Neuron*, *30*(2), 319–333. https://doi.org/10.1016/s0896-6273(01)00285-9.

Fuster, J. M. (2015). *The prefrontal cortex* (5th ed.). London, UK: Academic Press.

Geller, I., & Seifter, J. (1960). The effects of meprobamate, barbiturates, d-amphetamine and promazine on experimentally induced conflict in the rat. *Psychopharmacologia*, *1*(6), 482–492. https://doi.org/10.1007/BF00429273.

Goldman-Rakic, P. S. (1995). Cellular basis of working memory. *Neuron*, *14*(3), 477–485. https://doi.org/10.1016/0896-6273(95)90304-6.

Gourley, S. L., & Taylor, J. R. (2016). Going and stopping: Dichotomies in behavioral control by the prefrontal cortex. *Nature Neuroscience*, *19*(6), 656–664. https://doi.org/10.1038/nn.4275.

Granon, S., & Poucet, B. (1995). Medial prefrontal lesions in the rat and spatial navigation: Evidence for impaired planning. *Behavioral Neuroscience*, *109*(3), 474–484. https://doi.org/10.1037//0735-7044.109.3.474.

Grillon, C. (2008). Models and mechanisms of anxiety: Evidence from startle studies. *Psychopharmacology*, *199*(3), 421–437. https://doi.org/10.1007/s00213-007-1019-1.

Gruene, T. M., Flick, K., Stefano, A., Shea, S. D., & Shansky, R. M. (2015). Sexually divergent expression of active and passive conditioned fear responses in rats. *eLife*, *4*, 1–9. https://doi.org/10.7554/eLife.11352.

Halladay, L. R., Kocharian, A., Piantadosi, P. T., Authement, M. E., Lieberman, A. G., Spitz, N. A., et al. (2020). Prefrontal regulation of punished ethanol self-administration. *Biological Psychiatry*, *87*(11), 967–978. https://doi.org/10.1016/j.biopsych.2019.10.030.

Han, H. J., Jung, W. H., Yun, J. Y., Park, J. W., Cho, K. K., Hur, J. W., et al. (2016). Disruption of effective connectivity from the dorsolateral prefrontal cortex to the orbitofrontal cortex by negative emotional distraction in obsessive-compulsive disorder. *Psychological Medicine*, *46*(5), 921–932. https://doi.org/10.1017/s0033291715002391.

Hartley, C. A., & Phelps, E. A. (2012). Anxiety and decision-making. *Biological Psychiatry*, *72*(2), 113–118. https://doi.org/10.1016/j.biopsych.2011.12.027.

Holmes, A., & Wellman, C. L. (2009). Stress-induced prefrontal reorganization and executive dysfunction in rodents. *Neuroscience and Biobehavioral Reviews*, *33*(6), 773–783. https://doi.org/10.1016/j.neubiorev.2008.11.005.

Hoover, W. B., & Vertes, R. P. (2007). Anatomical analysis of afferent projections to the medial prefrontal cortex in the rat. *Brain Structure & Function*, *212*(2), 149–179. https://doi.org/10.1007/s00429-007-0150-4.

Jackson, M. E., & Moghaddam, B. (2006). Distinct patterns of plasticity in prefrontal cortex neurons that encode slow and fast responses to stress. *European Journal of Neuroscience*, *24*(6), 1702–1710.

Jacobs, D. S., & Moghaddam, B. (2020). Prefrontal cortex representation of learning of punishment probability during reward-motivated actions. *The Journal of Neuroscience*, *40*(26), 5063–5077. https://doi.org/10.1523/jneurosci.0310-20.2020.

Johansen, J. P., Cain, C. K., Ostroff, L. E., & LeDoux, J. E. (2011). Molecular mechanisms of fear learning and memory. *Cell*, *147*(3), 509–524. https://doi.org/10.1016/j.cell.2011.10.009.

Kesner, R. P., & Churchwell, J. C. (2011). An analysis of rat prefrontal cortex in mediating executive function. *Neurobiology of Learning and Memory*, *96*(3), 417–431.

Kesner, R. P., Hunt, M. E., Williams, J. M., & Long, J. M. (1996). Prefrontal cortex and working memory for spatial response, spatial location, and visual object information in the rat. *Cerebral Cortex*, *6*(2), 311–318.

Killcross, S., & Coutureau, E. (2003). Coordination of actions and habits in the medial prefrontal cortex of rats. *Cerebral Cortex*, *13*(4), 400–408. https://doi.org/10.1093/cercor/13.4.400.

Kim, J. J., Lee, H. J., Welday, A. C., Song, E., Cho, J., Sharp, P. E., et al. (2007). Stress-induced alterations in hippocampal plasticity, place cells, and spatial memory. *Proceedings of the National Academy of Sciences of the United States of America*, *104*(46), 18297–18302. https://doi.org/10.1073/pnas.0708644104.

Kim, C. K., Ye, L., Jennings, J. H., Pichamoorthy, N., Tang, D. D., Yoo, A. W., et al. (2017). Molecular and circuit-dynamical identification of top-down neural mechanisms for restraint of reward seeking. *Cell*, *170*(5), 1013–1027.e1014. https://doi.org/10.1016/j.cell.2017.07.020.

Kumar, S., Black, S. J., Hultman, R., Szabo, S. T., DeMaio, K. D., Du, J., et al. (2013). Cortical control of affective networks. *The Journal of Neuroscience*, *33*(3), 1116–1129. https://doi.org/10.1523/jneurosci.0092-12.2013.

Lacroix, L., Broersen, L., Weiner, I., & Feldon, J. (1998). The effects of excitotoxic lesion of the medial prefrontal cortex on latent inhibition, prepulse inhibition, food hoarding, elevated plus maze, active avoidance and locomotor activity in the rat. *Neuroscience*, *84*(2), 431–442.

Lammel, S., Tye, K., & Warden, M. (2014). Progress in understanding mood disorders: Optogenetic dissection of neural circuits. *Genes, Brain and Behavior*, *13*(1), 38–51.

Laubach, M., Amarante, L. M., Swanson, K., & White, S. R. (2018). What, if anything, is rodent prefrontal cortex? *eNeuro*, *5*(5), 1–14. https://doi.org/10.1523/eneuro.0315-18.2018.

Lezak, K. R., Missig, G., & Carlezon, W. A., Jr. (2017). Behavioral methods to study anxiety in rodents. *Dialogues in Clinical Neuroscience*, *19*(2), 181–191.

Liljequist, S., & Engel, J. A. (1984). The effects of GABA and benzodiazepine receptor antagonists on the anti-conflict actions of diazepam or ethanol. *Pharmacology Biochemistry and Behavior, 21*(4), 521–525.

Liston, C., Miller, M. M., Goldwater, D. S., Radley, J. J., Rocher, A. B., Hof, P. R., et al. (2006). Stress-induced alterations in prefrontal cortical dendritic morphology predict selective impairments in perceptual attentional set-shifting. *The Journal of Neuroscience, 26*(30), 7870–7874. https://doi.org/10.1523/jneurosci.1184-06.2006.

Loewke, A. C., Minerva, A. R., Nelson, A. B., Kreitzer, A. C., & Gunaydin, L. A. (2020). Fronto-striatal projections regulate approach-avoidance conflict. *bioRxiv*, 1–28. https://doi.org/10.1101/2020.03.05.979708. 2020.2003.2005.979708.

McEwen, B. S., Bowles, N. P., Gray, J. D., Hill, M. N., Hunter, R. G., Karatsoreos, I. N., et al. (2015). Mechanisms of stress in the brain. *Nature Neuroscience, 18*(10), 1353–1363.

McEwen, B. S., & Morrison, J. H. (2013). The brain on stress: Vulnerability and plasticity of the prefrontal cortex over the life course. *Neuron, 79*(1), 16–29. https://doi.org/10.1016/j.neuron.2013.06.028.

McEwen, B. S., Nasca, C., & Gray, J. D. (2016). Stress effects on neuronal structure: Hippocampus, amygdala, and prefrontal cortex. *Neuropsychopharmacology, 41*(1), 3–23.

McKlveen, J. M., Moloney, R. D., Scheimann, J. R., Myers, B., & Herman, J. P. (2019). "braking" the prefrontal cortex: The role of glucocorticoids and interneurons in stress adaptation and pathology. *Biological Psychiatry, 86*(9), 669–681. https://doi.org/10.1016/j.biopsych.2019.04.032.

McKlveen, J. M., Morano, R. L., Fitzgerald, M., Zoubovsky, S., Cassella, S. N., Scheimann, J. R., et al. (2016). Chronic stress increases prefrontal inhibition: A mechanism for stress-induced prefrontal dysfunction. *Biological Psychiatry, 80*(10), 754–764.

Milad, M. R., & Quirk, G. J. (2002). Neurons in medial prefrontal cortex signal memory for fear extinction. *Nature, 420*(6911), 70–74. https://doi.org/10.1038/nature01138.

Miller, E. K. (2000). The prefontral cortex and cognitive control. *Nature Reviews Neuroscience, 1*(1), 59–65.

Mizoguchi, K., Yuzurihara, M., Ishige, A., Sasaki, H., Chui, D. H., & Tabira, T. (2000). Chronic stress induces impairment of spatial working memory because of prefrontal dopaminergic dysfunction. *The Journal of Neuroscience, 20*(4), 1568–1574. https://doi.org/10.1523/jneurosci.20-04-01568.2000.

Moghaddam, B. (1993). Stress preferentially increases extraneuronal levels of excitatory amino acids in the prefrontal cortex: Comparison to hippocampus and basal ganglia. *Journal of Neurochemistry, 60*(5), 1650–1657. https://doi.org/10.1111/j.1471-4159.1993.tb13387.x.

Morrow, B. A., Elsworth, J. D., Lee, E. J., & Roth, R. H. (2000). Divergent effects of putative anxiolytics on stress-induced Fos expression in the mesoprefrontal system of the rat. *Synapse, 36*(2), 143–154.

Moscarello, J. M., & LeDoux, J. E. (2013). Active avoidance learning requires prefrontal suppression of amygdala-mediated defensive reactions. *The Journal of Neuroscience, 33*(9), 3815–3823. https://doi.org/10.1523/jneurosci.2596-12.2013.

Murphy, B. L., Arnsten, A. F., Jentsch, J. D., & Roth, R. H. (1996). Dopamine and spatial working memory in rats and monkeys: Pharmacological reversal of stress-induced impairment. *Journal of Neuroscience, 16*(23), 7768–7775.

Naqvi, N., Shiv, B., & Bechara, A. (2006). The role of emotion in decision making: A cognitive neuroscience perspective. *Current Directions in Psychological Science, 15*(5), 260–264.

Olson, E. A., Kaiser, R. H., Pizzagalli, D. A., Rauch, S. L., & Rosso, I. M. (2019). Regional prefrontal resting-state functional connectivity in posttraumatic stress disorder. *Biological Psychiatry: Cognitive Neuroscience and Neuroimaging, 4*(4), 390–398. https://doi.org/10.1016/j.bpsc.2018.09.012.

Ostrander, M. M., Richtand, N. M., & Herman, J. P. (2003). Stress and amphetamine induce Fos expression in medial prefrontal cortex neurons containing glucocorticoid receptors. *Brain Research, 990*(1–2), 209–214.

Park, J., & Moghaddam, B. (2017a). Impact of anxiety on prefrontal cortex encoding of cognitive flexibility. *Neuroscience, 345*, 193–202. https://doi.org/10.1016/j.neuroscience.2016.06.013.

Park, J., & Moghaddam, B. (2017b). Risk of punishment influences discrete and coordinated encoding of reward-guided actions by prefrontal cortex and VTA neurons. *eLife, 6*, 1–28. https://doi.org/10.7554/eLife.30056.

Park, J., Wood, J., Bondi, C., Del Arco, A., & Moghaddam, B. (2016). Anxiety evokes Hypofrontality and disrupts rule-relevant encoding by dorsomedial prefrontal cortex neurons. *The Journal of Neuroscience, 36*(11), 3322–3335. https://doi.org/10.1523/jneurosci.4250-15.2016.

Pati, S., Sood, A., Mukhopadhyay, S., & Vaidya, V. A. (2018). Acute pharmacogenetic activation of medial prefrontal cortex excitatory neurons regulates anxiety-like behaviour. *Journal of Biosciences, 43*(1), 85–95.

Paz, R., Bauer, E. P., & Paré, D. (2008). Theta synchronizes the activity of medial prefrontal neurons during learning. *Learning & Memory, 15*(7), 524–531. https://doi.org/10.1101/lm.932408.

Pellow, S., & File, S. E. (1986). Anxiolytic and anxiogenic drug effects on exploratory activity in an elevated plus-maze: A novel test of anxiety in the rat. *Pharmacology, Biochemistry, and Behavior, 24*(3), 525–529. https://doi.org/10.1016/0091-3057(86)90552-6.

Pizzagalli, D. A., Oakes, T. R., Fox, A. S., Chung, M. K., Larson, C. L., Abercrombie, H. C., et al. (2004). Functional but not structural subgenual prefrontal cortex abnormalities in melancholia. *Molecular Psychiatry, 9*(4), 393–405. 325 https://doi.org/10.1038/sj.mp.4001501.

Porcelli, A. J., Cruz, D., Wenberg, K., Patterson, M. D., Biswal, B. B., & Rypma, B. (2008). The effects of acute stress on human prefrontal working memory systems. *Physiology & Behavior, 95*(3), 282–289. https://doi.org/10.1016/j.physbeh.2008.04.027.

Preuss, T. M. (1995). Do rats have prefrontal cortex? The rose-Woolsey-akert program reconsidered. *Journal of Cognitive Neuroscience*, 7(1), 1–24. https://doi.org/10.1162/jocn.1995.7.1.1.

Quirk, G. J., Russo, G. K., Barron, J. L., & Lebron, K. (2000). The role of ventromedial prefrontal cortex in the recovery of extinguished fear. *The Journal of Neuroscience, 20*(16), 6225–6231. https://doi.org/10.1523/jneurosci.20-16-06225.2000.

Radley, J. J., Arias, C. M., & Sawchenko, P. E. (2006). Regional differentiation of the medial prefrontal cortex in regulating adaptive responses to acute emotional stress. *The Journal of Neuroscience, 26*(50), 12967–12976. https://doi.org/10.1523/jneurosci.4297-06.2006.

Radley, J. J., Rocher, A. B., Miller, M., Janssen, W. G., Liston, C., Hof, P. R., et al. (2006). Repeated stress induces dendritic spine loss in the rat medial prefrontal cortex. *Cerebral Cortex, 16*(3), 313–320. https://doi.org/10.1093/cercor/bhi104.

Radley, J. J., Williams, B., & Sawchenko, P. E. (2008). Noradrenergic innervation of the dorsal medial prefrontal cortex modulates hypothalamo-pituitary-adrenal responses to acute emotional stress. *The Journal of Neuroscience, 28*(22), 5806–5816. https://doi.org/10.1523/jneurosci.0552-08.2008.

Resstel, L., Souza, R., & Guimaraes, F. (2008). Anxiolytic-like effects induced by medial prefrontal cortex inhibition in rats submitted to the Vogel conflict test. *Physiology & Behavior, 93*(1–2), 200–205.

Roberts, A. C. (2020). Prefrontal regulation of threat-elicited behaviors: A pathway to translation. *Annual Review of Psychology, 71*, 357–387. https://doi.org/10.1146/annurev-psych-010419-050905.

Rushworth, M. F., & Owen, A. M. (1998). The functional organization of the lateral frontal cortex: Conjecture or conjuncture in the electrophysiology literature? *Trends in Cognitive Sciences*, *2*(2), 46–53. https://doi.org/10.1016/s1364-6613(98)01127-9.

Sandi, C. (2013). Stress and cognition. *Wiley Interdisciplinary Reviews: Cognitive Science*, *4*(3), 245–261.

Schmitz, A., & Grillon, C. (2012). Assessing fear and anxiety in humans using the threat of predictable and unpredictable aversive events (the NPU-threat test). *Nature Protocols*, 7(3), 527–532. https://doi.org/10.1038/nprot.2012.001.

Schwabe, L., & Wolf, O. T. (2011). Stress-induced modulation of instrumental behavior: From goal-directed to habitual control of action. *Behavioural Brain Research*, *219*(2), 321–328.

Shah, A. A., & Treit, D. (2003). Excitotoxic lesions of the medial prefrontal cortex attenuate fear responses in the elevated-plus maze, social interaction and shock probe burying tests. *Brain Research*, *969*(1–2), 183–194.

Shimizu, H., Kumasaka, Y., Tanaka, H., Hirose, A., & Nakamura, M. (1992). Anticonflict action of tandospirone in a modified Geller-Seifter conflict test in rats. *The Japanese Journal of Pharmacology*, *58*(3), 283–289.

Shin, L. M., & Liberzon, I. (2010). The neurocircuitry of fear, stress, and anxiety disorders. *Neuropsychopharmacology*, *35*(1), 169–191.

Sierra-Mercado, D., Padilla-Coreano, N., & Quirk, G. J. (2011). Dissociable roles of prelimbic and infralimbic cortices, ventral hippocampus, and basolateral amygdala in the expression and extinction of conditioned fear. *Neuropsychopharmacology*, *36*(2), 529–538. https://doi.org/10.1038/npp.2010.184.

Stefani, M. R., & Moghaddam, B. (2005). Systemic and prefrontal cortical NMDA receptor blockade differentially affect discrimination learning and set-shift ability in rats. *Behavioral Neuroscience*, *119*(2), 420–428. https://doi.org/10.1037/0735-7044.119.2.420.

Stern, C., Do Monte, F., Gazarini, L., Carobrez, A., & Bertoglio, L. (2010). Activity in prelimbic cortex is required for adjusting the anxiety response level during the elevated plus-maze retest. *Neuroscience*, *170*(1), 214–222.

Suzuki, S., Saitoh, A., Ohashi, M., Yamada, M., Oka, J.-I., & Yamada, M. (2016). The infralimbic and prelimbic medial prefrontal cortices have differential functions in the expression of anxiety-like behaviors in mice. *Behavioural Brain Research*, *304*, 120–124.

Uylings, H. B., Groenewegen, H. J., & Kolb, B. (2003). Do rats have a prefrontal cortex? *Behavioural Brain Research*, *146*(1–2), 3–17. https://doi.org/10.1016/j.bbr.2003.09.028.

Vertes, R. P. (2004). Differential projections of the infralimbic and prelimbic cortex in the rat. *Synapse*, *51*(1), 32–58. https://doi.org/10.1002/syn.10279.

Vogel, J. R., Beer, B., & Clody, D. E. (1971). A simple and reliable conflict procedure for testing anti-anxiety agents. *Psychopharmacologia*, *21*(1), 1–7. https://doi.org/10.1007/BF00403989.

Wager, T. D., Davidson, M. L., Hughes, B. L., Lindquist, M. A., & Ochsner, K. N. (2008). Prefrontal-subcortical pathways mediating successful emotion regulation. *Neuron*, *59*(6), 1037–1050.

Walker, D. L., Toufexis, D. J., & Davis, M. (2003). Role of the bed nucleus of the stria terminalis versus the amygdala in fear, stress, and anxiety. *European Journal of Pharmacology*, *463*(1–3), 199–216.

Yang, S. T., Shi, Y., Wang, Q., Peng, J. Y., & Li, B. M. (2014). Neuronal representation of working memory in the medial prefrontal cortex of rats. *Molecular Brain*, 7, 61. https://doi.org/10.1186/s13041-014-0061-2.

Yuen, E. Y., Wei, J., Liu, W., Zhong, P., Li, X., & Yan, Z. (2012). Repeated stress causes cognitive impairment by suppressing glutamate receptor expression and function in prefrontal cortex. *Neuron*, *73*(5), 962–977. https://doi.org/10.1016/j.neuron.2011.12.033.

CHAPTER THREE

Integration of value and action in medial prefrontal neural systems

Beata Kaminska[a], Jessica P. Caballero[a], and David E. Moorman[a,b,*]
[a]Neuroscience and Behavior Graduate Program, University of Massachusetts Amherst, Amherst, MA, United States
[b]Department of Psychological and Brain Sciences, University of Massachusetts Amherst, Amherst, MA, United States
*Corresponding author: e-mail address: moorman@umass.edu

Contents

Abstract

The rodent medial prefrontal cortex (mPFC) plays a key role in regulating cognition, emotion, and behavior. mPFC neurons are activated in diverse experimental paradigms, raising the questions of whether there are specific task elements or dimensions encoded by mPFC neurons, and whether these encoded parameters are selective to neurons in particular mPFC subregions or networks. Here, we consider the role of mPFC neurons in processing appetitive and aversive cues, outcomes, and related behaviors. mPFC neurons are strongly activated in tasks probing value and outcome-associated actions, but these responses vary across experimental paradigms. Can we identify specific categories of responses (e.g., positive or negative value), or do mPFC neurons exhibit response properties that are too heterogeneous/complex to cluster into distinct conceptual groups? Based on a review of relevant studies, we consider what has been done and what needs to be further explored in order to address these questions.

International Review of Neurobiology, Volume 158
ISSN 0074-7742
https://doi.org/10.1016/bs.irn.2020.11.007

1. Rodent medial prefrontal cortex

The prefrontal cortex (PFC) plays important roles in cognition, motivation, and behavior control (Cassaday, Nelson, & Pezze, 2014; Dalley, Cardinal, & Robbins, 2004; Euston, Gruber, & McNaughton, 2012; Miller & Cohen, 2001). However, the specifics and extent of what is being encoded in PFC circuit activity remain unclear. The PFC is highly interconnected with structures throughout the brain, and the resulting integrations produce complex computations at the circuit level. Although progress has been made in characterizing these computations in the service of understanding behavioral contributions of the PFC, the boundaries and intersection points of its functional roles are still poorly defined.

Although much of the research on PFC has been performed in human and non-human primates (Laubach, Amarante, Swanson, & White, 2018; Miller & Cohen, 2001; Passingham & Wise, 2012), there is a rich history of investigating PFC function in rodents, and interest has increased in recent years (Laubach et al., 2018). Despite concerns with rodent-primate homology (Preuss, 1995; Wise, 2008), anatomical parallels can guide cross-species comparisons (Heilbronner, Rodriguez-Romaguera, Quirk, Groenewegen, & Haber, 2016; Ongur & Price, 2000). Further, it is indisputable that rodents are capable of producing complex cognitive- and emotionally-driven behaviors, at least some of which activate and are impacted by manipulation of PFC. Thus, irrespective of specific anatomical homologies, there are functional equivalences across species that warrant investigating the rodent PFC as a key component of complex behavior (Brown & Bowman, 2002; Dalley et al., 2004; Kesner & Churchwell, 2011; Seamans, Lapish, & Durstewitz, 2008; Uylings, Groenewegen, & Kolb, 2003).

Most rodent studies of PFC focus on the medial PFC (mPFC), subdividing it into prelimbic (PL) and infralimbic (IL) cortex based on location, connectivity and function. Anterior cingulate cortex (ACC) is sometimes incorporated but more often separated based on anatomical and functional grounds, and orbitofrontal cortex (OFC) is typically considered separately (Gabbott, Warner, Jays, Salway, & Busby, 2005; Heidbreder & Groenewegen, 2003; Izquierdo, 2017; van Heukelum et al., 2020; Vertes, 2006). There is also some discussion about which regions constitute PFC (Laubach et al., 2018) and a growing interest in characterizing complex functions associated with frontal regions outside of ACC/PL/IL. Areas such as secondary motor cortex, frontal orienting fields, and frontal association cortex have been probed as being part of rodent prefrontal cortex (Barthas & Kwan, 2017; Ebbesen et al., 2018; Erlich, Bialek, & Brody, 2011). Although there is fascinating work being performed to elucidate the functions of these regions, here we will focus on the PL and IL "core" of mPFC.

2. Medial prefrontal neuronal representations of value and motivation

Here, we consider mPFC neural correlates of positive and negative value and how these signals interact with motivated action. There is clearly a relationship between mPFC neural function and outcome value as shown by the studies described below. Representations of outcomes, cues predicting outcomes, and actions needed to approach or avoid outcomes, are fundamental to almost all behaviors associated with mPFC function. Underlying cognition-probing tasks is the need for a motivated action. Outcome or value signals are frequently seen in these tasks—see, for example, action and outcome encoding during behavioral flexibility testing in (Del Arco, Park, Wood, Kim, & Moghaddam, 2017). Consideration of value also allows some cross-study comparisons. Although many studies of mPFC function are siloed in specific experimental paradigms or themes, a few have probed mPFC function associated with different values, bridging across research domains to understand the diversity of mPFC encoding. Finally, understanding value representations is of translational importance with potential implications for drug and alcohol dependence, eating disorders, ADHD and other disorders of impulsivity, depression, and other psychiatric diseases. A focus on studies investigating representation of value neglects much of the diversity of neural signaling in mPFC. However, given the pervasiveness of value in almost all behaviors, consideration of mPFC function in the context of value is worthwhile in understanding fundamental properties of this complex system.

3. Negative value: Punishment, fear, and aversion

Early studies of rodent mPFC neural activity described correlates of aversive learning (Peterson, 1986), supporting even earlier studies of cat and rabbit mPFC, e.g., (Gabriel & Orona, 1982). There has since been significant interest in mPFC activity related to learning and behaviors associated with negative outcomes (Gilmartin, Balderston, & Helmstetter, 2014; Giustino & Maren, 2015; Grunfeld & Likhtik, 2018; Maren & Quirk, 2004;

Rozeske & Herry, 2018; Sotres-Bayon & Quirk, 2010). This has been driven by at least two goals—first to understand neuronal representation of negatively-valenced emotions and outcomes, and second to understand neuronal mechanisms underlying learning and memory, as learning in Pavlovian fear conditioning paradigms is rapid and robust. This intersection between value/emotion representation and learning underscores the point that experimental paradigms incorporate multiple dimensions (outcomes, behaviors, learning), even when the primary research goal is focused on understanding one particular phenomenon.

Potentially reflecting these multiple dimensions, early studies of mPFC activity during fear conditioning identified that many neurons recorded responded to a shock-predicting cue, but that neuronal responses were diverse, lasting different durations and exhibiting both excitatory and inhibitory properties (Baeg et al., 2001; Garcia, Vouimba, Baudry, & Thompson, 1999; Peterson, 1986). Studies have repeatedly reported neuronal responses to aversive cues and outcomes, typically during fear conditioning involving shock stimuli, though studies have also reported mPFC neuronal responses to primary aversive stimuli (Jezzini, Mazzucato, La Camera, & Fontanini, 2013; Zhang, Tomida, Katayama, & Kawakami, 2004).

There is a relatively dichotomous role for PL vs IL in fear conditioning and expression vs extinction. PL neurons are generally activated by aversive outcomes and cues that predict them, and IL neurons are generally inhibited by aversive outcome-predicting cues and increase firing during extinction of fear-conditioned cues (Burgos-Robles, Vidal-Gonzalez, & Quirk, 2009; Gilmartin & McEchron, 2005; Giustino, Fitzgerald, Ressler, & Maren, 2019; Milad & Quirk, 2002; Sotres-Bayon, Sierra-Mercado, Pardilla-Delgado, & Quirk, 2012). These data support studies using PL or IL neuron manipulation, in which similar dichotomies have been observed (Courtin, Bienvenu, Einarsson, & Herry, 2013; Gilmartin et al., 2014; Giustino & Maren, 2015; Sotres-Bayon & Quirk, 2010). However, there is substantial heterogeneity in neuronal responses in these experiments, even within mPFC subregions (Baeg et al., 2001; Chang, Berke, & Maren, 2010; Fitzgerald, Giustino, Seemann, & Maren, 2015; Garcia et al., 1999; Giustino, Fitzgerald, & Maren, 2016; Peterson, 1986). Anxiety, as measured in elevated plus maze or open field testing or during restraint stress, also increases activation of PL neurons (Adhikari, Topiwala, & Gordon, 2010, 2011; Jackson & Moghaddam, 2006), and PL neurons encode the risk of punishment during a punished reward-seeking task (Park & Moghaddam, 2017).

Temporal dynamics of firing are important in aversive signaling. PL neurons fire at theta frequency during fear conditioning and anxiety (Adhikari et al., 2010, 2011; Dejean et al., 2016; Karalis et al., 2016), promoting the formation of task-relevant neuronal ensembles, and synchronizing activity with basolateral amygdala (BLA) and hippocampus (Adhikari et al., 2010; Dejean et al., 2016; Karalis et al., 2016; Padilla-Coreano et al., 2016). Synchronization is also associated with increased PL parvalbumin and somatostatin interneuron activity, which has been observed during fear conditioning, (Baeg et al., 2001; Courtin et al., 2014; Cummings & Clem, 2020). The interaction between PL and IL during fear conditioning and extinction is also important factor. PL and IL exhibit inhibitory interactions (Ji & Neugebauer, 2012; van Aerde, Heistek, & Mansvelder, 2008), though see (Marek et al., 2018). Some studies have shown that both PL and IL are activated during both fear conditioning and extinction, but the relative degrees of excitation and inhibition is important for determining the resulting emotional/behavioral response driven by mPFC outputs (Chang et al., 2010; Fitzgerald et al., 2015; Giustino et al., 2016).

Heterogeneity in PL and IL activity during fear conditioning and extinction may also result from the participation of these neurons in different afferent/efferent circuits. PL neurons projecting to BLA and periaqueductal gray (PAG) are activated during fear conditioning (Rozeske et al., 2018; Vander Weele et al., 2018). Connectivity between PL/IL and hippocampus and amygdala influences neuronal responses during anxiety and fear conditioning (Adhikari et al., 2010; Burgos-Robles et al., 2017; Dejean et al., 2016; Karalis et al., 2016; Klavir, Prigge, Sarel, Paz, & Yizhar, 2017; Padilla-Coreano et al., 2016). Neuromodulatory inputs from ventral tegmental area and locus coeruleus also influence neuronal responses and the behavioral output of mPFC during fear expression and/or extinction (Fitzgerald et al., 2015; Giustino et al., 2019; Vander Weele et al., 2018). These results indicate that continued analysis of circuit-level contributions may help resolve aspects of heterogeneous signaling observed related to aversive processing.

Neural activity associated with aversive processing may be particularly influenced by behavioral responses associated with the cues or outcomes. This has been shown during fear conditioning (Garcia et al., 1999; Halladay & Blair, 2015; Milad & Quirk, 2002), swim stress tests (Warden et al., 2012), and active avoidance (Diehl et al., 2018; Padilla-Coreano et al., 2019). Although most reports de-emphasize a purely motoric explanation of mPFC activity during fear conditioning, action may be a key factor that is integrated with emotional processing, memory consolidation, and recall, to produce the complex response properties seen at the single neuron level (Grunfeld & Likhtik, 2018). Given recent interest in characterizing active (e.g., avoidance) vs passive (e.g., freezing) responses to aversive stimuli (Fadok et al., 2017; Gruene, Flick, Stefano, Shea, & Shansky, 2015; LeDoux & Daw, 2018), understanding what role action plays in value-associated behaviors appears critical to our understanding of mPFC function.

4. Positive value: Reward and approach

The mPFC has also been well-characterized as a key brain region in signaling appetitive value and regulating appetitively-motivated behaviors (Hayes, Duncan, Xu, & Northoff, 2014). Appetitive signals in rodent mPFC have been described since at least the 1970s. Early reports demonstrated responses to delivered rewards (Ito & Olds, 1971) and cues conditioned to predict rewards (Olds, Disterhoft, Segal, Kornblith, & Hirsh, 1972; Pirch & Peterson, 1981). Interestingly, mPFC neurons responded to orally-consumed rewards such as food as well as to passively-received rewards such as brain stimulation. This suggests the possibility of positive value representation independent of its means of acquisition. Neurons recorded throughout the mPFC and ACC demonstrated selective responses to both intracranial stimulation and sucrose rewards, in some cases with the same neuron firing for both (Takenouchi et al., 1999). Along these lines, mPFC neurons, primarily those in PL, were active during outcome-alternating experiments where either food or water was presented after a delay (Miyazaki, Miyazaki, & Matsumoto, 2004). Approximately half of the neurons responded selectively to either food or water whereas the rest fired in anticipation of both. These results demonstrate the presence of both general positive valence signals and those encoding the identity of the rewarded outcome in mPFC neurons. Similar observations of reward- and reward-cue-related activity are found in both PL and IL subregions (Bouret & Sara, 2004; Le Merre et al., 2018; Otis et al., 2017; Valdes, Maldonado, Recabarren, Fuentes, & Torrealba, 2006). One possibility may be that value is represented in a graded fashion, with some neurons firing for more preferred rewards (e.g., sucrose) than others (e.g., water) (Petyko, Toth, Szabo, Galosi, & Lenard, 2009). Although variation in outcome magnitude, presumably correlating with value, has not been associated with mPFC firing

(Hong et al., 2019; Simon, Wood, & Moghaddam, 2015), neural correlates of value and individual preference have been observed under choice conditions (Sackett, Moschak, & Carelli, 2019). Comparably, mouse PL neurons recorded in a decision-making task signaled stimulus-driven value predictions, hypothesized to be compared with actual outcomes to promote adaptive behavior (Lak et al., 2020).

Studies of decision-making and spatial processing in the mPFC have reported high proportions of neurons responding during approach to or consumption of a food reward (Insel & Barnes, 2015; Lak et al., 2020; Pratt & Mizumori, 2001). However, multiple factors complicate interpretation of purely reward-related activity. One issue is the behavioral complexity associated with reward acquisition. During fear conditioning and other assays of aversive processing, negative outcomes are passively delivered, limiting explicit action demands (though see discussion of aversive outcome avoidance below). In contrast, rewards—particularly orally-ingested rewards—must be approached and consumed, raising the issue of whether neuronal activity changes represent value and/or action. This is more significant as behavioral complexity increases. Many studies demonstrating reward-related activity changes in mPFC do so during instrumental tasks where a lever-press or nosepoke delivers a food or liquid reward, which must then be acquired (e.g., Burgos-Robles, Bravo-Rivera, & Quirk, 2013; Homayoun & Moghaddam, 2006, 2009; Mulder, Nordquist, Orgut, & Pennartz, 2003; Peters, O'Donnell, & Carelli, 2005). Unsurprisingly, mPFC responses in these studies are complex, including those associated with predictive cues, instrumental responses, delay between response and outcome, reward approach, and reward consumption. Thus caution must be taken in interpreting neural signals related to appetitive value encoding as opposed to, e.g., behaviors associated with acquiring rewards.

As with aversive stimuli, there are differences in appetitive responses across PL and IL, although the exact nature of these differences has not been thoroughly characterized. Most studies investigating reward-related activity in mPFC subregions have focused on PL (or ACC), though reward-associated signals have been observed in IL (Burgos-Robles et al., 2013; Gentry & Roesch, 2018; Valdes et al., 2006). A few studies have compared PL and IL, reporting that PL responses are more associated with instrumental behaviors and IL responses are more related to reward anticipation/acquisition (Burgos-Robles et al., 2013; Takenouchi et al., 1999), although these differences may be more a matter of degree than category (Moorman & Aston-Jones, 2015). Response properties described across studies are heterogeneous. For example, PL neurons fire at the onset of reward-predicting actions such as lever presses (Peters et al., 2005), as well as during the time from action initiation to reward in a Go/NoGo task (Mulder et al., 2003). This diversity likely reflects a complex interaction of behavioral (task-dependence) and biological (specific region or neural circuit) variability.

Neuron phenotype also contributes to response heterogeneity. Most studies focus on characterizing pyramidal neuron activity or do not specify the neuron subtype recorded, but descriptions of interneuron activity have been reported. Some examples of interneuron responses include a relatively high proportion of sensory responses, reward approach and consumption, and reward/outcome evaluation (Insel & Barnes, 2015; Kvitsiani et al., 2013; Pinto & Dan, 2015). Importantly, the ability to target specific subtypes of interneurons (e.g., somatostatin vs parvalbumin) has revealed specific categories of response properties related to appetitive or aversive outcomes, described further below.

Neuronal connectivity is another parameter that plays a critical role in the response properties of mPFC neurons. As with neuronal responses to aversive outcomes, mPFC neurons projecting to different targets display different reward-related response properties. PL neurons projecting to NAc exhibited excitatory responses to a sucrose-predicting cue whereas PL projections to the paraventricular thalamus primarily responded with inhibition (Otis et al., 2017), revealing how commonly-observed diverse neuronal responses to the same cue or outcome are in part due signaling by separate populations defined by their anatomical projection target. These, and related results, e.g., (Ye et al., 2016), set the stage for future work that may explain functional heterogeneity seen in mPFC neuronal dynamics.

5. Drugs of abuse

mPFC responses during drug seeking are also diverse (Moorman, James, McGlinchey, & Aston-Jones, 2015). Early studies of cocaine and heroin self-administration characterized anticipatory responses, typically following a predictive cue or leading to the lever press, as well as post-drug delivery responses (Chang, Janak, & Woodward, 1998, 2000; Chang, Sawyer, Paris, Kirillov, & Woodward, 1997; Chang, Zhang, Janak, & Woodward, 1997; Rebec & Sun, 2005; Sun & Rebec, 2006). Activity in NAc-projecting IL neurons during cocaine seeking appeared to be inversely related to motivation for cocaine (Cameron, Murugan, Choi, Engel, & Witten, 2019), in line, to some degree, with the role of IL in facilitating extinction in fear conditioning and drug-seeking studies (Muller Ewald & LaLumiere, 2017; Peters, Kalivas, & Quirk, 2009). This suppressed activity lessened after an incubation period, arguing that one driver of increased motivation for cocaine is decreased inhibitory control by IL. A direct comparison of PL and IL neuronal responses to cocaine self-administration associated cues found responses in both PL and IL, but increased PL activation after a 30-day abstinence period, which produced increased motivation for cocaine (West, Saddoris, Kerfoot, & Carelli, 2014). These results indicate that PL and IL are both activated by cues predicting and behaviors directed toward drugs of abuse, but that the relative contributions of neurons in subregions or networks may shift over the course of abstinence/withdrawal.

mPFC activity has also been described in studies of alcohol use. The activity of dorsal PAG-projecting mPFC neurons recorded during alcohol consumption predicted future proclivity to compulsively consume alcohol under threat of punishment (Siciliano et al., 2019). Other studies have shown elevated overall mPFC neuronal activity, but blunted evoked responses, in alcohol-preferring "P" rats during cued alcohol drinking (Linsenbardt & Lapish, 2015; Linsenbardt, Timme, & Lapish, 2019; McCane et al., 2018), suggesting differential mPFC neuronal signaling based on alcohol preference. Subregion differences have also been observed in alcohol responses—although dorsal and ventral mPFC neurons were both activated during alcohol seeking, ventral, but not dorsal, neurons fired more during avoidance of alcohol-paired punishment (Halladay et al., 2020). The consistent finding of mPFC activation during drug and alcohol use indicates that a better understanding of the neural circuitry and dynamics of this system may improve our understanding and potential treatments of motivation-associated diseases such as drug and alcohol dependence, in addition to providing an additional perspective on value encoding in mPFC circuits.

6. Comparisons of positive and negative value

Most studies characterize either appetitive or aversive signaling by mPFC neurons in isolation, raising the question of whether mPFC neurons exhibit value selectivity when these outcomes are compared. In a small number of experiments, the same mPFC neurons have been recorded during presentation of both positive and negative cues and/or outcomes. During presentation of sucrose- or shock-predicting cues, rat PL neurons exhibited diverse response profiles, encoding most combinations of excitation and inhibition in response to both/either cues (Burgos-Robles et al., 2017). However, the predominant response was excitation to shock-predicting cues, and activity in this population of neurons was correlated with that in BLA neurons, suggesting a valence-selective population defined by its connectivity. Similar diverse responses in PL neurons were observed in response to unpredictable deliveries of either sucrose or tailpinch (Del Arco, Park, & Moghaddam, 2020). An approximately equal number of mPFC neurons responded to both sucrose and pinch, exhibiting both excitation and inhibition, and approximately 30% responded to both outcomes. Intraoral delivery of appetitive (sucrose or sodium chloride) or aversive (quinine or citric acid) solutions revealed a role for mPFC (broadly-defined) in encoding the intersection of palatability and tastant identity, with a bias in neural responses toward aversive tastants (Jezzini et al., 2013). Recordings from ACC neurons showed strong neuronal responses to food and shock, with some neurons firing under both conditions, hypothesized to reflect salience or attention signals (Schneider, Sciarillo, Nudelman, Cheer, & Roesch, 2020). In a separate study, a higher proportion of ACC neurons exhibited responses to shock vs reward, as well as to a neutral cue predicting no outcome (Caracheo, Grewal, & Seamans, 2018). Neurons also responded throughout blocks of trials, potentially revealing the use of valence to support representation of context (Hyman, Ma, Balaguer-Ballester, Durstewitz, & Seamans, 2012; Moorman & Aston-Jones, 2015). Emphasizing the role of network identity in response properties, PL neurons projecting to either NAc shell or dorsal PAG demonstrated differential signaling whereby most PAG-projecting neurons responded to shock and airpuff, and NAc-projecting neurons exhibited a mix of moderate responses to either shock or sucrose (Vander Weele et al., 2018). The lack of prominent PFC-NAc sucrose responding might be surprising given the results of (Otis et al., 2017), but may underscore the importance of context and behavioral components in shaping value-specific responses.

The cellular phenotype of mPFC neurons is also associated with response selectivity. In a reward/punishment go/no-go task, mouse PL neurons encoded multiple task parameters (Kamigaki & Dan, 2017; Pinto & Dan, 2015). Pyramidal neurons reflected diverse aspects of the task, with some response types differentially distributed across lamina, indicating an interaction with connectivity. Subtypes of GABAergic interneurons responded to different elements of the task (stimuli, outcomes, licking), indicating specific within-mPFC network correlates based on the intersection of cellular phenotype and extra-PFC connectivity. Differences across mPFC neuronal subtype were also observed in a reward foraging task, with somatostatin neurons encoding reward approach and parvalbumin neurons encoding disengagement (Kvitsiani et al., 2013). These data highlight the importance of cell-type identification to understand neural circuits associated with value encoding.

Other studies have compared mPFC activity during rewards and/or threats of punishments that can be avoided by withholding or performing an action. IL neurons exhibited a bias toward food-delivering, as opposed to shock-avoiding behaviors, even under extinction conditions, suggesting a privileged representation of positive reinforcement in IL (Gentry & Roesch, 2018). These data are aligned with previous work describing mPFC neuron activity during behaviors to acquire sucrose and ICSS, but not to avoid a shock (Takenouchi et al., 1999). Rostral PL neurons were inhibited during performance of platform avoidance of shock (Diehl et al., 2018), indicating that avoidance responses are regionally specific and may exhibit different dynamics than responses to inescapable shock-predicting stimuli. In a task where rats approached rewards or avoided/escaped from shock, PL neurons encoded complex combinations of stimulus identity, outcome valence, and behavioral response, in line with the concept of mPFC neurons exhibiting mixed selectivity (Fusi, Miller, & Rigotti, 2016; Rigotti et al., 2013), and in contrast with BLA neurons which showed stronger pure valence encoding (Kyriazi, Headley, & Pare, 2020). In a slightly different paradigm, in which rewarded outcome was associated with varying probabilities of shock, mPFC neurons were found to encode the risk of punishment though increases in baseline activity and decreases in action-associated responses and synchronous firing (Park & Moghaddam, 2017), demonstrating an intersection between valence and motivated action.

Together, these studies confirm that mPFC neurons respond to both positive and negative outcomes, even within the same task. Signals are heterogeneous and simultaneously encode value and action-associated parameters. Valence representation appears to be largely segregated, but occasionally overlaps at the level of the single neuron. Cellular identity and projection target may be key factors that identify which and how many variables are encoded. It will be of value in future studies to identify these correlates in tasks that can be directly compared, with matched actions and outcomes, in order to identify how combinations of factors are encoded at the level of neurons or populations (Fig. 1).

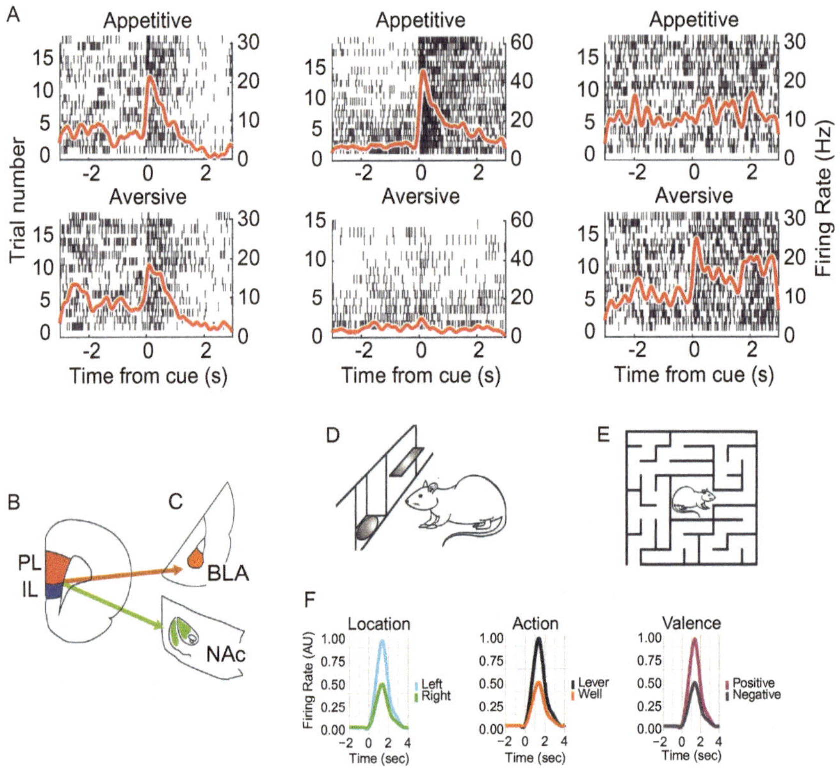

Fig. 1 What factors may explain apparent heterogeneity in mPFC neuron value signaling? (A) Three example mPFC neurons (from Kaminska and Moorman, in prep.) responsive to both appetitive and aversive outcome-predicting cues (left), only appetitive cues (middle), or only aversive cues (right). (B–F) Schematics of potential explanations for diversity in mPFC value encoding. (B) Subregion differences (e.g., different responses in PL vs IL neurons), (C) Differences in afferent projection target (e.g., different responses in neurons projecting to BLA vs NAc), (D) Differences in actions (e.g., neurons may respond to behaviors such as lever-press vs well-entry instead of or in addition to outcomes), (E) Potential cognitive factors (e.g., neurons may respond to variables associated with decision-making, learning, navigation, etc.), (F) Heterogeneous signaling (e.g., mixed selectivity) whereby individual neurons encode unique combinations of value, action, cognition, etc., at different strengths.

7. Integration of value and action

Not included above are numerous studies with goals beyond understanding direct relationships between mPFC neuronal activity and value. This includes studies such as those related to cognition, action control and social behavior. Reward or outcome related signals have been described in many of these studies, e.g., (Del Arco et al., 2017; Gruber et al., 2010), indicating that mPFC neurons encode some aspect of value, even when this is not the main focus of the experiment.

This highlights the point that signals related to multiple task parameters overlap during any behavioral test. Animal behavior is almost always motivated by potential rewards or punishments, producing a persistent motivational signal that may be hard to extract from neural signatures of cognition or behavior. Conversely, as demonstrated by many studies discussed above, it is challenging to extract pure value signals from mPFC neurons (Bissonette, Gentry, Padmala, Pessoa, & Roesch, 2014). Although mPFC neurons respond to positive or negative stimuli, it is challenging to identify value signals in the absence of any other factor driving activity. One major confounding factor is the behavior required to obtain or avoid an outcome. In some cases, what appear to be reward-related signals may actually be more closely associated to actions such as licking and consumption-related behavior, even when reward is not delivered (Amarante, Caetano, & Laubach, 2017; Horst & Laubach, 2013). These motor-related signals appear to be modulated by motivation levels (de Haan et al., 2018), suggesting that both value and action are represented in mPFC neuronal activity.

One framework for investigating the intersection of value and action is characterizing activity related to goal-directed vs habitual behaviors. Overtraining transitions animals from a goal-directed (outcome-devaluation sensitive) to habitual (devaluation insensitive) behavioral strategy. Over the course of overtraining, IL neurons gradually increased, and PL neurons decreased, activity (Smith & Graybiel, 2013). Using a lever-based operant task, changes in IL activity were also observed after extensive training on a random interval, habit-associated task (Barker, Glen, Linsenbardt, Lapish, & Chandler, 2017). The details of the neuronal responses varied between these studies (e.g., excitation vs inhibition, task epoch, etc.), but the results reflect differences in mPFC neuronal activity based on whether behaviors are goal-directed vs habitual.

The observation that mPFC neurons integrate action and outcome should not be surprising. Reward, fear, action, learning, attention, and other aspects of cognition and behavior are interlinked. Adaptive behavior, broadly construed, is a result of the seamless integration of these variables to allow responses to exogenous and endogenous stimuli. From an evolutionary and ethological perspective this is intuitive. Rewards (food, mates, shelter) must be acquired and punishers (noxious stimuli, threats, dangerous environments) must be avoided. This is dependent on calculating outcome costs and values, orienting to goals, performing actions, and learning from results. Actions themselves are costly and performance of actions (at least when goal-directed) must be evaluated in the context of the value of an outcome. So value computations are inherently built into almost all actions.

8. Summary and conclusions: Value, actions, both, or neither?

As is clear from the extensive studies described above, neuronal activity in mPFC appears to be strongly influenced by both positive and negative value. These value representations include responses to outcomes (e.g., food rewards, shocks), cues that predict outcomes, actions associated with acquiring or avoiding outcomes, and combinations of all of these. Based on the frequency of their observations, it seems logical to conclude that value signals make up a predominant component of mPFC function. However, a number of caveats suggest caution in this interpretation (see Box 1). The vast

BOX 1 Information needed to help refine understanding of value coding in mPFC.

Experimental design
- Account for behavioral variables (lever press vs freeze, etc.)
- Account for cognitive variables (salience/attention, decisions, learning, etc.)
- Precise measurement of behaviors to account for unanticipated behavioral correlates (video analysis, EMG, etc.)
- Parameterize outcomes (sucrose concentration, shock vs airpuff, etc.)
- Compare behaviors across studies

Biological sources of functional heterogeneity of value responsive neurons
- Anatomical location (PFC subregions)
- Connectivity (afferent & efferent)
- Cellular differences (pyramidal vs interneuron subtypes)
- Molecular and genetic differences (receptors and other signaling proteins, transcription factors and other genetic regulators)

Mechanisms of encoding
- Ensemble vs single-neuron selectivity
- Mixed selectivity (how heterogeneous? what are the parameter axes?)
- Integration of response properties with biological factors
- Contributions of dynamics (epochs of excitation or inhibition, temporally patterned activity, coherence with LFP)
- Encoding plasticity—do mPFC neurons always respond to valued outcomes the same way?

majority of the studies described above measured neural responses to one type of outcome in isolation. Most studies characterizing mPFC responses to footshocks do not also probe responses to sucrose, and vice versa. In the absence of parametric manipulation of value, it is challenging to conclude that a response associated with a particularly-valued outcome is in fact a value representation as opposed to, for example, a salience representation or the representation of an action associated with the outcome such as an approach or avoidance. The few studies directly comparing positive and negative outcomes clearly demonstrate neuronal responses to both with different biases (more positive or more negative) depending on the study. Although these studies advance our understanding of value representation by contrasting positive and negative outcomes they, too, are challenged by differential levels of salience and different actions, both of which may be main drivers of mPFC function. So, for example, delivery of a shock may be more salient or intensely negative than delivery of sucrose is positive. As such, differential representations within or across neurons may have alternate foundations than positive or negative value per se. Similarly, the actions associated with each outcome differ, as noted above. In order to make a clear association between mPFC neuronal activity, future studies need to incorporate parametric manipulations of outcomes and control for, or closely monitor, behaviors associated with different outcomes (Box 1). Forming conclusions based on these manipulations may even prove challenging, as there is a close relationship between outcome value, such as high vs low sucrose concentration, and value-associated actions, such as intensity of consumption (Amarante et al., 2017). For now, we are left with the overarching observation that mPFC neurons respond when outcomes of a particular valence (or predictors of such outcomes) are presented. But it is difficult, if not impossible, to say with confidence that such responses are specifically reflecting value as opposed to multiple other potential associated factors.

So should we consider an alternate interpretation—for example, that mPFC signals are best explained by actions associated with outcomes but not values? It is equally challenging to confidently make this conclusion, or related conclusions such as value representations being best described by the salience of cues or outcomes. In most of the studies described above, actions are motivated—freezing occurs in response to a shock-predicting cue, and approach or consumption is driven by an appetitive outcome. Thus, the value of an expected outcome is a frequent driver of action, even when the goal of a study is to characterize action-associated mPFC activity. In the same way that it is hard to isolate value in the absence of other factors, it is challenging to identify neural correlates of action in the absence of parameters driving such actions. In some sense, studies of goal-directed vs habitual behavior may permit some insight into this, under the assumption that habitual behaviors are actions in the absence of any expected outcome, though it is notable that mPFC neuronal activity is different under goal-directed vs habitual conditions, as described above. This indicates that actions, independent of their goals or the context surrounding them, are not the exclusive focus of mPFC activity and that some aspect of expected outcomes, or the absence of such expectations, influence mPFC neuronal activity.

The sections above describe a large number of studies in which neurons respond during situations in which outcomes have a valence. Few, if any, conclusively identify a "pure" value signal or, conversely, a "pure" signal related to action, salience, or other parameter that reliably explains variations in mPFC neuronal activity across studies. In some sense this is disappointing—it would be quite satisfying to align the research to date and conclude that the results together indicate a specific role for mPFC circuits in any one of these aspects of behavior. And yet we probably shouldn't be surprised that there is no single encoded variable in mPFC neuronal activity. Further, we should be encouraged that there are clear paths forward to refine our understanding of the presence or absence of value encoding in mPFC. For example, controlling for behavior across outcomes, varying salience within and across outcomes of different valence, and more (see Box 1). It would be disheartening to think that we have exhausted our potential manipulations and are still left with inconclusive results. Instead, there are obvious experiments, either underway or waiting to be done, that will answer some of these questions for us. When these behavioral controls are applied to studies incorporating biological heterogeneity across mPFC neurons (e.g., responses of NAc-projecting vs BLA-projecting neurons) we will have the opportunity to identify the extent of behavioral encoding with greater precision has been previously possible.

The intersection of value with so many other aspects of behavior raises an alternate perspective in which value is not directly encoded by individual neurons but is an emergent property of network activation. Neural responses related to value are distributed throughout the brain (Hunt & Hayden, 2017; Vickery, Chun, & Lee, 2011). mPFC may be a node within a larger value calculation network, and individual neurons may participate in the representation of value without being specifically dedicated to value encoding. Similarly, mPFC neurons may simultaneously encode other functions as well, as demonstrated by the large numbers of behavioral and cognitive factors that are associated with mPFC neuronal activity (Cassaday et al., 2014; Euston et al., 2012). This framework dovetails with the conceptualization of mPFC neurons exhibiting mixed selectivity, responding in a nonlinear fashion to multiple sensory, behavioral, or cognitive parameters (Fusi et al., 2016; Grunfeld & Likhtik, 2018; Rigotti et al., 2013). Although responses to value may be detected in a study in which it is probed (e.g., neural responses to sucrose presentation), a full characterization of the representational suite of each neuron may depend on multiple tests to identify the extent of response selectivity. An additional important future direction for understanding how mPFC neurons encode value with respect to other variables is to integrate functional, anatomical, and molecular properties of mPFC neurons (Ye et al., 2016), as subpopulations of neurons may exhibit more restricted selectivity. Some studies have characterized limits of mixed selectivity using model-free clustering and network identification of neuron subpopulations (Hirokawa, Vaughan, Masset, Ott, & Kepecs, 2019), suggesting that functional categories of prefrontal neurons may exist once properly classified.

Perhaps a more extreme conceptualization is that the representation of value is epiphenomenal at the level of the single neuron or even neuronal ensembles. This perspective is rooted in questions of whether the psychological concepts that we experience and test experimentally are actually represented at the level of neurons or neural networks (Buzsaki, 2020; Churchland, 1981; Hunt & Hayden, 2017). In some sense, this aligns with reframing value through the lens of a classical characterization of value inferred through action—reinforcers produce approach, and punishers produce avoidance. Positive value, for example, might simply be a term used to describe a stimulus that produces approach more strongly than another. By this logic, neural representation of value-associated stimuli (reward- or punishment-predicting cues in the lab) might be general association signals that could be equally applied to non-valued outcomes that produce similar behaviors, draw attention, or produce outcome-related learning.

Ultimately, animals are able to make preference-based decisions, presumably through assigning value to objects and action outcomes, and to execute and adjust their behaviors based on these decisions and outcomes. In the brain there must be some type of computation that permits these associations to be made (Vlaev, Chater, Stewart, & Brown, 2011). Given the studies described above, and despite the caveats above, neurons in the mPFC likely participate in this process. Instead of exclusively processing value information, mPFC neurons may encode value when it is critical to drive behavior, consistent with the region's role in decision-making and flexible (i.e., cost/benefit-driven) behaviors. mPFC may also use value to generate a representation of context (Hyman et al., 2012; Moorman & Aston-Jones, 2015), or to support learning and memory recall (Euston et al., 2012). One possible conclusion permitting a function for explicit value representations is that value signals may be present in a subset of mPFC neurons (defined, perhaps, by shared anatomical or other cellular features), in the service of complex representations at the ensemble level so that the maximum amount of world information can be used to guide behavior. Despite impressive studies revealing the presence of value-related signals in mPFC, exactly how mPFC neurons, within specific subregions and circuits, integrate this information is still unclear. The widespread, and growing, interest in mPFC coding befits its role as a highly connected structure where neurons exhibit complex response properties in the service of complex flexible value-motivated behavior. When this interest is paired with experiments designed to optimally extract value (or action, or other) signals, there is a significant chance of identifying the presence or absence of pure value signals and, in doing so, reveal core neuronal principles underlying behavioral control.

References

Adhikari, A., Topiwala, M. A., & Gordon, J. A. (2010). Synchronized activity between the ventral hippocampus and the medial prefrontal cortex during anxiety. *Neuron, 65*(2), 257–269. https://doi.org/10.1016/j.neuron.2009.12.002.

Adhikari, A., Topiwala, M. A., & Gordon, J. A. (2011). Single units in the medial prefrontal cortex with anxiety-related firing patterns are preferentially influenced by ventral hippocampal activity. *Neuron, 71*(5), 898–910. https://doi.org/10.1016/j.neuron.2011.07.027.

Amarante, L. M., Caetano, M. S., & Laubach, M. (2017). Medial frontal theta is entrained to rewarded actions. *The Journal of Neuroscience, 37*(44), 10757–10769. https://doi.org/10.1523/JNEUROSCI.1965-17.2017.

Baeg, E. H., Kim, Y. B., Jang, J., Kim, H. T., Mook-Jung, I., & Jung, M. W. (2001). Fast spiking and regular spiking neural correlates of fear conditioning in the medial prefrontal cortex of the rat. *Cerebral Cortex, 11*(5), 441–451.

Barker, J. M., Glen, W. B., Linsenbardt, D. N., Lapish, C. C., & Chandler, L. J. (2017). Habitual behavior is mediated by a shift in response-outcome encoding by infralimbic cortex. *eNeuro, 4*(6). https://doi.org/10.1523/ENEURO.0337-17.2017.

Barthas, F., & Kwan, A. C. (2017). Secondary motor cortex: Where 'sensory' meets 'motor' in the rodent frontal cortex. *Trends in Neurosciences, 40*(3), 181–193. https://doi.org/10.1016/j.tins.2016.11.006.

Bissonette, G. B., Gentry, R. N., Padmala, S., Pessoa, L., & Roesch, M. R. (2014). Impact of appetitive and aversive outcomes on brain responses: Linking the animal and human literatures. *Frontiers in Systems Neuroscience, 8*, 24. https://doi.org/10.3389/fnsys.2014.00024.

Bouret, S., & Sara, S. J. (2004). Reward expectation, orientation of attention and locus coeruleus-medial frontal cortex interplay during learning. *The European Journal of Neuroscience*, *20*(3), 791–802. https://doi.org/10.1111/j.1460-9568.2004.03526.x.

Brown, V. J., & Bowman, E. M. (2002). Rodent models of prefrontal cortical function. *Trends in Neurosciences*, *25*(7), 340–343. https://doi.org/10.1016/s0166-2236(02)02164-1.

Burgos-Robles, A., Bravo-Rivera, H., & Quirk, G. J. (2013). Prelimbic and infralimbic neurons signal distinct aspects of appetitive instrumental behavior. *PLoS One*, *8*, 1–7. https://doi.org/10.1371/journal.pone.0057575.

Burgos-Robles, A., Kimchi, E. Y., Izadmehr, E. M., Porzenheim, M. J., Ramos-Guasp, W. A., Nieh, E. H., et al. (2017). Amygdala inputs to prefrontal cortex guide behavior amid conflicting cues of reward and punishment. *Nature Neuroscience*, *20*(6), 824–835. https://doi.org/10.1038/nn.4553.

Burgos-Robles, A., Vidal-Gonzalez, I., & Quirk, G. J. (2009). Sustained conditioned responses in prelimbic prefrontal neurons are correlated with fear expression and extinction failure. *The Journal of Neuroscience*, *29*(26), 8474–8482. https://doi.org/10.1523/JNEUROSCI.0378-09.2009.

Buzsaki, G. (2020). The brain-cognitive behavior problem: A retrospective. *eNeuro*, 7(4). https://doi.org/10.1523/ENEURO.0069-20.2020.

Cameron, C. M., Murugan, M., Choi, J. Y., Engel, E. A., & Witten, I. B. (2019). Increased cocaine motivation is associated with degraded spatial and temporal representations in IL-NAc neurons. *Neuron*, *103*(1), 80–91.e87. https://doi.org/10.1016/j.neuron.2019.04.015.

Caracheo, B. F., Grewal, J. J. S., & Seamans, J. K. (2018). Persistent valence representations by ensembles of anterior cingulate cortex neurons. *Frontiers in Systems Neuroscience*, *12*, 51. https://doi.org/10.3389/fnsys.2018.00051.

Cassaday, H. J., Nelson, A. J., & Pezze, M. A. (2014). From attention to memory along the dorsal-ventral axis of the medial prefrontal cortex: Some methodological considerations. *Frontiers in Systems Neuroscience*, *8*, 160. https://doi.org/10.3389/fnsys.2014.00160.

Chang, C. H., Berke, J. D., & Maren, S. (2010). Single-unit activity in the medial prefrontal cortex during immediate and delayed extinction of fear in rats. *PLoS One*, *5*(8), e11971. https://doi.org/10.1371/journal.pone.0011971.

Chang, J. Y., Janak, P. H., & Woodward, D. J. (1998). Comparison of mesocorticolimbic neuronal responses during cocaine and heroin self-administration in freely moving rats. *The Journal of Neuroscience*, *18*(8), 3098–3115.

Chang, J. Y., Janak, P. H., & Woodward, D. J. (2000). Neuronal and behavioral correlations in the medial prefrontal cortex and nucleus accumbens during cocaine self-administration by rats. *Neuroscience*, *99*(3), 433–443.

Chang, J. Y., Sawyer, S. F., Paris, J. M., Kirillov, A., & Woodward, D. J. (1997). Single neuronal responses in medial prefrontal cortex during cocaine self-administration in freely moving rats. *Synapse*, *26*(1), 22–35. https://doi.org/10.1002/(SICI)1098-2396(199705)26:1<22::AID-SYN3>3.0.CO;2-G. (pii) 10.1002/(SICI)1098-2396(199705)26:1<22::AID-SYN3>3.0.CO;2-G.

Chang, J. Y., Zhang, L., Janak, P. H., & Woodward, D. J. (1997). Neuronal responses in prefrontal cortex and nucleus accumbens during heroin self-administration in freely moving rats. *Brain Research*, *754*(1–2), 12–20. https://doi.org/10.1016/s0006-8993(97)00012-7.

Churchland, P. M. (1981). Eliminative materialism and propositional attitudes. *Journal of Philosophy*, *78*, 67–90.

Courtin, J., Bienvenu, T. C., Einarsson, E. O., & Herry, C. (2013). Medial prefrontal cortex neuronal circuits in fear behavior. *Neuroscience*, *240*, 219–242. https://doi.org/10.1016/j.neuroscience.2013.03.001.

Courtin, J., Chaudun, F., Rozeske, R. R., Karalis, N., Gonzalez-Campo, C., Wurtz, H., et al. (2014). Prefrontal parvalbumin interneurons shape neuronal activity to drive fear expression. *Nature, 505*(7481), 92–96. https://doi.org/10.1038/nature12755.

Cummings, K. A., & Clem, R. L. (2020). Prefrontal somatostatin interneurons encode fear memory. *Nature Neuroscience, 23*(1), 61–74. https://doi.org/10.1038/s41593-019-0552-7.

Dalley, J. W., Cardinal, R. N., & Robbins, T. W. (2004). Prefrontal executive and cognitive functions in rodents: Neural and neurochemical substrates. *Neuroscience and Biobehavioral Reviews, 28*(7), 771–784. https://doi.org/10.1016/j.neubiorev.2004.09.006.

de Haan, R., Lim, J., van der Burg, S. A., Pieneman, A. W., Nigade, V., Mansvelder, H. D., et al. (2018). Neural representation of motor output, context and behavioral adaptation in rat medial prefrontal cortex during learned behavior. *Frontiers in Neural Circuits, 12*, 75. https://doi.org/10.3389/fncir.2018.00075.

Dejean, C., Courtin, J., Karalis, N., Chaudun, F., Wurtz, H., Bienvenu, T. C., et al. (2016). Prefrontal neuronal assemblies temporally control fear behaviour. *Nature, 535*(7612), 420–424. https://doi.org/10.1038/nature18630.

Del Arco, A., Park, J., & Moghaddam, B. (2020). Unanticipated stressful and rewarding experiences engage the same prefrontal cortex and ventral tegmental area neuronal populations. *eNeuro*, 7(3). https://doi.org/10.1523/ENEURO.0029-20.2020.

Del Arco, A., Park, J., Wood, J., Kim, Y., & Moghaddam, B. (2017). Adaptive encoding of outcome prediction by prefrontal cortex ensembles supports behavioral flexibility. *The Journal of Neuroscience, 37*(35), 8363–8373. https://doi.org/10.1523/JNEUROSCI.0450-17.2017.

Diehl, M. M., Bravo-Rivera, C., Rodriguez-Romaguera, J., Pagan-Rivera, P. A., Burgos-Robles, A., Roman-Ortiz, C., et al. (2018). Active avoidance requires inhibitory signaling in the rodent prelimbic prefrontal cortex. *eLife*, 7, e34657. https://doi.org/10.7554/eLife.34657.

Ebbesen, C. L., Insanally, M. N., Kopec, C. D., Murakami, M., Saiki, A., & Erlich, J. C. (2018). More than just a "motor": Recent surprises from the frontal cortex. *The Journal of Neuroscience, 38*(44), 9402–9413. https://doi.org/10.1523/JNEUROSCI.1671-18.2018.

Erlich, J. C., Bialek, M., & Brody, C. D. (2011). A cortical substrate for memory-guided orienting in the rat. *Neuron, 72*(2), 330–343. https://doi.org/10.1016/j.neuron.2011.07.010.

Euston, D. R., Gruber, A. J., & McNaughton, B. L. (2012). The role of medial prefrontal cortex in memory and decision making. *Neuron, 76*(6), 1057–1070. https://doi.org/10.1016/j.neuron.2012.12.002.

Fadok, J. P., Krabbe, S., Markovic, M., Courtin, J., Xu, C., Massi, L., et al. (2017). A competitive inhibitory circuit for selection of active and passive fear responses. *Nature, 542*(7639), 96–100. https://doi.org/10.1038/nature21047.

Fitzgerald, P. J., Giustino, T. F., Seemann, J. R., & Maren, S. (2015). Noradrenergic blockade stabilizes prefrontal activity and enables fear extinction under stress. *Proceedings of the National Academy of Sciences of the United States of America, 112*(28), E3729–E3737. https://doi.org/10.1073/pnas.1500682112.

Fusi, S., Miller, E. K., & Rigotti, M. (2016). Why neurons mix: High dimensionality for higher cognition. *Current Opinion in Neurobiology, 37*, 66–74. https://doi.org/10.1016/j.conb.2016.01.010.

Gabbott, P. L., Warner, T. A., Jays, P. R., Salway, P., & Busby, S. J. (2005). Prefrontal cortex in the rat: Projections to subcortical autonomic, motor, and limbic centers. *The Journal of Comparative Neurology, 492*(2), 145–177. https://doi.org/10.1002/cne.20738.

Gabriel, M., & Orona, E. (1982). Parallel and serial processes of the prefrontal and cingulate cortical systems during behavioral learning. *Brain Research Bulletin, 8*(6), 781–785. https://doi.org/10.1016/0361-9230(82)90107-1.

Garcia, R., Vouimba, R. M., Baudry, M., & Thompson, R. F. (1999). The amygdala modulates prefrontal cortex activity relative to conditioned fear. *Nature, 402*(6759), 294–296. https://doi.org/10.1038/46286.

Gentry, R. N., & Roesch, M. R. (2018). Neural activity in ventral medial prefrontal cortex is modulated more before approach than avoidance during reinforced and extinction trial blocks. *The Journal of Neuroscience, 38*(19), 4584–4597. https://doi.org/10.1523/JNEUROSCI.2579-17.2018.

Gilmartin, M. R., Balderston, N. L., & Helmstetter, F. J. (2014). Prefrontal cortical regulation of fear learning. *Trends in Neurosciences, 37*(8), 455–464. https://doi.org/10.1016/j.tins.2014.05.004.

Gilmartin, M. R., & McEchron, M. D. (2005). Single neurons in the medial prefrontal cortex of the rat exhibit tonic and phasic coding during trace fear conditioning. *Behavioral Neuroscience, 119*(6), 1496–1510. https://doi.org/10.1037/0735-7044.119.6.1496.

Giustino, T. F., Fitzgerald, P. J., & Maren, S. (2016). Fear expression suppresses medial prefrontal cortical firing in rats. *PLoS One, 11*(10). https://doi.org/10.1371/journal.pone.0165256, e0165256.

Giustino, T. F., Fitzgerald, P. J., Ressler, R. L., & Maren, S. (2019). Locus coeruleus toggles reciprocal prefrontal firing to reinstate fear. *Proceedings of the National Academy of Sciences of the United States of America, 116*(17), 8570–8575. https://doi.org/10.1073/pnas.1814278116.

Giustino, T. F., & Maren, S. (2015). The role of the medial prefrontal cortex in the conditioning and extinction of fear. *Frontiers in Behavioral Neuroscience, 9*, 298. https://doi.org/10.3389/fnbeh.2015.00298.

Gruber, A. J., Calhoon, G. G., Shusterman, I., Schoenbaum, G., Roesch, M. R., & O'Donnell, P. (2010). More is less: A disinhibited prefrontal cortex impairs cognitive flexibility. *The Journal of Neuroscience, 30*(50), 17102–17110. https://doi.org/10.1523/JNEUROSCI.4623-10.2010.

Gruene, T. M., Flick, K., Stefano, A., Shea, S. D., & Shansky, R. M. (2015). Sexually divergent expression of active and passive conditioned fear responses in rats. *eLife, 4*, e11352. https://doi.org/10.7554/eLife.11352.

Grunfeld, I. S., & Likhtik, E. (2018). Mixed selectivity encoding and action selection in the prefrontal cortex during threat assessment. *Current Opinion in Neurobiology, 49*, 108–115. https://doi.org/10.1016/j.conb.2018.01.008.

Halladay, L. R., & Blair, H. T. (2015). Distinct ensembles of medial prefrontal cortex neurons are activated by threatening stimuli that elicit excitation vs. inhibition of movement. *Journal of Neurophysiology, 114*(2), 793–807. https://doi.org/10.1152/jn.00656.2014.

Halladay, L. R., Kocharian, A., Piantadosi, P. T., Authement, M. E., Lieberman, A. G., Spitz, N. A., et al. (2020). Prefrontal regulation of punished ethanol self-administration. *Biological Psychiatry, 87*(11), 967–978. https://doi.org/10.1016/j.biopsych.2019.10.030.

Hayes, D. J., Duncan, N. W., Xu, J., & Northoff, G. (2014). A comparison of neural responses to appetitive and aversive stimuli in humans and other mammals. *Neuroscience and Biobehavioral Reviews, 45*, 350–368. https://doi.org/10.1016/j.neubiorev.2014.06.018.

Heidbreder, C. A., & Groenewegen, H. J. (2003). The medial prefrontal cortex in the rat: Evidence for a dorso-ventral distinction based upon functional and anatomical characteristics. *Neuroscience and Biobehavioral Reviews, 27*(6), 555–579. https://doi.org/10.1016/j.neubiorev.2003.09.003.

Heilbronner, S. R., Rodriguez-Romaguera, J., Quirk, G. J., Groenewegen, H. J., & Haber, S. N. (2016). Circuit-based corticostriatal homologies between rat and primate. *Biological Psychiatry, 80*(7), 509–521. https://doi.org/10.1016/j.biopsych.2016.05.012.

Hirokawa, J., Vaughan, A., Masset, P., Ott, T., & Kepecs, A. (2019). Frontal cortex neuron types categorically encode single decision variables. *Nature, 576*(7787), 446–451. https://doi.org/10.1038/s41586-019-1816-9.

Homayoun, H., & Moghaddam, B. (2006). Progression of cellular adaptations in medial prefrontal and orbitofrontal cortex in response to repeated amphetamine. *The Journal of Neuroscience, 26*(31), 8025–8039. https://doi.org/10.1523/JNEUROSCI.0842-06.2006. 26/31/8025 [pii].

Homayoun, H., & Moghaddam, B. (2009). Differential representation of Pavlovian-instrumental transfer by prefrontal cortex subregions and striatum. *The European Journal of Neuroscience, 29*(7), 1461–1476. https://doi.org/10.1111/j.1460-9568.2009.06679.x.

Hong, D. D., Huang, W. Q., Ji, A. A., Yang, S. S., Xu, H., Sun, K. Y., et al. (2019). Neurons in rat orbitofrontal cortex and medial prefrontal cortex exhibit distinct responses in reward and strategy-update in a risk-based decision-making task. *Metabolic Brain Disease, 34*(2), 417–429. https://doi.org/10.1007/s11011-018-0360-x.

Horst, N. K., & Laubach, M. (2013). Reward-related activity in the medial prefrontal cortex is driven by consumption. *Frontiers in Neuroscience*, 7, 56. https://doi.org/10.3389/fnins.2013.00056.

Hunt, L. T., & Hayden, B. Y. (2017). A distributed, hierarchical and recurrent framework for reward-based choice. *Nature Reviews. Neuroscience, 18*(3), 172–182. https://doi.org/10.1038/nrn.2017.7.

Hyman, J. M., Ma, L., Balaguer-Ballester, E., Durstewitz, D., & Seamans, J. K. (2012). Contextual encoding by ensembles of medial prefrontal cortex neurons. *Proceedings of the National Academy of Sciences of the United States of America, 109*(13), 5086–5091. https://doi.org/10.1073/pnas.1114415109.

Insel, N., & Barnes, C. A. (2015). Differential activation of fast-spiking and regular-firing neuron populations during movement and reward in the dorsal medial frontal cortex. *Cerebral Cortex, 25*(9), 2631–2647. https://doi.org/10.1093/cercor/bhu062.

Ito, M., & Olds, J. (1971). Unit activity during self-stimulation behavior. *Journal of Neurophysiology, 34*(2), 263–273. https://doi.org/10.1152/jn.1971.34.2.263.

Izquierdo, A. (2017). Functional heterogeneity within rat orbitofrontal cortex in reward learning and decision making. *The Journal of Neuroscience, 37*(44), 10529–10540. https://doi.org/10.1523/JNEUROSCI.1678-17.2017.

Jackson, M. E., & Moghaddam, B. (2006). Distinct patterns of plasticity in prefrontal cortex neurons that encode slow and fast responses to stress. *The European Journal of Neuroscience, 24*(6), 1702–1710. https://doi.org/10.1111/j.1460-9568.2006.05054.x.

Jezzini, A., Mazzucato, L., La Camera, G., & Fontanini, A. (2013). Processing of hedonic and chemosensory features of taste in medial prefrontal and insular networks. *The Journal of Neuroscience, 33*(48), 18966–18978. https://doi.org/10.1523/JNEUROSCI.2974-13.2013.

Ji, G., & Neugebauer, V. (2012). Modulation of medial prefrontal cortical activity using in vivo recordings and optogenetics. *Molecular Brain, 5*, 36. https://doi.org/10.1186/1756-6606-5-36.

Kamigaki, T., & Dan, Y. (2017). Delay activity of specific prefrontal interneuron subtypes modulates memory-guided behavior. *Nature Neuroscience, 20*(6), 854–863. https://doi.org/10.1038/nn.4554.

Karalis, N., Dejean, C., Chaudun, F., Khoder, S., Rozeske, R. R., Wurtz, H., et al. (2016). 4-Hz oscillations synchronize prefrontal-amygdala circuits during fear behavior. *Nature Neuroscience, 19*(4), 605–612. https://doi.org/10.1038/nn.4251.

Kesner, R. P., & Churchwell, J. C. (2011). An analysis of rat prefrontal cortex in mediating executive function. *Neurobiology of Learning and Memory, 96*(3), 417–431. https://doi.org/10.1016/j.nlm.2011.07.002.

Klavir, O., Prigge, M., Sarel, A., Paz, R., & Yizhar, O. (2017). Manipulating fear associations via optogenetic modulation of amygdala inputs to prefrontal cortex. *Nature Neuroscience, 20*(6), 836–844. https://doi.org/10.1038/nn.4523.

Kvitsiani, D., Ranade, S., Hangya, B., Taniguchi, H., Huang, J. Z., & Kepecs, A. (2013). Distinct behavioural and network correlates of two interneuron types in prefrontal cortex. *Nature*, *498*, 363–366. https://doi.org/10.1038/nature12176.

Kyriazi, P., Headley, D. B., & Pare, D. (2020). Different multidimensional representations across the Amygdalo-Prefrontal network during an approach-avoidance task. *Neuron*, *107*(4), 717–730.e715. https://doi.org/10.1016/j.neuron.2020.05.039.

Lak, A., Okun, M., Moss, M. M., Gurnani, H., Farrell, K., Wells, M. J., et al. (2020). Dopaminergic and prefrontal basis of learning from sensory confidence and reward value. *Neuron*, *105*(4), 700–711.e706. https://doi.org/10.1016/j.neuron.2019.11.018.

Laubach, M., Amarante, L. M., Swanson, K., & White, S. R. (2018). What, if anything, is rodent prefrontal cortex? *eNeuro*, *5*(5). https://doi.org/10.1523/ENEURO.0315-18.2018.

Le Merre, P., Esmaeili, V., Charriere, E., Galan, K., Salin, P. A., Petersen, C. C. H., et al. (2018). Reward-based learning drives rapid sensory signals in medial prefrontal cortex and dorsal hippocampus necessary for goal-directed behavior. *Neuron*, *97*(1), 83–91.e85. https://doi.org/10.1016/j.neuron.2017.11.031.

LeDoux, J., & Daw, N. D. (2018). Surviving threats: Neural circuit and computational implications of a new taxonomy of defensive behaviour. *Nature Reviews. Neuroscience*, *19*(5), 269–282. https://doi.org/10.1038/nrn.2018.22.

Linsenbardt, D. N., & Lapish, C. C. (2015). Neural firing in the prefrontal cortex during alcohol intake in alcohol-preferring "P" versus Wistar rats. *Alcoholism, Clinical and Experimental Research*, *39*(9), 1642–1653. https://doi.org/10.1111/acer.12804.

Linsenbardt, D. N., Timme, N. M., & Lapish, C. C. (2019). Encoding of the intent to drink alcohol by the prefrontal cortex is blunted in rats with a family history of excessive drinking. *eNeuro*, *6*(4). https://doi.org/10.1523/ENEURO.0489-18.2019.

Marek, R., Jin, J., Goode, T. D., Giustino, T. F., Wang, Q., Acca, G. M., et al. (2018). Hippocampus-driven feed-forward inhibition of the prefrontal cortex mediates relapse of extinguished fear. *Nature Neuroscience*, *21*(3), 384–392. https://doi.org/10.1038/s41593-018-0073-9.

Maren, S., & Quirk, G. J. (2004). Neuronal signalling of fear memory. *Nature Reviews. Neuroscience*, *5*(11), 844–852. https://doi.org/10.1038/nrn1535.

McCane, A. M., DeLory, M. J., Timm, M. M., Janetsian-Fritz, S. S., Lapish, C. C., & Czachowski, C. L. (2018). Differential COMT expression and behavioral effects of COMT inhibition in male and female Wistar and alcohol preferring rats. *Alcohol*, *67*, 15–22. https://doi.org/10.1016/j.alcohol.2017.08.007.

Milad, M. R., & Quirk, G. J. (2002). Neurons in medial prefrontal cortex signal memory for fear extinction. *Nature*, *420*(6911), 70–74. https://doi.org/10.1038/nature01138.

Miller, E. K., & Cohen, J. D. (2001). An integrative theory of prefrontal cortex function. *Annual Review of Neuroscience*, *24*, 167–202. https://doi.org/10.1146/annurev.neuro.24.1.167.

Miyazaki, K., Miyazaki, K. W., & Matsumoto, G. (2004). Different representation of forthcoming reward in nucleus accumbens and medial prefrontal cortex. *Neuroreport*, *15*(4), 721–726. https://doi.org/10.1097/00001756-200403220-00030.

Moorman, D. E., & Aston-Jones, G. (2015). Prefrontal neurons encode context-based response execution and inhibition in reward seeking and extinction. *Proceedings of the National Academy of Sciences of the United States of America*, *112*(30), 9472–9477. https://doi.org/10.1073/pnas.1507611112.

Moorman, D. E., James, M. H., McGlinchey, E. M., & Aston-Jones, G. (2015). Differential roles of medial prefrontal subregions in the regulation of drug seeking. *Brain Research*, *1628*(Pt. A), 130–146. https://doi.org/10.1016/j.brainres.2014.12.024.

Mulder, A. B., Nordquist, R. E., Orgut, O., & Pennartz, C. M. (2003). Learning-related changes in response patterns of prefrontal neurons during instrumental conditioning. *Behavioural Brain Research*, *146*(1–2), 77–88.

Muller Ewald, V. A., & LaLumiere, R. T. (2017). Neural systems mediating the inhibition of cocaine-seeking behaviors. *Pharmacology, Biochemistry, and Behavior, 174*, 53–63. https://doi.org/10.1016/j.pbb.2017.07.006.

Olds, J., Disterhoft, J. F., Segal, M., Kornblith, C. L., & Hirsh, R. (1972). Learning centers of rat brain mapped by measuring latencies of conditioned unit responses. *Journal of Neurophysiology, 35*(2), 202–219. https://doi.org/10.1152/jn.1972.35.2.202.

Ongur, D., & Price, J. L. (2000). The organization of networks within the orbital and medial prefrontal cortex of rats, monkeys and humans. *Cerebral Cortex, 10*(3), 206–219.

Otis, J. M., Namboodiri, V. M., Matan, A. M., Voets, E. S., Mohorn, E. P., Kosyk, O., et al. (2017). Prefrontal cortex output circuits guide reward seeking through divergent cue encoding. *Nature, 543*(7643), 103–107. https://doi.org/10.1038/nature21376.

Padilla-Coreano, N., Bolkan, S. S., Pierce, G. M., Blackman, D. R., Hardin, W. D., Garcia-Garcia, A. L., et al. (2016). Direct ventral hippocampal-prefrontal input is required for anxiety-related neural activity and behavior. *Neuron, 89*(4), 857–866. https://doi.org/10.1016/j.neuron.2016.01.011.

Padilla-Coreano, N., Canetta, S., Mikofsky, R. M., Alway, E., Passecker, J., Myroshnychenko, M. V., et al. (2019). Hippocampal-prefrontal Theta transmission regulates avoidance behavior. *Neuron, 104*(3), 601–610.e604. https://doi.org/10.1016/j.neuron.2019.08.006.

Park, J., & Moghaddam, B. (2017). Risk of punishment influences discrete and coordinated encoding of reward-guided actions by prefrontal cortex and VTA neurons. *eLife, 6*, e30056. https://doi.org/10.7554/eLife.30056.

Passingham, R. E., & Wise, S. P. (2012). *The neurobiology of the prefrontal cortex: Anatomy, evolution, and the origin of insight* (1st ed.). Oxford, United Kingdom: Oxford University Press.

Peters, J., Kalivas, P. W., & Quirk, G. J. (2009). Extinction circuits for fear and addiction overlap in prefrontal cortex. *Learning & Memory, 16*(5), 279–288. https://doi.org/10.1101/lm.1041309.

Peters, Y. M., O'Donnell, P., & Carelli, R. M. (2005). Prefrontal cortical cell firing during maintenance, extinction, and reinstatement of goal-directed behavior for natural reward. *Synapse, 56*(2), 74–83. https://doi.org/10.1002/syn.20129.

Peterson, S. L. (1986). Prefrontal cortex neuron activity during a discriminative conditioning paradigm in unanesthetized rats. *The International Journal of Neuroscience, 29*(3–4), 245–254. https://doi.org/10.3109/00207458608986154.

Petyko, Z., Toth, A., Szabo, I., Galosi, R., & Lenard, L. (2009). Neuronal activity in rat medial prefrontal cortex during sucrose solution intake. *Neuroreport, 20*(14), 1235–1239. https://doi.org/10.1097/WNR.0b013e32832fbf30.

Pinto, L., & Dan, Y. (2015). Cell-type-specific activity in prefrontal cortex during goal-directed behavior. *Neuron, 87*(2), 437–450. https://doi.org/10.1016/j.neuron.2015.06.021.

Pirch, J. H., & Peterson, S. L. (1981). Event-related slow potentials and activity of singly neurons in rat frontal cortex. *The International Journal of Neuroscience, 15*(3), 141–146. https://doi.org/10.3109/00207458108985906.

Pratt, W. E., & Mizumori, S. J. (2001). Neurons in rat medial prefrontal cortex show anticipatory rate changes to predictable differential rewards in a spatial memory task. *Behavioural Brain Research, 123*(2), 165–183.

Preuss, T. M. (1995). Do rats have prefrontal cortex? The Rose-Woolsey-Akert program reconsidered. *Journal of Cognitive Neuroscience,* 7(1), 1–24. https://doi.org/10.1162/jocn.1995.7.1.1.

Rebec, G. V., & Sun, W. (2005). Neuronal substrates of relapse to cocaine-seeking behavior: Role of prefrontal cortex. *Journal of the Experimental Analysis of Behavior, 84*(3), 653–666.

Rigotti, M., Barak, O., Warden, M. R., Wang, X. J., Daw, N. D., Miller, E. K., et al. (2013). The importance of mixed selectivity in complex cognitive tasks. *Nature*, *497*(7451), 585–590. https://doi.org/10.1038/nature12160.

Rozeske, R. R., & Herry, C. (2018). Neuronal coding mechanisms mediating fear behavior. *Current Opinion in Neurobiology*, *52*, 60–64. https://doi.org/10.1016/j.conb.2018.04.017.

Rozeske, R. R., Jercog, D., Karalis, N., Chaudun, F., Khoder, S., Girard, D., et al. (2018). Prefrontal-periaqueductal gray-projecting neurons mediate context fear discrimination. *Neuron*, *97*(4), 898–910.e896. https://doi.org/10.1016/j.neuron.2017.12.044.

Sackett, D. A., Moschak, T. M., & Carelli, R. M. (2019). Prelimbic cortical neurons track preferred reward value and reflect impulsive choice during delay discounting behavior. *The Journal of Neuroscience*, *39*(16), 3108–3118. https://doi.org/10.1523/JNEUROSCI.2532-18.2019.

Schneider, K. N., Sciarillo, X. A., Nudelman, J. L., Cheer, J. F., & Roesch, M. R. (2020). Anterior cingulate cortex signals attention in a social paradigm that manipulates reward and shock. *Current Biology*, *30*, 3724–3735. https://doi.org/10.1016/j.cub.2020.07.039.

Seamans, J. K., Lapish, C. C., & Durstewitz, D. (2008). Comparing the prefrontal cortex of rats and primates: Insights from electrophysiology. *Neurotoxicity Research*, *14*(2–3), 249–262. https://doi.org/10.1007/BF03033814.

Siciliano, C. A., Noamany, H., Chang, C. J., Brown, A. R., Chen, X., Leible, D., et al. (2019). A cortical-brainstem circuit predicts and governs compulsive alcohol drinking. *Science*, *366*(6468), 1008–1012. https://doi.org/10.1126/science.aay1186.

Simon, N. W., Wood, J., & Moghaddam, B. (2015). Action-outcome relationships are represented differently by medial prefrontal and orbitofrontal cortex neurons during action execution. *Journal of Neurophysiology*, *114*(6), 3374–3385. https://doi.org/10.1152/jn.00884.2015.

Smith, K. S., & Graybiel, A. M. (2013). A dual operator view of habitual behavior reflecting cortical and striatal dynamics. *Neuron*, *79*, 361–374. https://doi.org/10.1016/j.neuron.2013.05.038.

Sotres-Bayon, F., & Quirk, G. J. (2010). Prefrontal control of fear: More than just extinction. *Current Opinion in Neurobiology*, *20*(2), 231–235. https://doi.org/10.1016/j.conb.2010.02.005.

Sotres-Bayon, F., Sierra-Mercado, D., Pardilla-Delgado, E., & Quirk, G. J. (2012). Gating of fear in prelimbic cortex by hippocampal and amygdala inputs. *Neuron*, *76*(4), 804–812. https://doi.org/10.1016/j.neuron.2012.09.028.

Sun, W., & Rebec, G. V. (2006). Repeated cocaine self-administration alters processing of cocaine-related information in rat prefrontal cortex. *The Journal of Neuroscience*, *26*(30), 8004–8008. https://doi.org/10.1523/JNEUROSCI.1413-06.2006. 26/30/8004 [pii].

Takenouchi, K., Nishijo, H., Uwano, T., Tamura, R., Takigawa, M., & Ono, T. (1999). Emotional and behavioral correlates of the anterior cingulate cortex during associative learning in rats. *Neuroscience*, *93*(4), 1271–1287. https://doi.org/10.1016/s0306-4522(99)00216-x.

Uylings, H. B., Groenewegen, H. J., & Kolb, B. (2003). Do rats have a prefrontal cortex? *Behavioural Brain Research*, *146*(1–2), 3–17. https://doi.org/10.1016/j.bbr.2003.09.028.

Valdes, J. L., Maldonado, P., Recabarren, M., Fuentes, R., & Torrealba, F. (2006). The infralimbic cortical area commands the behavioral and vegetative arousal during appetitive behavior in the rat. *The European Journal of Neuroscience*, *23*(5), 1352–1364. https://doi.org/10.1111/j.1460-9568.2006.04659.x.

van Aerde, K. I., Heistek, T. S., & Mansvelder, H. D. (2008). Prelimbic and infralimbic prefrontal cortex interact during fast network oscillations. *PLoS One*, *3*(7), e2725. https://doi.org/10.1371/journal.pone.0002725.

van Heukelum, S., Mars, R. B., Guthrie, M., Buitelaar, J. K., Beckmann, C. F., Tiesinga, P. H. E., et al. (2020). Where is cingulate cortex? A cross-species view. *Trends in Neurosciences*, *43*(5), 285–299. https://doi.org/10.1016/j.tins.2020.03.007.

Vander Weele, C. M., Siciliano, C. A., Matthews, G. A., Namburi, P., Izadmehr, E. M., Espinel, I. C., et al. (2018). Dopamine enhances signal-to-noise ratio in cortical-brainstem encoding of aversive stimuli. *Nature*, *563*(7731), 397–401. https://doi.org/10.1038/s41586-018-0682-1.

Vertes, R. P. (2006). Interactions among the medial prefrontal cortex, hippocampus and midline thalamus in emotional and cognitive processing in the rat. *Neuroscience*, *142*, 1–20. https://doi.org/10.1016/j.neuroscience.2006.06.027.

Vickery, T. J., Chun, M. M., & Lee, D. (2011). Ubiquity and specificity of reinforcement signals throughout the human brain. *Neuron*, *72*(1), 166–177. https://doi.org/10.1016/j.neuron.2011.08.011.

Vlaev, I., Chater, N., Stewart, N., & Brown, G. D. (2011). Does the brain calculate value? *Trends in Cognitive Sciences*, *15*(11), 546–554. https://doi.org/10.1016/j.tics.2011.09.008.

Warden, M. R., Selimbeyoglu, A., Mirzabekov, J. J., Lo, M., Thompson, K. R., Kim, S. Y., et al. (2012). A prefrontal cortex-brainstem neuronal projection that controls response to behavioural challenge. *Nature*, *492*(7429), 428–432. https://doi.org/10.1038/nature11617.

West, E. A., Saddoris, M. P., Kerfoot, E. C., & Carelli, R. M. (2014). Prelimbic and infralimbic cortical regions differentially encode cocaine-associated stimuli and cocaine-seeking before and following abstinence. *The European Journal of Neuroscience*, *39*(11), 1891–1902. https://doi.org/10.1111/ejn.12578.

Wise, S. P. (2008). Forward frontal fields: Phylogeny and fundamental function. *Trends in Neurosciences*, *31*(12), 599–608. https://doi.org/10.1016/j.tins.2008.08.008.

Ye, L., Allen, W. E., Thompson, K. R., Tian, Q., Hsueh, B., Ramakrishnan, C., et al. (2016). Wiring and molecular features of prefrontal ensembles representing distinct experiences. *Cell*, *165*(7), 1776–1788. https://doi.org/10.1016/j.cell.2016.05.010.

Zhang, R., Tomida, M., Katayama, Y., & Kawakami, Y. (2004). Response durations encode nociceptive stimulus intensity in the rat medial prefrontal cortex. *Neuroscience*, *125*(3), 777–785. https://doi.org/10.1016/j.neuroscience.2004.01.055.

CHAPTER FOUR

Dynamic decision making and value computations in medial frontal cortex

Bilal A. Bari and Jeremiah Y. Cohen*
The Solomon H. Snyder Department of Neuroscience, Brain Science Institute, Kavli Neuroscience Discovery Institute, Johns Hopkins University, Baltimore, MD, United States
*Corresponding author: e-mail address: Jeremiah.cohen@jhmi.edu

Contents

Abstract

Dynamic decision making requires an intact medial frontal cortex. Recent work has combined theory and single-neuron measurements in frontal cortex to advance models of decision making. We review behavioral tasks that have been used to study dynamic

International Review of Neurobiology, Volume 158
ISSN 0074-7742
https://doi.org/10.1016/bs.irn.2020.12.001

decision making and algorithmic models of these tasks using reinforcement learning theory. We discuss studies linking neurophysiology and quantitative decision variables. We conclude with hypotheses about the role of other cortical and subcortical structures in dynamic decision making, including ascending neuromodulatory systems.

1. Introduction

Nervous systems evolved in highly dynamic environments. Adaptive behavior in the natural world requires not just learning which actions improve survival, but also changing behavior as the environment changes. This describes an active feedback process in which organisms interact with the environment through actions, learn from feedback, and adjust future actions adaptively. The ability to behave flexibly is a ubiquitous feature of life, ranging from flies (Ofstad, Zuker, & Reiser, 2011) to humans.[a] Despite the ubiquity of flexible decision making, it is a historically understudied problem in systems neuroscience. Progress in reinforcement learning over the past two decades has provided a biologically plausible and mathematically sophisticated framework for studying these problems (Bertsekas & Tsitsiklis, 1996; Sutton & Barto, 1998). Parallel progress in tool development in mice has enabled the dissection of neural circuits needed for detailed biological insight (Luo, Callaway, & Svoboda, 2018). The intersection of these two fields—reinforcement learning and neural circuit dissection—holds promise to further our algorithmic- and implementation-level understanding of cognition.

This chapter reviews behavioral assays for investigating value-guided behavior, explores biologically plausible algorithms of value-based decision making, and ends with an overview of the neural systems thought to instantiate these functions. We highlight recent evidence demonstrating value computations in the medial prefrontal cortex (mPFC) and cortico-basal-ganglia loops. Given the myriad decision making dysfunctions seen in patients with mental illnesses, understanding its neural basis on multiple levels is crucial for developing targeted therapies (Cáceda, Nemeroff, & Harvey, 2014).

[a] It has even been observed in organisms without nervous systems (pea plants; Dener, Kacelnik, & Shemesh, 2016).

2. Value-based decision making

Decision making is at the heart of many fields, including economics, political science, psychology, engineering, medicine, and neuroscience. A decision is defined as a deliberative process that results in commitment to a categorical proposition (Gold & Shadlen, 2007). Decisions are the result of integrating both external (e.g., sensory) as well as internal evidence (e.g., predictions). These two domains have largely been studied in the context of perceptual (i.e., external evidence) and value-based (i.e., internally generated evidence) decision-making tasks (Gold & Shadlen, 2007; Yu, 2014). Within each domain, quantitative mathematical frameworks have provided semantically meaningful interpretations of neural activity in key brain structures. A key concept is the decision variable, which "*represents the accrual of all sources of priors, evidence, and value into a quantity that is interpreted by the decision rule to produce a choice*" (Gold & Shadlen, 2007). The decision variable represents the common currency used by the brain to generate a single categorical action among all possibilities.

Value-based decision making tasks require subjects to choose on the basis of expected utility or subjective value (Glimcher, Dorris, & Bayer, 2005; Sanfey, Loewenstein, McClure, & Cohen, 2006; Sugrue, Corrado, & Newsome, 2005). Typically, sensory stimuli are salient, which minimizes perceptual uncertainty. Decisions are made on the basis of values learned over long time periods (Morris, Nevet, Arkadir, Vaadia, & Bergman, 2006) or values learned over short timescales (Lau & Glimcher, 2005; Sugrue, Corrado, & Newsome, 2004; Tsutsui, Grabenhorst, Kobayashi, & Schultz, 2016). Importantly, these values are not sensory properties of stimuli (like brightness or contrast) but are internal variables that must be learned through experience. Compared to perceptual tasks, the subjective nature of value-based tasks makes them much more difficult to control. However, rigorous formalisms from the fields of economics (Rangel, Camerer, & Montague, 2008) and reinforcement learning (Sutton & Barto, 1998) have greatly benefited the study of value-based decisions. As such, these tasks are well-poised to address questions about representations of cognitive information.

2.1 Pavlovian systems

Pavlovian behavior describes innate, reflexive behavioral responses to stimuli that have been assigned value (Rescorla, 1988). The phenomenon was discovered by Ivan Pavlov while conducting studies of the digestive system in dogs (Pavlov & Anrep, 1928).[b] In the case of reward, the Pavlovian response is preparatory (e.g., approach) and consummatory. In the case of punishment, the Pavlovian response is avoidance (Rangel et al., 2008). These responses can be innate (e.g., avoidance of predator odors) or learned through experience. Owing to their simplicity, Pavlovian behavioral tasks have been invaluable in studying the neural basis of reward learning. One of the most celebrated examples at the intersection of reinforcement learning and neuroscience—that midbrain dopamine neurons encode reward prediction errors—was discovered in monkeys performing Pavlovian behaviors (Schultz, Dayan, & Montague, 1997). The simplicity of Pavlovian tasks is also limiting, particularly if there is a mismatch between the innate Pavlovian behavior and the response needed to obtain reward (e.g., withhold an action to obtain reward; Dayan, Niv, Seymour, & Daw, 2006).

2.2 Habitual systems

Habitual behavior describes the mapping of a large range of arbitrary motor responses to stimuli, through repeated reinforcement. Habit systems in the brain learn through trial-and-error over relatively long timescales (Balleine & O'Doherty, 2010). Once learned, values are thought to be cached and behavior can be carried out "automatically." As such, habits are thought to be computationally cheap but inflexible (Daw, Niv, & Dayan, 2005). It can also be difficult to unlearn habitual behavior. This particular feature is the rationale for outcome devaluation, a gold-standard test of habitual behavior. In outcome devaluation experiments, animals are first trained to associate a stimulus with a response (e.g., pressing a lever in response to a tone) to receive a reward. Following training, the outcome is devalued either by pairing with sickness or making the outcome freely available before the task. Animals are then tested to see if they respond to the stimulus. Continued behavioral responses are consistent with habitual behavior (Balleine & O'Doherty, 2010).

2.3 Goal-directed systems

Goal-directed systems are responsible for flexible decision making. They are thought to compute the outcomes associated with particular actions on fast timescales. As such, they are sensitive to changes in environmental contingencies. Outcome devaluation is also used to test the contribution of goal-directed systems to behavior (Balleine & O'Doherty, 2010).

[b] Although it is commonly stated that Ivan Pavlov used a bell as a stimulus, this story appears to be apocryphal (Cambiaghi & Sacchetti, 2015).

A reduction in responding is taken as evidence that behavior was driven largely by goal-directed systems. This chapter focuses on delineating the contributions of goal-directed systems to flexible decision making. We will focus largely on behaviors, primarily because, in contrast to studies of sensory decision making, it is paramount to have a controlled behavior, in which the experimenter can quantify hidden variables, such as value.

3. Matching behavior

Several behavioral tasks have been used to study flexible decision making: outcome devaluation, reversal learning, set shifting, foraging, "mixed-strategy" games, and matching behavior. Matching behavior (or the matching law) is a type of behavior that we will argue provides excellent conditions to study continual value-based learning. In matching tasks, animals freely choose among two or more options to harvest reward (Herrnstein & Heyman, 1979). Matching describes the tendency of animals to "match" the fraction of choices to a particular option with the fraction of rewards received from that option. Mathematically, matching describes the following relationship:

$$\frac{c_i}{\sum_{i=1}^{N} c_i} = \frac{r_i}{\sum_{i=1}^{N} r_i} \tag{1}$$

where c_i is the number of choices allocated to option i and r_i is the number of rewards obtained from option i, given N possible options. Matching behavior was first observed in pigeons (Herrnstein, 1961) and has since been observed in mice (Bari et al., 2019; Fonseca, Murakami, & Mainen, 2015), rats (Gallistel, Mark, King, & Latham, 2001; Graft, Lea, & Whitworth, 1977), monkeys (Lau & Glimcher, 2005; Sugrue et al., 2004; Tsutsui et al., 2016), and humans (Pierce & Epling, 1983). Matching behavior is typically highly dynamic, with animals switching between options on fast timescales. In trial-based tasks, animals typically switch from one option to another with a mode of 1 trial. However, animals remain reward sensitive, and repeat recently rewarded choices. As such, it can be considered a form of goal-directed behavior.

3.1 Task conditions

Experimental psychologists use tasks in which reward delivery is contingent on schedules of reinforcement (Ferster & Skinner, 1957). Two commonly used schedules are called "variable ratio" and "variable interval." Matching behavior is classically observed in tasks with variable interval schedules.

Variable ratio schedules are intuitive—reward is simply delivered with a fixed probability, much like flipping a (biased) coin. If given the choice between two variable ratio schedules, the optimal policy is to choose the higher probability option exclusively. Note that exclusively choosing one option is trivially consistent with matching behavior since all choices are allocated to one option, and all rewards are received from that option. Variable-ratio-schedule tasks are typically called "two-armed bandit" tasks and tasks with changing probabilities are called "dynamic two-armed bandit" tasks or, if the probabilities reverse, "probabilistic reversal learning" tasks. In variable interval schedules, reward is delivered after a fixed time has elapsed. Once the time has elapsed, that option is "baited," guaranteeing reward delivery once chosen. This feature is thought to make these tasks ethologically relevant to study foraging. Although this seems like a trivial change, it changes the optimal policy significantly, which will be expanded on below. Intuitively, if given the choice between two variable interval schedules, it does not make sense to choose one option exclusively. Instead, one should occasionally probe the lower-probability option, to not miss out on a baited reward. In these circumstances, matching behavior emerges. In modern, trial-based tasks, discrete versions of variable interval schedules are used, to allow for independent control over inter-trial intervals.

Variable ratio and variable interval schedules are not necessarily different categories but may be thought of as two extremes of a competitive foraging environment (Sakai & Fukai, 2008a). Under this interpretation, "baited" rewards are withdrawn with a particular probability. Variable interval schedules describe environments where the withdrawal probability is 0, imitating herbivore foraging environments without competitors. Intermediate withdrawal probabilities between 0 and 1 imitate competitive foraging environments where food may be intercepted by competitors. Variable ratio schedules are ones where the withdrawal probability is 1, which may resemble the foraging of carnivores.

3.2 Matching behavior is generally a suboptimal probabilistic policy

Given the task conditions that engender matching behavior, a natural question to ask is whether animals are matching because they are aware of baiting (i.e., accounting for environmental statistics) or are they agnostic to it? A key insight into this problem was developed by Sakai and Fukai (2008b), which we expand on below.

For simplicity, we will assume two choices (options a and b). Matching Eq. (1) then reduces to the following form

$$\frac{c_a}{c_a + c_b} = \frac{r_a}{r_a + r_b} \text{ and } \frac{c_b}{c_a + c_b} = \frac{r_b}{r_a + r_b} \tag{2}$$

Matching occurs when the relative fraction of choices to option i, c_i, "matches" the relative fraction of rewards, r_i, from option i. Rearranging either equation gives $\frac{r_a}{c_a} = \frac{r_b}{c_b}$. In other words, matching occurs when the mean reward from all options is equated. Written compactly,

$$\hat{r}_a = \hat{r}_b \tag{3}$$

The average reward from both options can be written as

$$\hat{r} = \hat{r}_a\pi_a + \hat{r}_b\pi_b \tag{4}$$

where π_i is the probability of choosing option i. Because we have two options, $\pi_b = 1 - \pi_a$. We assume that the policy is controlled by a parameter x, yielding[c]

$$\hat{r}(x) = \hat{r}_a(x)\pi_a(x) + \hat{r}_b(x)\pi_b(x) \tag{5}$$

The parameter x explicitly influences the policy and implicitly influences the reward probabilities. Under variable-interval schedules, the longer one has stayed away from an option, the higher the probability of reward when that option is eventually selected[d]

[c] For example, assume $\pi_a(x)$ is a softmax function: $\pi_a(x) = \frac{1}{1+e^{-x}}$.

[d] The mean reward obtained is a function of x. In variable-interval tasks, this is correct—the animal's policy influences whether it is able to take advantage of the baiting rule. The longer the animal abstains from choosing an option, the greater the probability of reward when the animal chooses it. To express the baiting rule mathematically, we write the probability of reward from option i as

$$P_i(t) = 1 - (1 - p_i)^{t+1} \tag{6}$$

where p_i is the base reward probability and t is the number of consecutive trials since that option was last chosen. This equation states that the probability of reward asymptotically grows from p_i to 1 the longer option i is unchosen. The mean return from option i can be written as

$$\begin{aligned}\hat{r}_i(x) &= \sum_{t=0}^{\infty}(1 - \pi_i(x))^t\pi_i(x)P_i(t) = \sum_{t=0}^{\infty}(1 - \pi_i(x))^t\pi_i(x)\left(1 - (1 - p_i)^{t+1}\right)\\ &= \frac{p_i}{\pi_i(x) + p_i(1 - \pi_i(x))}\end{aligned} \tag{7}$$

These equations can be used to test whether matching is optimal, in closed form.

To maximize reward, we take the derivative of $\hat{r}(x)$ with respect to x and set it equal to 0.

$$\begin{aligned}\frac{d\hat{r}(x)}{dx} &= 0\\ &= \left(\hat{r}_a(x)\frac{d\pi_a(x)}{dx} + \hat{r}_b(x)\frac{d\pi_b(x)}{dx}\right)\\ &+ \left(\frac{d\hat{r}_a(x)}{dx}\pi_a(x) + \frac{d\hat{r}_b(x)}{dx}\pi_b(x)\right)\end{aligned} \tag{8}$$

This equation defines the optimal probabilistic policy. The first set of terms in parentheses is the explicit change in behavior when x is changed. The second set of terms is the implicit change in the environment brought about by the animal's policy. One may hypothesize that this first computation (explicit change in behavior) is easy for the brain to perform while the second computation (implicit change in environment) is much more difficult. Sakai and Fukai's critical insight was to recognize that the brain might ignore this second computation, which yields the reduced form

$$\frac{d\hat{r}(x)}{dx} = 0 = \hat{r}_a(x)\frac{d\pi_a(x)}{dx} + \hat{r}_b(x)\frac{d\pi_b(x)}{dx} \tag{9}$$

It can be shown that solving Eq. (9) yields matching behavior. In other words, the solution is when $\hat{r}_a = \hat{r}_b$. Under what conditions is matching optimal?

Fig. 1 provides a geometric intuition for how the policy influences reward in standard matching paradigms. Both options follow variable interval schedules ($p_a = 0.4; p_b = 0.1$). Light and dark blue illustrate $\hat{r}_a$ and $\hat{r}_b$, respectively, as a function of the policy. The consequences of baiting are clear. The less often option *a* is chosen, the greater the probability of reward when it is chosen. The orange trace illustrates $\hat{r}$ which, from Eq. (4), is the reward rate from both options, weighted by the probability of choosing each option. The optimal reward is at the maximum of this function, which occurs at *P*(choice to option a) ≈ 0.86. Matching behavior, when $\hat{r}_a = \hat{r}_b$ (Eq. 3), occurs when the two blue traces intersect one another. In standard matching paradigms, matching *is* the optimal probabilistic policy. However, in this circumstance matching can occur either if animals are aware of environmental statistics (Eq. 8) or if they ignore it (Eq. 9).

There is a strong-inference experiment that can test whether animals will continue to exhibit matching behavior, even when it is suboptimal. If they

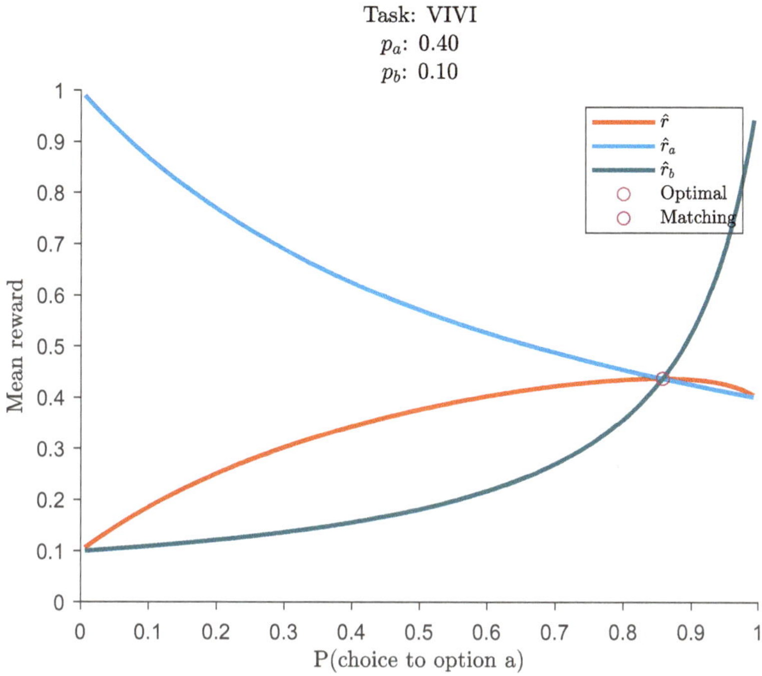

Fig. 1 Matching is the optimal probabilistic policy in variable-interval/variable-interval tasks.

continue to show matching behavior, this is evidence that they likely do not take into account environmental statistics to make decisions. If they show optimal behavior, then they likely are. The key experiment is to only let one option follow a variable-interval schedule. The second option should deliver reward with a fixed reward probability (variable-ratio schedule). This example is illustrated in Fig. 2 ($p_a = 0.3$, variable-ratio option; $p_b = 0.24$, variable-interval option). It is clear that option *a*, the variable ratio option, does not benefit from baiting. No matter how often it is chosen, the probability of reward is fixed. In this task, matching is not the optimal solution. When these types of tasks have been tested, matching behavior has been observed (Herrnstein & Heyman, 1979; Vyse & Belke, 1992; Williams, 1985), including in humans (Savastano & Fantino, 1994). Under the theory proposed by Sakai and Fukai, these findings indicate that animals behave as if they are not aware of environmental statistics. This should not be taken as evidence that animals are *unable* to calculate these environmental statistics—simply that under these task conditions, they behave as if they

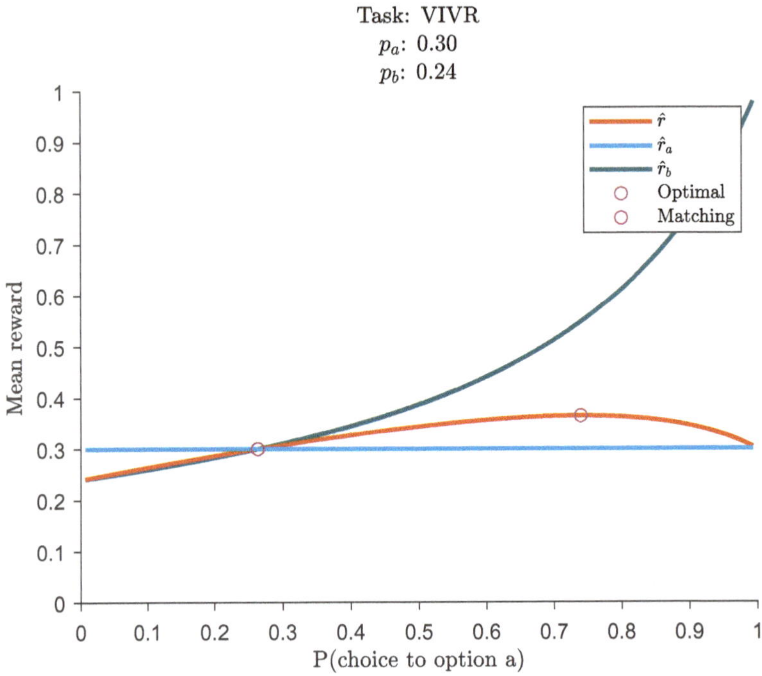

Fig. 2 Matching is not optimal in variable-interval/variable-ratio tasks.

do not. Manipulations designed to encourage optimal behavioral can be successful—for example, rewards that differ in magnitude rather than probability, and allowing practice without reinforcement (Tunney & Shanks, 2002).

3.2.1 Animals do not adopt deterministic switching policies

One limitation of the argument above is that it is limited to probabilistic policies. Deterministic switching policies (i.e., sample the other arm every n choices) are the true optimal policies. For example, if given two variable-interval options in which the base probabilities are $p_a = 0.4$ and $p_b = 0.1$, the optimal policy is to choose option a four times and choose option b once. This is because, according to Eq. (6), after not being selected for four consecutive trials, the probability of reward from option b has climbed from 0.1 to 0.4095, at which point the probability of reward is greater than option a, and it should be chosen. After option b has been chosen, the probability of reward drops again to 0.1, the probability of reward on option a is now 0.64 (since it has not been chosen for one trial), and option a should be chosen.

These policies are easy to diagnose, since stay duration histograms will not display a characteristic exponential shape. Under the hypothesis that animals are adopting a deterministic switching policy, stay durations should instead display a bimodal shape. We are not aware of any human or animal studies demonstrating such behavior, although a computational study trained artificial neural networks which exhibited these switching policies in certain conditions (Wang et al., 2018).

3.2.2 Matching behavior vs probability matching

Matching behavior should *not* be confused with probability matching. Matching behavior and probability matching are observed in very different task conditions. Probability matching refers to the tendency of subjects to match the relative fraction of choices to the probability of reward in two-armed bandit tasks (Mongillo, Shteingart, & Loewenstein, 2014). In two-armed bandit tasks, unlike matching behavior tasks, reward probability does *not* depend on past choices—the options follow variable-ratio schedules. For example, if $p_a = 0.75$ and $p_b = 0.25$, probability matching occurs when subjects choose option a 75% of the time and option b 25% of the time. This is clearly a suboptimal policy, since the subject should choose option a 100% of the time.

Mathematically, probability matching can be written as

$$\frac{c_a}{c_a + c_b} = \frac{p_a}{p_a + p_b} \tag{10}$$

where p_i is the probability of reward associated with option i. Unlike matching behavior, there is no circumstance in which probability matching is optimal. Intuitively, in a two-armed bandit task, the optimal policy is to exclusively choose the high-probability option. To see more rigorously how matching behavior and probability matching are incompatible, note that the actual reward received from option i is $r_i = p_i \cdot c_i$. Matching behavior is therefore

$$\frac{c_a}{c_a + c_b} = \frac{r_a}{r_a + r_b} = \frac{p_a \cdot c_a}{p_a \cdot c_a + p_b \cdot c_b} \tag{11}$$

With fixed reward probabilities, matching behavior is obtained when $c_a = 0$ or $c_b = 0$, which is incompatible with probability matching. In two-armed bandit tasks, matching behavior is (trivially) to exclusively choose one option - the optimal policy.

Interestingly, probability matching and *undermatching*, the tendency to behave more randomly than perfect matching behavior, may share the same underlying mechanism. If subjects believe the world is more unstable or prone to change than it truly is, both of these phenomena can emerge (Shanks, Tunney, & McCarthy, 2002; Yu & Cohen, 2009; Yu & Huang, 2014).

3.3 Dynamic foraging tasks

In addition to the ethological relevance of matching behavior, a useful feature is the tendency of animals to exhibit highly flexible behavior, switching from one option to another on short timescales. However, one limitation of classic matching paradigms is reward contingencies are typically fixed within sessions and varied across sessions. Practically, both trial-by-trial algorithmic studies of behavior and neurophysiologic studies of flexible behavior benefit from task designs that better capture the full dynamic range of matching within session. Take the example in Fig. 3. Matching behavior is typically illustrated in these types of plots, with the relative reward ratio on the x-axis and choice ratio on the y-axis. Imagine a single session with fixed reward

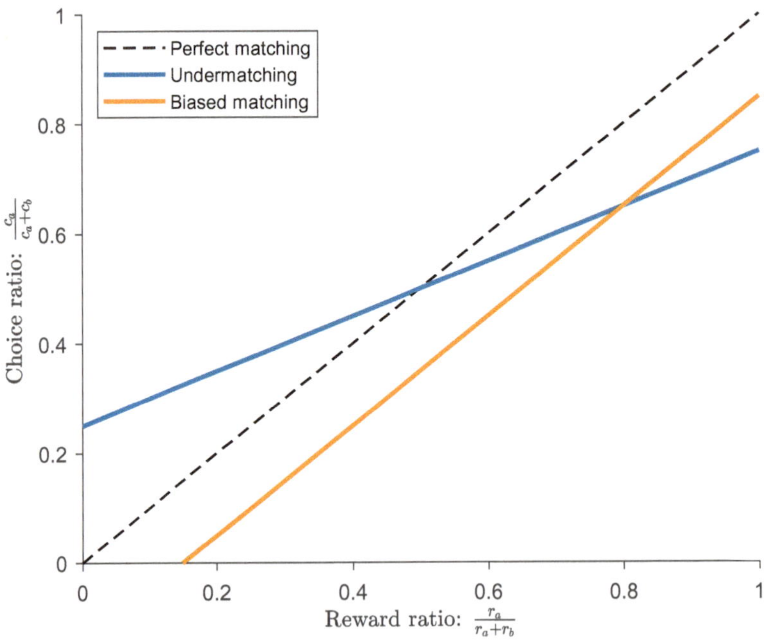

Fig. 3 Dynamic foraging tasks allow for better characterization of behavior.

contingencies that yield a reward ratio of ~0.8 and a choice ratio of ~0.6. This single data point would be consistent with both undermatching (the blue curve; tendency to behave more randomly than perfect matching) and biased matching (the orange curve; tendency to prefer one choice more than another). Since these two hypotheses have very different algorithmic and neural underpinnings, it is useful to use a task design that allows one to measure along multiple choice/reward ratios. Dynamic foraging tasks are variable-interval/variable-interval tasks with multiple reward contingency changes in one session (Gallistel et al., 2001; Lau & Glimcher, 2005; Sugrue et al., 2004; Tsutsui et al., 2016). This task variant elegantly solves this problem.

4. Algorithms underlying matching behavior

Matching is a description of macroscopic, averaged behavior. An important question therefore is how does matching behavior emerge from trial-by-trial behavior?

4.1 Melioration

Melioration is among the earliest trial-by-trial algorithms developed to solve this problem (Herrnstein & Vaughan, 1980). This algorithm states that behavior should tend toward the option with the highest local rate of reinforcement (the highest $\frac{r_i}{c_i}$ ratio), which yields matching behavior in the limit. This algorithm was studied extensively by Herrnstein and others, largely to contrast with theories of optimal decision making (Herrnstein, 2000).

4.2 Local matching

Local matching is another algorithm designed to yield matching behavior (Sugrue et al., 2004). Under local matching, the agent exponentially integrates reward from each option and maps the *local* reward ratio to a probability of choice.

$$\frac{\hat{r}_a}{\hat{r}_a + \hat{r}_b} = P(c_a) \tag{12}$$

where $\hat{r}_i$ is the local estimate of reward. For example, $\hat{r}_i$ can be updated as

$$\hat{r}_i = \hat{r}_i + \alpha(R - \hat{r}_i) \tag{13}$$

each time option i is chosen (where R is reward). Although melioration and local matching both do a reasonable job describing behavior, deep insight is limited since both of these algorithms were designed to exhibit matching behavior.

4.3 Covariance-based update rules

A general insight into this problem was developed by Loewenstein and Seung (2006). Assume a change in synaptic weights ΔW is given by the following form

$$\Delta W = \alpha \cdot \mathrm{cov}(R, N) \tag{14}$$

where α is a plasticity rate, R is reward, and N is neural activity. It can be shown that synaptic plasticity update rules of this form converge to matching behavior. Examples of these update rules include

$$\Delta W = \alpha \cdot (R - \mathbf{E}(R))N \tag{15}$$

$$\Delta W = \alpha \cdot R(N - \mathbf{E}(N)) \tag{16}$$

$$\Delta W = \alpha \cdot (R - \mathbf{E}(R))(N - \mathbf{E}(N)) \tag{17}$$

where $\mathbf{E(X)}$ is the expected value. These particular update rules are equivalent to the learning rules in the direct actor and actor-critic reinforcement learning algorithms (Dayan & Abbott, 2001; Sakai & Fukai, 2008a, 2008b). These results demonstrate that matching behavior can be the outcome of very simple learning rules, a remarkably deep insight and one that makes it possible to more confidently interpret neural correlates of value-based decision making in these behaviors.

4.4 Logistic regressions

One of the most common means of analyzing behavior in dynamic foraging tasks is to calculate logistic regressions to predict choice as a function of reward history and choice history (Fonseca et al., 2015; Lau & Glimcher, 2005; Sul, Jo, Lee, & Jung, 2011; Tsutsui et al., 2016). These models take the following form

$$\log\left(\frac{P(c_a(t))}{1 - P(c_a(t))}\right) = \sum_{i=1}^{N} \beta_i^R (R_a(t-i) - R_b(t-i)) + \sum_{i=1}^{N} \beta_i^c (c_a(t-i) - c_b(t-i)) + \beta_0 \tag{18}$$

where

$$R_a(t) = \begin{cases} 1, & \text{if option } a \text{ was rewarded} \\ 0, & \text{if either option was not rewarded} \end{cases} \tag{19}$$

$$c_a(t) = \begin{cases} 1, & \text{if option } a \text{ was chosen} \\ 0, & \text{if option } b \text{ was chosen} \end{cases} \tag{20}$$

and vice versa for $R_b(t)$ and $c_b(t)$.

These particular models are powerful since they can capture arbitrary linear combinations of reward and choice history (up to N trials into the past) to predict upcoming choices. A frequent observation is that choices have a positive dependence on reward history (β_i^R coefficients), often multiple trials into the past. The interpretation is that previous rewarded choices reinforce future choices to that option. Previous choices (generally 1–2 trials) tend to have negative coefficients (β_i^c coefficients), meaning animals tend to switch their choices over short timescales, regardless of reward history. Interestingly, negative β_i^c coefficients are generally not seen in tasks without baiting (Parker et al., 2016). Given the arguments above suggesting that animals behave as if they are *not* aware of the baiting rule, one potential explanation is that animals are implementing a simple "switch" heuristic to increase reward rate.

A major limitation of the logistic regression formulation is that only linear combinations of the input can be used. An immediate consequence of this can be seen by observing that when $R_a(t) - R_b(t) = 0$, the model cannot discriminate whether it was option a that was not rewarded or option b. A solution to this problem may exist, since it has been shown that the logistic regression model is identical to an action-value reinforcement learning model with identical learning and forgetting rates (Katahira, 2015). This suggests that the action-value reinforcement learning model may be a suitable template to build interpretable algorithms that capture the essence of matching behavior.

4.5 Action-value reinforcement learning algorithm (*Q*-learning)

The action-value reinforcement learning (or Q-learning) algorithm is a general-purpose learning algorithm that keeps track of the values (i.e., expected future reward) of available actions and makes decisions based on the difference between action values (Watkins & Dayan, 1992). It has been widely applied both in neuroscience (Akam et al., 2017; Li & Daw, 2011;

Samejima, Ueda, Doya, & Kimura, 2005; Sul, Kim, Huh, Lee, & Jung, 2010) and machine learning (Mnih et al., 2015). If we assume a task with two actions, a common implementation is to assume a single state and update the Q-values as follows. If action on trial $t = a$

$$Q_{t+1}(a) = Q_t(a) + \alpha(r - Q_t(a)) \tag{21}$$

$$Q_{t+1}(b) = Q_t(b) \tag{22}$$

The difference between Q-values is used as an input into a softmax function to produce the probability of a choice.

$$P(\text{action on trial } t = a) = \frac{1}{1 + e^{-\beta(Q_t(a) - Q_t(b))}} \tag{23}$$

$$P(\text{action on trial } t = b) = 1 - P(\text{action on trial } t = a) \tag{24}$$

Forgetting can be introduced as (where action on trial $t = a$)

$$Q_{t+1}(a) = \zeta Q_t(a) + \alpha(r - Q_t(a)) \tag{25}$$

$$Q_{t+1}(b) = \zeta Q_t(b) \tag{26}$$

Forgetting can be applied to just the unchosen action or both actions (Bari et al., 2019; Farashahi, Rowe, Aslami, Gobbini, & Soltani, 2018; Hattori, Danskin, Babic, Mlynaryk, & Komiyama, 2019; Katahira, 2015).

4.6 Bayesian inference

Bayesian inference algorithms use Bayes rule to iteratively update estimates of reward probability. They are powerful since they estimate full probability distributions, allowing them to make decisions that take into account uncertainty in their estimates, as well as higher-order moments. For example, if the algorithm estimates both options to have a mean reward probability $p_a = p_b = 0.5$, but the uncertainty of $p_a > p_b$, then it is adaptive to choose p_a since the true mean reward rate might be higher, increasing reward in the long term (Sutton & Barto, 1998). These algorithms have been used to argue that matching might occur due to uncertainty about changing reward dynamics (Yu & Huang, 2014). In general, they have seen limited use for quantifying matching behavior, since simpler models often do well enough.

5. Movement vigor during flexible decision making

Decisions are much more complex than just discrete choices. Behavior occurs in real-time and the nervous system must finely calibrate the vigor of movements. Vigor, often defined as reaction time plus speed of movements, has long been studied in the context of motor control (Choi, Vaswani, & Shadmehr, 2014; Reppert et al., 2018; Rigas, Komogortsev, & Shadmehr, 2016). Recently, vigor has been appreciated as a reflection of value (Shadmehr, Reppert, Summerside, Yoon, & Ahmed, 2019). Vigor is increased by increasing reward (Summerside, Shadmehr, & Ahmed, 2018), decreased by increasing effort (Stelmach & Worringham, 1988), and modulated on short, individual-decision timescales (Reppert, Lempert, Glimcher, & Shadmehr, 2015).

From a normative perspective, modulating vigor in relation to value/effort is an appropriate computation (Niv, Daw, Joel, & Dayan, 2007; Yoon, Geary, Ahmed, & Shadmehr, 2018). Because increased vigor requires greater energy (Hoyt & Taylor, 1981; Ralston, 1958; Selinger, O'Connor, Wong, & Donelan, 2015), it is not always appropriate to move with high vigor. However, in the context of a highly rewarding environment, it can be worth increasing vigor to increase reward rate, since slow movements necessitate a longer time between movement initiation and receipt of reward. It is clear that the brain modulates movement vigor to maximize reward rates (Haith, Reppert, & Shadmehr, 2012).

In the context of flexible decision making, moment-to-moment vigor has been shown to be as flexible as choice-based behavior. One study employed a dynamic two-armed bandit task in rats and found that latency of task initiation was highly correlated with instantaneous probability of reward (Hamid et al., 2016). Movement vigor was most strongly modulated by the reward rate of the environment (how much reward can be expected regardless of action), and much less so by relative reward rates (how much better one option is relative to another). A number of studies have demonstrated that recent reward history modulates movement vigor (Bari et al., 2019; Del Arco, Park, Wood, Kim, & Moghaddam, 2017; Ottenheimer et al., 2020; Simon, Wood, & Moghaddam, 2015).

6. Brain structures underlying flexible behavior

Extensive work, most of it in humans, has correlated changes in activity in multiple brain regions with variables from reinforcement-learning models (e.g., Daw, O'Doherty, Dayan, Seymour, & Dolan, 2006; Doya, 2008). The vast majority of these studies have focused on brief changes in activity of neurons or fMRI signal (seconds or less) and model variables that change on similar timescales.

One key variable, common to many algorithms, is the reward prediction error, the difference between actual and predicted reward. We know much about these signals, especially in the context of foraging (Morris et al., 2006; Parker et al., 2016). However, we know much less about the representations of the decision variables updated by these reward prediction errors. In particular, all algorithms of flexible behavior require memory: a summary of previous interactions with the environment that allows for adaptive behavior in the future. Whereas this memory signal—typically in the form of action values or their arithmetic combinations—has been observed in several brain structures, most notably the dorsal striatum (e.g., Samejima et al., 2005), it is largely transient and occurs around the time of cues and actions. It is less clear where these memory signals reside in between bouts of interaction with the environment.

6.1 Cortico-basal-ganglia loops: Medial prefrontal cortex and dorsomedial striatum

The medial prefrontal cortex (mPFC) and its downstream target, the dorsomedial striatum, have long been studied as critical components for generating flexible behavior[e] One view of this circuit is that cortex provides signals to bias action selection (Murakami, Shteingart, Loewenstein, & Mainen, 2017) and the striatum is responsible for action selection itself. Cortical circuitry is predominantly recurrent, which is hypothesized to allow for integration of information which can be routed to striatum to bias action selection. Striatal circuitry, in contrast, is predominantly inhibitory and weak, with lateral inhibition motifs, facilitating a winner-take-all operation (Morita, Jitsev, & Morrison, 2016). The classic view of the striatum considers two parallel cortico-striato-cortical loops, operating to initiate and inhibit actions (Fig. 4). The "Direct" or "Go" pathway, consists of cortical inputs to D1-expressing medium spiny neurons in the striatum, which synapse onto globus pallidus pars interna (entopeduncular nucleus in the rodent (Grillner & Robertson, 2016; Wallace et al., 2017)) and the substantia nigra pars reticulata. These in turn synapse onto the thalamus, and back to the cortex, completing the loop. The "Indirect" or "No-Go" pathway consists of cortical inputs synapsing onto D2-expressing medium spiny neurons,

[e] In rodents, mPFC typically refers to (from dorsal to ventral) the anterior cingulate cortex, the prelimbic cortex, the infralimbic cortex, and the medial orbital cortex. Dorsal mPFC usually means anterior cingulate cortex and prelimbic cortex, and ventral mPFC usually means infralimbic cortex and medial orbital cortex. In general, since the study of mPFC is still nascent, especially in mice, stereotactic coordinates are useful for comparing studies.

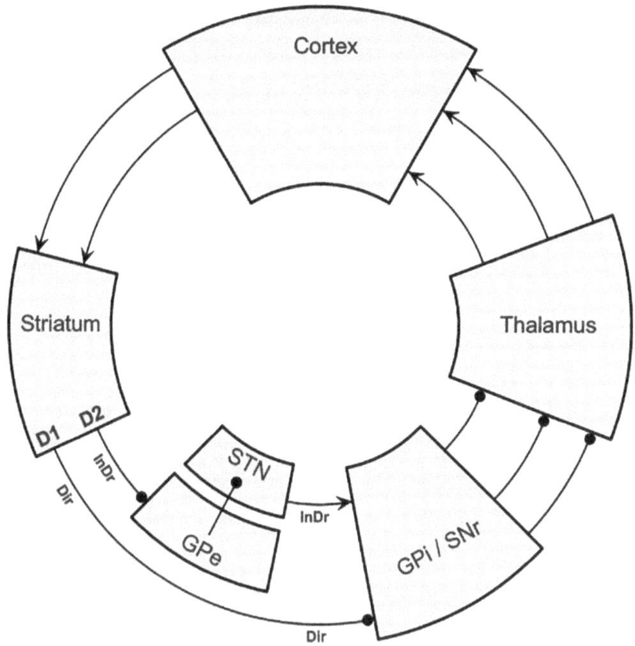

Fig. 4 Classic view of the "Direct" and "Indirect" striatal pathways. *Adapted from Shipp, S. (2017). The functional logic of corticostriatal connections. Brain Structure and Function, 222(2), 669–706. https://doi.org/10.1007/s00429-016-1250-9.*

which in turn synapse onto the globus pallidus pars externa → subthalamis nucleus → globus pallidus pars interna and substantia nigra pars reticulata → thalamus, and finally back to cortex (Shipp, 2017).

Lesions of either the medial prefrontal cortex or the dorsomedial striatum are known to abolish goal-directed behavior and render behavior under the control of sensorimotor associations (Balleine, Delgado, & Hikosaka, 2007; Kennerley, Walton, Behrens, Buckley, & Rushworth, 2006). Neurons in rat dorsal mPFC predict upcoming outcomes, before they have been presented (Del Arco et al., 2017), and monitor action/outcome contingencies (Hyman, Whitman, Emberly, Woodward, & Seamans, 2013; Simon et al., 2015; Sul et al., 2010). Neurons in primate medial frontal cortex signal prediction errors of action values (Matsumoto, Matsumoto, Abe, & Tanaka, 2007) and those in anterior cingulate cortex encode reward history as well as reward prediction errors[f] (Seo & Lee, 2007). In humans, fMRI studies

[f] Comparisons between rodent and primate frontal structures should be taken with a grain of salt, as anatomical homologies between these orders is weak (Uylings, Groenewegen, & Kolb, 2003).

have revealed a role for medial frontal regions in encoding reward magnitude and value of chosen actions (Daw et al., 2006). Higher network coordination between cortex and striatum predicts changes in learning and decision making (Gerraty et al., 2018).

Recordings in the primate caudate, a homolog of the rodent dorsomedial striatum, have demonstrated encoding of action values and chosen values, key signals in reinforcement learning models of behavior (Lau & Glimcher, 2008; Samejima et al., 2005). Later studies discovered preferential encoding of the difference of temporally discounted values, and encoding of future actions (Cai, Kim, & Lee, 2011). Recordings in the dorsomedial striatum of the rat have confirmed these findings (Ito & Doya, 2009, 2015; Kim, Lee, & Jung, 2013; Seo, Lee, & Averbeck, 2012), and extended them by demonstrating encoding of total value, necessary for modulating vigor (Wang, Miura, & Uchida, 2013). Manipulation of striatal subpopulations modulates choice-based behavior in a manner consistent with changes in action values (Tai, Lee, Benavidez, Bonci, & Wilbrecht, 2012). Unilateral activation of D1-expressing medium spiny neurons in the dorsomedial striatum bias animals to make more contralateral actions. Conversely, unilateral activation of D2-expressing neurons in the dorsomedial striatum bias animals to make more ipsilateral actions. Similar results were obtained with pharmacological manipulation of D1/D2 receptors in the primate putamen (Ueda et al., 2017). These findings align with the classic view that D1-expressing neurons are organized into the "Go" pathway to initiate behavior, and D2-expressing neurons are organized into a "No-Go" pathway to inhibit behavior.

Recent work has focused on mouse mPFC, a structure known to have persistent working-memory-like neural correlates (Liu et al., 2014), as the potential site where decision variables are maintained in between bouts of interaction with the environment. Persistent activity is a viable network mechanism for maintaining representations of decision variables in the times between choices. Persistent activity, defined here as activity that lasts longer than the time constants of synaptic inputs, was first described in the prefrontal cortex and mediodorsal thalamus of primates and is thought to be a critical component of working memory (Fuster & Alexander, 1971).

We recently recorded from mPFC neurons, including those that projected to dorsomedial striatum, in mice performing a matching task (Bari et al., 2019). Individual mPFC neurons showed persistent representations of two key decision variables for matching behavior: relative value—used to bias choices—and total value—used to bias response time (or "vigor").

Remarkably, these two forms of persistent activity showed different rates of decay, that matched the behavior they supported. Relative-value activity did not appear to decay during long inter-choice intervals (tens of seconds); neither did the mouse's memory of its choice policy (Fig. 5). In contrast, total-value activity decayed slowly over long inter-choice intervals; likewise, mice made slower choices after long waiting times. These variables did not appear to be robustly encoded by tongue premotor neurons. These data suggest that cortico-basal-ganglia-thalamo-cortical loops maintain value-based decision variables, and that information flow in the circuit is not a simple linear flow of computations.

Tool development has allowed for precise pathway-specific modulation. In our recent study, we found that inactivation of the mPFC → dorsomedial striatal pathway disrupted choice behavior and slowed vigor, consistent with a disruption of cognitive variables necessary for value-based decision making (Bari et al., 2019). This is similar to a recent study which demonstrated that the mPFC → striatum and mPFC → thalamus pathways were necessary for choice behavior, but not the mPFC → mPFC pathway (Nakayama, Ibañez-Tallon, & Heintz, 2018). Another group demonstrated that the mediodorsal → mPFC pathway, but not the mPFC → mediodorsal thalamus, is necessary for updating understanding about the causal structure of actions, although both were necessary for goal-directed behavior (Alcaraz et al., 2018). This is consistent with the notion that the cortico-basal ganglia system is critical for flexible behavior, since the former, but not the latter, pathway is part of this system.

Taken together, these results highlight the key concept that the brain has dedicated circuitry for different behavioral strategies, and that the rodent dorsal medial prefrontal cortex and downstream dorsomedial striatum are important structures for driving flexible behavior.

6.2 Neuromodulatory systems

Neuromodulatory systems are remarkably unique. These systems are each composed of very small numbers of cells, yet have outputs that arborize to span large volumes of tissue. This feature makes them poised to exert global control over neural states and computations.

Midbrain dopamine is a particularly well-studied neuromodulator. Dopamine neurons in the midbrain encode reward prediction errors, a critical variable in reinforcement learning models of flexible behavior (Bayer & Glimcher, 2005; Cohen, Haesler, Vong, Lowell, & Uchida, 2012;

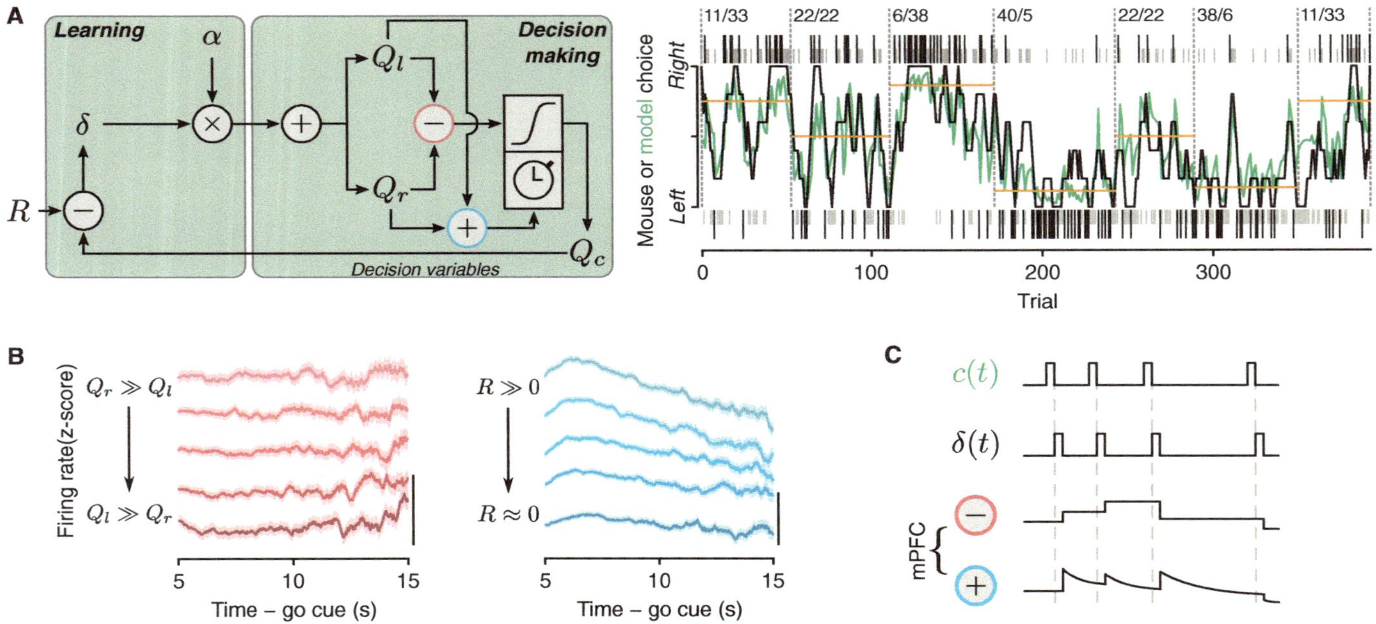

Fig. 5 (A) Reinforcement-learning model (*left*) and fit to example mouse behavior (*right*). (B) Persistent activity representing relative (*left panel*) and total (*right panel*) value. (C) Schematic of variables from the model represented in mPFC. *Adapted from Bari, B. A., Grossman, C. D., Lubin, E. E., Rajagopalan, A. E., Cressy, J. I., & Cohen, J. Y. (2019). Stable representations of decision variables for flexible behavior. Neuron, 103(5), 922–933.*

Morris et al., 2006; Roesch, Stalnaker, & Schoenbaum, 2007; Schultz et al., 1997). Manipulation of the dopamine system profoundly affects learning, for both Pavlovian associations (Steinberg et al., 2013) and for flexible decisions (Hamid et al., 2016; Parker et al., 2016). Dopamine densely innervates the striatum, where it bidirectionally modulates the plasticity of corticostriatal synapses, depending on the downstream receptor (Reynolds & Wickens, 2002). This feature makes dopamine an integral component that enables the cortico-basal ganglia system to modulate flexible behavior.

Serotonin is comparatively less-well understood, in large part to substantial heterogeneity in this system relative to midbrain dopamine. In large thanks to tool development (Lima, Hromádka, Znamenskiy, & Zador, 2009), extracellular recording of single cell-type-identified serotonin neurons has allowed for inroads to be made. Recent work has shown that a subset of serotonin neurons in the dorsal raphe nucleus encodes value over very long timescales (Cohen, Amoroso, & Uchida, 2015), which may relate to the encoding of background reward rate, a key variable in models of optimal foraging behavior (Charnov et al., 1976). Activation of dorsal raphe serotonin neurons in mice performing a dynamic foraging task leads to increases in learning rates (Iigaya, Fonseca, Murakami, Mainen, & Dayan, 2018) and promotes persistence in mice performing a patch foraging task (Lottem et al., 2018). These findings must be interpreted with caution, however, since serotonin manipulation can have opposite effects on behavior, depending on the outputs of distinct subpopulations of neurons (Ren et al., 2018).

Norepinephrine produced by neurons in the locus coeruleus is thought to be critical for behavioral flexibility. These neurons respond to salient events, which has led to the hypothesis that norepinephrine is critical for arousal and attention (Aston-Jones, Rajkowski, & Cohen, 2000; Carter et al., 2010; Harley, 1987; Kety, 1970). Classic work demonstrated that activation of locus coeruleus norepinephrine can alleviate forgetting in a complex maze task (Devauges & Sara, 1991) and facilitate attentional shifts (Devauges & Sara, 1990). Formal theories of norepinephrine suggest that the system encodes unexpected uncertainty (Dayan & Yu, 2006; Yu & Dayan, 2003, 2005) and facilitate exploitation/exploration of task contingencies (Aston-Jones & Cohen, 2005).

6.3 Other structures

Although we have focused heavily on cortico-basal ganglia systems and neuromodulatory systems, flexible behavior relies on a much larger network of structures. The orbitofrontal cortex, and its upstream/downstream structures, are critical for flexible behavior. Examples include the orbitofrontal cortex → submedius nucleus pathway (Fresno, Parkes, Faugère, Coutureau, & Wolff, 2019) and the amygdala → orbitofrontal cortex pathway (Fiuzat, Rhodes, & Murray, 2017). The orbitofrontal cortex encodes reward magnitude (Simon et al., 2015) and codes for chosen value and reward prediction errors more strongly than mPFC (Sul et al., 2010). Other important structures include the posterior parietal cortex and posteromedial cortex (Funamizu, Kuhn, & Doya, 2016) and the ventral hippocampus (Yoshida, Drew, Mimura, & Tanaka, 2019). This list is by no means exhaustive and serves simply to indicate that large portions of the brain are critical for flexible decision making.

7. Future directions

Despite sophisticated theoretical and modeling insight about flexible decision making, there remains a gap in our knowledge of how the entire loop—sensory signal to discrete action—is instantiated in neural circuitry. One direction that we believe is due for further study is expanding action space to more than two discrete actions, especially using continuous action space. This is likely to enhance our theories of value representation in the mPFC, extending beyond the relatively impoverished regime of binary choices.

References

Akam, T., Rodrigues-Vaz, I., Zhang, X., Pereira, M., Oliveira, R., Dayan, P., & Costa, R. M. (2017). Single-trial inhibition of anterior cingulate disrupts model-based reinforcement learning in a two-step decision task. *bioRxiv*, 126292.

Alcaraz, F., Fresno, V., Marchand, A. R., Kremer, E. J., Coutureau, E., & Wolff, M. (2018). Thalamocortical and corticothalamic pathways differentially contribute to goal-directed behaviors in the rat. *Elife*, 7, e32517.

Aston-Jones, G., & Cohen, J. D. (2005). An integrative theory of locus coeruleus-norepinephrine function: Adaptive gain and optimal performance. *Annual Review of Neuroscience*, *28*, 403–450. https://doi.org/10.1146/annurev.neuro.28.061604.135709.

Aston-Jones, G., Rajkowski, J., & Cohen, J. D. (2000). Locus coeruleus and regulation of behavioral flexibility and attention. *Progress in Brain Research*, *126*, 165–182.

Balleine, B. W., Delgado, M. R., & Hikosaka, O. (2007). The role of the dorsal striatum in reward and decision-making. *The Journal of Neuroscience: The Official Journal of the Society for Neuroscience*, *27*, 8161–8165. https://doi.org/10.1523/JNEUROSCI.1554-07.2007.

Balleine, B. W., & O'Doherty, J. P. (2010). Human and rodent homologies in action control: Corticostriatal determinants of goal-directed and habitual action. *Neuropsychopharmacology*, *35*(1), 48–69. https://doi.org/10.1038/npp.2009.131.

Bari, B. A., Grossman, C. D., Lubin, E. E., Rajagopalan, A. E., Cressy, J. I., & Cohen, J. Y. (2019). Stable representations of decision variables for flexible behavior. *Neuron*, *103*(5), 922–933.

Bayer, H. M., & Glimcher, P. W. (2005). Midbrain dopamine neurons encode a quantitative reward prediction error signal. *Neuron*, *47*(1), 129–141. https://doi.org/10.1016/j.neuron.2005.05.020.

Bertsekas, D. P., & Tsitsiklis, J. N. (1996). *Neuro-dynamic programming*. Athena Scientific Belmont.

Cáceda, R., Nemeroff, C. B., & Harvey, P. D. (2014). Toward an understanding of decision making in severe mental illness. *The Journal of Neuropsychiatry and Clinical Neurosciences*, *26*(3), 196–213. https://doi.org/10.1176/appi.neuropsych.12110268.

Cai, X., Kim, S., & Lee, D. (2011). Heterogeneous coding of temporally discounted values in the dorsal and ventral striatum during intertemporal choice. *Neuron*, *69*, 170–182. https://doi.org/10.1016/j.neuron.2010.11.041.

Cambiaghi, M., & Sacchetti, B. (2015). Ivan Petrovich Pavlov (1849–1936). *Journal of Neurology*, *262*(6), 1599–1600. https://doi.org/10.1007/s00415-015-7743-2.

Carter, M. E., Yizhar, O., Chikahisa, S., Nguyen, H., Adamantidis, A., Nishino, S., … de Lecea, L. (2010). Tuning arousal with optogenetic modulation of locus coeruleus neurons. *Nature Neuroscience*, *13*(12), 1526–1533. https://doi.org/10.1038/nn.2682.

Charnov, E. L. (1976). Optimal foraging, the marginal value theorem. *Theoretical Population Biology*, *9*(2), 129–136.

Choi, J. E., Vaswani, P. A., & Shadmehr, R. (2014). Vigor of movements and the cost of time in decision making. *Journal of Neuroscience*, *34*(4), 1212–1223.

Cohen, J. Y., Amoroso, M. W., & Uchida, N. (2015). Serotonergic neurons signal reward and punishment on multiple timescales. *eLife*, *4*. https://doi.org/10.7554/eLife.06346.

Cohen, J. Y., Haesler, S., Vong, L., Lowell, B. B., & Uchida, N. (2012). Neuron-type-specific signals for reward and punishment in the ventral tegmental area. *Nature*, *482*(7383), 85–88. https://doi.org/10.1038/nature10754.

Daw, N. D., Niv, Y., & Dayan, P. (2005). Uncertainty-based competition between prefrontal and dorsolateral striatal systems for behavioral control. *Nature Neuroscience*, *8*, 1704–1711. https://doi.org/10.1038/nn1560.

Daw, N. D., O'Doherty, J. P., Dayan, P., Seymour, B., & Dolan, R. J. (2006). Cortical substrates for exploratory decisions in humans. *Nature*, *441*(7095), 876–879. https://doi.org/10.1038/nature04766.

Dayan, P., & Abbott, L. F. (2001). *Theoretical neuroscience: Computational and mathematical modeling of neural systems*. MIT Press.

Dayan, P., Niv, Y., Seymour, B., & Daw, N. D. (2006). The misbehavior of value and the discipline of the will. *Neural Networks: The Official Journal of the International Neural Network Society*, *19*, 1153–1160. https://doi.org/10.1016/j.neunet.2006.03.002.

Dayan, P., & Yu, A. J. (2006). Phasic norepinephrine: A neural interrupt signal for unexpected events. *Network*, *17*(4), 335–350. https://doi.org/10.1080/09548980601004024.

Del Arco, A., Park, J., Wood, J., Kim, Y., & Moghaddam, B. (2017). Adaptive encoding of outcome prediction by prefrontal cortex ensembles supports behavioral flexibility. *The Journal of Neuroscience: The Official Journal of the Society for Neuroscience*, *37*, 8363–8373. https://doi.org/10.1523/JNEUROSCI.0450-17.2017.

Dener, E., Kacelnik, A., & Shemesh, H. (2016). Pea Plants Show Risk Sensitivity. *Current Biology*, *26*(13), 1763–1767. https://doi.org/10.1016/j.cub.2016.05.008.

Devauges, V., & Sara, S. J. (1990). Activation of the noradrenergic system facilitates an attentional shift in the rat. *Behavioural Brain Research*, *39*(1), 19–28. https://doi.org/10.1016/0166-4328(90)90118-X.

Devauges, V., & Sara, S. J. (1991). Memory retrieval enhancement by locus coeruleus stimulation: Evidence for mediation by beta-receptors. *Behavioural Brain Research*, *43*(1), 93–97. https://doi.org/10.1016/S0166-4328(05)80056-7.

Doya, K. (2008). Modulators of decision making. *Nature Neuroscience, 11*(4), 410–416. https://doi.org/10.1038/nn2077.

Farashahi, S., Rowe, K., Aslami, Z., Gobbini, M. I., & Soltani, A. (2018). Influence of learning strategy on response time during complex value-based learning and choice. *PLoS One, 13*(5), e0197263.

Ferster, C. B., & Skinner, B. F. (1957). *Schedules of reinforcement.* East Norwalk: Appleton-Century-Crofts. https://doi.org/10.1037/10627-000.

Fiuzat, E. C., Rhodes, S. E. V., & Murray, E. A. (2017). The role of orbitofrontal-amygdala interactions in updating action-outcome valuations in macaques. *The Journal of Neuroscience: The Official Journal of the Society for Neuroscience, 37*(9), 2463–2470. https://doi.org/10.1523/JNEUROSCI.1839-16.2017.

Fonseca, M. S., Murakami, M., & Mainen, Z. F. (2015). Activation of dorsal raphe serotonergic neurons promotes waiting but is not reinforcing. *Current Biology, 25*(3), 306–315. https://doi.org/10.1016/j.cub.2014.12.002.

Fresno, V., Parkes, S. L., Faugère, A., Coutureau, E., & Wolff, M. (2019). A thalamocortical circuit for updating action-outcome associations. *eLife, 8,* 1–13. https://doi.org/10.7554/eLife.46187.

Funamizu, A., Kuhn, B., & Doya, K. (2016). Neural substrate of dynamic Bayesian inference in the cerebral cortex. *Nature Neuroscience, 19*(12), 1682–1689. https://doi.org/10.1038/nn.4390.

Fuster, J. M., & Alexander, G. E. (1971). Neuron activity related to short-term memory. *Science, 173*(3997), 652–654.

Gallistel, C. R., Mark, T. A., King, A. P., & Latham, P. E. (2001). The rat approximates an ideal detector of changes in rates of reward: Implications for the law of effect. *Journal of Experimental Psychology. Animal Behavior Processes, 27*(4), 354–372.

Gerraty, R. T., Davidow, J. Y., Foerde, K., Galvan, A., Bassett, D. S., & Shohamy, D. (2018). Dynamic flexibility in striatal-cortical circuits supports reinforcement learning. *The Journal of Neuroscience: The Official Journal of the Society for Neuroscience.* https://doi.org/10.1523/JNEUROSCI.2084-17.2018.

Glimcher, P. W., Dorris, M. C., & Bayer, H. M. (2005). Physiological utility theory and the neuroeconomics of choice. *Games and Economic Behavior, 52*(2), 213–256. https://doi.org/10.1016/j.geb.2004.06.011.

Gold, J. I., & Shadlen, M. N. (2007). The neural basis of decision making. *Annual Review of Neuroscience, 30,* 535–574. https://doi.org/10.1146/annurev.neuro.29.051605.113038.

Graft, D. A., Lea, S. E. G., & Whitworth, T. L. (1977). The matching law in and within groups of rats. *Journal of the Experimental Analysis of Behavior, 27*(1), 1333563. https://doi.org/10.1901/jeab.1977.27-183.

Grillner, S., & Robertson, B. (2016). The basal ganglia over 500 million years. *Current Biology, 26*(20), R1088–R1100.

Haith, A. M., Reppert, T. R., & Shadmehr, R. (2012). Evidence for hyperbolic temporal discounting of reward in control of movements. *The Journal of Neuroscience: The Official Journal of the Society for Neuroscience, 32*(34), 11727–11736. https://doi.org/10.1523/JNEUROSCI.0424-12.2012.

Hamid, A. A., Pettibone, J. R., Mabrouk, O. S., Hetrick, V. L., Schmidt, R., Vander Weele, C. M., … Berke, J. D. (2016). Mesolimbic dopamine signals the value of work. *Nature Neuroscience, 19*(1), 117–126. https://doi.org/10.1038/nn.4173.

Harley, C. W. (1987). A role for norepinephrine in arousal, emotion and learning?: Limbic modulation by norepinephrine and the Kety hypothesis. *Progress in Neuro-Psychopharmacology & Biological Psychiatry, 11,* 419–458. https://doi.org/10.1016/0278-5846(87)90015-7.

Hattori, R., Danskin, B., Babic, Z., Mlynaryk, N., & Komiyama, T. (2019). Area-specificity and plasticity of history-dependent value coding during learning. *Cell, 177*(7), 1858–1872.

Herrnstein, R. J. (1961). Relative and absolute strength of response as a function of frequency of reinforcement. *Journal of the Experimental Analysis of Behavior, 4*, 267–272. https://doi.org/10.1901/jeab.1961.4-267.

Herrnstein, R. J. (2000). *The Matching Law: Papers in Psychology and economics*. Harvard University Press.

Herrnstein, R. J., & Heyman, G. M. (1979). Is matching compatible with reinforcement maximization on concurrent variable interval variable ratio? *Journal of the Experimental Analysis of Behavior, 31*(2), 209–223. https://doi.org/10.1901/jeab.1979.31-209.

Herrnstein, R. J., & Vaughan, W. (1980). Melioration and behavioral allocation. *Limits to Action: The Allocation of Individual Behavior, 1*, 143–176.

Hoyt, D. F., & Taylor, C. R. (1981). Gait and the energetics of locomotion in horses. *Nature, 292*(5820), 239.

Hyman, J. M., Whitman, J., Emberly, E., Woodward, T. S., & Seamans, J. K. (2013). Action and outcome activity state patterns in the anterior cingulate cortex. *Cerebral Cortex, 23*(6), 1257–1268. https://doi.org/10.1093/cercor/bhs104.

Iigaya, K., Fonseca, M. S., Murakami, M., Mainen, Z. F., & Dayan, P. (2018). An effect of serotonergic stimulation on learning rates for rewards apparent after long intertrial intervals. *Nature Communications, 9*(1), 2477. https://doi.org/10.1038/s41467-018-04840-2.

Ito, M., & Doya, K. (2009). Validation of decision-making models and analysis of decision variables in the rat basal ganglia. *The Journal of Neuroscience: The Official Journal of the Society for Neuroscience, 29*(31), 9861–9874. https://doi.org/10.1523/JNEUROSCI.6157-08.2009.

Ito, M., & Doya, K. (2015). Distinct neural representation in the dorsolateral, dorsomedial, and ventral parts of the striatum during fixed- and free-choice tasks. *The Journal of Neuroscience: The Official Journal of the Society for Neuroscience, 35*(8), 3499–3514. https://doi.org/10.1523/JNEUROSCI.1962-14.2015.

Katahira, K. (2015). The relation between reinforcement learning parameters and the influence of reinforcement history on choice behavior. *Journal of Mathematical Psychology, 66*, 59–69. https://doi.org/10.1016/j.jmp.2015.03.006.

Kennerley, S. W., Walton, M. E., Behrens, T. E. J., Buckley, M. J., & Rushworth, M. F. S. (2006). Optimal decision making and the anterior cingulate cortex. *Nature Neuroscience, 9*(7), 940–947. https://doi.org/10.1038/nn1724.

Kety, S. S. (1970). The biogenic amines in the central nervous system: Their possible roles in arousal, emotion, and learning. In F. O. Schmitt (Ed.), *The neurosciences: Second study program* (pp. 324–336). New York: Rockefeller University Press.

Kim, H., Lee, D., & Jung, M. W. (2013). Signals for previous goal choice persist in the dorsomedial, but not dorsolateral striatum of rats. *The Journal of Neuroscience: The Official Journal of the Society for Neuroscience, 33*(1), 52–63. https://doi.org/10.1523/JNEUROSCI.2422-12.2013.

Lau, B., & Glimcher, P. W. (2005). Dynamic response-by-response models of matching behavior in rhesus monkeys. *Journal of the Experimental Analysis of Behavior, 84*(3), 555–579.

Lau, B., & Glimcher, P. W. (2008). Value representations in the primate striatum during matching behavior. *Neuron, 58*(3), 451–463. https://doi.org/10.1016/j.neuron.2008.02.021.

Li, J., & Daw, N. D. (2011). Signals in human striatum are appropriate for policy update rather than value prediction. *The Journal of Neuroscience: The Official Journal of the Society for Neuroscience, 31*(14), 5504–5511. https://doi.org/10.1523/JNEUROSCI.6316-10.2011.

Lima, S. Q., Hromádka, T., Znamenskiy, P., & Zador, A. M. (2009). PINP: A new method of tagging neuronal populations for identification during in vivo electrophysiological recording. *PLoS One, 4*(7), e6099. https://doi.org/10.1371/journal.pone.0006099.

Liu, D., Gu, X., Zhu, J., Zhang, X., Han, Z., Yan, W., ... Li, C. T. (2014). Medial prefrontal activity during delay period contributes to learning of a working memory task. *Science, 346*(6208), 458–463. https://doi.org/10.1126/science.1256573.

Loewenstein, Y., & Seung, H. S. (2006). Operant matching is a generic outcome of synaptic plasticity based on the covariance between reward and neural activity. *Proceedings of the National Academy of Sciences of the United States of America, 103*(41), 15224–15229. https://doi.org/10.1073/pnas.0505220103.

Lottem, E., Banerjee, D., Vertechi, P., Sarra, D., Lohuis, M. O., & Mainen, Z. F. (2018). Activation of serotonin neurons promotes active persistence in a probabilistic foraging task. *Nature Communications, 9*(1), 1000. https://doi.org/10.1038/s41467-018-03438-y.

Luo, L., Callaway, E. M., & Svoboda, K. (2018). Genetic dissection of neural circuits: A decade of progress. *Neuron, 98*(2), 256–281. https://doi.org/10.1016/j.neuron.2018.03.040.

Matsumoto, M., Matsumoto, K., Abe, H., & Tanaka, K. (2007). Medial prefrontal cell activity signaling prediction errors of action values. *Nature Neuroscience, 10*(5), 647–656. https://doi.org/10.1038/nn1890.

Mnih, V., Kavukcuoglu, K., Silver, D., Rusu, A. A., Veness, J., Bellemare, M. G., ... Hassabis, D. (2015). Human-level control through deep reinforcement learning. *Nature, 518*(7540), 529.

Mongillo, G., Shteingart, H., & Loewenstein, Y. (2014). The misbehavior of reinforcement learning. *Proceedings of the IEEE, 102*(4), 528–541. https://doi.org/10.1109/JPROC.2014.2307022.

Morita, K., Jitsev, J., & Morrison, A. (2016). Corticostriatal circuit mechanisms of value-based action selection: Implementation of reinforcement learning algorithms and beyond. *Behavioural Brain Research, 311*, 110–121. https://doi.org/10.1016/j.bbr.2016.05.017.

Morris, G., Nevet, A., Arkadir, D., Vaadia, E., & Bergman, H. (2006). Midbrain dopamine neurons encode decisions for future action. *Nature Neuroscience, 9*(8), 1057–1063. https://doi.org/10.1038/nn1743.

Murakami, M., Shteingart, H., Loewenstein, Y., & Mainen, Z. F. (2017). Distinct sources of deterministic and stochastic components of action timing decisions in rodent frontal cortex. *Neuron, 94*(4), 908–919.e7. https://doi.org/10.1016/j.neuron.2017.04.040.

Nakayama, H., Ibañez-Tallon, I., & Heintz, N. (2018). Cell-type-specific contributions of medial prefrontal neurons to flexible behaviors. *The Journal of Neuroscience: The Official Journal of the Society for Neuroscience, 38*(19), 4490–4504. https://doi.org/10.1523/JNEUROSCI.3537-17.2018.

Niv, Y., Daw, N. D., Joel, D., & Dayan, P. (2007). Tonic dopamine: Opportunity costs and the control of response vigor. *Psychopharmacology, 191*(3), 507–520. https://doi.org/10.1007/s00213-006-0502-4.

Ofstad, T. A., Zuker, C. S., & Reiser, M. B. (2011). Visual place learning in Drosophila melanogaster. *Nature, 474*(7350), 204–209. https://doi.org/10.1038/nature10131.

Ottenheimer, D. J., Bari, B. A., Sutlief, E., Fraser, K. M., Kim, T. H., Richard, J. M., ... Janak, P. H. (2020). A quantitative reward prediction error signal in the ventral pallidum. *Nature Neuroscience, 23*, 1267–1276.

Parker, N. F., Cameron, C. M., Taliaferro, J. P., Lee, J., Choi, J. Y., Davidson, T. J., ... Witten, I. B. (2016). Reward and choice encoding in terminals of midbrain dopamine neurons depends on striatal target. *Nature Neuroscience, 19*(6), 845–854. https://doi.org/10.1038/nn.4287.

Pavlov, I. P., & Anrep, G. V. (1928). Conditioned reflexes. *Journal of Philosophical Studies, 3*(11), 380–383.

Pierce, W. D., & Epling, W. F. (1983). Choice, matching, and human behavior: A review of the literature. *Behavior Analyst, 6*(1), 57–76. https://doi.org/10.1007/BF03391874.

Ralston, H. J. (1958). Energy-speed relation and optimal speed during level walking. *Internationale Zeitschrift für Angewandte Physiologie Einschliesslich Arbeitsphysiologie, 17*(4), 277–283.

Rangel, A., Camerer, C., & Montague, P. R. (2008). A framework for studying the neurobiology of value-based decision making. *Nature Reviews Neuroscience, 9*(7), 545–556. https://doi.org/10.1038/nrn2357.

Ren, J., Friedmann, D., Xiong, J., Liu, C. D., Ferguson, B. R., Weerakkody, T., ... Luo, L. (2018). Anatomically defined and functionally distinct dorsal raphe serotonin sub-systems. *Cell, 175*(2), 472–487.

Reppert, T. R., Lempert, K. M., Glimcher, P. W., & Shadmehr, R. (2015). Modulation of saccade vigor during value-based decision making. *The Journal of Neuroscience: The Official Journal of the Society for Neuroscience, 35*(46), 15369–15378. https://doi.org/10.1523/JNEUROSCI.2621-15.2015.

Reppert, T. R., Rigas, I., Herzfeld, D. J., Sedaghat-Nejad, E., Komogortsev, O., & Shadmehr, R. (2018). Movement vigor as a traitlike attribute of individuality. *Journal of Neurophysiology, 120*(2), 741–757.

Rescorla, R. A. (1988). Behavioral studies of Pavlovian conditioning. *Annual Review of Neuroscience, 11*(1), 329–352. https://doi.org/10.1146/annurev.neuro.11.1.329.

Reynolds, J. N. J., & Wickens, J. R. (2002). Dopamine-dependent plasticity of corticostriatal synapses. *Neural Networks, 15*(4–6), 507–521.

Rigas, I., Komogortsev, O., & Shadmehr, R. (2016). Biometric recognition via eye movements: Saccadic vigor and acceleration cues. *ACM Transactions on Applied Perception (TAP), 13*(2), 6.

Roesch, M. R., Stalnaker, T. A., & Schoenbaum, G. (2007). Associative encoding in anterior piriform cortex versus orbitofrontal cortex during odor discrimination and reversal learning. *Cerebral Cortex (New York, N.Y. : 1991), 17*(3), 643–652. https://doi.org/10.1093/cercor/bhk009.

Sakai, Y., & Fukai, T. (2008a). The Actor-Critic Learning Is Behind the Matching Law: Matching Versus Optimal Behaviors. *Neural Computation, 20*(1), 227–251. https://doi.org/10.1162/neco.2008.20.1.227.

Sakai, Y., & Fukai, T. (2008b). When does reward maximization lead to matching law? *PLoS ONE, 3*(11), e3795. https://doi.org/10.1371/journal.pone.0003795.

Samejima, K., Ueda, Y., Doya, K., & Kimura, M. (2005). Representation of action-specific reward values in the striatum. *Science, 310*(5752), 1337–1340. https://doi.org/10.1126/science.1115270.

Sanfey, A. G., Loewenstein, G., McClure, S. M., & Cohen, J. D. (2006). Neuroeconomics: Cross-currents in research on decision-making. *Trends in Cognitive Sciences, 10*(3), 108–116. https://doi.org/10.1016/j.tics.2006.01.009.

Savastano, H. I., & Fantino, E. (1994). Human choice in concurrent ratio-interval schedules of reinforcement. *Journal of the Experimental Analysis of Behavior, 61*(3), 453–463. https://doi.org/10.1901/jeab.1994.61-453.

Schultz, W., Dayan, P., & Montague, P. R. (1997). A neural substrate of prediction and reward. *Science, 275*(5306), 1593–1599.

Selinger, J. C., O'Connor, S. M., Wong, J. D., & Donelan, J. M. (2015). Humans can continuously optimize energetic cost during walking. *Current Biology, 25*(18), 2452–2456.

Seo, H., & Lee, D. (2007). Temporal filtering of reward signals in the dorsal anterior cingulate cortex during a mixed-strategy game. *The Journal of Neuroscience: The Official Journal of the Society for Neuroscience, 27*(31), 8366–8377. https://doi.org/10.1523/JNEUROSCI.2369-07.2007.

Seo, M., Lee, E., & Averbeck, B. B. (2012). Action selection and action value in frontal-striatal circuits. *Neuron, 74*(5), 947–960. https://doi.org/10.1016/j.neuron.2012.03.037.

Shadmehr, R., Reppert, T. R., Summerside, E. M., Yoon, T., & Ahmed, A. A. (2019). Movement vigor as a reflection of subjective economic utility. *Trends in Neurosciences*, *42*.

Shanks, D. R., Tunney, R. J., & McCarthy, J. D. (2002). A re-examination of probability matching and rational choice. *Journal of Behavioral Decision Making*, *15*(3), 233–250. https://doi.org/10.1002/bdm.413.

Shipp, S. (2017). The functional logic of corticostriatal connections. *Brain Structure and Function*, *222*(2), 669–706. https://doi.org/10.1007/s00429-016-1250-9.

Simon, N. W., Wood, J., & Moghaddam, B. (2015). Action-outcome relationships are represented differently by medial prefrontal and orbitofrontal cortex neurons during action execution. *Journal of Neurophysiology*, *114*(6), 3374–3385. https://doi.org/10.1152/jn.00884.2015.

Steinberg, E. E., Keiflin, R., Boivin, J. R., Witten, I. B., Deisseroth, K., & Janak, P. H. (2013). A causal link between prediction errors, dopamine neurons and learning. *Nature Neuroscience*, *16*(7), 966.

Stelmach, G. E., & Worringham, C. J. (1988). The preparation and production of isometric force in Parkinson's disease. *Neuropsychologia*, *26*(1), 93–103.

Sugrue, L. P., Corrado, G. S., & Newsome, W. T. (2004). Matching behavior and the representation of value in the parietal cortex. *Science*, *304*(5678), 1782–1787. https://doi.org/10.1126/science.1094765.

Sugrue, L. P., Corrado, G. S., & Newsome, W. T. (2005). Choosing the greater of two goods: Neural currencies for valuation and decision making. *Nature Reviews Neuroscience*, *6*(5), 363–375. https://doi.org/10.1038/nrn1666.

Sul, J. H., Jo, S., Lee, D., & Jung, M. W. (2011). Role of rodent secondary motor cortex in value-based action selection. *Nature Neuroscience*, *14*(9), 1202–1208. https://doi.org/10.1038/nn.2881.

Sul, J. H., Kim, H., Huh, N., Lee, D., & Jung, M. W. (2010). Distinct roles of rodent orbitofrontal and medial prefrontal cortex in decision making. *Neuron*, *66*(3), 449–460. https://doi.org/10.1016/j.neuron.2010.03.033.

Summerside, E. M., Shadmehr, R., & Ahmed, A. A. (2018). Vigor of reaching movements: Reward discounts the cost of effort. *Journal of Neurophysiology*, *119*(6), 2347–2357.

Sutton, R. S., & Barto, A. G. (1998). *Reinforcement learning: An introduction*. Cambridge: MIT Press.

Tai, L.-H., Lee, A. M., Benavidez, N., Bonci, A., & Wilbrecht, L. (2012). Transient stimulation of distinct subpopulations of striatal neurons mimics changes in action value. *Nature Neuroscience*, *15*(9), 1281–1289. https://doi.org/10.1038/nn.3188.

Tsutsui, K.-I., Grabenhorst, F., Kobayashi, S., & Schultz, W. (2016). A dynamic code for economic object valuation in prefrontal cortex neurons. *Nature Communications*, *7*, 12554. https://doi.org/10.1038/ncomms12554.

Tunney, R. J., & Shanks, D. R. (2002). A re-examination of melioration and rational choice. *Journal of Behavioral Decision Making*, *15*(4), 291–311.

Ueda, Y., Yamanaka, K., Noritake, A., Enomoto, K., Matsumoto, N., Yamada, H., … Kimura, M. (2017). Distinct functions of the primate putamen direct and indirect pathways in adaptive outcome-based action selection. *Frontiers in Neuroanatomy*, *11*, 66. https://doi.org/10.3389/fnana.2017.00066.

Uylings, H. B., Groenewegen, H. J., & Kolb, B. (2003). Do rats have a prefrontal cortex? *Behavioural Brain Research*, *146*(1–2), 3–17.

Vyse, S. A., & Belke, T. W. (1992). Maximizing versus matching on concurrent variable-interval schedules. *Journal of the Experimental Analysis of Behavior*, *58*(2), 325–334. https://doi.org/10.1901/jeab.1992.58-325.

Wallace, M. L., Saunders, A., Huang, K. W., Philson, A. C., Goldman, M., Macosko, E. Z., … Sabatini, B. L. (2017). Genetically distinct parallel pathways in the entopeduncular nucleus for limbic and sensorimotor output of the basal ganglia. *Neuron*, *94*(1), 138–152.

Wang, A. Y., Miura, K., & Uchida, N. (2013). The dorsomedial striatum encodes net expected return, critical for energizing performance vigor. *Nature Neuroscience*, *16*(5), 639–647. https://doi.org/10.1038/nn.3377.

Wang, J. X., Kurth-Nelson, Z., Kumaran, D., Tirumala, D., Soyer, H., Leibo, J. Z., … Botvinick, M. (2018). Prefrontal cortex as a meta-reinforcement learning system. *Nature Neuroscience*, *21*, 860–868. https://doi.org/10.1038/s41593-018-0147-8.

Watkins, C. J. C. H., & Dayan, P. (1992). Q-learning. *Machine Learning*, *8*(3–4), 279–292.

Williams, B. A. (1985). Choice behavior in a discrete-trial concurrent VI-VR: A test of maximizing theories of matching. *Learning and Motivation*, *16*(4), 423–443. https://doi.org/10.1016/0023-9690(85)90025-6.

Yoon, T., Geary, R. B., Ahmed, A. A., & Shadmehr, R. (2018). Control of movement vigor and decision making during foraging. *Proceedings of the National Academy of Sciences of the United States of America*, *115*(44), E10476–E10485. https://doi.org/10.1073/pnas.1812979115.

Yoshida, K., Drew, M. R., Mimura, M., & Tanaka, K. F. (2019). Serotonin-mediated inhibition of ventral hippocampus is required for sustained goal-directed behavior. *Nature Neuroscience*, 1.

Yu, A. J. (2014). Decision-making tasks. *Encyclopedia of computational neuroscience*, *2*, 931–937.

Yu, A. J., & Cohen, J. D. (2009). Sequential effects: Superstition or rational behavior? *Advances in Neural Information Processing Systems*, *21*, 1873–1880.

Yu, A. J., & Dayan, P. (2003). Expected and unexpected uncertainty: ACh and NE in the neocortex. In *Advances in neural information processing systems 15: Proceedings of the 2002 conference* (pp. 157–164).

Yu, A. J., & Dayan, P. (2005). Uncertainty, neuromodulation, and attention. *Neuron*, *46*(4), 681–692. https://doi.org/10.1016/j.neuron.2005.04.026.

Yu, A. J., & Huang, H. (2014). Maximizing masquerading as matching in human visual search choice behavior. *Decision*, *1*(4), 275–287. https://doi.org/10.1037/dec0000013.

CHAPTER FIVE

Reward signaling by the rodent medial frontal cortex

Mark Laubach[a,*], Linda M. Amarante[b], Marcelo S. Caetano[c], and Nicole K. Horst[d]

[a]Department of Neuroscience, American University, Washington, DC, United States
[b]Department of Neuroscience, Johns Hopkins University, Baltimore, MD, United States
[c]Center for Mathematics, Computing, and Cognition, Universidade Federal do ABC (UFABC), Santo André, Brazil
[d]Centre for Teaching and Learning and Postdoc Academy, University of Cambridge, Cambridge, England
*Corresponding author: e-mail address: mark.laubach@american.edu

Contents

Abstract

The anatomical relevance and functional significance of medial parts of the rodent frontal cortex have been intensely debated over the modern history of neuroscience. Early studies emphasized common functions among medial frontal regions in rodents and the dorsolateral prefrontal cortex of primates. Behavioral tasks emphasized memory-guided performance and persistent neural activity as a marker of working memory. Over time, it became clear that long-standing concerns about cross-species homology were justified and the view emerged that rodents are useful for understanding medial parts of the frontal cortex in primates, and not the dorsolateral prefrontal cortex. Here, we summarize a series of studies on the rodent medial frontal cortex that began with an interest in studying working memory in the perigenual prelimbic area and ended up studying reward processing in the medial orbital region. Our experiments revealed a role for a 4–8 Hz "theta" rhythm in tracking engagement in the consumption of rewarding fluids and denoting the value of a given reward. Evidence for a functional

International Review of Neurobiology, Volume 158
ISSN 0074-7742
https://doi.org/10.1016/bs.irn.2020.11.012

differentiation between the rostral and caudal medial frontal cortex and its relationship to other frontal cortical areas is also discussed with the hope of motivating future work on this part of the cerebral cortex.

1. What is the rodent medial frontal cortex?

The anatomical relevance and functional significance of medial parts of the rodent frontal cortex have been intensely debated over the modern history of neuroscience (Carlén, 2017; Laubach, Amarante, Swanson, & White, 2018). This part of the cerebral cortex runs along the medial wall of the cortex, between the cortical surface and the corpus callosum caudally and the rhinal sulcus rostrally. The region between the genu of the corpus callosum and the rhinal sulcus contains three distinct cortical regions, the anterior cingulate, prelimbic, and infralimbic cortices. The areas from the rhinal sulcus to the frontal edge of the brain were at one time denoted as the prelimbic and medial orbital areas, and have also been referred to as medial orbital cortex (Fuchs, Evans, Parker, & See, 2004; Izquierdo, 2017). The locations of these areas, in rats, are denoted in recent editions of the popular Paxinos brain atlas (Paxinos & Watson, 2014), and the criteria for each region and the terms used to denote them (e.g., Cg3, PL, PrL, area 32) are described at length in several publications (van Heukelum et al., 2020; Vogt et al., 2013; Vogt & Paxinos, 2014). (For an alternative perspective, based on functional imaging methods, see Balsters, Zerbi, Sallet, Wenderoth, and Mars (2020) and Schaeffer et al. (2020).) Comparable areas with similar cyto- and chemo-architecture and connectivity exist in primates (Preuss, 1995), and have commonly been referred to as medial frontal cortex (MFC) or medial prefrontal cortex (mPFC) in the recent literature. While there has been considerable confusion and debate about what to call these regions and how they relate to each other across species (Laubach et al., 2018), the functional significance of the medial frontal cortex has been just as confusing, especially in the early 2000s when our lab started to work on this part of the brain. Our journey to understand the MFC, at least with regards to reward processing is described below.

2. Dorsolateral prefrontal cortex, working memory, and persistent neural activity

Early neuroscience studies on the frontal cortex targeted a dorsolateral region, located around the principal sulcus, where lesions suggested a role for this area in spatial working memory (Goldman, Rosvold, Vest, & Galkin, 1971). Some of the earliest single unit recording studies supported this idea by the discovery of persistent activity during the delay periods that are common in working memory tasks (Fuster & Alexander, 1971; Kubota & Niki, 1971). These early findings led to the view that persistent activity in the dlPFC supports working memory function (Goldman-Rakic, 1995); however, a number of competing views emerged (Lebedev, Messinger, Kralik, & Wise, 2004; Miller & Cohen, 2001; Miller, Erickson, & Desimone, 1996; Quintana & Fuster, 1992; Rowe, Toni, Josephs, Frackowiak, & Passingham, 2000). In recent years, it has been established that items can be encoded in working memory without persistent activity (Wasmuht, Spaak, Buschman, Miller, & Stokes, 2018), and a view emerged that rhythmic activity, beyond spiking, is crucial for working memory and any other cognitive process that depends on the dlPFC (Miller, Lundqvist, & Bastos, 2018).

The early primate work on the dlPFC inspired research on the frontal cortex of rodents. Rats were almost exclusively used in these earliest studies on this part of the brain (Laubach et al., 2018). Given the predominance of the working memory paradigm in the 1990s, most early studies examined the role of the rat frontal cortex in spatial memory using delayed alternation tasks and the radial arm maze. (See (Brito, Yamasaki, Paumgartten, & Brito, 1987) for critical commentary on the use of delayed alternation for the study of working memory.) Some studies argued for a role of the rodent MFC in spatial memory (e.g., Brito, Thomas, Davis, & Gingold, 1982; Dunnett, Nathwani, & Brasted, 1999; van Haaren et al., 1988), but others disputed this idea (Delatour & Gisquet-Verrier, 1996, 1999; Delatour & Gisquet-Verrier, 2001; Gisquet-Verrier & Delatour, 2006; Horst & Laubach, 2009). Single unit recordings from rats performing memory tasks emerged in the late 1990s. One study found no evidence for an encoding of working memory information in single units from the MFC (Jung, Qin, McNaughton, & Barnes, 1998), specially from the caudal MFC as described above. A subsequent machine learning analysis of groups of simultaneously recorded neurons suggested that populations of neurons could encode spatial information over delay periods (Baeg et al., 2003). However, it was unclear how such population activity could emerge in the absence of coding at the single neuron level, and several additional studies emerged arguing that the apparent spatial memory encoding reflects behavioral variability over delay periods (Chudasama & Muir, 1997; Cowen & McNaughton, 2007; Euston & McNaughton, 2006).

These studies, together with others that found signals in common with primate medial PFC activity within other kinds of tasks (Narayanan & Laubach, 2006; Niki & Watanabe, 1979), led to a shift in the community away from working memory coding and towards consideration of other functions associated with MFC processing in primates including humans, such as cognitive control and performance monitoring (Ridderinkhof, Ullsperger, Crone, & Nieuwenhuis, 2004). For many researchers, it became clear that long-standing concerns about cross-species homology (Preuss, 1995) were justified and most researchers stopped referring to the rodent MFC as the "rodent homolog of the dlPFC."

3. Medial frontal cortex, performance monitoring, and persistent neural activity

MFC research in humans was largely driven by a debate among cognitive neuroscientists in the 1990s about whether the anterior cingulate cortex (ACC) is involved in processing errors or task difficulty. Early EEG recordings and fMRI led to the view that the ACC serves for monitoring errors in the performance of cognitive tasks (Gehring, Goss, Coles, Meyer, & Donchin, 1993). A theory emerged based on early fMRI that instead suggested that the ACC was crucial for monitoring conflict about what to do when a task is difficult and not processing outcomes per se (Carter et al., 1998). Single unit recordings in the ACC did not support this idea and instead found the initial evidence for the encoding of behavioral outcomes by single neurons in the ACC (Ito, Stuphorn, Brown, & Schall, 2003). (This same group found evidence in support of the conflict monitoring theory in an adjacent medial frontal region, the supplementary motor area (Stuphorn, Taylor, & Schall, 2000).) Over time, the debate about what the MFC, specifically the ACC, does has lingered within the human research community (Kolling et al., 2016; Shenhav, Cohen, & Botvinick, 2016). Animal research groups have largely focused on the role of MFC in performance monitoring and adjustment, following the findings that MFC neurons also track the need to adjust performance in tasks that have changing action-outcome contingencies (Shima & Tanji, 1998), encode the reward expected on a given trial (Roesch & Olson, 2004), show error-related activity that depends on reward predictability (Amiez, Joseph, & Procyk, 2005), and emit rhythmic activity in the 4–8 Hz range (theta) when animals must adapt to changes in behavioral tasks (Womelsdorf, Johnston, Vinck, & Everling, 2010).

Our lab started to use reversible inactivation and multi-electrode recording methods to study the rodent MFC in the early 2000s, as the debate over conflict monitoring and performance monitoring and adjustment was raging. The lab started out with an interest in resolving how the rodent motor cortex becomes increasingly aware of behavioral outcomes as rats learn to perform a simple reaction time task (Laubach, Wessberg, & Nicolelis, 2000). Following a lesion study by another group (Risterucci, Terramorsi, Nieoullon, & Amalric, 2003), we set out to record in the perigenual MFC to determine if cells there would be active during delay periods and track the animal's successes and failures in performing the simple reaction time task. We also wanted to know if MFC activity varied with the animal's temporal expectancies, based on prevailing psychological theories about timing and reaction times, e.g. (Los, Knol, & Boers, 2001) and classic single unit recordings in the monkey ACC in a similar task (Niki & Watanabe, 1979). Our initial recordings found cells in the MFC that were quite distinct from the motor cortex (Narayanan & Laubach, 2006). They became active when trials were initiated, fired persistently over the delay period, and/or fired during the reaction time period. A subset of neurons showed what we later came to call prospective or retrospective outcome encoding (Laubach, Caetano, & Narayanan, 2015), in which the cells fired persistently and at distinct rates before (prospective) or after (retrospective) animals responded correctly or incorrectly. These activity patterns were associated with post-error slowing (Narayanan & Laubach, 2008), a classic form of performance adjustment in the simple reaction time task. Our findings, together with reversible inactivation experiments (Narayanan, Horst, & Laubach, 2006; Narayanan & Laubach, 2006, 2008; Smith, Horst, Liu, Caetano, & Laubach, 2010) led to the view that the rat MFC, similar to the monkey MFC, has a crucial role in performance monitoring and adjustment (Laubach, 2011).

Computational analysis was then carried out to understand the dynamics of performance monitoring and adjustment by the MFC. Multivariate statistical analysis (PCA) was used to understand coordinated changes in MFC firing rates, and led to the idea that persistent activity in MFC during the simple reaction time task tracks the animal's instrumental action and terminates when the action is completed on time and leads to a reward (Narayanan & Laubach, 2009). By contrast on error trials, the lack of the rewarding stimulus resulted in a subset of the neuronal population remaining in an elevated or reduced level of activity. The computational significance of these activity patterns was then evaluated using spiking network models, in collaboration with Chris Eliasmith and Trevor Bekolay. This led to the idea that reward feedback alters the flow of neuronal activity along a neural manifold such that neurons experience unique synaptic states following errors and thus could learn from unexpected outcomes (errors) (Bekolay, Laubach, & Eliasmith, 2014; Laubach et al., 2015).

A major test of the cross-species significance of these concepts was then carried out in collaboration with Michael Frank and Jim Cavanagh (Narayanan, Cavanagh, Frank, & Laubach, 2013). They ran human studies, including with EEG recordings, using the same simple reaction time tasks used in our lab. They found markers of correct and error trials in human EEG that were similar to our LFP recordings in rats, and these markers were notably associated with rhythmic activity in the theta range (4–8 Hz). This same theta frequency had been suggested as a key signal for cognitive processing in earlier studies by Dr. Cavanagh (Cavanagh, Cohen, & Allen, 2009; Cavanagh, Zambrano-Vazquez, & Allen, 2012), and our collaboration was the first cross-species demonstration of MFC theta with a common functional interpretation in rats and humans. (See Hyman, Holroyd, & Seamans, 2017 for another demonstration of these cross-species signals.) That is, the MFC generates theta signals to synchronize brain activity to guide performance and is a key neural component of the ability to adapt to unexpected outcomes.

4. From working memory to reward processing

During this same period of time, our lab was also studying how neural ensembles in the MFC are involved in working memory. At the time, working memory, as measured via classic working memory paradigms, was still considered to be a function of rodent MFC, and we were studying how neurons in the rat MFC might track spatial choices and maintain spatial information over delay periods. We developed an operant version of the classic T-maze based delayed spatial alternation task and showed that performance depended on processing in the rat MFC (Caetano, Jin, et al., 2012; Horst & Laubach, 2009). The task was then used for multi-electrode recordings. In two separate studies, we found abundant evidence for spatial coding by MFC neurons and many neurons that fired persistently over the delay period (Caetano, Horst, et al., 2012; Horst & Laubach, 2012). However, none of the neurons with persistent activity fired selectively to signal past or future spatial choices during the delay period.

A key aspect of this study was the manipulation of reward delivery duration to create a delay period in between spatial responses. Rats learned to maintain licking behavior at the reward port with their backs to the spatial response ports until, after an unpredictable amount of time, a cue indicated that it was time to perform the next trial. The reward port was placed within a narrow alleyway and behavior was tracked using multiple cameras. The idea behind this design was to avoid issues with behavioral variability (Chudasama & Muir, 1997; Cowen & McNaughton, 2007; Euston & McNaughton, 2006). When the timing of the animals' licks was plotted against the timing of persistent firing by the MFC neurons, it became clear that some of the cells were firing in direct proportion to the rats' licking behavior. In fact, further analysis revealed that the spikes were temporally aligned with the licks and the onset of the rewarding fluid triggered a major change in MFC theta (Horst & Laubach, 2012, 2013). Rather than reflecting spatial information, MFC spike and field activity seemed to be related to the rewarding fluid. This was a major unexpected result at the time. In addition, we also observed subsets of MFC neurons that fired in a similar way to the neurons reported in our reaction time studies (Narayanan & Laubach, 2006). They tracked prospective and retrospective information about the behavioral outcomes (Horst & Laubach, 2012). Based on these studies, we set out on a new direction of research to understand how reward consumption engages the rodent MFC.

5. Theta range activity is synchronized to reward consumption

As a result of the findings described above, we decided to use a simple behavioral task to assess how licking behavior and fluid delivery influence neural activity in the rat MFC (Horst & Laubach, 2013). A variable interval procedure was used, in which rats behaved in the same behavioral chamber as in the working memory studies. Now they simply had to lick on a spout to receive fluid, and the first lick after a randomly chosen interval expired would activate a syringe pump and deliver a liquid reward. The rats were maintained with regulated access to fluid, and received water as a reward in the task. There were three new findings from this study. First, engaging in persistent licking activated many MFC neurons. The neurons were especially driven when rats licked over sustained bouts, and the depth of modulation around the beginnings and ends of the licking bouts was as strong as any modulation associated with task events in other behavioral studies in our lab. Second, periods with sustained licking were associated with elevated theta-band activity in MFC LFPs, and the LFPs were phase locked to the rats' lick cycle. Third, the majority of recording sites with licking-modulated activity were localized to what was, at the time, called the rostral part of the prelimbic cortex (figure 8 in Horst & Laubach, 2013). This cortical region is now commonly referred to as medial orbital cortex, aka medial orbitofrontal cortex or mOFC (Izquierdo, 2017). It is heavily interconnected with the gustatory insular cortex (Gabbott, Warner, Jays, & Bacon, 2003) and projects to subcortical feeding-related centers (Gabbott, Warner, Jays, Salway, & Busby, 2005), as described below. Together, these findings led us to propose that reward-related activity in MFC is driven by consummatory behavior.

6. Theta range activity tracks the value of consummatory actions

At this point in our research, we implemented behavioral procedures that have been commonly used to assess reward value using liquid sucrose rewards. We developed an operant version of a "simultaneous contrast" procedure using liquid sucrose rewards (Parent, Amarante, Liu, Weikum, & Laubach, 2015), based on classic studies by Flaherty (1982) and others. The procedure was designed to provide access to rats of two levels of liquid sucrose, a relatively high concentration (20% or 16%, weight by volume)—called the high value option—and a relatively low concentration (4%)—called the low value option. Rats will readily consume both levels of sucrose when offered either option alone in a test session. The two solutions are delivered from a common reward port, and the only behavior in the procedure is to stand at the port and lick for the fluids. The two fluids are presented in alternating order for periods of 30s. Over days, rats learn to lick persistently when the high value option is available and to suppress licking when the low value option is available. Importantly, they come to lick more for the high value option and less for the low value option when the options are presented in alternation compared to when only one level of reward is available. This behavior is typical of incentive contrast procedures. We eventually came to refer to this task as the "Shifting Values Licking Task" or SVLT (Amarante, Caetano, & Laubach, 2017).

We used reversible inactivation methods to determine if specific MFC regions are crucial for the performance of the SVLT. (Parent, Amarante, Liu, et al., 2015). Based on our studies in the reaction time task (Narayanan et al., 2006), which suggested a role for the MFC in behavioral inhibition (specifically, the perigenual prelimbic area), we expected to find reductions in persistent licking and over-consumption of the lower value option following inactivation of the MFC. While we did find that reversible inactivation of the MFC (using muscimol and archaerhodopsin, in separate experiments) led to dramatic reductions in persistent licking, we were surprised to find reduced licking for the higher, not the lower, value option. These effects were especially apparent with inactivation of the rostral MFC, in the medial orbital area and adjacent frontal agranular cortex anterior to the rhinal sulcus (figures 3 and 6 in Parent, Amarante, Liu, et al., 2015). This was the same rostral location that showed strong spike and LFP synchronization to licking in the Horst and Laubach (2013) study (see figure 8 from that paper). Rather than affecting impulsivity, which should have increased licking for the lower value option, reversible inactivation of the MFC in the SVLT revealed effects on what might best be called the expression of "incentive contrast," the increased rate of consumption of the higher value option when it is available in alternating sequence with a lower value reward. That is, the MFC is crucial for the expression of learned value-guided behavioral strategies.

Following on from this work, multi-electrode recordings were then made in the rat MFC during the SVLT (Amarante et al., 2017). The recordings focused on the same rostral part of the MFC as in the study by Horst and Laubach (2013) (see figure 2 in Amarante et al. (2017)). The recordings revealed that same kind of lick entrained rhythmic activity in the SVLT. The strength of phase locking, measured with inter-trial phase coherence, varied with the value of the currently available reward. That is, phase locking was strongest when the rats infested the higher value liquid sucrose compared to when they consumed the lower value fluid. Phase locking developed with experience and in proportion to the increase in licking for the higher value option. Therefore, the signals seem to track the animals' learning about the incentive properties (aka subjective value) of the sucrose rewards in the SVLT. Importantly, by reversibly inactivating the MFC in one hemisphere and recording neural activity in the other, we were able to show that the effects of incentive value on lick-entrained neural activity depended on cells in the MFC.

Most recently, neural recordings were made in the same rostral part of the MFC and simultaneously in lateral frontal regions that are densely interconnected with the MFC, i.e., lateral orbital and agranular insular cortices (Gabbott et al., 2003). Paired recordings showed common signaling of reward value in the SVLT (Amarante & Laubach, 2020), and extended earlier results by using a novel open-source syringe pump that delivered different volumes of fluid over common periods of time (Amarante, Newport, Mitchell, Wilson, & Laubach, 2019). The study showed evidence for a common signaling of differences in sucrose concentration and fluid volume, suggesting that the lick-entrained theta signaling described above is not specific to processing differences in taste (sucrose concentration) and instead might track reward magnitude in a more general manner. By using several variations on the design of the SVLT it was possible to determine that the signals in the MFC, but not the OFC, vary with response vigor (licking rate) and shifts in absolute, but not relative, reward value (Amarante & Laubach, 2020). New studies are ongoing in our lab to further investigate these issues and extend the findings to the period of consumption in decision making tasks.

The mechanisms of these rhythmic signals have remained unclear. Understanding the neurochemical basis of MFC theta activity might lead to new therapeutics for treating a variety of neurological and psychiatric illnesses associated with MFC function. One clue to understanding the mechanisms of the rhythms is available from another study from our lab (Parent, Amarante, Swanson, & Laubach, 2015). It examined the role of a class of voltage-dependent potassium channels (KCNQ, aka Kv7) on licking behavior and incentive contrast in the SVLT. This class of ion channels has been shown to be crucial for neuronal rhythms in the frequency range of licking and the theta rhythm (Guan, Higgs, Horton, Spain, & Foehring, 2011; Pafundo, Miyamae, Lewis, & Gonzalez-Burgos, 2013). The channels are regulated by the cholinergic system, and mediate the so-called M (muscarinic) current (Brown & Passmore, 2009) and also by the hunger-related peptide grehlin (Shi et al., 2013). Intra-cortical infusions of drugs that selectively act on the ion channels and receptors for the two transmitter systems had dramatic effects on behavior in the SVLT (Parent, Amarante, Swanson, et al., 2015). Similar to inactivation of the MFC, the drugs had specific effects on consumption of the higher, but not the lower, value liquid sucrose rewards. The neuronal basis of these effects warrants future investigation, and might be the key to understanding how reward value modulates MFC function in general. For example, another transmitter system, the mu opioid system, has been shown to directly modulate value-guided consumption in the MFC (Mena, Sadeghian, & Baldo, 2011), and we are currently working on understanding how this transmitter system impacts behavior and neural activity in the SVLT.

7. Anatomical basis for the rostral-caudal gradient of reward signaling

Anatomical studies over the years have consistently revealed that the MFC is not a monolithic cortical region. Clear differences exist between the rostral medial orbital and caudal prelimbic areas, and for the most part these differences have not been emphasized in the behavioral and electrophysiological literature. Evidence in support of functional differences between the rostral and caudal halves of the MFC, specifically with regards to consummatory behavior, was reported in several studies described in this chapter (Horst & Laubach, 2012, 2013; Parent, Amarante, Liu, et al., 2015; Amarante et al., 2017), and in another recent study that used reversible inactivation methods to study the role of the medial orbital cortex in responding in a progressive ratio licking task (Swanson, Goldbach, & Laubach, 2019). Fig. 1 summarizes differences in connectivity along the rostral to caudal axis of the MFC. These anatomical connections might mediate different behavioral functions, with the rostral MFC participating in a "feed and breed" circuit that is distinct from a more caudal "fight-or-flight" circuit linked to the caudal MFC. Curiously, stimulation of neurons in the neighboring agranular cortex leads to movements by different body parts, with the rostral agranular cortex producing oral and facial movements and the caudal cortex producing neck and shoulder movements (Neafsey et al., 1986). Functionally, as these cortical networks share local intercortical connections with the adjacent MFC, they could mediate switching between states of foraging and vigilance, a phenomenon that is well known in the field of behavioral ecology

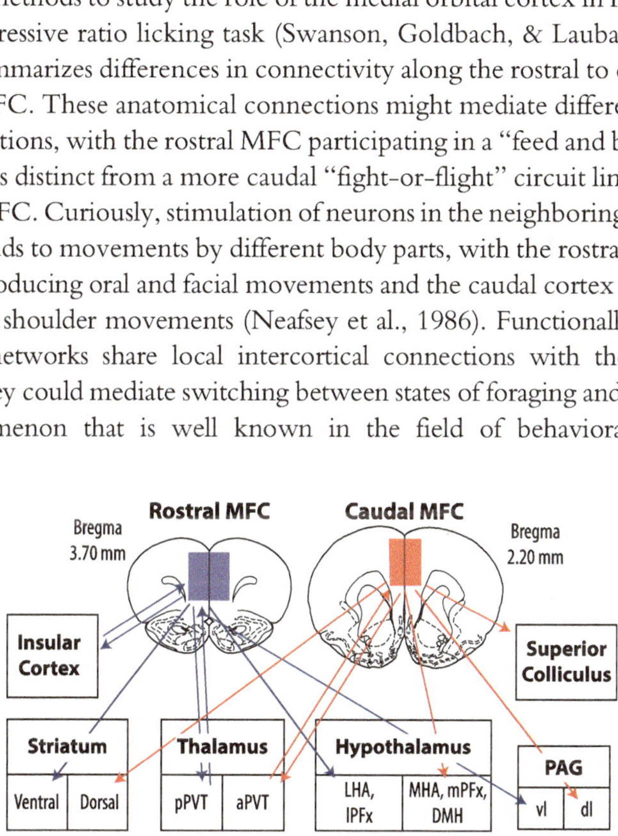

Fig. 1 Divergent connections of the rostral and caudal MFC. Anterograde tract-tracing studies suggest that the rostral and caudal MFC send diverging connections to a number of brain structures involved in reward and homeostasis (Floyd, Price, Ferry, Keay, & Bandler, 2000, 2001; Gabbott et al., 2003, 2005; Li & Kirouac, 2012). The rostral MFC projects to "feed-or-breed" sites in the thalamus (Li & Kirouac, 2012), hypothalamus (Floyd et al., 2001) and periacqueductal gray, or PAG (Floyd et al., 2000), as well as the gustatory insular cortex (Gabbott et al., 2003). By contrast, the caudal MFC projects to "fight-or-flight" sites in these same areas (Floyd et al., 2000, 2001; Li & Kirouac, 2012) and also the superior colliculus (Gabbott et al., 2005). Abbreviations: PVT = paraventricular nucleus of the thalamus, posterior and anterior; LHA = lateral hypothalamus; MHA = medial hypothalamus; DMH = dosomedial hypothalamus; Pfx = perifornical area, medial and lateral; vlPAG and dlPAG = ventrolateral and dorsolateral PAG.

(Beauchamp, 2015). With the recent increase in interest in the MFC by researchers studying animal models of fear and anxiety (Roberts, 2020), it is likely that these issues will become more apparent and eventually resolved in the literature. We would like to suggest that researchers keep in mind the well established role of medial frontal theta (e.g., Narayanan et al., 2013) and expectancy signals (e.g., Hayden, Heilbronner, Pearson, & Platt, 2011; Hyman et al., 2017) in future studies.

In between the well studied MFC and OFC regions is an area described commonly as frontal agranular cortex (Fig. 2). The region has received considerable attention in mice (Barthas & Kwan, 2017). The most rostral part of this region contains neurons that have been shown to modulate orolingual actions (Neafsey et al., 1986; Terreberry & Neafsey, 1983; Yoshida et al., 2009) and to receive orolingual input (Haque et al., 2010). The dorsal part of this intermediate cortical region might be homologous to a frontal motor region studied extensively in mice, and referred to as the anterior lateral motor cortex or ALM (Guo et al., 2017). The most ventral portion of the intermediate cortex is granular and borders the lateral OFC regions that have been studied by many groups (e.g., Gutierrez, Carmena, Nicolelis, & Simon, 2006). This cortical region receives input from the lingual nerve (innervating the tongue) by way of the rostrodorsomedial trigeminal nucleus and ventromedial and ventrolateral thalamus (Haque et al., 2010).

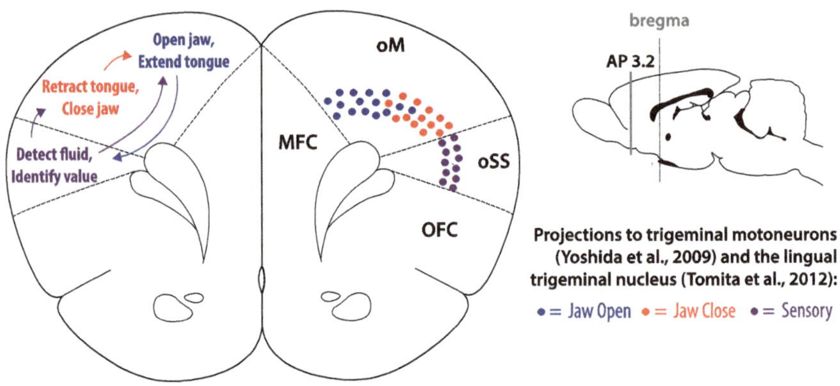

Fig. 2 Hierarchical processing of orolingual actions by the rodent frontal cortex. Two orolingual regions, referred to here as the oral motor [oM] and oral somatosensory [oSS] areas, are depicted in the left hemisphere. Anatomical data from two key studies (Tomita et al., 2012; Yoshida et al., 2009), which define the boundaries of these areas, are depicted in the right hemisphere. These cortical regions are between the two most commonly studied cortical regions in the rodent brain and yet they have received virtually no attention with regard to their roles in the control of feeding or reward processing.

It sends "top-down" projections back to the trigeminal sensory nucleus (Chang et al., 2009; Tomita et al., 2012) and also to the more dorsal oral motor area (Zhang & Sasamoto, 1990). The functional significance of these motor and sensory regions in processing reward information and their connectivity with the MFC and OFC have not been examined in detail. In Fig. 2, we propose a form of hierarchical processing among these areas that could enable monitoring and adjustments of orolingual actions during the process of reward consumption. Together with studies on how the regions mediate tradeoffs between foraging and vigilance described in the paragraph above, studies that examine functional interactions across the entire medial to lateral scope of the rodent frontal cortex seem warranted.

8. Summary and conclusions

The recent history of research on the rodent medial frontal cortex suggests that we are only beginning to understand its role in the control of goal-directed behavior. A lot more work will be needed to understand this interesting part of the brain. Studies reviewed here started from an ill posed classic view of the rodent frontal cortex, that it is similar to the dorsolateral prefrontal cortex of primates. The studies led to discoveries of performance monitoring and adjustment signals and the idea that reward signaling is driven by the actions of the animals as they consume rewarding fluids. As modern circuit methods are increasingly applied to the MFC (e.g., Ährlund-Richter et al., 2019; Otis et al., 2017), we are learning a lot more about how the wiring of the MFC may determine its ultimate role in the control of behavior. We hope to see cleverly designed behavioral studies combined with new tools for imaging and manipulating specific classes of MFC neurons emerge on this part of the brain in the coming years, including from our own lab.

Acknowledgments

We thank Drs Nandakumar Narayanan, Eyal Kimchi, Benjamin Yixing Xu, Linda Harenberg, James Cavanagh, Michael Frank, Chris Eliasmith, Trevor Bekolay, Marc Parent, Kim Stachenfeld, Kyra Swanson, and Wambura Fobbs and Benjamine Liu, Damian Weikum, and Samantha White for intellectual and experimental contributions to some of the research that is reviewed in this manuscript.

Conflict of interest

None.

Financial support

NIH DA046375 to ML and a NSF Graduate Research Fellowship to LMA.

References

Ährlund-Richter, S., Xuan, Y., van Lunteren, J. A., Kim, H., Ortiz, C., Pollak Dorocic, I., et al. (2019). A whole-brain atlas of monosynaptic input targeting four different cell types in the medial prefrontal cortex of the mouse. *Nature Neuroscience*, *22*, 657–668.

Amarante, L. M., Caetano, M. S., & Laubach, M. (2017). Medial frontal theta is entrained to rewarded actions. *The Journal of Neuroscience*, *37*, 10757–10769. https://doi.org/10.1523/JNEUROSCI.1965-17.2017. 17.

Amarante, L. M., & Laubach, M. (2020). Rhythmic activity in the medial and orbital frontal cortices tracks reward value and the vigor of consummatory behavior. *Biorxiv*. 2020.09.22.308809.

Amarante, L. M., Newport, J., Mitchell, M., Wilson, J., & Laubach, M. (2019). An open source syringe pump controller for fluid delivery of multiple volumes. *ENeuro*, *6*.

Amiez, C., Joseph, J.-P., & Procyk, E. (2005). Anterior cingulate error-related activity is modulated by predicted reward. *The European Journal of Neuroscience*, *21*, 3447–3452.

Baeg, E. H., Kim, Y. B., Huh, K., Mook-Jung, I., Kim, H. T., & Jung, M. W. (2003). Dynamics of population code for working memory in the prefrontal cortex. *Neuron*, *40*, 177–188.

Balsters, J. H., Zerbi, V., Sallet, J., Wenderoth, N., & Mars, R. B. (2020). Primate homologs of mouse cortico-striatal circuits. *eLife*, *9*, e53680. https://doi.org/10.7554/eLife.53680.

Barthas, F., & Kwan, A. C. (2017). Secondary motor cortex: Where "sensory" meets "motor" in the rodent frontal cortex. *Trends in Neurosciences*, *40*, 181–193.

Beauchamp, G. (2015). Overview of animal vigilance. In *Animal vigilance* (pp. 1–21). Elsevier. Available at: https://linkinghub.elsevier.com/retrieve/pii/B9780128019832000012. Accessed October 6, 2020.

Bekolay, T., Laubach, M., & Eliasmith, C. (2014). A spiking neural integrator model of the adaptive control of action by the medial prefrontal cortex. *Journal of Neuroscience: The Official Journal of the Society for Neuroscience*, *34*, 1892–1902.

Brito, G. N., Thomas, G. J., Davis, B. J., & Gingold, S. I. (1982). Prelimbic cortex, mediodorsal thalamus, septum, and delayed alternation in rats. *Experimental Brain Research*, *46*, 52–58.

Brito, L. S., Yamasaki, E. N., Paumgartten, F. J., & Brito, G. N. (1987). Continuous and discrete T-maze alternation in rats: Effects of intertrial and interrun intervals. *Brazilian Journal of Medical and Biological Research*, *20*, 125–135.

Brown, D. A., & Passmore, G. M. (2009). Neural KCNQ (Kv7) channels. *British Journal of Pharmacology*, *156*, 1185–1195.

Caetano, M. S., Horst, N. K., Harenberg, L., Liu, B., Arnsten, A. F. T., & Laubach, M. (2012). Lost in transition: Aging-related changes in executive control by the medial prefrontal cortex. *Journal of Neuroscience: The Official Journal of the Society for Neuroscience*, *32*, 3765–3777.

Caetano, M. S., Jin, L. E., Harenberg, L., Stachenfeld, K. L., Arnsten, A. F. T., & Laubach, M. (2012). Noradrenergic control of error perseveration in medial prefrontal cortex. *Frontiers in Integrative Neuroscience*, *6*, 125.

Carlén, M. (2017). What constitutes the prefrontal cortex? *Science*, *358*, 478–482.

Carter, C. S., Braver, T. S., Barch, D. M., Botvinick, M. M., Noll, D., & Cohen, J. D. (1998). Anterior cingulate cortex, error detection, and the online monitoring of performance. *Science*, *280*, 747–749.

Cavanagh, J. F., Cohen, M. X., & Allen, J. J. B. (2009). Prelude to and resolution of an error: EEG phase synchrony reveals cognitive control dynamics during action monitoring. *Journal of Neuroscience: The Official Journal of the Society for Neuroscience, 29*, 98–105.

Cavanagh, J. F., Zambrano-Vazquez, L., & Allen, J. J. B. (2012). Theta lingua franca: A common mid-frontal substrate for action monitoring processes. *Psychophysiology, 49*, 220–238.

Chang, Z., Haque, T., Iida, C., Seki, S., Sato, F., Kato, T., et al. (2009). Distribution of premotoneurons for jaw-closing and jaw-opening motor nucleus receiving contacts from axon terminals of primary somatosensory cortical neurons in rats. *Brain Research, 1275*, 43–53.

Chudasama, Y., & Muir, J. L. (1997). A behavioural analysis of the delayed non-matching to position task: The effects of scopolamine, lesions of the fornix and of the prelimbic region on mediating behaviours by rats. *Psychopharmacology, 134*, 73–82.

Cowen, S. L., & McNaughton, B. L. (2007). Selective delay activity in the medial prefrontal cortex of the rat: Contribution of sensorimotor information and contingency. *Journal of Neurophysiology, 98*, 303–316.

Delatour, B., & Gisquet-Verrier, P. (1996). Prelimbic cortex specific lesions disrupt delayed-variable response tasks in the rat. *Behavioral Neuroscience, 110*, 1282–1298.

Delatour, B., & Gisquet-Verrier, P. (1999). Lesions of the prelimbic-infralimbic cortices in rats do not disrupt response selection processes but induce delay-dependent deficits: Evidence for a role in working memory? *Behavioral Neuroscience, 113*, 941–955.

Delatour, B., & Gisquet-Verrier, P. (2001). Involvement of the dorsal anterior cingulate cortex in temporal behavioral sequencing: Subregional analysis of the medial prefrontal cortex in rat. *Behavioural Brain Research, 126*, 105–114.

Dunnett, S. B., Nathwani, F., & Brasted, P. J. (1999). Medial prefrontal and neostriatal lesions disrupt performance in an operant delayed alternation task in rats. *Behavioural Brain Research, 106*, 13–28.

Euston, D. R., & McNaughton, B. L. (2006). Apparent encoding of sequential context in rat medial prefrontal cortex is accounted for by behavioral variability. *Journal of Neuroscience: The Official Journal of the Society for Neuroscience, 26*, 13143–13155.

Flaherty, C. F. (1982). Incentive contrast: A review of behavioral changes following shifts in reward. *Animal Learning & Behavior, 10*, 409–440.

Floyd, N. S., Price, J. L., Ferry, A. T., Keay, K. A., & Bandler, R. (2000). Orbitomedial prefrontal cortical projections to distinct longitudinal columns of the periaqueductal gray in the rat. *The Journal of Comparative Neurology, 422*, 556–578.

Floyd, N. S., Price, J. L., Ferry, A. T., Keay, K. A., & Bandler, R. (2001). Orbitomedial prefrontal cortical projections to hypothalamus in the rat. *The Journal of Comparative Neurology, 432*, 307–328.

Fuchs, R. A., Evans, K. A., Parker, M. P., & See, R. E. (2004). Differential involvement of orbitofrontal cortex subregions in conditioned cue-induced and cocaine-primed reinstatement of cocaine seeking in rats. *Journal of Neuroscience: The Official Journal of the Society for Neuroscience, 24*, 6600–6610.

Fuster, J. M., & Alexander, G. E. (1971). Neuron activity related to short-term memory. *Science, 173*, 652–654.

Gabbott, P. L. A., Warner, T. A., Jays, P. R. L., & Bacon, S. J. (2003). Areal and synaptic interconnectivity of prelimbic (area 32), infralimbic (area 25) and insular cortices in the rat. *Brain Research, 993*, 59–71.

Gabbott, P. L. A., Warner, T. A., Jays, P. R. L., Salway, P., & Busby, S. J. (2005). Prefrontal cortex in the rat: Projections to subcortical autonomic, motor, and limbic centers. *The Journal of Comparative Neurology, 492*, 145–177.

Gehring, W. J., Goss, B., Coles, M. G. H., Meyer, D. E., & Donchin, E. (1993). A neural system for error detection and compensation. *Psychological Science, 4*, 385–390.

Gisquet-Verrier, P., & Delatour, B. (2006). The role of the rat prelimbic/infralimbic cortex in working memory: Not involved in the short-term maintenance but in monitoring and processing functions. *Neuroscience, 141*, 585–596.

Goldman, P. S., Rosvold, H. E., Vest, B., & Galkin, T. W. (1971). Analysis of the delayed-alternation deficit produced by dorsolateral prefrontal lesions in the rhesus monkey. *Journal of Comparative and Physiological Psychology*, 77, 212–220.

Goldman-Rakic, P. S. (1995). Cellular basis of working memory. *Neuron, 14*, 477–485.

Guan, D., Higgs, M. H., Horton, L. R., Spain, W. J., & Foehring, R. C. (2011). Contributions of Kv7-mediated potassium current to sub- and suprathreshold responses of rat layer II/III neocortical pyramidal neurons. *Journal of Neurophysiology, 106*, 1722–1733.

Guo, Z. V., Inagaki, H. K., Daie, K., Druckmann, S., Gerfen, C. R., & Svoboda, K. (2017). Maintenance of persistent activity in a frontal thalamocortical loop. *Nature, 545*, 181–186.

Gutierrez, R., Carmena, J. M., Nicolelis, M. A. L., & Simon, S. A. (2006). Orbitofrontal ensemble activity monitors licking and distinguishes among natural rewards. *Journal of Neurophysiology, 95*, 119–133.

Haque, T., Yamamoto, S., Masuda, Y., Kato, T., Sato, F., Uchino, K., et al. (2010). Thalamic afferent and efferent connectivity to cerebral cortical areas with direct projections to identified subgroups of trigeminal premotoneurons in the rat. *Brain Research, 1346*, 69–82.

Hayden, B. Y., Heilbronner, S. R., Pearson, J. M., & Platt, M. L. (2011). Surprise signals in anterior cingulate cortex: Neuronal encoding of unsigned reward prediction errors driving adjustment in behavior. *Journal of Neuroscience: The Official Journal of the Society for Neuroscience, 31*, 4178–4187.

Horst, N. K., & Laubach, M. (2009). The role of rat dorsomedial prefrontal cortex in spatial working memory. *Neuroscience, 164*, 444–456.

Horst, N. K., & Laubach, M. (2012). Working with memory: Evidence for a role for the medial prefrontal cortex in performance monitoring during spatial delayed alternation. *Journal of Neurophysiology, 108*, 3276–3288.

Horst, N. K., & Laubach, M. (2013). Reward-related activity in the medial prefrontal cortex is driven by consumption. *Frontiers in Neuroscience*, 7, 56.

Hyman, J. M., Holroyd, C. B., & Seamans, J. K. (2017). A novel neural prediction error found in anterior cingulate cortex ensembles. *Neuron, 95*, 447–456. e3.

Ito, S., Stuphorn, V., Brown, J. W., & Schall, J. D. (2003). Performance monitoring by the anterior cingulate cortex during saccade countermanding. *Science, 302*, 120–122.

Izquierdo, A. (2017). Functional heterogeneity within rat orbitofrontal cortex in reward learning and decision making. *The Journal of Neuroscience, 37*, 10529–10540.

Jung, M. W., Qin, Y., McNaughton, B. L., & Barnes, C. A. (1998). Firing characteristics of deep layer neurons in prefrontal cortex in rats performing spatial working memory tasks. *Cerebral cortex (New York, N.Y.: 1991), 8*, 437–450.

Kolling, N., Wittmann, M. K., Behrens, T. E. J., Boorman, E. D., Mars, R. B., & Rushworth, M. F. S. (2016). Value, search, persistence and model updating in anterior cingulate cortex. *Nature Neuroscience, 19*, 1280–1285.

Kubota, K., & Niki, H. (1971). Prefrontal cortical unit activity and delayed alternation performance in monkeys. *Journal of Neurophysiology, 34*, 337–347.

Laubach, M., Amarante, L. M., Swanson, K., & White, S. R. (2018). What, if anything, is rodent prefrontal cortex? *eNeuro, 5*.

Laubach, M., Caetano, M. S., & Narayanan, N. S. (2015). Mistakes were made: Neural mechanisms for the adaptive control of action initiation by the medial prefrontal cortex. *Journal of Physiology, Paris, 109*, 104–117.

Laubach, M., Wessberg, J., & Nicolelis, M. A. (2000). Cortical ensemble activity increasingly predicts behaviour outcomes during learning of a motor task. *Nature*, *405*, 567–571.
Lebedev, M. A., Messinger, A., Kralik, J. D., & Wise, S. P. (2004). Representation of attended versus remembered locations in prefrontal cortex. *PLoS Biology*, *2*, e365.
Li, S., & Kirouac, G. J. (2012). Sources of inputs to the anterior and posterior aspects of the paraventricular nucleus of the thalamus. *Brain Structure & Function*, *217*, 257–273.
Los, S. A., Knol, D. L., & Boers, R. M. (2001). The foreperiod effect revisited: Conditioning as a basis for nonspecific preparation. *Acta Psychologica*, *106*, 121–145.
Mena, J. D., Sadeghian, K., & Baldo, B. A. (2011). Induction of hyperphagia and carbohydrate intake by μ-opioid receptor stimulation in circumscribed regions of frontal cortex. *Journal of Neuroscience: The Official Journal of the Society for Neuroscience*, *31*, 3249–3260.
Miller, E. K., & Cohen, J. D. (2001). An integrative theory of prefrontal cortex function. *Annual Review of Neuroscience*, *24*, 167–202.
Miller, E. K., Erickson, C. A., & Desimone, R. (1996). Neural mechanisms of visual working memory in prefrontal cortex of the macaque. *The Journal of Neuroscience*, *16*, 5154–5167.
Miller, E. K., Lundqvist, M., & Bastos, A. M. (2018). Working memory 2.0. *Neuron*, *100*, 463–475.
Narayanan, N. S., Cavanagh, J. F., Frank, M. J., & Laubach, M. (2013). Common medial frontal mechanisms of adaptive control in humans and rodents. *Nature Neuroscience*, *16*, 1888–1895.
Narayanan, N. S., Horst, N. K., & Laubach, M. (2006). Reversible inactivations of rat medial prefrontal cortex impair the ability to wait for a stimulus. *Neuroscience*, *139*, 865–876.
Narayanan, N. S., & Laubach, M. (2006). Top-down control of motor cortex ensembles by dorsomedial prefrontal cortex. *Neuron*, *52*, 921–931.
Narayanan, N. S., & Laubach, M. (2008). Neuronal correlates of post-error slowing in the rat dorsomedial prefrontal cortex. *Journal of Neurophysiology*, *100*, 520–525.
Narayanan, N. S., & Laubach, M. (2009). Delay activity in rodent frontal cortex during a simple reaction time task. *Journal of Neurophysiology*, *101*, 2859–2871.
Neafsey, E. J., Bold, E. L., Haas, G., Hurley-Gius, K. M., Quirk, G., Sievert, C. F., et al. (1986). The organization of the rat motor cortex: A microstimulation mapping study. *Brain Research*, *396*, 77–96.
Niki, H., & Watanabe, M. (1979). Prefrontal and cingulate unit activity during timing behavior in the monkey. *Brain Research*, *171*, 213–224.
Otis, J. M., Namboodiri, V. M. K., Matan, A. M., Voets, E. S., Mohorn, E. P., Kosyk, O., et al. (2017). Prefrontal cortex output circuits guide reward seeking through divergent cue encoding. *Nature*, *543*, 103–107.
Pafundo, D. E., Miyamae, T., Lewis, D. A., & Gonzalez-Burgos, G. (2013). Cholinergic modulation of neuronal excitability and recurrent excitation-inhibition in prefrontal cortex circuits: Implications for gamma oscillations. *The Journal of Physiology*, *591*, 4725–4748.
Parent, M. A., Amarante, L. M., Liu, B., Weikum, D., & Laubach, M. (2015). The medial prefrontal cortex is crucial for the maintenance of persistent licking and the expression of incentive contrast. *Frontiers in Integrative Neuroscience*, *9*, 23.
Parent, M. A., Amarante, L. M., Swanson, K., & Laubach, M. (2015). Cholinergic and ghrelinergic receptors and KCNQ channels in the medial PFC regulate the expression of palatability. *Frontiers in Behavioral Neuroscience*, *9*, 284.
Paxinos, G., & Watson, C. (2014). *The rat brain in stereotaxic coordinates* (7th ed.). San Diego, CA: Academic Press.
Preuss, T. (1995). Do rats have prefrontal cortex? The rose-Woolsey-Akert program reconsidered. *Journal of Cognitive Neuroscience*, *7*, 1–24.

Quintana, J., & Fuster, J. M. (1992). Mnemonic and predictive functions of cortical neurons in a memory task. *Neuroreport: An International Journal for the Rapid Communication of Research in Neuroscience*, *3*, 721–724.

Ridderinkhof, K. R., Ullsperger, M., Crone, E. A., & Nieuwenhuis, S. (2004). The role of the medial frontal cortex in cognitive control. *Science*, *306*, 443–447.

Risterucci, C., Terramorsi, D., Nieoullon, A., & Amalric, M. (2003). Excitotoxic lesions of the prelimbic-infralimbic areas of the rodent prefrontal cortex disrupt motor preparatory processes. *The European Journal of Neuroscience*, *17*, 1498–1508.

Roberts, A. C. (2020). Prefrontal regulation of threat-elicited behaviors: A pathway to translation. *Annual Review of Psychology*, *71*, 357–387.

Roesch, M. R., & Olson, C. R. (2004). Neuronal activity related to reward value and motivation in primate frontal cortex. *Science*, *304*, 307–310.

Rowe, J. B., Toni, I., Josephs, O., Frackowiak, R. S., & Passingham, R. E. (2000). The prefrontal cortex: Response selection or maintenance within working memory? *Science*, *288*, 1656–1660.

Schaeffer, D. J., Hori, Y., Gilbert, K. M., Gati, J. S., Menon, R. S., & Everling, S. (2020). Divergence of rodent and primate medial frontal cortex functional connectivity. *Proceedings of the National Academy of Sciences of the United States of America*, *117*, 21681–21689.

Shenhav, A., Cohen, J. D., & Botvinick, M. M. (2016). Dorsal anterior cingulate cortex and the value of control. *Nature Neuroscience*, *19*, 1286–1291.

Shi, L., Bian, X., Qu, Z., Ma, Z., Zhou, Y., Wang, K., et al. (2013). Peptide hormone ghrelin enhances neuronal excitability by inhibition of Kv7/KCNQ channels. *Nature Communications*, *4*, 1435.

Shima, K., & Tanji, J. (1998). Role for cingulate motor area cells in voluntary movement selection based on reward. *Science*, *282*, 1335–1338.

Smith, N. J., Horst, N. K., Liu, B., Caetano, M. S., & Laubach, M. (2010). Reversible inactivation of rat premotor cortex impairs temporal preparation, but not inhibitory control, during simple reaction-time performance. *Frontiers in Integrative Neuroscience*, *4*, 124.

Stuphorn, V., Taylor, T. L., & Schall, J. D. (2000). Performance monitoring by the supplementary eye field. *Nature*, *408*, 857–860.

Swanson, K., Goldbach, H. C., & Laubach, M. (2019). The rat medial frontal cortex controls pace, but not breakpoint, in a progressive ratio licking task. *Behavioral Neuroscience*, *133*, 385–397.

Terreberry, R. R., & Neafsey, E. J. (1983). Rat medial frontal cortex: A visceral motor region with a direct projection to the solitary nucleus. *Brain Research*, *278*, 245–249.

Tomita, A., Kato, T., Sato, F., Haque, T., Oka, A., Yamamoto, M., et al. (2012). Somatotopic direct projections from orofacial areas of primary somatosensory cortex to pons and medulla, especially to trigeminal sensory nuclear complex, in rats. *Neuroscience*, *200*, 166–185.

van Haaren, F., van Zijderveld, G., van Hest, A., de Bruin, J. P., van Eden, C. G., & van de Poll, N. E. (1988). Acquisition of conditional associations and operant delayed spatial response alternation: Effects of lesions in the medial prefrontal cortex. *Behavioral Neuroscience*, *102*, 481–488.

van Heukelum, S., Mars, R. B., Guthrie, M., Buitelaar, J. K., Beckmann, C. F., Tiesinga, P. H. E., et al. (2020). Where is cingulate cortex? A cross-species view. *Trends in Neurosciences*, *43*, 285–299.

Vogt, B. A., Hof, P. R., Zilles, K., Vogt, L. J., Herold, C., & Palomero-Gallagher, N. (2013). Cingulate area 32 homologies in mouse, rat, macaque and human: Cytoarchitecture and receptor architecture. *The Journal of Comparative Neurology*, *521*, 4189–4204.

Vogt, B. A., & Paxinos, G. (2014). Cytoarchitecture of mouse and rat cingulate cortex with human homologies. *Brain Structure & Function*, *219*, 185–192.

Wasmuht, D. F., Spaak, E., Buschman, T. J., Miller, E. K., & Stokes, M. G. (2018). Intrinsic neuronal dynamics predict distinct functional roles during working memory. *Nature Communications*, *9*, 3499.

Womelsdorf, T., Johnston, K., Vinck, M., & Everling, S. (2010). Theta-activity in anterior cingulate cortex predicts task rules and their adjustments following errors. *Proceedings of the National Academy of Sciences*, *107*, 5248–5253.

Yoshida, A., Taki, I., Chang, Z., Iida, C., Haque, T., Tomita, A., et al. (2009). Corticofugal projections to trigeminal motoneurons innervating antagonistic jaw muscles in rats as demonstrated by anterograde and retrograde tract tracing. *The Journal of Comparative Neurology*, *514*, 368–386.

Zhang, G., & Sasamoto, K. (1990). Projections of two separate cortical areas for rhythmical jaw movements in the rat. *Brain Research Bulletin*, *24*, 221–230.

Laubach, M. (2011). A comparative perspective on executive and motivational control by the medial prefrontal cortex. In RB Mars, J Sallet, MFS Rushworth, & N Yeung (Eds.), *Neural basis of motivational and cognitive control*. Cambridge, MA: Massachusetts Institute of Technology.

CHAPTER SIX

The anterior cingulate cortex and event-based modulation of autonomic states

Jeremy K. Seamans*
Department of Psychiatry, University of British Columbia, Vancouver, BC, Canada
*Corresponding author: e-mail address: jeremyseamans@gmail.com

Contents

Abstract

In spite of being an intensive area of research focus, the anterior cingulate cortex (ACC) remains somewhat of an enigma. Many theories have focused on its role in various aspects of cognition yet surgically precise lesions of the ACC, used to treat severe emotional disorders in human patients, typically have no lasting effects on cognition. An alternative view is that the ACC has a prominent role in regulating autonomic states. This view is consistent with anatomical data showing that a main target of the ACC are regions involved in autonomic control and with the observation that stimulation of the ACC evokes changes in autonomic states in both animals and humans. From an electrophysiological perspective, ACC neurons appear able to represent virtually any event or internal state, even though there is not always a strong link between these representations and behavior. Ensembles of neurons form robust contextual representations that strongly influence how specific events are encoded. The activity patterns associated with these contextually-based event representations presumably impact activity in downstream regions that control autonomic state. As a result, the ACC may regulate the autonomic and perhaps emotional reactions to events it is representing. This event-based control of autonomic tone by the ACC would likely arise during all types of cognitive and affective processes, without necessarily being critical for any of them.

International Review of Neurobiology, Volume 158
ISSN 0074-7742
https://doi.org/10.1016/bs.irn.2020.11.006

A main goal of this chapter is to provide a mechanistic perspective on the possible functions of the anterior cingulate cortex (ACC). This perspective arises out of rodent electrophysiology experiments where it has been feasible to study the behavior of ACC ensembles across a variety of different tasks and situations. I will also attempt to integrate this work with observations at the other end of the spectrum, the effects of surgically precise lesions of the ACC (anterior cingulotomies) in human patients. In spite of the obvious differences, the conclusions about the ACC drawn from these two literatures seem quite similar.

1. Homology of the ACC from rat to human

In humans, the ACC encompasses subdivisions of area 24/24c and 32 rostrally (Devinsky, Morrell, & Vogt, 1995; Vogt, 2005; Vogt, Hof, Zilles, Vogt, & Herold, 2013). Immediately posterior to the ACC is the anterior midcingulate cortex (aMCC), which includes extensions of area 24′ and 32′. There are important considerations in regards to the exact boundaries between these subregions (Vogt, Berger, & Derbyshire, 2003; Vogt, 2016), some of which will be discussed in further detail below. van Heukelum et al. (2020) described the distinction between the ACC and aMCC as follows "Human ACC encompasses the frontal third of cingulate cortex, surrounding the rostral part of the corpus callosum…, while MCC occupies the middle third of the cingulate cortex" (Fig. 1).

There is reasonably good homology between the rat and human ACC, unlike many other subregions of the frontal cortex (Laubach, Amarante, Swanson, & White, 2018; van Heukelum et al., 2020; Vogt et al., 2013). In the rat, the ACC is located on the medial wall, dorsal and rostral to the genu of the corpus callosum. According to the classic terminology of the Paxinos and Watson (1986) atlas, area 24 of the rat was referred to as the ACC while area 32 was prelimbic (PL) and area 25 was infralimbic (IL) cortex. These areas have often collectively been referred to as the rodent prefrontal cortex (PFC). Vogt and Paxinos (2014) updated this terminology and considered the ACC to be the area from Bregma to the frontal pole. The portion from Bregma to ~3 mm rostrally, includes areas 24a (ventral) and 24b (dorsal), while the region from ~3 mm to the frontal pole is area 32, which also has dorsal and ventral subdivisions. Areas 25 and 33 constitute a thin band of the ACC surrounding the corpus callosum. For the purpose of this review,

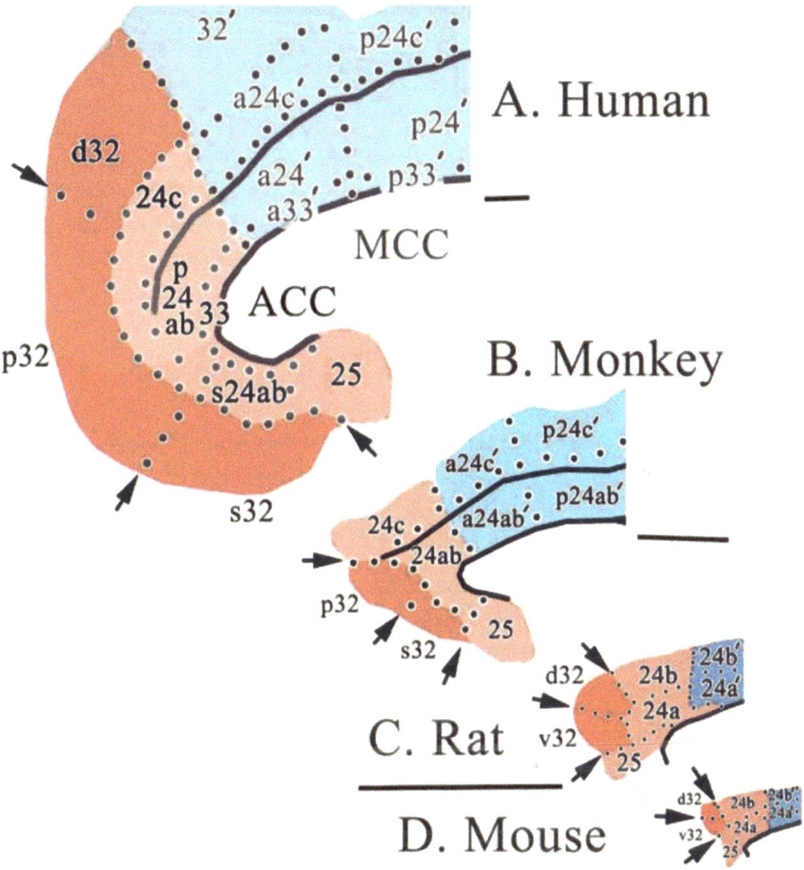

Fig. 1 A–D: *"Flat maps of the medial surfaces with anterior cingulate cortex (ACC) and mid-cingulate cortex (MCC) color-coded for four species."* The ACC is shaded in light pink and is centered on area 24ab (human and primate) or area 24a (rodent). The aMCC is shaded in light blue and is centered on a24′ and a24ab (humans and primates) and 24a′ (rodent). *Taken from Vogt, B.A., Hof, P.R., Zilles, K., Vogt, L.J., Herold, C., Palomero-Gallagher, N. (2013). Cingulate area 32 homologies in mouse, rat, macaque and human: cytoarchitecture and receptor architecture.* The Journal of Comparative Neurology, 521*(18), 4189–204.*

I will consider the portions of areas 24a and dorsal 25, rostral to the corpus callosum as the ACC (Fig. 1), even though the studies I discuss may have referred to these areas as the rat PFC, medial PFC (mPFC), or simply as frontal cortex. The electrophysiological data described in sections to follow were collected mainly from area 24a, which would correspond to areas p24 and s24 in humans.

2. Popular theories of ACC function

There is no shortage of theories about the ACC but they tend to fall into a few main categories that are not mutually exclusive. The most popular category focuses on some aspect of control, with the main premise being that the ACC and/or aMCC is important whenever the subject must monitor cognitive processes, typically during situations that require prepotent, habitual, or unconscious responses to be over ridden. Prominent theories in this category emphasize the role of the ACC and/or aMCC in the control of attention (Bush, Luu, & Posner, 2000; Paus, 2001); the current value of control (Shenhav, Cohen, & Botvinick, 2016), the application of cognitive control in contextually adaptive ways (Narayanan, Cavanagh, Frank, & Laubach, 2013; Shackman et al., 2011), the ability to monitor multiple sensory streams for potential conflicts (Botvinick, Braver, Barch, Carter, & Cohen, 2001; Carter, Botvinick, & Cohen, 1999), the detection of errors (Bernstein, Scheffers, & Coles, 1995; Ito, Stuphorn, Brown, & Schall, 2003), or the likelihood of committing an error (Alexander & Brown, 2011; Holroyd & Coles, 2002). A somewhat related category of theories emphasizes a role for the ACC in the monitoring and control of emotional and autonomic states, including the monitoring and regulation of pain and negative affect (Critchley, Mathias, Josephs, et al., 2003; Critchley, Wiens, Rotshtein, Ohman, & Dolan, 2004; Papez, 1995; Robinson et al., 2009; Vogt, 2005; Zhuo, 2016). As discussed in detail below, this group of theories is arguably most consistent with the anatomical profile of the ACC and the effects of ACC stimulation. A third category of theories emphasizes the role of the ACC in processing outcomes including the tracking of recent rewards and reward histories (Matsumoto, Suzuki, & Tanaka, 2003; Rushworth, Noonan, Boorman, Walton, & Behrens, 2011) as well as the reward value of the current task context (Holroyd & McClure, 2015). By tracking the history of outcomes, the ACC may assign a relative value to actions (Kennerley & Wallis, 2009) thereby altering choice policies. A fourth category of theories posits that the ACC is particularly involved in movement planning or initiation (Devinsky et al., 1995; Mesulam, 1981), especially of self-generated or "willed" actions (Passingham, 1996; Paus, 2001). Finally, studies in rodents suggest that the ACC may be involved in establishing remote memories (Kitamura, Ogawa, Roy, et al., 2017; Teixeira, Pomedli, Maei, Kee, & Frankland, 2006).

With so many theories it is difficult to make sense of what is going on. For this reason, it seems important to establish an anchor point or standard from which to evaluate theories and data. I would argue that the study of surgically precise lesions of the ACC in human patients might be the best candidate.

3. Anterior cingulotomies

Since the late 1940s, targeted lesions of the ACC and aMCC have been performed in patients with severe major depressive disorder (MDD), obsessive compulsive disorder (OCD) or chronic pain that are unresponsive to traditional treatments. The exact procedure differs between centers which may deliver one, two or three bilateral lesions all at once or separated over years, depending on clinical response (Steele, Christmas, Eljamel, & Matthews, 2008; Yang, Ginat, Dougherty, Makris, & Eskandar, 2014; Richter et al., 2004). The exact site of the lesion also differs, but the target is typically just posterior to the anterior extent of the frontal horn (Hassenbusch, Pillay, & Barnett, 1990; Lapidus, Kopell, Ben-Haim, Rezai, & Goodman, 2013). Lesion sites were initially quite anterior and were contained within the ACC (Brown & Lighthill, 1968) but the targeted region has since moved more posteriorly into the aMCC (Shields et al., 2008), even though imaging data from patients suggests more anteriorly placed lesions are more effective in some cases (Steele et al., 2008).

Across studies dating back over 60 years, clinically efficacious outcomes are achieved in 30–60% of the patients receiving anterior cingulotomies for emotional disorders, chronic pain and OCD (Brotis, Kapsalaki, Paterakis, Smith, & Fountas, 2009; Brown et al., 2016; Lapidus et al., 2013; McGovern & Sheth, 2017; Sharim & Pouratian, 2016; Volpini et al., 2017; Wilkinson, 2009). While effective in these seemingly diverse disorders, the lesions are not effective for all psychiatric conditions as attempts to use targeted anterior cingulotomies to treat for example, psychosis were unsuccessful (Mettler, 1949).

Regardless of whether or not the lesions are clinically efficacious, lasting side effects are rare. There can be some level of apathy, decreases in emotional control and urinary incontinence after surgery in subsets of patients. However, there is a striking lack of consistent or lasting side-effects on cognition or indeed any other aspect of the patient's lives in the months to decades after the surgery (Brotis et al., 2009; Brown et al., 2016; McGovern & Sheth, 2017; Teuber, Corkin, & Twitchell, 1977). Numerous studies employing either standard neuropsychological test batteries or the Cambridge Neuropsychological Test Automated Battery have found no adverse effects on intelligence quotient (IQ), forward/reverse digit span, paired associates learning, spatial working memory, delayed retention (3h), visual-verbal/nonverbal recurring figures test, block design, Rey-Osterrieth complex figure test, Stockings of Cambridge task, Wisconsin Card Sorting test, trail making test, porteous maze, verbal fluency and the Go-NoGo task (Brotis et al., 2009; Cohen, Kaplan, Moser, Jenkins, & Wilkinson, 1999; Faillace, Allen, McQueen, & Northrup, 1971; Fodstad, Strandman, Karlsson, & West, 1982; Long, Pueschel, & Hunter, 1978; Sharim & Pouratian, 2016; Steele et al., 2008; Strauss, Berger, Ben Moshe, et al., 2017; Teuber et al., 1977; Yen, Kuan, Sheehan, et al., 2009).

Even though performance of these tasks is not affected, lesioned patients do tend to be less vigorous in their responding. This is also true of performance on the Stroop task, where patients have slowed reaction times and, in some cases, perform more poorly. Teuber et al. (1977) failed to find deficits in lesioned patients on the regular Stroop task while Ochsner, Kosslyn, Cosgrove, et al. (2001) subsequently reported on a single patient that was impaired. Cohen et al. (1999) and Janer and Pardo (1991) also reported Stroop deficits in a group of patients that largely resolved within 12 months after surgery. Finally, Tolomeo, Christmas, Jentzsch, et al. (2016) found a deficit on the Stroop task but it was restricted to a subset of patients (see below for further discussion). It is also important to note that non-surgical lesions of the ACC do not appear to impair Stroop performance (Fellows & Farah, 2005; Stuss, Floden, Alexander, Levine, & Katz, 2001).

The most straightforward conclusion from all these studies would be that the ACC plays an non-essential role in most aspects of cognition, with the possible exception of focused attention as assessed by the Stroop task. There are of course a number of caveats to consider however. First, it could be argued that the lesions were simply not large enough to produce a cognitive deficit. In response to this concern it is important to bear in mind that the lesions were large enough to be clinically efficacious and while they do not work in every patient, when they do work, the beneficial effects on the patient's lives are obvious. Furthermore, the lesions are typically quite large and can span almost the entire cingulate gyrus in some cases (Steele et al., 2008; Yang et al., 2014). It is also worth noting that more tissue damage does not always equal a greater task deficit. As noted above, Tolomeo et al. (2016) reported that some anterior cingulotomy patients failed to perform optimally on the Stroop task. However, the critical determinant was not the absolute size of the lesion but rather the extent to which the lesion overlapped with a region localized within the aMCC that a meta-analysis of functional imaging studies had shown to be consistently activated by negative affect and cognitive control (Shackman et al., 2011). The more the lesion overlapped with the aMCC sites described by Shackman et al. (2011), the greater the effect on the Stroop task. Interestingly, neurons that represent conflict-related information necessary for Stroop performance were found within this region of the aMCC in human patients (Sheth, Mian, Patel, et al., 2012). Even though Tolomeo et al. (2016) found that some patients performed more poorly on the Stroop tasks, they also stated that "interpersonal deficits are not apparent clinically in our patients". Therefore, it would seem that as long as the lesion avoids certain parts of the aMCC, lasting cognitive or behavioral deficits are exceedingly rare.

Another issue to consider is that the lesions are typically performed in patients with an underlying psychiatric condition. As a consequence, any improvement in disease symptoms might mask cognitive deficits caused by the lesion. However, anterior cingulotomies also fail to produce lasting cognitive side-effects in patients that have no pre-existing psychiatric condition that received the lesions for emotional relief from severe chronic pain (Cohen et al., 1999; Sharim & Pouratian, 2016; Strauss et al., 2017; Teuber et al., 1977; Viswanathan, Harsh, Pereira, & Aziz, 2013; Yen et al., 2009).

4. Why are anterior cingulotomies clinically efficacious?

More important than the question of why these lesions do not produce more side-effects is the question of why they have a therapeutic effect. This is a difficult question, but consider these quotes from case reports of patients who received an anterior cingulotomy to deal with chronic pain. *"Although cingulotomy produced only mild improvement in pain severity...most patients (66.6%) reported being less bothered by pain", "The perception of pain as such does not appear to be modified, but the patient's total reaction to pain...is modified markedly", "the pain was still present but did not concern her now", "the patients stated they continued to have pain but it 'was not distressing, not particularly bothersome, doesn't worry me anymore'"* (Ballantine, Cassidy, Flanagan, & Marino, 1967; Foltz & White, 1962; Hurt & Ballantine, 1974). Prior to surgery these patients had tremendous anxiety about their pain and it was this preoccupation, rather than the pain itself, that was removed by the surgery. In fact, the psychological impact of the pain could be so lessened that some patients no longer expressed a need for narcotics and apparently exhibited no obvious signs of withdrawal (Sharma, 1974; Teuber et al., 1977; Wilkinson, 2009).

Case studies of patients who received anterior cingulotomies for other conditions are less abundant. However, early case studies of patients who received an entire ablation of area 24 for 'mental disease' noted that the most significant psychological change was once again, a reduction in tension and distress (Laitinen, 1972; Whitty, Duffield, Tov, & Cairns, 1952). Likewise, modern studies have observed that anterior cingulotomies reduce scores on the 'tension' subscale of the Profile of Mood states (Cohen et al., 2001) and on the Hamilton Anxiety Rating Scale (Brown et al., 2016).

While lesion studies can be immensely valuable, further insights into ACC function can be gained by considering what happens when it is electrically stimulated. In both animals and humans, ACC stimulation alters autonomic tone, changing aerterial blood pressure, heart rate, respiration, circulating corticosterone levels, pupillary dilation, muscle tension and gastric motility (Buchanan & Powell, 1982; Caruana, Gerbella, Avanzini, et al., 2018; Delgado, 1960; Feldman & Conforti, 1985; Lewin & Whitty, 1960; Neafsey, 1990; Pool & Ransohoff, 1949; Scopinho, Tavares, & Corrêa, 2009; Smith, 1945; Verberne & Owens, 1998). Interestingly, the changes can be in opposing directions, suggesting that the ACC modulates autonomic function rather than simply inducing a sympathetic or parasympathetic state. In addition to these objectively measurable changes, human subjects report that ACC stimulation can produce the sensation of an empty stomach or heat/shivers. It can also evoke more fully developed emotional responses such as fear, agitation and anxiety or pleasure, laughter and euphoria (Caruana et al., 2015; Foltz & White, 1962; Meyer, McElhaney, Martin, & McGraw, 1973; Talairach et al., 1973). The finding that ACC stimulation can produce feelings that are interpreted positively or negatively is consistent with the results of functional neuroimaging studies in normal subjects, as the ACC is strongly activated by the experience of negative affective states and pain (Etkin, Egner, & Kalisch, 2011; Fullana, Albajes-Eizagirre, Soriano-Mas, et al., 2018; Kulkarni, Bentley, Elliott, et al., 2005; Palermo, Benedetti, Costa, & Amanzio, 2015; Shackman et al., 2011; Wiech & Tracey, 2009; Yoshino, Okamoto, Onoda, et al., 2010) as well as by the experience of pleasure and happiness (Lindgren et al., 2012; Matsunaga, Kawamichi, Koike, et al., 2016; Rolls, Kringelbach, & de Araujo, 2003; Suardi, Sotgiu, Costa, Cauda, & Rusconi, 2016).

Recently Caruana et al. (2018) reviewed the effects of brief (5 s) stimulation of nearly 1800 sites across the cingulate cortex in a cohort of 329 drug-resistant epileptic patients. Caudal sites in the aMCC evoked motor-related

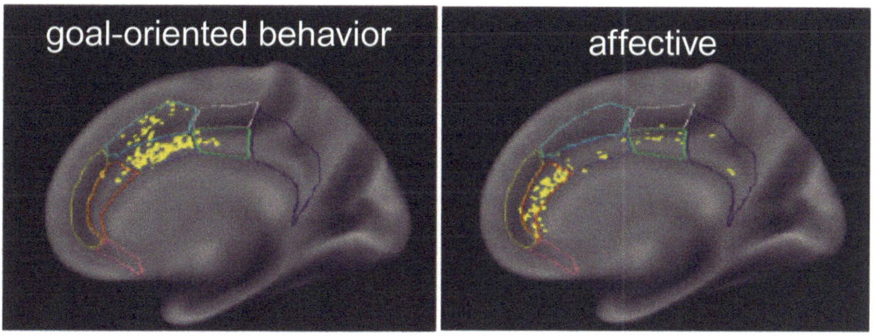

Fig. 2 Anatomical sites where stimulation elicits goal-oriented behaviors (e.g., grasping, exploratory eye movements) are mainly located in the aMCC (Left), whereas sites that evoke affective and autonomic reactions (e.g., laughter, shivers, tachycardia), are located more anteriorly in the ACC (Right). *Taken from Caruana, F., Gerbella, M., Avanzini, P., et al. (2018). Motor and emotional behaviours elicited by electrical stimulation of the human cingulate cortex.* Brain, 141*(10), 3035–3051.*

responses of a goal-oriented nature whereas ACC sites centered around area p24ab evoked a variety of affective and autonomic responses (Fig. 2). The patients often interpreted these responses as feelings of fear or anxiety. When Parvizi, Rangarajan, Shirer, Desai, and Greicius (2013) stimulated sites within the ACC of awake patients, the effects were described as creating "the will to persevere" or the sense of rising to a challenge in the face of worry or anxiety about a forthcoming event (perhaps "the *willingness* to persevere" might be even more appropriate).

To summarize the clinical evidence reviewed above, bilateral, surgical destruction of the ACC can be therapeutically efficacious while having little effect on the ability of patients to perform various laboratory tests or to live normal lives. The main therapeutic benefit of the lesion appears to be a reduction in concern/pre-occupation/anxiety about an emotional state (in patients with MDD), a physical condition (chronic pain), or a troubling thought or action (in patients with OCD). Conversely, the main effect of ACC stimulation is an autonomic reaction that in some cases is interpreted as pleasant and in others as anxiety, perhaps in the context of preparation for a forthcoming challenge. From this it would seem that the ACC is primarily involved in modulating the strength of emotional or autonomic reactions to specific events, thoughts or internal states.

This interpretation is consistent with the anatomical profile of the ACC given that a sizeable portion of its efferents synapse either in limbic regions such as the amygdala or in subcortical regions that control autonomic tone including the lateral, posterior and central parts of the anterior hypothalamus, midbrain monamine neuromodulatory nuclei, the periaqueductal gray (PAG), deep mesencephalic nucleus, the ventral / lateroposterior recticular formation and spinal cord (Barbas, Saha, Rempel-Clower, & Ghashghaei, 2003; Calejesan, Kim, & Zhuo, 2000; Carr & Sesack, 2000; Gabbott, Warner, Jays, Salway, & Busby, 2005; Fillinger, Yalcin, Barrot, & Veinante, 2018; Groenewegen & Uylings, 2000; Groenewegen, Wright, & Uylings, 1997; Sesack, Deutch, Roth, & Bunney, 1989; Terreberry & Neafsey, 1983; Van Eden & Buijs, 2000; van Heukelum et al., 2020; Warden, Selimbeyoglu, Mirzabekov, et al., 2012). These and related data have prompted multiple groups to conclude that the control of autonomic state might be the major output function of the ACC (Buchanan & Powell, 1982; Critchley et al., 2003, 2004; Devinsky et al., 1995; Medford & Critchley, 2010; Neafsey, Terreberry, Hurley, Ruit, & Frysztak, 1993; Papez, 1995).

The ability to modulate autonomic states might seem like a rather primitive function for a region located in the frontal cortex. It is nevertheless important. In animals, it could mean the difference between fleeing too early or not at all when faced with a threat. In humans, the ability to control autonomic tone determines how we will react in most situations, and especially in social situations. Although such a function is not inherently cognitive in nature, it would arise whenever a subject performs a task that they are invested in, regardless of whether or not the changes in autonomic state have any effect on task performance.

5. Properties of single ACC neurons: Flexibility, multi-responsivity and contextualization

It is often said that one can find a neuronal correlate for everything in frontal cortex (Duncan, 2001) and the ACC is no exception. One of the first in vivo recording studies of rat mPFC categorized 78 unique correlates during maze-based tasks (Jung, Qin, McNaughton, & Barnes, 1998). Since then, an endless variety of correlates have been reported. In rodents, mPFC neurons have been found to respond to (among other things); actions, action sequences, action planning, movements of individual limbs, body positions, allocentric locations, contexts, present and future choices, effort, movement trajectory, stopping, stranger approach, pup interaction, errors, prediction errors, post-error slowing, conflict, cost-benefit ratio, timing, spatial goals, progress towards a goal, rewards, anticipated reward, reward consumption, reward-related feedback, drug-related cues/stimuli, valence, working memory, remote memory, mneumonic replay, rules/strategies, decisions, fear, threats, stress, noxious mechanical/electrical/thermal stimuli to the skin or sciatic nerve, formalin injections into the paw/tail, colon distension, and the pain of other rodents (Baeg et al., 2001; Balaguer-Ballester, Lapish, Seamans, & Durstewitz, 2011; Bissonette & Roesch, 2015; Caracheo, Emberly, Hadizadeh, Hyman, & Seamans, 2013; Caracheo, Grewal, & Seamans, 2018; Cowen & McNaughton, 2007; Durstewitz, Vittoz, Floresco, & Seamans, 2010; Euston & McNaughton, 2006; Febo, 2012; Halladay & Blair, 2015; Hillman & Bilkey, 2010; Hok, Save, Lenck-Santini, & Poucet, 2005; Horst & Laubach, 2012, 2013; Huang et al., 2006; Hyman, Holroyd, & Seamans, 2017; Hyman, Ma, Balaguer-Ballester, Durstewitz, & Seamans, 2012; Insel & Barnes, 2015; Kim, Jeong, Lee, et al., 2016; Kitamura et al., 2017; Lapish, Durstewitz, Chandler, & Seamans, 2008; Lindsay, Caracheo, Grewal, Leibovitz, & Seamans, 2018; Ma, Hyman, Durstewitz, Phillips, & Seamans, 2016; Ma, Hyman, Phillips, & Seamans, 2014; Mashhoori, Hashemnia, McNaughton, Euston, & Gruber, 2018; Narayanan, 2016; Narayanan & Laubach, 2008; Passecker, Mikus, Malagon-Vina, et al., 2019; Powell & Redish, 2014; Pratt & Mizumori, 2001; Rich & Shapiro, 2009; Sikes & Vogt, 1992; Sikes, Vogt, & Vogt, 2008; Struik, Marchant, de Haan, et al., 2019; Wang et al., 2008).

It would be extremely unlikely that in each of these studies the experimenters happened to find the subset of neurons that were inherently tuned to their particular domain of interest. Instead one would assume that these neurons are so multi-modal that they are capable of exhibiting apparently selective responses on any task. As a consequence, seemingly specialized neuronal correlates would be produced by neurons that are themselves not inherently specialized. Or to put it another way, you get out what you train in. For this reason, one must be very careful about the conclusions that can be drawn from responses recorded during a single task. Any theory of ACC must take this important characteristic into account.

Individual ACC neurons are also not inherently tuned to a specific motivational valence. Numerous studies conducted in rats and monkeys have shown that ACC neurons respond to appetitive cues and outcomes (Amiez, Joseph, & Procyk, 2006; Cai & Padoa-Schioppa, 2012; Caracheo et al., 2013; Hayden, Heilbronner, Pearson, & Platt, 2011; Kennerley & Wallis, 2009; Khamassi, Quilodran, Enel, Dominey, & Procyk, 2015; Petykó, Tóth, Szabó, Gálosi, & Lénárd, 2009, Petykó et al., 2015; Pratt & Mizumori, 2001; Shidara & Richmond, 2002; Toda et al., 2012) while an equally large number of studies have found that ACC neurons respond robustly to aversive or painful cues and outcomes (Baeg et al., 2001;

Burgos-Robles, Vidal-Gonzalez, & Quirk, 2009; Dejean, Courtin, Karalis, et al., 2016; Gao, Ren, Zhang, & Zhao, 2004; Halladay & Blair, 2015; Powell, Buchanan, & Gibbs, 1990; Powell & Ginsberg, 2005; Powell, Maxwell, & Penney, 1996; Sikes et al., 2008; Yang, Tan, Cheng, et al., 2018). When tested in the same animal, individual ACC neurons can be activated by both pain and sexual attraction (Lu, Chen, Zhou, Inokuchi, & Zhuo, 2018). Recently, we recorded from the same set of ACC neurons during a 3-tone task in which rats were exposed to auditory cues signaling appetitive (food pellet), aversive (footshock) or neutral outcomes (Caracheo et al., 2018) and found few if any neurons that responded only to cues and outcomes of a single affective valence. Instead, most neurons responded to mixtures of different cues and outcomes and individual neurons could respond to a conditioned cue and an outcome of an opposing valence (Fig. 3).

Another complicating factor is that there is often a lack of tight correspondence between neuronal responses and behavior. For instance, even though action correlates are commonly observed in the ACC, using detailed motion analysis and machine learning approaches we found few ACC neurons that reliably encoded discrete limb movements (Lindsay et al., 2018). Likewise, in spite of the robust responses of some ACC neurons during fear conditioning, it has been difficult to find a clear link to fear-related movements. Rats typically exhibit the defensive behavior of freezing in response to a footshock or a cue previously paired with a footshock and in some studies the activity of mPFC neurons was correlated with these freezing responses (Burgos-Robles et al., 2009; Dejean et al., 2016). Yet, ACC lesions do not always eliminate freezing behavior (Bissière, Plachta, Hoyer, et al., 2008; Gao et al., 2004). Furthermore, after carefully monitoring EMG activity, Steenland, Li, and Zhuo (2012) were unable to find a positive correlation between the activity of ACC neurons and freezing responses. Therefore, fearful events activate ACC neurons and rats freeze when frightened but the fear-related responses of ACC neurons probably do not directly drive the freezing behavior.

Given all these complications, we reasoned that perhaps rather than focusing on what information is encoded, it might be better to focus on how information is encoded. In particular how neuronal selectivity develops and evolves. To address this issue, we tracked the responses of ACC neurons while rats performed a 3-action/2-context task that required them to execute three different actions (a nose-poke (NP), lever press (LP) and wheel turn (WT)) in different sequences or environments (Fig. 4; Ma et al., 2016). The spike rasters and peri-event time histograms for three representative

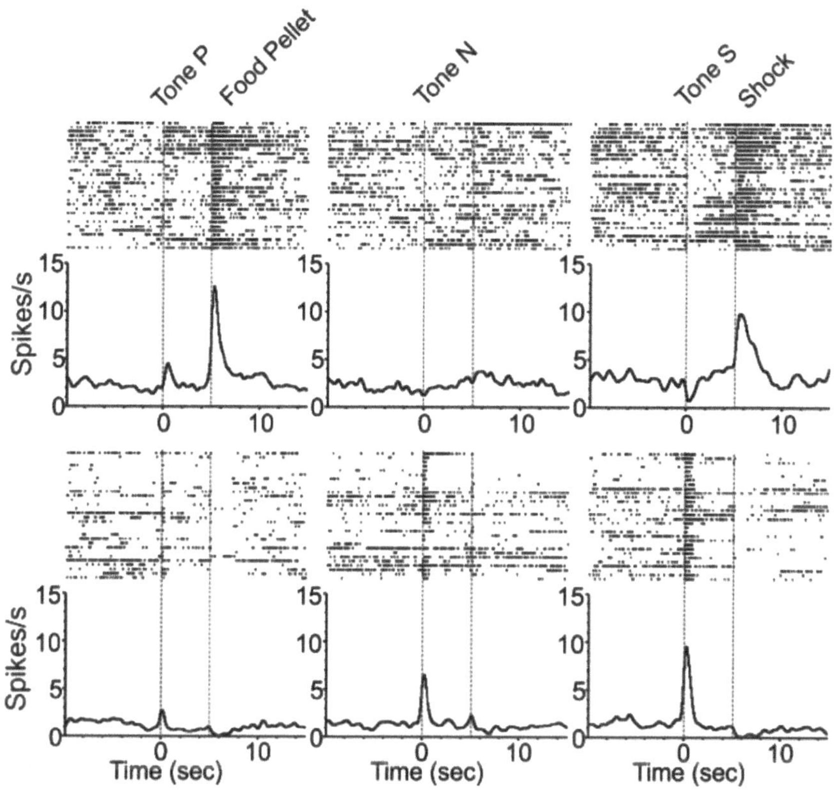

Fig. 3 Examples of 2 ACC neurons recorded during the 3-tone task. (Top) This neuron exhibited a weak excitatory response to "Tone P" and a weak inhibitory response to "Tone S," but responded strongly to both the food pellet and footshock. (Bottom) This neuron exhibited a response to all three tones but no outcomes. In the six individual panels, the spike rasters (each tiny line is a spike and each row of lines is a trial) and peri-event histograms (centered at tone onset) are shown on the top and bottom, respectively. Tone P = tone paired with food pellet; Tone N = tone followed by no outcome; Tone S = tone paired with footshock.

ACC neurons are shown in Fig. 4. Of the three example neurons, one responded consistently to WTs in both contexts (left), another to NPs only in Context 1 (middle) and the third only to LPs in Context 2. This was typical of the action correlates recorded during this task as the neurons tended to respond in varying degrees to all three actions and the mixtures of response preferences changed when the animals performed the same actions in a different environment (Ma et al., 2016).

To gain a more global perspective on how the action selectivties changed, we plotted the activity of all neurons in a multiple single-unit activity

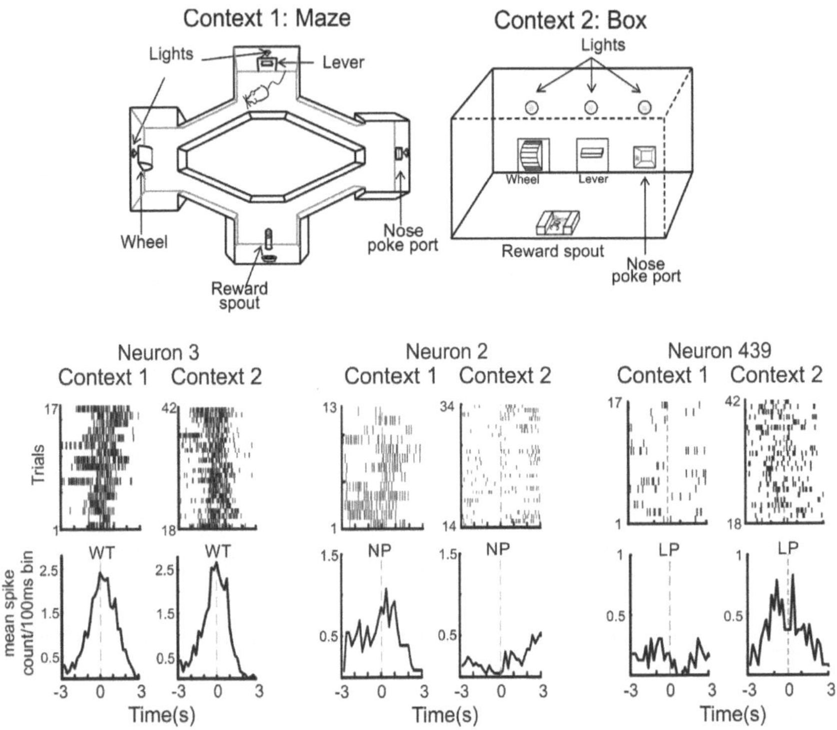

Fig. 4 (Top) A schematic of the 3-action/2-context task. Rats would traverse the maze (Context 1) or Box (Context 2) performing the three actions before receiving a reward on each trial. (Bottom) Rasters (top panels) and peri-event histograms (bottom panels) for three neurons recorded during this task. The neuron on the left responded to WTs in both contexts, the neuron in the middle responded to NPs only in Context 1 while the neuron on the right responded to LPs only in Context 2. WT=wheel turn, NP=nose poke, LP=lever press. *Adapted from Ma, L., Hyman, J.M., Durstewitz, D., Phillips, A.G., Seamans, J.K. (2016). A quantitative analysis of context-dependent remapping of medial frontal cortex neurons and ensembles.* The Journal of Neuroscience, *36(31), 8258–8272.*

(MSUA) space that was constructed as follows. Imagine there were 100 simultaneously recorded neurons and the spike counts of each neuron were plotted on individual axes. This plot would have 100 axis or dimensions and the activity of all the neurons during a single time bin would be represented by a single dot in the space (since it is not possible to visualize 100 dimensions, dimensionality reduction techniques are typically used to reduce the space to 3 axes). Therefore, if the spike count data was binned at 0.2 s and we plotted the 1 s period a lever press, the MSUA space would contain 5 dots. If we plotted the 1 s period around 10 LPs, it would have 50 dots and so on.

We quantified the differences in ensemble activity-states in the two contexts by calculating the (Mahalanobis, D_{Mah}) distance between the corresponding clusters of dots in the full, non-reduced MSUA space. Although we had already observed that changing contexts caused a significant shift in ensemble activity(Hyman et al. 2012), we did not expect to find that changing contexts would cause the three action clusters to immediately move with them (Fig. 5C). In spite of the shift, action differentiation remained consistent in both contexts because the action clusters all shifted by the same amount in the MSUA space. As a result, the ensembles were able to encode both the differences between the two contexts as well as the fact that the actions performed in the two contexts were the same. Collectively, this suggested that the ACC may encode actions within a larger contextual framework.

The contextual framework need not be defined by the physical cues in the environment, but could also be defined by the internal state of the animal. We explored this issue in Caracheo et al. (2018) using the 3-tone task mentioned above, where the rats received blocks of trials in which a unique tone was followed by either a footshock, a food pellet or nothing, all within a single enclosed chamber. The MSUA space from one of these sessions is shown in Fig. 6. This session involved 64 simultaneously recorded neurons (for visualization the MSUA space was reduced to three dimensions using Principle Components Analysis, PCA). Even though most individual neurons did not respond in a valence specific manner (Fig. 3), the ensemble space was strongly organized according to valence. Specifically, the activity pattern associated with the delivery of the shock-paired tone (red squares) and footshocks (red triangles), the food-paired tone (green squares) and food consumption (green triangles) and the null-tone (blue squares) formed unique and highly separable clusters in the space. In addition, the circular dots, which represented the activity during the inter-trial intervals ($\sim$50 s), were located within the clusters of similarly colored squares and triangles. This meant that the ensemble remained in the same state during and long after the tone and outcome epochs had ended. Importantly, the ensembles were in a more similar state (i.e., had smaller Mahalanobis distances) during all the epochs of a given block (i.e., tone, outcome and ITI) than during any matched epoch across blocks (Fig. 6). This was quite remarkable considering that the similarities between the three tones were obviously far greater than the similarities between a tone, outcome and ITI within a given block.

Once evoked, the block-specific ensemble states persisted until the next block of tones/outcomes occurred. The persistent nature of the responses

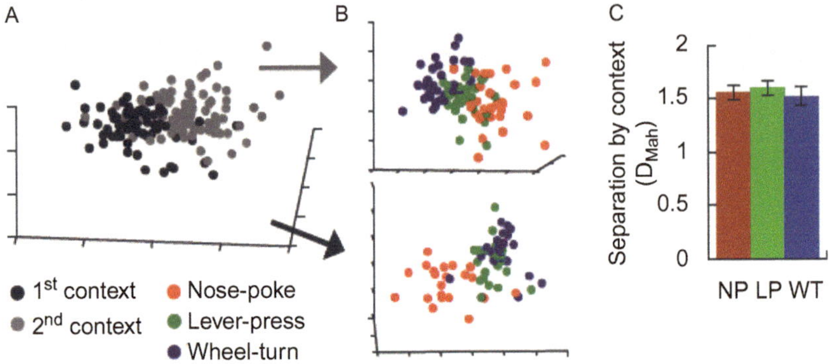

Fig. 5 Ensemble dynamics during the 3-action/2-context task. (A) A reduced MSUA space constructed from a single session of the task (reduced to three dimensions using multi-dimensional scaling). All time bins recorded while the rat was in Contexts 1 or 2 are shown in black or gray, respectively. (B) The same data as in (A) but plotted separately for each context. All time bins during the 1 s period surrounding the nose poke (red) lever press(green) or wheel turn(blue) formed separate clusters in each context space. (C) Quantification of the cluster differences based on the Mahalanobis distances (D_{Mah}) in the full, non-reduced MSUA spaces across all sessions. The D_{Mah} in the MSUA space between the activity state clusters for NPs in the first context versus NPs in the second context (red) was the same as the distance between the activity state clusters for LPs in the first and second contexts (green) and the distance between the activity state clusters of WTs in the first and second contexts (blue). *Adapted from Ma, L., Hyman, J.M., Durstewitz, D., Phillips, A.G., Seamans, J.K. (2016). A quantitative analysis of context-dependent remapping of medial frontal cortex neurons and ensembles.* The Journal of Neuroscience, 36*(31), 8258–8272.*

were also obvious in the rasters of individual neurons (Fig. 7). Since the ensemble states persisted well into the ITI, long after the tone/outcome had ended, they were not representing the sensory aspects of the tones/outcomes. Instead, we argued that through learning, each tone/outcome pair came to evoke a unique physiological/autonomic reaction that was represented by the ensemble patterns we had observed (Caracheo et al., 2018).

The results of these studies suggested that ACC ensembles can represent diverse events in a contextualized manner. By "context" I mean something similar to what has been advocated by Heilbronner and Hayden (2016) or the totality of information about the event, the physical environment, the task and the current emotional or autonomic state. The highly multi-modal nature of these contextual representations is likely a direct byproduct of anatomy. The ACC receives input from a wide variety of cortical, thalamic, striatal, limbic, hypothalamic, midbrain, and medullary brain regions (Condé, Maire-Lepoivre, Audinat, & Crépel, 1995; Fillinger et al., 2018;

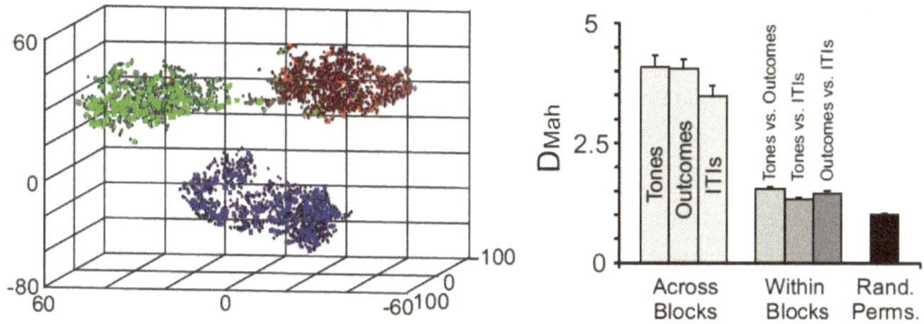

Fig. 6 Encoding of 'emotional contexts' by ACC ensembles. (Left) A MSUA space from a single session was reduced to three dimensions for visualization. Each point in the space represents the activity of the ensemble during a single 1 s time bin. The gray diamonds are derived from time bins when tones were presented, the squares from time bins when the outcomes were delivered (specifically, the 0–5 s following the tones) and the black dots from time bins associated with the ITIs. Each symbol is filled based on the block from which it was derived (Food block = green, Neutral block = blue, Shock block = red). (Right) The average D_{Mah} (and sem) between the clusters of points corresponding to time bins associated with tone, outcome or ITI periods derived from different blocks (light gray bars), from within a single block (dark gray bars) or from randomly chosen time bins taken at random from throughout the task (black bar; Rand. Perms. = random permutations). Each light gray bar involved an averaging over all combinations of block pairs while all darker gray bars averaged over all blocks. *Adapted from Caracheo, B.F., Grewal, J.J.S., Seamans, J.K. (2018). Persistent valence representations by ensembles of anterior Cingulate cortex neurons.* Frontiers in Systems Neuroscience, *12, 51.*

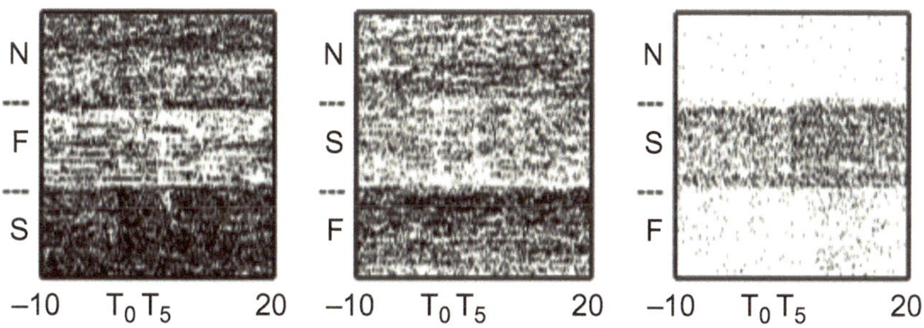

Fig. 7 Block specific shifts in the activity of single ACC neurons. Sample raster plots from three neurons illustrating the changes in activity during trial and ITI periods of different blocks. Note how the neurons changed their firing throughout both the trial (black dotted bars) and/or ITI periods of the blocks. The N,F and S denotes the neutral, food and shock blocks, respectively. In each panel, every dot is a spike and every line is a trial. The rasters are centered around the time the 5 s tone that preceded each outcome started (To) and ended (T5). Time (x-axis) is in seconds. *Adapted from Caracheo, B.F., Grewal, J.J.S., Seamans, J.K. (2018). Persistent valence representations by ensembles of anterior cingulate cortex neurons.* Frontiers in Systems Neuroscience, *12, 51.*

Hoover & Vertes, 2007; Sesack et al., 1989) and the activity in each of these afferent regions would contribute to different aspects of the overall contextual representation. For example, input from the hippocampus may attempt to organize ACC activity according to a "where-based" pattern, input from the amygdala according to a "threat-based" pattern, input from intralaminar thalamic nuclei according to a "pain-based" pattern, input from hypothalamus according to a "hunger-based" pattern and so on. Input from all active and relevant regions would combine to form an integrated representation of the overall context and all the events that occur within it. This integrated representation would fluctuate based on the prevailing mixture of afferent inputs but if an afferent input remained consistent, that aspect of the integrated representation would presumably also remain consistent. This could explain why the ensemble patterns shift when moving into a new context yet still maintain a stable representation of the actions (Fig. 5) or cues (Fig. 6) common to the both the old and new contexts. It may also explain the highly variable nature of single ACC neuron responses as these responses would be organized according to some over-arching contextual framework that may or may not be obvious to the experimenter.

This is not to say that the main function of the ACC is to represent contexts, external events or internal states. If this were the case anterior cingulotomies would alter how events are perceived, which they do not. It is also important to emphasize that the lesions do not remove the ability to generate normal emotional or autonomic reactions. Rather, as noted above, the key therapeutic effect is a reduction in the excessively strong reactions to a limited set of external events (e.g., dirty hands in the case of OCD) or internal states (e.g., sadness or pain in the case of MDD and chronic pain, respectively). Therefore, I would argue that the ACC does not simply create contextualized representations of events but it allows these representations to gain control over emotional or autonomic reactions.

The contextual representations are activity patterns that are created by relative increases/decreases in the firing of individual ACC neurons. Such changes in the relative activity of large groups of ACC neurons would be felt in downstream regions, including those that control autonomic state. Therefore, the activity patterns may exert effects similar to those evoked by artificial electrical stimulation discussed above. The critical difference being that electrical stimulation synchronously drives all neurons as a collective, whereas naturally occurring patterns would be a diverse mixture of activity across the population. As a consequence, the patterns would likely only alter or bias activity in downstream regions, rather than turning an area "on" or "off." However, because each contextually-based pattern is statistically unique, they could alter the autonomic state in slightly different ways.

6. Neuromodulation of the ACC: Lessons from the dopamine system

If we assume that the patterns represent contexts and in turn alter the autonomic states, then it is critically important to understand how are these patterns created, modified and fined-tuned. The ensemble activity patterns can be thought of as quasi-attractor states (Balaguer-Ballester et al., 2011; Lapish, Balaguer-Ballester, Seamans, Phillips, & Durstewitz, 2015) and are likely initiated by glutamatergic input from the cortical, thalamic and limbic regions mentioned above. However, once attractor states becomes established, they are subjected to modulation by monoamines (Durstewitz & Seamans, 2002, 2008; Durstewitz, Seamans, & Sejnowski, 2000; Rolls & Deco, 2011; Ueltzhöffer, Armbruster-Genç, & Fiebach, 2015). In particular, my colleague Daniel Durstewitz described how dopamine (DA) could modulate attractor dynamics by incorporating the known biophysical effects of DA on intrinsic and synaptic currents into computational models of realistic mPFC networks (Durstewitz et al., 2000; Durstewitz & Seamans, 2008). When the biophysical effects of D1 receptor activation were simulated, the attractors became stronger and more stable (i.e., the attractor basins deepened). In contrast, when the known biophysical effects of D2 receptor activation were simulated, the attractors became weaker and less stable. Even though the receptors acted on a variety of intrinsic and synaptic currents in both pyramidal and interneurons, in each case the effects converged to produce the final results.

Both states in the model have advantages and disadvantages. The deep attractor basins that emerge in the D1 state are stable and resistant to distractors and noise. However, these attractor patterns could potentially become so strong and stable that they cause the subject to perseverate on information that is no longer contextually appropriate. The opposite is true of the shallow attractor basins that emerge in the D2 state, as ensemble patterns are highly flexible and switches between patterns occur easily and frequently. As a consequence, in an extreme D2-dominated state the subject may be flexible but easily distracted. These and other predictions of the model have been largely supported by experiments performed in animals (Ellwood, Patel, Wadia, et al., 2017; Kroener, Chandler, Phillips, & Seamans, 2009; Lapish et al., 2015; Ott & Nieder, 2017; St Onge, Abhari, & Floresco, 2011; Thurley, Senn, & Lüscher, 2008) and humans (Bensmann, Zink, Arning, Beste, & Stock, 2020;

Cameron, Wallace, Al-Zughoul, Kayser, & D'Esposito, 2018; Cools & D'Esposito, 2011; Fagerlund, Pinborg, Mortensen, et al., 2013; Fallon, Williams-Gray, Barker, Owen, & Hampshire, 2013; Kahnt, Weber, Haker, Robbins, & Tobler, 2015; Kuo, Paulus, & Nitsche, 2008; Meyer-Lindenberg, Kohn, Kolachana, et al., 2005; Rosa, Dickinson, Apud, Weinberger, & Elvevåg, 2010; Shafiei, Zeighami, Clark, et al., 2019; Stelzel, Fiebach, Cools, Tafazoli, & D'Esposito, 2013; van Holstein, Aarts, van der Schaaf, et al., 2011).

We have argued that the relative [DA] is a critical determinant of whether D1 versus D2 receptors are preferentially activated and therefore whether a D1- or a D2-state emerges. Moderate DA levels were proposed to preferentially activate D1 receptors whereas very high or very low DA levels would preferentially activate D2 receptors (Durstewitz & Seamans, 2008; Seamans & Robbins, 2009; Seamans & Yang, 2004). However, recent data suggest that [DA] is not uniformly distributed, at least in other projection fields (i.e., there are hotspots of higher [DA]; Siciliano, Calipari, & Jones, 2014; Peters & Michael, 2000). Hence there would likely always be some mixture of D1 and D2 states present.

The ability of DA to either enhance or collapse attractor states arises out of its biophysical actions on intrinsic and synaptic currents. Obviously, DA is not the only neuromodulator present in frontal cortex nor is it unique in its biophysical effects. Since other monoamines like serotonin and noradrenaline also alter the intrinsic and synaptic currents in frontal cortex neurons (Andrews & Lavin, 2006; Celada, Puig, & Artigas, 2013; Di Pietro & Seamans, 2011; Koga, Yamada, Song, et al., 2020; Satake, Mitani, Nakagome, & Kaneko, 2008; Stephens, Avesar, & Gulledge, 2014; Tian, Schmidt, & Lambe, 2016; Yin et al., 2017; Yuen et al., 2014), they too would modulate attractor dynamics, as would other non-monoamine neuromodulators. Therefore, the prevailing mixture of neuromodulators present at any given time would ultimately determine the overall attractor dynamics. But because the effects of other neuromodulators on attractor dynamics have not been studied in the same detail as DA, I'll assume that certain combinations of neuromodulators produce something akin to a "D1-like" or "D2-like" state for the present purposes.

Our original theory was framed in terms DA modulation of working memory processes mediated by the dorsolateral PFC, but the model would hold regardless of the information represented by the attractor patterns. When reframed in terms of the theorized function of the ACC noted above, in a D1-like state, the events represented by ACC ensembles would tend to evoke strong autonomic reactions that would be persistent and difficult to switch out of. Conversely, if the neuromodulatory tone promoted something like a D2-state, the events represented by ACC ensembles would evoke weaker autonomic reactions that would not be as noticeable or persistent but could be more dynamically tailored to fit the current situation.

Of course, it is also important to acknowledge that monoamine neuromodulators do not only act in the ACC but serve various functions throughout the brain. Multiple authors have proposed that they may also act collectively to produce, or at least regulate, affective states (Goschke & Bolte, 2014; Gruber & McDonald, 2012; Gu et al., 2018; Kringelbach & Berridge, 2017; Lövheim, 2012). This idea is captured in the Lövheim (2012) cube theory of emotion, where various combinations of the 3 monoamine neurotransmitters (dopamine, serotonin and noradrenaline) act throughout the brain to shape the 8 basic emotions. Combining this theory with the current hypothesis, monoamine neuromodulators in the ACC might determine the strength of the autonomic component of the emotional response created by these same monoamines acting throughout the brain.

7. A model for event-specific control of autonomic states

The discussion above can be summarized in the following model. Contexts, events or thoughts evoke emotional reactions created by neuromodulatory systems acting throughout the brain. Both the initiating event and the emotional response would be encoded by ACC ensemble activity patterns. These patterns would then be shaped by the local mixture of neuromodulators present within the ACC. The modulated patterns would act on downstream regions to create the autonomic component of the emotional response. The change in autonomic state as well as the outcome of any actions taken would subsequently incorporated into the ACC ensemble patterns and in turn subjected to local neuromodulation. The newly modulated patterns would again act on downstream regions to further alter the autonomic response and so on.

This iterative process could help ensure that the body is best able to deal with the evolving demands of emotionally laden events. However, like any iterative process, it is important to ensure that is tightly regulated. The iterative cooperativity between neuromodulatory systems and the ACC could be controlled by way of the reciprocal connections between the ACC and the nuclei containing the monoamine cell bodies (Celada, Puig, Casanovas,

Guillazo, & Artigas, 2001; Elston & Bilkey, 2017; Elston, Croy, & Bilkey, 2019; Fillinger et al., 2018; Warden et al., 2012). For example, if the ACC neurons that project to the ventral tegmental area (the midbrain nuclei that contains DA cell bodies) participate in an ensemble pattern, the strength of the pattern would in turn affect DA neuron activity and the subsequent release of DA.

To be clear, I am not claiming that the ACC is required to experience an emotion nor is it required for the subject to know where they are, what stimuli are currently relevant or which actions to perform. However, it may be required for the subject to know how *strongly* to react, given their current emotional state, where they are and what they are doing. In the present model, ACC ensemble representations do not directly produce a specific behavior (or any behavior at all), but instead alter autonomic states according to the current context.

The present theory has obvious parallels to the somatic marker hypothesis of Damasio (2008) in that the autonomic reactions mentioned above would be the equivalent of the somatic markers in that theory. One difference is that the somatic marker hypothesis focuses on regions ventral to the ACC that interpret somatic markers in the service of decision making. Whereas in the current hypothesis, a somatic marker is not used by the ACC to inform the subject which emotion they are feeling but rather the ACC shapes somatic markers in order to determine the strength of the overall emotional reaction.

The present theory is also closely related to that of Critchley et al. (Critchley et al. 2004, 2005; Medford & Critchley, 2010) who argued that the main function of the ACC is the control of autonomic state. However, these authors assigned the integration of external/internal information to the anterior insular cortex and the modulation of autonomic states to the ACC, whereas the current theory assigns both functions to the ACC (although the insula is likely also involved). In addition, the current theory provides a neuronal mechanism for context-specific regulation of autonomic states.

8. The modulation of attractor dynamics in psychiatric conditions

One way to conceive of OCD, MDD and the distress caused by chronic pain is as an excessive amplification of the physiological reactions to a specific event or internal state (Devinsky et al., 1995). In the current theory, pathologies in how ACC ensembles represent contexts/events or how the representations are modulated might contribute to the development of these disorders. In our experimental data, clear individual differences existed in terms of how the contextual representations were formed. For example, every rat in the 3-tone experiments expressed a unique ensemble pattern in each of the three blocks (Fig. 6) but the relative strengths of the three ensemble patterns (i.e., the distances between the clusters in the MSUA spaces) varied across rats. In a similar way, ACC ensembles in humans might normally differentiate between the physiological states associated with joy and sadness but the relative strengths of these patterns may vary across the human population. Perhaps in some individuals the ensemble patterns that represent the emotional state of sadness are so strong that they dominate the activity in downstream regions for prolonged periods. As a result, whenever sadness is experienced, it is experienced with a stronger and more persistent autonomic reaction than in non-depressed subjects. MDD may arise if this continues over a period of months or years.

The current theory can also provide insights into why anterior cingulotomies work as they do. MDD, OCD and chronic pain are characterized by an exaggerated and persistent autonomic reaction to an anxiety-provoking event or internal state. In the case of MDD an anterior cingulotomy removes this exaggerated response to sadness without removing the ability to feel sadness or any other emotion. In OCD, the lesion removes the exaggerated reaction to a specific event (e.g., unclean hands), without necessarily affecting the normal response to other events. In patients suffering from severe chronic pain, the lesions diminish the exaggerated reaction to the pain without removing the pain itself or any other sensation. I am claiming that a unique feature of the ACC is that it can represent any event or internal state and that it allows these representations to modulate autonomic tone. Because each pattern is distinct from every other pattern, even if one representation has gained a pathologically high level of control over autonomic tone, this should not impact how other representations affect autonomic tone. It is also important to bear in mind that in the current theory, the ACC does not generate autonomic states directly but rather exerts a biasing influence on the downstream regions that do. Therefore, anterior cingulotomies may work in part because they remove the controlling influence that certain event-based representations have over autonomic tone.

Although highly speculative, it is hoped that this theory generates predictions that can be tested in experimental animals and human subjects. Ensemble-based attractor states might be difficult to detect and quantify using functional magnetic resonance imaging techniques in humans because they are created by both increases and decreases in firing of single neurons and can arise without a change in overall activity levels (Ma et al., 2016). Nevertheless, each attractor state may possess a unique macroscopic signature that could be detected using novel analysis methods (Deco & Jirsa, 2012; Shine, Breakspear, Bell, et al., 2019). The current theory would predict that moving human subjects between virtual environments that differ in terms of sensory cues and the emotions they evoke, should produce widespread shifts in voxel patterns within the ACC. These shifts should be stronger or weaker in different patient populations. For example, in patients with MDD, the signatures of attractor dynamics should be stronger and/or more persistent in environments that evoke sadness.

These signatures should also be modifiable by manipulations of monoaminergic systems. As a case in point, we found that doses of amphetamine that would be predicted to evoke a "D1-state" in cortex strengthened the differentiation of task epoch representations during a rodent working memory task whereas doses of amphetamine that would be predicted to evoke a "D2-state," weakened the differentiation (Lapish et al., 2015). An optogenetic approach to manipulation of monoamine levels in animals would also be informative as it would make it possible to precisely stimulate or inhibit each monoamine system alone or in combination at precise points during a task.

The current theory also generates predictions that could be tested in the absence of multiple-single unit recordings. For instance, modifications of monoamine systems in the ACC should alter context-dependent changes in autonomic variables such as heart-rate, breathing and blood pressure and these effects should be reduced by inhibiting the axons of ACC neurons in downstream targets. The results of these and other experiments could help refine the theory while providing insights into the function of the ACC in normal and disease states.

References

Alexander, W. H., & Brown, J. W. (2011). Medial prefrontal cortex as an action-outcome predictor. *Nature Neuroscience*, *14*(10), 1338–1344.

Amiez, C., Joseph, J. P., & Procyk, E. (2006). Reward encoding in the monkey anterior cingulate cortex. *Cerebral Cortex*, *16*(7), 1040–1055.

Andrews, G. D., & Lavin, A. (2006). Methylphenidate increases cortical excitability via activation of alpha-2 noradrenergic receptors. *Neuropsychopharmacology*, *31*(3), 594–601.

Baeg, E. H., Kim, Y. B., Jang, J., Kim, H. T., Mook-Jung, I., & Jung, M. W. (2001). Fast spiking and regular spiking neural correlates of fear conditioning in the medial prefrontal cortex of the rat. *Cerebral Cortex*, *11*(5), 441–451.

Balaguer-Ballester, E., Lapish, C. C., Seamans, J. K., & Durstewitz, D. (2011). Attracting dynamics of frontal cortex ensembles during memory-guided decision-making. *PLoS Computational Biology*, 7(5), e1002057.

Ballantine, H. T., Jr., Cassidy, W. L., Flanagan, N. B., & Marino, R., Jr. (1967). Stereotaxic anterior cingulotomy for neuropsychiatric illness and intractable pain. *Journal of Neurosurgery*, *26*(5), 488–495.

Barbas, H., Saha, S., Rempel-Clower, N., & Ghashghaei, T. (2003). Serial pathways from primate prefrontal cortex to autonomic areas may influence emotional expression. *BMC Neuroscience*, *4*, 25.

Bensmann, W., Zink, N., Arning, L., Beste, C., & Stock, A. K. (2020). Dopamine D1, but not D2, signaling protects mental representations from distracting bottom-up influences. *NeuroImage*, *204*, 116243.

Bernstein, P. S., Scheffers, M. K., & Coles, M. G. H. (1995). "Where did I go wrong?" A psychophysiological analysis of error detection. *Journal of Experimental Psychology. Human Perception and Performance*, *21*(6), 1312–1322.

Bissière, S., Plachta, N., Hoyer, D., et al. (2008). The rostral anterior cingulate cortex modulates the efficiency of amygdala-dependent fear learning. *Biological Psychiatry*, *63*(9), 821–831.

Bissonette, G. B., & Roesch, M. R. (2015). Neural correlates of rules and conflict in medial prefrontal cortex during decision and feedback epochs. *Frontiers in Behavioral Neuroscience*, *9*, 266.

Botvinick, M. M., Braver, T. S., Barch, D. M., Carter, C. S., & Cohen, J. D. (2001). Conflict monitoring and cognitive control. *Psychological Review*, *108*(3), 624–652.

Brotis, A. G., Kapsalaki, E. Z., Paterakis, K., Smith, J. R., & Fountas, K. N. (2009). Historic evolution of open cingulectomy and stereotactic cingulotomy in the management of medically intractable psychiatric disorders, pain and drug addiction. *Stereotactic and Functional Neurosurgery*, *87*(5), 271–291.

Brown, H. M., & Lighthill, J. A. (1968). Selective anterior cingulotomy: A psycho-surgical evaluation. *Journal of Neurosurgery*, *29*, 513–519.

Brown, L. T., Mikell, C. B., Youngerman, B. E., Zhang, Y., McKhann, G. M., 2nd, & Sheth, S. A. (2016). Dorsal anterior cingulotomy and anterior capsulotomy for severe, refractory obsessive-compulsive disorder: a systematic review of observational studies. *Journal of Neurosurgery*, *124*(1), 77–89.

Buchanan, S. L., & Powell, D. A. (1982). Cingulate cortex: its role in Pavlovian conditioning. *Journal of Comparative and Physiological Psychology*, *96*(5), 755–774.

Burgos-Robles, A., Vidal-Gonzalez, I., & Quirk, G. J. (2009). Sustained conditioned responses in prelimbic prefrontal neurons are correlated with fear expression and extinction failure. *The Journal of Neuroscience*, *29*(26), 8474–8482.

Bush, G., Luu, P., & Posner, M. I. (2000). Cognitive and emotional influences in anterior cingulate cortex. *Trends in Cognitive Sciences*, *4*(6), 215–222.

Cai, X., & Padoa-Schioppa, C. (2012). Neuronal encoding of subjective value in dorsal and ventral anterior cingulate cortex. *The Journal of Neuroscience*, *32*(11), 3791–3808.

Calejesan, A. A., Kim, S. J., & Zhuo, M. (2000). Descending facilitatory modulation of a behavioral nociceptive response by stimulation in the adult rat anterior cingulate cortex. *European Journal of Pain*, *4*(1), 83–96.

Cameron, I. G. M., Wallace, D. L., Al-Zughoul, A., Kayser, A. S., & D'Esposito, M. (2018). Effects of tolcapone and bromocriptine on cognitive stability and flexibility. *Psychopharmacology*, *235*(4), 1295–1305.

Caracheo, B. F., Emberly, E., Hadizadeh, S., Hyman, J. M., & Seamans, J. K. (2013). Abrupt changes in the patterns and complexity of anterior cingulate cortex activity when food is introduced into an environment. *Frontiers in Neuroscience*, *7*, 74.

Caracheo, B. F., Grewal, J. J. S., & Seamans, J. K. (2018). Persistent valence representations by ensembles of anterior Cingulate cortex neurons. *Frontiers in Systems Neuroscience*, *12*, 51.

Carr, D. B., & Sesack, S. R. (2000). Projections from the rat prefrontal cortex to the ventral tegmental area: target specificity in the synaptic associations with mesoaccumbens and mesocortical neurons. *The Journal of Neuroscience*, *20*(10), 3864–3873.

Carter, C. S., Botvinick, M. M., & Cohen, J. D. (1999). The contribution of the anterior cingulate cortex to executive processes in cognition. *Reviews in the Neurosciences*, *10*(1), 49–57.

Caruana, F., Avanzini, P., Gozzo, F., Francione, S., Cardinale, F., & Rizzolatti, G. (2015). Mirth and laughter elicited by electrical stimulation of the human anterior cingulate cortex. *Cortex*, *71*, 323–331.

Caruana, F., Gerbella, M., Avanzini, P., et al. (2018). Motor and emotional behaviours elicited by electrical stimulation of the human cingulate cortex. *Brain*, *141*(10), 3035–3051.

Celada, P., Puig, M. V., & Artigas, F. (2013). Serotonin modulation of cortical neurons and networks. *Frontiers in Integrative Neuroscience*, 7, 25.

Celada, P., Puig, M. V., Casanovas, J. M., Guillazo, G., & Artigas, F. (2001). Control of dorsal raphe serotonergic neurons by the medial prefrontal cortex: Involvement of serotonin-1A, GABA(A), and glutamate receptors. *The Journal of Neuroscience*, *21*(24), 9917–9929.

Cohen, R. A., Kaplan, R. F., Moser, D. J., Jenkins, M. A., & Wilkinson, H. (1999). Impairments of attention after cingulotomy. *Neurology*, *53*(4), 819–824.

Cohen, R. A., Paul, R., Zawacki, T. M., Moser, D. J., Sweet, L., & Wilkinson, H. (2001). Emotional and personality changes following cingulotomy. *Emotion*, *1*(1), 38–50.

Condé, F., Maire-Lepoivre, E., Audinat, E., & Crépel, F. (1995). Afferent connections of the medial frontal cortex of the rat. II. Cortical and subcortical afferents. *The Journal of Comparative Neurology*, *352*(4), 567–593.

Cools, R., & D'Esposito, M. (2011). Inverted-U-shaped dopamine actions on human working memory and cognitive control. *Biological Psychiatry*, *69*(12), e113–e125.

Cowen, S. L., & McNaughton, B. L. (2007). Selective delay activity in the medial prefrontal cortex of the rat: contribution of sensorimotor information and contingency. *Journal of Neurophysiology*, *98*(1), 303–316.

Critchley, H. D. (2005). Neural mechanisms of autonomic, affective, and cognitive integration. *The Journal of Comparative Neurology*, *493*(1), 154–166.

Critchley, H. D., Mathias, C. J., Josephs, O., et al. (2003). Human cingulate cortex and autonomic control: converging neuroimaging and clinical evidence. *Brain*, *126*(Pt 10), 2139–2152.

Critchley, H. D., Wiens, S., Rotshtein, P., Ohman, A., & Dolan, R. J. (2004). Neural systems supporting interoceptive awareness. *Nature Neuroscience*, 7(2), 189–195.

Damasio, A. R. (2008). *Descartes' error: Emotion, reason and the human brain*. Random House.

Deco, G., & Jirsa, V. K. (2012). Ongoing cortical activity at rest: criticality, multistability, and ghost attractors. *The Journal of Neuroscience*, *32*(10), 3366–3375.

Dejean, C., Courtin, J., Karalis, N., et al. (2016). Prefrontal neuronal assemblies temporally control fear behaviour. *Nature*, *535*(7612), 420–424.

Delgado, J. M. (1960). Circulatory effects of cortical stimulation. *Physiological Reviews. Supplement*, *4*, 146–178.

Devinsky, O., Morrell, M. J., & Vogt, B. A. (1995). Contributions of anterior cingulate cortex to behaviour. *Brain*, *118*(Pt 1), 279–306.

Di Pietro, N. C., & Seamans, J. K. (2011). Dopamine and serotonin interactively modulate prefrontal cortex neurons in vitro. *Biological Psychiatry*, *69*(12), 1204–1211.

Duncan, J. (2001). An adaptive coding model of neural function in prefrontal cortex. *Nature Reviews. Neuroscience*, *2*(11), 820–829.

Durstewitz, D., & Seamans, J. K. (2002). The computational role of dopamine D1 receptors in working memory. *Neural Networks*, *15*(4–6), 561–572.

Durstewitz, D., & Seamans, J. K. (2008). The dual-state theory of prefrontal cortex dopamine function with relevance to catechol-o-methyltransferase genotypes and schizophrenia. *Biological Psychiatry*, *64*(9), 739–749.

Durstewitz, D., Seamans, J. K., & Sejnowski, T. J. (2000). Dopamine-mediated stabilization of delay-period activity in a network model of prefrontal cortex. *Journal of Neurophysiology*, *83*(3), 1733–1750.

Durstewitz, D., Vittoz, N. M., Floresco, S. B., & Seamans, J. K. (2010). Abrupt transitions between prefrontal neural ensemble states accompany behavioral transitions during rule learning. *Neuron, 66*(3), 438–448.

Ellwood, I. T., Patel, T., Wadia, V., et al. (2017). Tonic or phasic stimulation of dopaminergic projections to prefrontal cortex causes mice to maintain or deviate from previously learned behavioral strategies. *The Journal of Neuroscience, 37*(35), 8315–8329.

Elston, T. W., & Bilkey, D. K. (2017). Anterior cingulate cortex modulation of the ventral tegmental area in an effort task. *Cell Reports, 19*(11), 2220–2230.

Elston, T. W., Croy, E., & Bilkey, D. K. (2019). Communication between the anterior cingulate cortex and ventral tegmental area during a cost-benefit reversal task. *Cell Reports, 26*(9), 2353–2361.

Etkin, A., Egner, T., & Kalisch, R. (2011). Emotional processing in anterior cingulate and medial prefrontal cortex. *Trends in Cognitive Sciences, 15*(2), 85–93.

Euston, D. R., & McNaughton, B. L. (2006). Apparent encoding of sequential context in rat medial prefrontal cortex is accounted for by behavioral variability. *The Journal of Neuroscience, 26*(51), 13143–13155.

Fagerlund, B., Pinborg, L. H., Mortensen, E. L., et al. (2013). Relationship of frontal D(2/3) binding potentials to cognition: a study of antipsychotic-naive schizophrenia patients. *The International Journal of Neuropsychopharmacology, 16*(1), 23–36.

Faillace, L. A., Allen, R. P., McQueen, J. D., & Northrup, B. (1971). Cognitive deficits from bilateral cingulotomy for intractable pain in man. *Diseases of the Nervous System, 32*(3), 171–175.

Fallon, S. J., Williams-Gray, C. H., Barker, R. A., Owen, A. M., & Hampshire, A. (2013). Prefrontal dopamine levels determine the balance between cognitive stability and flexibility. *Cerebral Cortex, 23*(2), 361–369.

Febo, M. (2012). Firing patterns of maternal rat prelimbic neurons during spontaneous contact with pups. *Brain Research Bulletin, 88*(5), 534–542.

Feldman, S., & Conforti, N. (1985). Modifications of adrenocortical responses following frontal cortex simulation in rats with hypothalamic deafferentations and medial forebrain bundle lesions. *Neuroscience, 15*(4), 1045–1047.

Fellows, L. K., & Farah, M. J. (2005). Is anterior cingulate cortex necessary for cognitive control? *Brain, 128*(Pt 4), 788–796.

Fillinger, C., Yalcin, I., Barrot, M., & Veinante, P. (2018). Efferents of anterior cingulate areas 24a and 24b and midcingulate areas 24a' and 24b' in the mouse. *Brain Structure & Function, 223*(4), 1747–1778.

Fodstad, H., Strandman, E., Karlsson, B., & West, K. A. (1982). Treatment of chronic obsessive compulsive states with stereotactic anterior capsulotomy or cingulotomy. *Acta Neurochirurgica, 62*(1–2), 1–23.

Foltz, E. L., & White, L. E., Jr. (1962). Pain "relief" by frontal cingulumotomy. *Journal of Neurosurgery, 19*, 89–100.

Fullana, M. A., Albajes-Eizagirre, A., Soriano-Mas, C., et al. (2018). Fear extinction in the human brain: A meta-analysis of fMRI studies in healthy participants. *Neuroscience and Biobehavioral Reviews, 88*, 16–25.

Gabbott, P. L., Warner, T. A., Jays, P. R., Salway, P., & Busby, S. J. (2005). Prefrontal cortex in the rat: Projections to subcortical autonomic, motor, and limbic centers. *The Journal of Comparative Neurology, 492*(2), 145–177.

Gao, Y. J., Ren, W. H., Zhang, Y. Q., & Zhao, Z. Q. (2004). Contributions of the anterior cingulate cortex and amygdala to pain- and fear-conditioned place avoidance in rats. *Pain, 110*(1–2), 343–353.

Goschke, T., & Bolte, A. (2014). Emotional modulation of control dilemmas: the role of positive affect, reward, and dopamine in cognitive stability and flexibility. *Neuropsychologia, 62*, 403–423.

Groenewegen, H. J., & Uylings, H. B. (2000). The prefrontal cortex and the integration of sensory, limbic and autonomic information. *Progress in Brain Research, 126*, 3–28.

Groenewegen, H. J., Wright, C. I., & Uylings, H. B. (1997). The anatomical relationships of the prefrontal cortex with limbic structures and the basal ganglia. *Journal of Psychopharmacology, 11*(2), 99–106.

Gruber, A. J., & McDonald, R. J. (2012). Context, emotion, and the strategic pursuit of goals: interactions among multiple brain systems controlling motivated behavior. *Frontiers in Behavioral Neuroscience, 6*, 50.

Gu, S., Gao, M., Yan, Y., Wang, F., Tang, Y. Y., & Huang, J. H. (2018). The neural mechanism underlying cognitive and emotional processes in creativity. *Frontiers in Psychology, 9*, 1924.

Halladay, L. R., & Blair, H. T. (2015). Distinct ensembles of medial prefrontal cortex neurons are activated by threatening stimuli that elicit excitation vs. inhibition of movement. *Journal of Neurophysiology, 114*(2), 793–807.

Hassenbusch, S. J., Pillay, P. K., & Barnett, G. H. (1990). Radiofrequency cingulotomy for intractable cancer pain using stereotaxis guided by magnetic resonance imaging. *Neurosurgery, 27*(2), 220–223.

Hayden, B. Y., Heilbronner, S. R., Pearson, J. M., & Platt, M. L. (2011). Surprise signals in anterior cingulate cortex: neuronal encoding of unsigned reward prediction errors driving adjustment in behavior. *The Journal of Neuroscience, 31*(11), 4178–4187.

Heilbronner, S. R., & Hayden, B. Y. (2016). Dorsal anterior cingulate cortex: A bottom-up view. *Annual Review of Neuroscience, 39*, 149–170.

Hillman, K. L., & Bilkey, D. K. (2010). Neurons in the rat anterior cingulate cortex dynamically encode cost-benefit in a spatial decision-making task. *The Journal of Neuroscience, 30*(22), 7705–7713.

Hok, V., Save, E., Lenck-Santini, P. P., & Poucet, B. (2005). Coding for spatial goals in the prelimbic/infralimbic area of the rat frontal cortex. *Proceedings of the National Academy of Sciences of the United States of America, 102*(12), 4602–4607.

Holroyd, C. B., & Coles, M. G. H. (2002). The neural basis of human error processing: reinforcement learning, dopamine, and the error-related negativity. *Psychological Review, 109*(4), 679–709.

Holroyd, C. B., & McClure, S. M. (2015). Hierarchical control over effortful behavior by rodent medial frontal cortex: A computational model. *Psychological Review, 122*(1), 54–83.

Hoover, W. B., & Vertes, R. P. (2007). Anatomical analysis of afferent projections to the medial prefrontal cortex in the rat. *Brain Structure & Function, 212*(2), 149–179.

Horst, N. K., & Laubach, M. (2012). Working with memory: evidence for a role for the medial prefrontal cortex in performance monitoring during spatial delayed alternation. *Journal of Neurophysiology, 108*(12), 3276–3288.

Horst, N. K., & Laubach, M. (2013). Reward-related activity in the medial prefrontal cortex is driven by consumption. *Frontiers in Neuroscience, 7*, 56.

Huang, J., Chang, J. Y., Woodward, D. J., Baccalá, L. A., Han, J. S., Wang, J. Y., et al. (2006). Dynamic neuronal responses in cortical and thalamic areas during different phases of formalin test in rats. *Experimental Neurology, 200*(1), 124–134.

Hurt, R. W., & Ballantine, H. T., Jr. (1974). Stereotactic anterior cingulate lesions for persistent pain: a report on 68 cases. *Clinical Neurosurgery, 21*, 334–351.

Hyman, J. M., Holroyd, C. B., & Seamans, J. K. (2017). A novel neural prediction error found in anterior cingulate cortex ensembles. *Neuron, 95*(2), 447–456.e3.

Hyman, J. M., Ma, L., Balaguer-Ballester, E., Durstewitz, D., & Seamans, J. K. (2012). Contextual encoding by ensembles of medial prefrontal cortex neurons. *Proceedings of the National Academy of Sciences of the United States of America, 109*(13), 5086–5091.

Insel, N., & Barnes, C. A. (2015). Differential activation of fast-spiking and regular-firing neuron populations during movement and reward in the dorsal medial frontal cortex. *Cerebral Cortex*, *25*(9), 2631–2647.

Ito, S., Stuphorn, V., Brown, J. W., & Schall, J. D. (2003). Performance monitoring by the anterior cingulate cortex during saccade countermanding. *Science*, *302*(5642), 120–122.

Janer, K. W., & Pardo, J. V. (1991). Deficits in selective attention following bilateral anterior cingulotomy. *Journal of Cognitive Neuroscience*, *3*(3), 231–241.

Jung, M. W., Qin, Y., McNaughton, B. L., & Barnes, C. A. (1998). Firing characteristics of deep layer neurons in prefrontal cortex in rats performing spatial working memory tasks. *Cerebral Cortex*, *8*(5), 437–450.

Kahnt, T., Weber, S. C., Haker, H., Robbins, T. W., & Tobler, P. N. (2015). Dopamine D2-receptor blockade enhances decoding of prefrontal signals in humans. *The Journal of Neuroscience*, *35*(9), 4104–4111.

Kennerley, S. W., & Wallis, J. D. (2009). Encoding of reward and space during a working memory task in the orbitofrontal cortex and anterior cingulate sulcus. *Journal of Neurophysiology*, *102*(6), 3352–3364.

Khamassi, M., Quilodran, R., Enel, P., Dominey, P. F., & Procyk, E. (2015). Behavioral regulation and the modulation of information coding in the lateral prefrontal and cingulate cortex. *Cerebral Cortex*, *25*(9), 3197–3218.

Kim, D., Jeong, H., Lee, J., et al. (2016). Distinct roles of parvalbumin- and somatostatin-expressing interneurons in working memory. *Neuron*, *92*(4), 902–915.

Kitamura, T., Ogawa, S. K., Roy, D. S., et al. (2017). Engrams and circuits crucial for systems consolidation of a memory. *Science*, *356*(6333), 73–78.

Koga, K., Yamada, A., Song, Q., et al. (2020). Ascending noradrenergic excitation from the locus coeruleus to the anterior cingulate cortex. *Molecular Brain*, *13*(1), 49.

Kringelbach, M. L., & Berridge, K. C. (2017). The affective core of emotion: Linking pleasure, subjective well-being, and optimal metastability in the brain. *Emotion Review*, *9*(3), 191–199.

Kroener, S., Chandler, L. J., Phillips, P. E., & Seamans, J. K. (2009). Dopamine modulates persistent synaptic activity and enhances the signal-to-noise ratio in the prefrontal cortex. *PLoS One*, *4*(8), e6507.

Kulkarni, B., Bentley, D. E., Elliott, R., et al. (2005). Attention to pain localization and unpleasantness discriminates the functions of the medial and lateral pain systems. *The European Journal of Neuroscience*, *21*(11), 3133–3142.

Kuo, M. F., Paulus, W., & Nitsche, M. A. (2008). Boosting focally-induced brain plasticity by dopamine. *Cerebral Cortex*, *18*(3), 648–651.

Laitinen, L. V. (1972). Stereotactic lesions in the knee of the corpus callosum in the treatment of emotional disorders. *Lancet*, *I*, 472–475.

Lapidus, K. A., Kopell, B. H., Ben-Haim, S., Rezai, A. R., & Goodman, W. K. (2013). History of psychosurgery: a psychiatrist's perspective. *World Neurosurgery*, *80*(3–4), S27.e1–S27.e16.

Lapish, C. C., Balaguer-Ballester, E., Seamans, J. K., Phillips, A. G., & Durstewitz, D. (2015). Amphetamine exerts dose-dependent changes in prefrontal cortex attractor dynamics during Working memory. *The Journal of Neuroscience*, *35*(28), 10172–10187.

Lapish, C. C., Durstewitz, D., Chandler, L. J., & Seamans, J. K. (2008). Successful choice behavior is associated with distinct and coherent network states in anterior cingulate cortex. *Proceedings of the National Academy of Sciences of the United States of America*, *105*(33), 11963–11968.

Laubach, M., Amarante, L. M., Swanson, K., & White, S. R. (2018). What, if anything, is rodent prefrontal cortex? *eNeuro*, *5*(5), 1–14.

Lewin, W., & Whitty, C. W. (1960). Effects of anterior cingulate stimulation in conscious human subjects. *Journal of Neurophysiology*, *23*, 445–447.

Lindgren, L., Westling, G., Brulin, C., Lehtipalo, S., Andersson, M., & Nyberg, L. (2012). Pleasant human touch is represented in pregenual anterior cingulate cortex. *NeuroImage*, *59*(4), 3427–3432.

Lindsay, A. J., Caracheo, B. F., Grewal, J. J. S., Leibovitz, D., & Seamans, J. K. (2018). How much does movement and location encoding impact prefrontal cortex activity? An algorithmic decoding approach in freely moving rats. *eNeuro*, *5*(2), ENEURO.0023-18.2018.

Long, C. J., Pueschel, K., & Hunter, S. E. (1978). Assessment of the effects of cinglulate gyrus lesions by neuropsychological techniques. *Journal of Neurosurgery*, *49*(2), 264–271.

Lövheim, H. (2012). A new three-dimensional model for emotions and monoamine neurotransmitters. *Medical Hypotheses*, *78*(2), 341–348.

Lu, J. S., Chen, Q. Y., Zhou, S., Inokuchi, K., & Zhuo, M. (2018). Dual roles of anterior cingulate cortex neurons in pain and pleasure in adult mice. *Molecular Brain*, *11*(1), 72.

Ma, L., Hyman, J. M., Durstewitz, D., Phillips, A. G., & Seamans, J. K. (2016). A quantitative analysis of context-dependent remapping of medial frontal cortex neurons and ensembles. *The Journal of Neuroscience*, *36*(31), 8258–8272.

Ma, L., Hyman, J. M., Phillips, A. G., & Seamans, J. K. (2014). Tracking progress toward a goal in corticostriatal ensembles. *The Journal of Neuroscience*, *34*(6), 2244–2253.

Mashhoori, A., Hashemnia, S., McNaughton, B. L., Euston, D. R., & Gruber, A. J. (2018). Rat anterior cingulate cortex recalls features of remote reward locations after disfavoured reinforcements. *eLife*, 7, e29793.

Matsumoto, K., Suzuki, W., & Tanaka, K. (2003). Neuronal correlates of goal-based motor selection in the prefrontal cortex. *Science*, *301*(5630), 229–232.

Matsunaga, M., Kawamichi, H., Koike, T., et al. (2016). Structural and functional associations of the rostral anterior cingulate cortex with subjective happiness. *NeuroImage*, *134*, 132–141.

McGovern, R. A., & Sheth, S. A. (2017). Role of the dorsal anterior cingulate cortex in obsessive-compulsive disorder: converging evidence from cognitive neuroscience and psychiatric neurosurgery. *Journal of Neurosurgery*, *126*(1), 132–147.

Medford, N., & Critchley, H. D. (2010). Conjoint activity of anterior insular and anterior cingulate cortex: awareness and response. *Brain Structure & Function*, *214*(5–6), 535–549.

Mesulam, M. M. (1981). A cortical network for directed attention and unilateral neglect. *Annals of Neurology*, *10*(4), 309–325.

Mettler, F. A. (1949). Anatomy and physiology. In F. A. Mettler (Ed.), *Selective partial ablation of the frontal cortex* (pp. 477–491). New York: Hoeber.

Meyer, G., McElhaney, M., Martin, W., & McGraw, C. P. (1973). Stereotactic cingulotomy with results of acute stimulation and serial psychological testing. In L. V. Laitinen, & K. E. Livingston (Eds.), *Surgical approaches in psychiatry* (pp. 39–58). Baltimore, MD: University Park Press.

Meyer-Lindenberg, A., Kohn, P. D., Kolachana, B., et al. (2005). Midbrain dopamine and prefrontal function in humans: interaction and modulation by COMT genotype. *Nature Neuroscience*, *8*(5), 594–596.

Narayanan, N. S. (2016). Ramping activity is a cortical mechanism of temporal control of action. *Current Opinion in Behavioral Sciences*, *8*, 226–230.

Narayanan, N. S., Cavanagh, J. F., Frank, M. J., & Laubach, M. (2013). Common medial frontal mechanisms of adaptive control in humans and rodents. *Nature Neuroscience*, *16*(12), 1888–1895.

Narayanan, N. S., & Laubach, M. (2008). Neuronal correlates of post-error slowing in the rat dorsomedial prefrontal cortex. *Journal of Neurophysiology*, *100*(1), 520–525.

Neafsey, E. J. (1990). Prefrontal cortical control of the autonomic nervous system: anatomical and physiological observations. *Progress in Brain Research*, *85*, 147–166.

Neafsey, E. J., Terreberry, R. R., Hurley, K. M., Ruit, K. G., & Frysztak, R. J. (1993). Anterior cingulate cortex in rodents: Connections, visceral control functions, and implications for emotion. In B. A. Vogt, & M. Gabriel (Eds.), *Neurobiology of cingulate cortex and limbic thalamus: A comprehensive handbook* (pp. 206–223).

Ochsner, K. N., Kosslyn, S. M., Cosgrove, G. R., et al. (2001). Deficits in visual cognition and attention following bilateral anterior cingulotomy. *Neuropsychologia*, *39*(3), 219–230.

Ott, T., & Nieder, A. (2017). Dopamine D2 receptors enhance population dynamics in primate prefrontal working memory circuits. *Cerebral Cortex*, *27*(9), 4423–4435.

Palermo, S., Benedetti, F., Costa, T., & Amanzio, M. (2015). Pain anticipation: an activation likelihood estimation meta-analysis of brain imaging studies. *Human Brain Mapping*, *36*(5), 1648–1661.

Papez, J. W. (1995). A proposed mechanism of emotion. 1937. *The Journal of Neuropsychiatry and Clinical Neurosciences*, 7(1), 103–112.

Parvizi, J., Rangarajan, V., Shirer, W. R., Desai, N., & Greicius, M. D. (2013). The will to persevere induced by electrical stimulation of the human cingulate gyrus. *Neuron*, *80*(6), 1359–1367.

Passecker, J., Mikus, N., Malagon-Vina, H., et al. (2019). Activity of prefrontal neurons predict future choices during gambling. *Neuron*, *101*(1), 152–164.e7.

Passingham, R. E. (1996). Attention to action. *Philosophical Transactions of the Royal Society of London. Series B, Biological Sciences*, *351*(1346), 1473–1479.

Paus, T. (2001). Primate anterior cingulate cortex: where motor control, drive and cognition interface. *Nature Reviews. Neuroscience*, *2*(6), 417–424.

Paxinos, G., & Watson, C. (1986). *The rat brain in stereotaxic coordinates*. New York: Academic Press.

Peters, J. L., & Michael, A. C. (2000). Changes in the kinetics of dopamine release and uptake have differential effects on the spatial distribution of extracellular dopamine concentration in rat striatum. *Journal of Neurochemistry*, *74*(4), 1563–1573.

Petykó, Z., Gálosi, R., Tóth, A., Máté, K., Szabó, I., Karádi, Z., et al. (2015). Responses of rat medial prefrontal cortical neurons to Pavlovian conditioned stimuli and to delivery of appetitive reward. *Behavioural Brain Research*, *287*, 109–119.

Petykó, Z., Tóth, A., Szabó, I., Gálosi, R., & Lénárd, L. (2009). Neuronal activity in rat medial prefrontal cortex during sucrose solution intake. *Neuroreport*, *20*(14), 1235–1239.

Pool, J. L., & Ransohoff, J. (1949). Autonomic effects on stimulating rostral portion of cingulate gyri in man. *Journal of Neurophysiology*, *12*(6), 385–392.

Powell, D. A., Buchanan, S. L., & Gibbs, C. M. (1990). Role of the prefrontal- -thalamic axis in classical conditioning. *Progress in Brain Research*, *85*, 433–466.

Powell, D. A., & Ginsberg, J. P. (2005). Single unit activity in the medial prefrontal cortex during Pavlovian heart rate conditioning: Effects of peripheral autonomic blockade. *Neurobiology of Learning and Memory*, *84*(3), 200–213.

Powell, D. A., Maxwell, B., & Penney, J. (1996). Neuronal activity in the medial prefrontal cortex during Pavlovian eyeblink and nictitating membrane conditioning. *The Journal of Neuroscience*, *16*(19), 6296–6306.

Powell, N. J., & Redish, A. D. (2014). Complex neural codes in rat prelimbic cortex are stable across days on a spatial decision task. *Frontiers in Behavioral Neuroscience*, *8*, 120.

Pratt, W. E., & Mizumori, S. J. (2001). Neurons in rat medial prefrontal cortex show anticipatory rate changes to predictable differential rewards in a spatial memory task. *Behavioural Brain Research*, *123*(2), 165–183.

Rich, E. L., & Shapiro, M. (2009). Rat prefrontal cortical neurons selectively code strategy switches. *The Journal of Neuroscience*, *29*(22), 7208–7219.

Richter, E. O., Davis, K. D., Hamani, C., Hutchison, W. D., Dostrovsky, J. O., & Lozano, A. M. (2004). Cingulotomy for psychiatric disease: Microelectrode guidance, a callosal reference system for documenting lesion location, and clinical results. *Neurosurgery*, *54*(3), 622–630.

Robinson, M. J., Edwards, S. E., Iyengar, S., Bymaster, F., Clark, M., & Katon, W. (2009). Depression and pain. *Frontiers in Bioscience (Landmark Ed)*, *14*, 5031–5051.

Rolls, E. T., & Deco, G. (2011). A computational neuroscience approach to schizophrenia and its onset. *Neuroscience and Biobehavioral Reviews*, *35*(8), 1644–1653.

Rolls, E. T., Kringelbach, M. L., & de Araujo, I. E. (2003). Different representations of pleasant and unpleasant odours in the human brain. *The European Journal of Neuroscience*, *18*(3), 695–703.

Rosa, E. C., Dickinson, D., Apud, J., Weinberger, D. R., & Elvevåg, B. (2010). COMT Val158Met polymorphism, cognitive stability and cognitive flexibility: an experimental examination. *Behavioral and Brain Functions*, *6*, 53.

Rushworth, M. F., Noonan, M. P., Boorman, E. D., Walton, M. E., & Behrens, T. E. (2011). Frontal cortex and reward-guided learning and decision-making. *Neuron*, *70*(6), 1054–1069.

Satake, T., Mitani, H., Nakagome, K., & Kaneko, K. (2008). Individual and additive effects of neuromodulators on the slow components of afterhyperpolarization currents in layer V pyramidal cells of the rat medial prefrontal cortex. *Brain Research*, *1229*, 47–60.

Scopinho, A. A., Tavares, R. F., & Corrêa, F. M. (2009). The medial forebrain bundle mediates cardiovascular responses to electrical stimulation of the medial prefrontal cortex. *Autonomic Neuroscience*, *147*(1–2), 38–47.

Seamans, J. K., & Robbins, T. W. (2009). Dopamine modulation of the prefrontal cortex and cognitive function. In K. Neve (Ed.), *Dopamine receptors*. Totowa, NJ: Humana.

Seamans, J. K., & Yang, C. R. (2004). The principal features and mechanisms of dopamine modulation in the prefrontal cortex. *Progress in Neurobiology*, *74*(1), 1–58.

Sesack, S. R., Deutch, A. Y., Roth, R. H., & Bunney, B. S. (1989). Topographical organization of the efferent projections of the medial prefrontal cortex in the rat: an anterograde tract-tracing study with Phaseolus vulgaris leucoagglutinin. *The Journal of Comparative Neurology*, *290*(2), 213–242.

Shackman, A. J., Salomons, T. V., Slagter, H. A., Fox, A. S., Winter, J. J., & Davidson, R. J. (2011). The integration of negative affect, pain and cognitive control in the cingulate cortex. *Nature Reviews. Neuroscience*, *12*(3), 154–167.

Shafiei, G., Zeighami, Y., Clark, C. A., et al. (2019). Dopamine signaling modulates the stability and integration of intrinsic brain networks. *Cerebral Cortex*, *29*(1), 397–409.

Sharim, J., & Pouratian, N. (2016). Anterior cingulotomy for the treatment of chronic intractable pain: A systematic review. *Pain Physician*, *19*(8), 537–550.

Sharma, T. (1974). Abolition of opiate hunger in humans following bilateral anterior cingulotomy. *Texas Medicine*, *70*(10), 49–52.

Shenhav, A., Cohen, J. D., & Botvinick, M. M. (2016). Dorsal anterior cingulate cortex and the value of control. *Nature Neuroscience*, *19*(10), 1286–1289.

Sheth, S. A., Mian, M. K., Patel, S. R., et al. (2012). Human dorsal anterior cingulate cortex neurons mediate ongoing behavioural adaptation. *Nature*, *488*(7410), 218–221.

Shidara, M., & Richmond, B. J. (2002). Anterior cingulate: single neuronal signals related to degree of reward expectancy. *Science*, *296*(5573), 1709–1711.

Shields, D. C., Asaad, W., Eskandar, E. N., Jain, F. A., Cosgrove, G. R., Flaherty, A. W., et al. (2008). Prospective assessment of stereotactic ablative surgery for intractable major depression. *Biological Psychiatry*, *64*(6), 449–454.

Shine, J. M., Breakspear, M., Bell, P. T., et al. (2019). Human cognition involves the dynamic integration of neural activity and neuromodulatory systems. *Nature Neuroscience*, *22*(2), 289–296.

Siciliano, C. A., Calipari, E. S., & Jones, S. R. (2014). Amphetamine potency varies with dopamine uptake rate across striatal subregions. *Journal of Neurochemistry*, *131*(3), 348–355.

Sikes, R. W., & Vogt, B. A. (1992). Nociceptive neurons in area 24 of rabbit cingulate cortex. *Journal of Neurophysiology*, *68*(5), 1720–1732.

Sikes, R. W., Vogt, L. J., & Vogt, B. A. (2008). Distribution and properties of visceral nociceptive neurons in rabbit cingulate cortex. *Pain*, *135*(1–2), 160–174.

Smith, W. K. (1945). The functional significance of the rostral cingular cortex as revealed by its responses to electrical excitation. *Journal of Neurophysiology*, *8*(4), 241–255.

St Onge, J. R., Abhari, H., & Floresco, S. B. (2011 Jun 8). Dissociable contributions by prefrontal D1 and D2 receptors to risk-based decision making. *The Journal of Neuroscience*, *31*(23), 8625–8633.

Steele, J. D., Christmas, D., Eljamel, M. S., & Matthews, K. (2008). Anterior cingulotomy for major depression: clinical outcome and relationship to lesion characteristics. *Biological Psychiatry*, *63*(7), 670–677.

Steenland, H. W., Li, X. Y., & Zhuo, M. (2012). Predicting aversive events and terminating fear in the mouse anterior cingulate cortex during trace fear conditioning. *The Journal of Neuroscience*, *32*(3), 1082–1095.

Stelzel, C., Fiebach, C. J., Cools, R., Tafazoli, S., & D'Esposito, M. (2013). Dissociable fronto-striatal effects of dopamine D2 receptor stimulation on cognitive versus motor flexibility. *Cortex*, *49*(10), 2799–2811.

Stephens, E. K., Avesar, D., & Gulledge, A. T. (2014). Activity-dependent serotonergic excitation of callosal projection neurons in the mouse prefrontal cortex. *Frontiers in Neural Circuits*, *8*, 97.

Strauss, I., Berger, A., Ben Moshe, S., et al. (2017). Double anterior stereotactic cingulotomy for intractable oncological pain. *Stereotactic and Functional Neurosurgery*, *95*(6), 400–408.

Struik, R. F., Marchant, N. J., de Haan, R., et al. (2019). Dorsomedial prefrontal cortex neurons encode nicotine-cue associations. *Neuropsychopharmacology*, *44*(12), 2011–2021.

Stuss, D. T., Floden, D., Alexander, M. P., Levine, B., & Katz, D. (2001). Stroop performance in focal lesion patients: dissociation of processes and frontal lobe lesion location. *Neuropsychologia*, *39*(8), 771–786.

Suardi, A., Sotgiu, I., Costa, T., Cauda, F., & Rusconi, M. (2016). The neural correlates of happiness: A review of PET and fMRI studies using autobiographical recall methods. *Cognitive, Affective, & Behavioral Neuroscience*, *16*(3), 383–392.

Talairach, J., Bancaud, J., Geier, S., Bordas-Ferrer, M., Bonis, A., Szikla, G., et al. (1973). The cingulate gyrus and human behavior. *Electroencephalography and Clinical Neurophysiology*, *34*, 45–52.

Teixeira, C. M., Pomedli, S. R., Maei, H. R., Kee, N., & Frankland, P. W. (2006). Involvement of the anterior cingulate cortex in the expression of remote spatial memory. *The Journal of Neuroscience*, *26*(29), 7555–7564.

Terreberry, R. R., & Neafsey, E. J. (1983). Rat medial frontal cortex: a visceral motor region with a direct projection to the solitary nucleus. *Brain Research*, *278*(1–2), 245–249.

Teuber, H. L., Corkin, S., & Twitchell, T. E. (1977). A study of cingulotomy in man. In *Appendix. Psychosurgery. The national commission for the protection of human subjects of biomedical and behavioral research* (pp. 1–115). U.S. Department of Health, Education and Welfare. Dhew Publication No. (os) 77–0002, Contract No. NO1-HU-6-2116.

Thurley, K., Senn, W., & Lüscher, H. R. (2008). Dopamine increases the gain of the input-output response of rat prefrontal pyramidal neurons. *Journal of Neurophysiology*, *99*(6), 2985–2997.

Tian, M. K., Schmidt, E. F., & Lambe, E. K. (2016). Serotonergic suppression of mouse prefrontal circuits implicated in task attention. *eNeuro*, *3*(5), ENEURO.0269–16.2016.

Toda, K., Sugase-Miyamoto, Y., Mizuhiki, T., Inaba, K., Richmond, B. J., & Shidara, M. (2012). Differential encoding of factors influencing predicted reward value in monkey rostral anterior cingulate cortex. *PLoS One*, 7(1), e30190.

Tolomeo, S., Christmas, D., Jentzsch, I., et al. (2016). A causal role for the anterior mid-cingulate cortex in negative affect and cognitive control. *Brain*, *139*(Pt 6), 1844–1854.

Ueltzhöffer, K., Armbruster-Genç, D. J., & Fiebach, C. J. (2015). Stochastic dynamics underlying cognitive stability and flexibility. *PLoS Computational Biology*, *11*(6), e1004331.

Van Eden, C. G., & Buijs, R. M. (2000). Functional neuroanatomy of the prefrontal cortex: autonomic interactions. *Progress in Brain Research*, *126*, 49–62.

van Heukelum, S., Mars, R. B., Guthrie, M., Buitelaar, J. K., Beckmann, C. F., Tiesinga, P. H. E., et al. (2020). Where is cingulate cortex? A cross-species view. *Trends in Neurosciences*, *43*(5), 285–299.

van Holstein, M., Aarts, E., van der Schaaf, M. E., et al. (2011). Human cognitive flexibility depends on dopamine D2 receptor signaling. *Psychopharmacology*, *218*(3), 567–578.

Verberne, A. J., & Owens, N. C. (1998). Cortical modulation of the cardiovascular system. *Progress in Neurobiology*, *54*(2), 149–168.

Viswanathan, A., Harsh, V., Pereira, E. A., & Aziz, T. Z. (2013). Cingulotomy for medically refractory cancer pain. *Neurosurgical Focus*, *35*(3), E1.

Vogt, B. A. (2005). Pain and emotion interactions in subregions of the cingulate gyrus. *Nature Reviews. Neuroscience*, *6*(7), 533–544.

Vogt, B. A. (2016). Midcingulate cortex: Structure, connections, homologies, functions and diseases. *Journal of Chemical Neuroanatomy*, *74*, 28–46.

Vogt, B. A., Berger, G. R., & Derbyshire, S. W. (2003). Structural and functional dichotomy of human midcingulate cortex. *The European Journal of Neuroscience*, *18*(11), 3134–3144.

Vogt, B. A., Hof, P. R., Zilles, K., Vogt, L. J., & Herold, C. (2013). Palomero-Gallagher N Cingulate area 32 homologies in mouse, rat, macaque and human: cytoarchitecture and receptor architecture. *The Journal of Comparative Neurology*, *521*(18), 4189–4204.

Vogt, B. A., & Paxinos, G. (2014). Cytoarchitecture of mouse and rat cingulate cortex with human homologies. *Brain Structure & Function*, *219*(1), 185–192.

Volpini, M., Giacobbe, P., Cosgrove, G. R., Levitt, A., Lozano, A. M., & Lipsman, N. (2017). The history and future of ablative neurosurgery for major depressive disorder. *Stereotactic and Functional Neurosurgery*, *95*(4), 216–228.

Wang, J. Y., Zhang, H. T., Chang, J. Y., Woodward, D. J., Baccalá, L. A., & Luo, F. (2008). Anticipation of pain enhances the nociceptive transmission and functional connectivity within pain network in rats. *Molecular Pain*, *4*, 34.

Warden, M. R., Selimbeyoglu, A., Mirzabekov, J. J., et al. (2012). A prefrontal cortex-brainstem neuronal projection that controls response to behavioural challenge. *Nature*, *492*(7429), 428–432.

Whitty, C. W., Duffield, J. E., Tov, P. M., & Cairns, H. (1952). Anterior cingulectomy in the treatment of mental disease. *Lancet*, *1*(6706), 475–481.

Wiech, K., & Tracey, I. (2009). The influence of negative emotions on pain: behavioral effects and neural mechanisms. *NeuroImage*, *47*(3), 987–994.

Wilkinson, H. A. (2009). Cingulotomy. *Journal of Neurosurgery*, *110*(3), 607.

Yang, J. C., Ginat, D. T., Dougherty, D. D., Makris, N., & Eskandar, E. N. (2014). Lesion analysis for cingulotomy and limbic leucotomy: comparison and correlation with clinical outcomes. *Journal of Neurosurgery*, *120*(1), 152–163.

Yang, Z., Tan, Q., Cheng, D., et al. (2018). The changes of intrinsic excitability of pyramidal neurons in anterior cingulate Cortex in neuropathic pain. *Frontiers in Cellular Neuroscience*, *12*, 436.

Yen, C. P., Kuan, C. Y., Sheehan, J., et al. (2009). Impact of bilateral anterior cingulotomy on neurocognitive function in patients with intractable pain. *Journal of Clinical Neuroscience*, *16*(2), 214–219.

Yin, L., Rasch, M. J., He, Q., Wu, S., Dou, F., & Shu, Y. (2017). Selective modulation of axonal sodium channel subtypes by 5-HT1A receptor in cortical pyramidal neuron. *Cerebral Cortex*, *27*(1), 509–521.
Yoshino, A., Okamoto, Y., Onoda, K., et al. (2010). Sadness enhances the experience of pain via neural activation in the anterior cingulate cortex and amygdala: an fMRI study. *NeuroImage*, *50*(3), 1194–1201.
Yuen, E. Y., Qin, L., Wei, J., Liu, W., Liu, A., & Yan, Z. (2014). Synergistic regulation of glutamatergic transmission by serotonin and norepinephrine reuptake inhibitors in prefrontal cortical neurons. *The Journal of Biological Chemistry*, *289*(36), 25177–25185.
Zhuo, M. (2016). Neural mechanisms underlying anxiety-chronic pain interactions. *Trends in Neurosciences*, *39*(3), 136–145.

CHAPTER SEVEN

Valence processing in the PFC: Reconciling circuit-level and systems-level views

Austin A. Coley[†], Nancy Padilla-Coreano[†], Reesha Patel[†], and Kay M. Tye*

Salk Institute for Biological Studies, La Jolla, CA, United States

*Corresponding author: e-mail address: tye@salk.edu

Contents

[†] These authors contributed equally.

International Review of Neurobiology, Volume 158
ISSN 0074-7742
https://doi.org/10.1016/bs.irn.2020.12.002

Abstract

An essential component in animal behavior is the ability to process emotion and dissociate among positive and negative valence in response to a rewarding or aversive stimulus. The medial prefrontal cortex (mPFC)—responsible for higher order executive functions that include cognition, learning, and working memory; and is also involved in sociability—plays a major role in emotional processing and control. Although the amygdala is widely regarded as the "emotional hub," the mPFC encodes for context-specific salience and elicits top-down control over limbic circuitry. The mPFC can then conduct behavioral responses, via cortico-striatal and cortico-brainstem pathways, that correspond to emotional stimuli. Evidence shows that abnormalities within the mPFC lead to sociability deficits, working memory impairments, and drug-seeking behavior that include addiction and compulsive disorders; as well as conditions such as anhedonia. Recent studies investigate the effects of aberrant salience processing on cortical circuitry and neuronal populations associated with these behaviors. In this chapter, we discuss mPFC valence processing, neuroanatomical connections, and physiological substrates involved in mPFC-associated behavior. We review neurocomputational and theoretical models such as "mixed selectivity," that describe cognitive control, attentiveness, and motivational drives. Using this knowledge, we describe the effects of valence imbalances and its influence on mPFC neural pathways that contribute to deficits in social cognition, while understanding the effects in addiction/compulsive behaviors and anhedonia.

1. Introduction

The medial prefrontal cortex (mPFC), a prominent brain region located within the frontal lobe, is primarily responsible for functions that include decision-making processes and judgement (Bechara, Damasio, & Damasio, 2000); attentional processing (Miller & Buschman, 2013), learning and memory consolidation (Euston, Gruber, & McNaughton, 2012); as well as working memory (Goldman-Rakic, 1996; Sawaguchi & Goldman-Rakic, 1991), social cognition (Forbes & Grafman, 2010) and emotional processing (Etkin, Egner, & Kalisch, 2011). However, this chapter will highlight the critical role of mPFC circuits and neuronal ensembles in emotional processing in the context of mood disorders, drug addiction and social cognition due to their direct relevance to emotional regulation, and provide a more harmonious and unified view of mPFC function incorporating both circuit-level and systems-level insights and conceptual frameworks.

An important feature in emotional processing is the ability to integrate sensory information—via external inputs and homeostatic internal states—to trigger a specific emotional and behavioral response. In animals, emotions can impact survival, reproduction, motivation, and decision-making processes.

From a basic science perspective, the urgency is necessitated by the rapid proliferation of literature in two diametrically-opposed conceptual frameworks underpinning PFC function. We have witnessed the paradigm-shifting emergence of a systems-level conceptual framework highlighting (nonlinear) "mixed selectivity" of PFC neurons, which not only suggests that individual neurons may be selectively responsive to many parameters under different contexts and are thereby recruited to multiple, seemingly-distinct functions in different contexts, but also provides a systems-level explanation for how the mPFC is endowed with high functioning processing power (Allsop, Vander Weele, Wichmann, & Tye, 2014; Johnston, Palmer, & Freedman, 2020; Muir, Lopez, & Bagot, 2019; Riga et al., 2014; Vander Weele et al., 2018; Vander Weele, Siciliano, & Tye, 2018). The concept of flexibility in function is juxtaposed with accumulating evidence—and canonical circuit diagrams—depicting PFC circuits as for diverse, but implicitly fixed, functions ascribed to specific components by the input-output ("hardwired") architecture of the circuit. (Allsop et al., 2014; Muir et al., 2019; Riga et al., 2014; Siciliano et al., 2019; Vander Weele et al., 2018; Vander Weele, Siciliano, & Tye, 2018).

The often-overlooked function of mPFC valence processing is essential in distinguishing among motivational drives of reward and aversion. In the two-dimensional model of valence and arousal (Lang, 1995), valence is assigned as a positive (pleasant) or negative (unpleasant) hedonic value, whereas arousal illustrates the intensity or degree of the emotional state (Lang, Bradley, & Cuthbert, 1990; Russell, 1980; Tye, 2018). Indeed, the mPFC provides context-dependent valence processing that influences behaviors.

The mPFC elicits top-down control over subcortical regions such as the basal ganglia and limbic structures that affect memory, motivation, and emotional behaviors (Miller & Cohen, 2001). Consequently, mPFC dysfunction alters behavioral output. For instance, lesions to the mPFC and orbital frontal cortex (OFC) reveal a loss of emotional inhibitory control, impaired attentional processing (Dias, Robbins, & Roberts, 1996), reactive aggression (Anderson et al., 2007; Grafman et al., 1996; Pennington & Bennetto,

1993), and a reduction in social cognition (Blair & Cipolotti, 2000; Hornak, Rolls, & Wade, 1996). Other studies showed the mPFC is implicated in context-dependent drug-reward seeking behavior and compulsive disorders (Bossert et al., 2011, 2012), and displays disrupted processing during anhedonia due to dopamine (DA) imbalances (Callicott et al., 2003; Ferenczi et al., 2016; Hamani et al., 2011).

In an effort to better understand how seemingly static circuits mediate these dynamic behaviors, we review neurocomputational models of mixed selectivity, that describe complex and prominent features of the mPFC, and integrate circuit-level function into a conceptual framework in which to understand valence processing mPFC circuits across a wide-range of behaviors.

We highlight the role of the mPFC in emotional valence processing with a focus on mPFC neural circuits mediating reward and social behaviors. Although the mPFC is involved in myriad of emotional behaviors that include fear expression/extinction and response to threat (Sotres-Bayon & Quirk, 2010), we focus on the mPFC circuits encoding valence and its role in social behaviors, and how valence processing is disrupted during addiction and anhedonia.

2. The mPFC encodes context-specific valence processing

2.1 Emotional processing in the mPFC vs the amygdala

While the limbic system has been largely conserved across evolution (Janak & Tye, 2015), the PFC represents the most dramatic and significant evolutionary changes in the brain (Fig. 1). Among these species, the PFC is considerably larger in primates, and largest in absolute size within humans (Brodmann, 1909; Semendeferi, Damasio, Frank, & Van Hoesen, 1997; Semendeferi, Lu, Schenker, & Damasio, 2002), suggesting a major role for higher-order functional processing in this region. The human prefrontal cortex (PFC) can be divided into dorsolateral (dlPFC), ventrolateral (vlPFC), dorsomedial (dmPFC), ventromedial (vmPFC) and orbitofrontal (OFC) regions (Passingham & Wise, 2014); although neuroanatomists argue the topographic maps of subdivisions within the PFC (Carmichael & Price, 1994; Ongür, Ferry, & Price, 2003; Petrides, 1995). The mPFC in rodents, including mouse and rats, can be subdivided into three primary regions: the anterior cingulate cortex (ACC), the prelimbic cortex (PL) and infralimbic cortex (IL) (Heidbreder & Groenewegen, 2003; Ongür & Price, 2000).

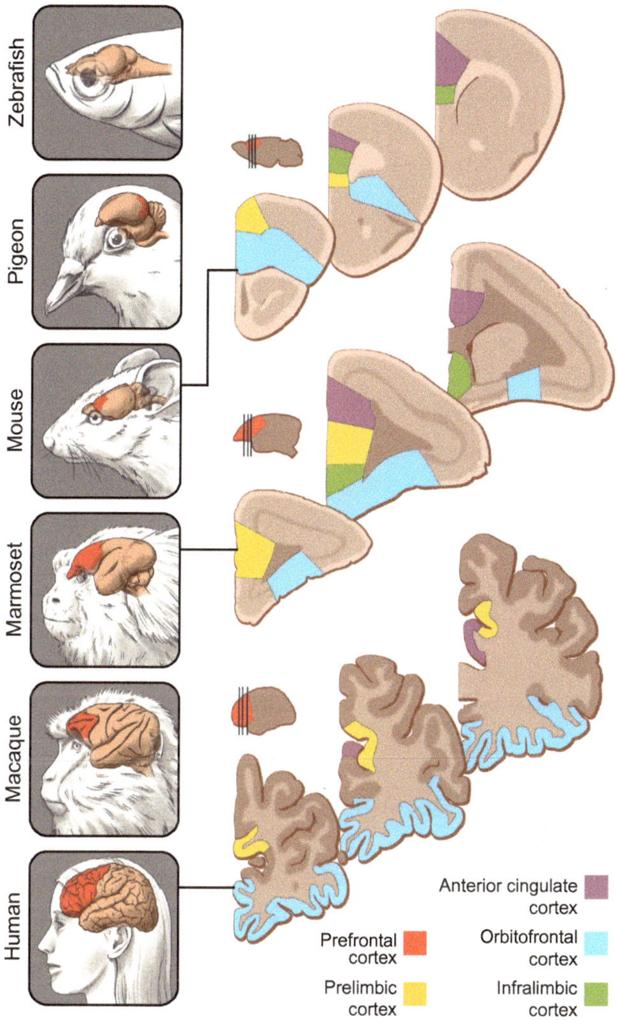

Fig. 1 Evolutionary changes of the PFC across species. The relative size and location of the prefrontal cortex (PFC; red) significantly differs among species: zebrafish (none), pigeon, mouse, marmoset, macaque, and human. Analogous PFC subregions are represented in coronal sections in the mouse, marmoset and human. Prelimbic cortex (yellow), anterior cingulate cortex (purple), orbitofrontal cortex (blue), and infralimbic (green).

However, some studies include the secondary motor cortex (M2) in the mPFC (Donoghue & Wise, 1982; Gabbott, Warner, Jays, Salway, & Busby, 2005), which remains debatable (Laubach, Amarante, Swanson, & White, 2018).

The mPFC plays an essential role in regulating emotions and salience processing by integrating multimodal sensory inputs and facilitating behavioral outputs dependent on context (Miller, 1999; Miller & Cohen, 2001; Ridderinkhof, Ullsperger, Crone, & Nieuwenhuis, 2004). In this respect, the mPFC elicits top-down control of limbic and brainstem regions that influence emotional behavior (Grace, Floresco, Goto, & Lodge, 2007; Lee & D'Esposito, 2012; Ochsner et al., 2009; Ochsner, Bunge, Gross, & Gabrieli, 2002; Quirk & Beer, 2006). Indeed, there are multiple circuit nodes that have been implicated in emotional processing, including the amygdala (Allsop et al., 2018; Burgos-Robles et al., 2017; Yizhar & Klavir, 2018), ventral hippocampus (vHP) (Fanselow & Dong, 2010), and ventromedial hypothalamus (VMH) (Kunwar et al., 2015); whereas the amygdala and a subdivision of the VMH has direct reciprocal connections with the mPFC (Lo et al., 2019; Yizhar & Klavir, 2018).

In contrast to the mPFC, the amygdala—widely considered the hub for emotional processing (Brown & Sharpey-Schafer III, 1888; Janak & Tye, 2015; Klüver & Bucy, 1937; LeDoux, 1996; Maren & Quirk, 2004; Tye, 2018; Weiskrantz, 1956)—provides bottom-up control in limbic circuitry, thereby facilitating rapid (LeDoux, 1996; Méndez-Bértolo et al., 2016; Tamietto & de Gelder, 2010), and automatic processing of emotions in response to salient stimuli, in addition to modulating attention and motivated-behavior (Davis & Whalen, 2001; LeDoux, 2000; Ochsner et al., 2009; Phelps & LeDoux, 2005). The amygdala is also involved in regulating fear responses (Fanselow & LeDoux, 1999; Maren, 2001). For instance, in the historical case study that examined patient S.M., diagnosed with Urbach-Wiethe disease, Adolph et al showed bilateral damage of the amygdala severely impairs fear and anger recognition in facial expressions. However, patient S.M. revealed no deficits in facial identity recognition, indicating the amygdala's role in facial affects and discerning emotional associations (Adolphs, Tranel, Damasio, & Damasio, 1994).

Indeed, the basolateral amygdala complex (BLA), that includes the lateral, basolateral, and basal amygdala nuclei, identified as an essential area in fear conditioning consists of specific neuronal populations involved in the acquisition of associative memories as well as retrieval (Clem & Huganir, 2010; Fanselow & Dong, 2010; Han et al., 2007, 2009; Namburi et al., 2015; Reijmers, Perkins, Matsuo, & Mayford, 2007; Rumpel, LeDoux, Zador, & Malinow, 2005). Whereas within the PFC, the acquisition, encoding and retrieval systems appear to be in distributed neuronal populations (Otis et al., 2017; Siciliano et al., 2019; Vander Weele et al., 2018). Although, within other emotional nodes such as the ventromedial hypothalamus, require specific neuronal populations activation that differentiate aggression and mating behaviors (Lin et al., 2011). These studies highlight key distinctions in neuronal circuits as well as cell properties among limbic regions and the PFC in emotional behavior.

2.2 The mPFC and emotional control

Another facet of mPFC valence processing is its involvement in emotional control and the ability to suppress impulse behavior. Previous studies in non-human primates showed damage to the OFC and lateral PFC elicits deficits in affective processing as well as attentional selection (Brozoski, Brown, Rosvold, & Goldman, 1979; Dias et al., 1996). Specifically, the mPFC, along with other circuits, play a role in inhibiting rage and aggression (Aleyasin, Flanigan, & Russo, 2018). In perhaps the earliest study examining the effects of frontal lobe damage, patient Phineas Gage, suffered severe prefrontal cortex damage during a work accident. Following the incident, Gage experienced significant cognitive and emotional changes that included learning deficits as well as irritable, hostile, capricious, and irrational behavior (Harlow, 1868, 1993). Thus, the PFC was targeted as a region for experimentation to control anomalous behavior. Early procedures, i.e., the psychosurgical interventions of the prefrontal leucotomy, now described as the "lobotomy," was developed to control human emotional and psychotic behavior in mental health patients (Moniz, 1937). The technique was designed to reduce aggressive tendencies by removing white matter that undercut afferent and efferent connections to the PFC. Although highly controversial and inconsistent, the PFC was highlighted as a region implicated in emotional inhibition (Abimbola & Awolowo, 2006; Tierney & Egas, 2000).

Historical evidence derived from human lesion studies corroborate the effects of aberrant emotional behavior sustained by prefrontal cortex damage (Papez, 1937). For example, human studies showed that patients exhibiting ventrolateral and OFC damage display increased risk-taking and maladaptive behaviors related to reward-based aspects in decision-making processes (Floden, Alexander, Kubu, Katz, & Stuss, 2008). Specifically, lesions within the OFC (which often includes the ventromedial PFC; vmPFC), a region involved in assessing reward-outcome and value (Wallis & Orbitofrontal, 2007), lead to increased risk-behavior regardless of the outcome or consequence (Bechara, Damasio, Damasio, & Anderson, 1994; Clark et al., 2008).

It was also shown that vmPFC lesions lead to emotional deficits that included irritability, anxiety, as well as social impairments (Anderson, Barrash, Bechara, & Tranel, 2006).

Bilateral lesions to the OFC lead to impairments in identifying voice and facial expressions, along with deficits in subjective emotional states and awareness (Hornak et al., 2003; Tsuchida & Fellows, 2012). Unilateral lesions to the anterior cingulate cortex showed similar deficits in voice and facial expressions identification and changes in subjective emotional states (Hornak et al., 2003). A subtle distinction in these studies compared to the bilateral amygdala lesions, as observed in patient S.M., is the specific absence of fear and anger in facial recognition (Adolphs et al., 1994; Forbes & Grafman, 2010; Fusi, Miller, & Rigotti, 2016; Rigotti et al., 2013) .These human lesion studies highlight the effects of PFC damage as it relates to emotional control, as well as the assessment of affective states in reward processing and motivation.

2.3 The mPFC and context-specificity

The mPFC is principally involved in executive functions, that includes cognitive processing and working memory, and is highly influenced depending on contextual information (Bechara et al., 2000; Euston et al., 2012; Sawaguchi & Goldman-Rakic, 1991). Context can be defined as a situation or circumstance describing an event and can be delineated into multiple aspects that include spatial, temporal, physiological, social, and/or cognitive contexts (Maren, 2001). Context can provide information regarding location and objects surrounding an environment, or actions and thoughts of an event occurring during time. Context can also be described as the internal or physiological state, for example hunger or stress; and depict the social environment that influences behavioral outcomes. Furthermore, a cognitive context is critical for encoding or processing information, and the retrieval of memories which is necessary for adaptive behavior. As a higher-order cortical region, the mPFC utilizes context during emotional valence processing that affect goal-directed behaviors and decision-making processes (Kennerley & Walton, 2011).

2.4 Mixed selectivity as a framework to flexibly encode valence in mPFC circuits

Neurons that respond to stimuli in a specific condition or context or that respond to a combination of task variables are referred to as 'mixed selectivity' cells (Fusi et al., 2016; Grunfeld & Likhtik, 2018; Parathasarathy et al., 2017;

Ramirez-Cardenas & Viswanathan, 2016; Rigotti et al., 2013). We propose that mixed selectivity is a model that can explain how mPFC circuits flexibly encode valence across contexts and situations. In the wild, animals encounter a range of contexts with varying degrees of risk and reward probabilities. Imagine the toy example of two contexts, one with high probability of reward and another with high probability of risk (Fig 2). mPFC neurons could respond to specific valence cues in a context-specific manner, thus encoding a combination of valence and risk level. Mixed selectivity is computationally advantageous because it increases the dimensionality of the population activity which allows for more efficient encoding of complex tasks, thus highlighting the ensemble of neurons, and not an individual cell, as the coding unit (Fusi et al., 2016; Johnston et al., 2020; Rigotti et al., 2013).

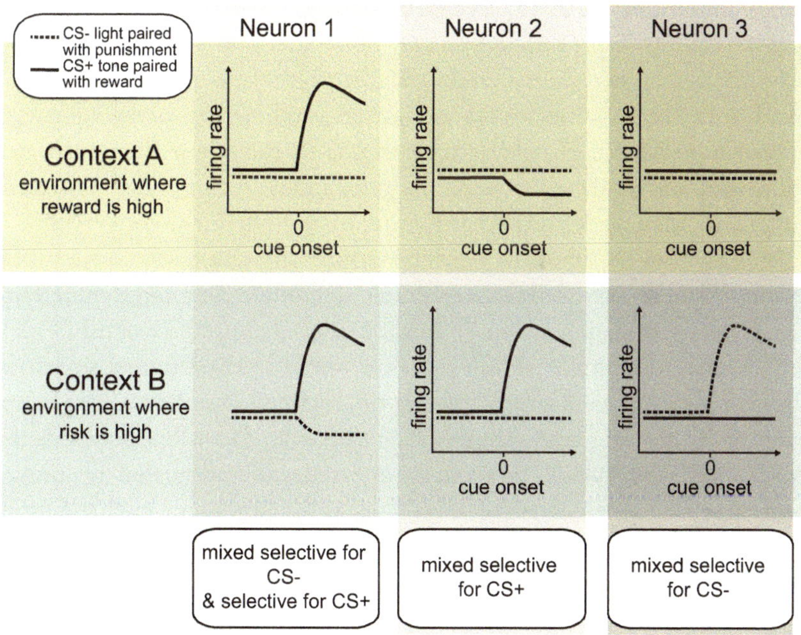

Fig. 2 Mixed selectivity in valence encoding cells. Toy example using cue and contexts as parameters and showing mixed selectivity for neurons that encode a positive CS+ cue that predicts a reward or a negative CS- cue that predicts a punishment. Top row shows firing rate to CS+ and CS- for 3 hypothetical neurons in context A which has high probability of reward. Middle row shows firing rate for the same 3 neurons in context B which has a high-risk probability. Bottom row indicates if neurons display nonlinear mixed selectivity. No individual neuron represents the 4 conditions, but the combination of the firing rate from the 3 neurons can be used to decode condition.

In both primates and rodents, the prefrontal cortex shows mixed selectivity during decision making and working memory assays (Machens, Romo, & Brody, 2010; Mante, Sussillo, Shenoy, & Newsome, 2013; Naya, Chen, Yang, & Suzuki, 2017; Powell & Redish, 2014). These types of PFC mixed selective cells are easiest to identify when using intricately-designed, model-driven behavioral experiments that are common in systems-level investigations of non-human primates. On the other hand, most circuit-specific studies use a data-driven approach, employing reductionist assays for their reproducibility and use manipulations complicating or even obscuring the identification of mixed selective cells. In addition, valence encoding literature has historically centered around single-cell (Kyriazi, Headley, & Pare, 2018; Paton, Belova, Morrison, & Salzman, 2006; Shabel & Janak, 2009) or populations of cells that are anatomically defined (Beyeler et al., 2016, 2018; Ciocchi et al., 2010; Janak & Tye, 2015; Namburi et al., 2015; Tye et al., 2011) encoding, thus overlooking what the population is encoding as a whole.

Rigotti and colleagues demonstrated that the encoding advantage of mixed selectivity is due to the nonlinear combination of firing rate to two parameters. We propose circuit motifs to provide a neurobiological mechanism for non-linear mixed selectivity and help consolidate this useful neural code scheme with the hardwiring of circuits. First, convergence of functionally diverse inputs of varying synaptic strength to the same mPFC neurons could lead to nonlinear encoding of multiple parameters underlying mixed selectivity (Fig. 3; motif 1). In addition, microcircuitry of mPFC cells with diverse functions and interneurons could lead to mixed selectivity (Fig. 3; motif 2). An example of this microcircuit motif was seen in the amygdala, as a recent study showed that two amygdala subpopulations that encode opposing valence are interconnected and modulated by hunger (Calhoon et al., 2018). Internal states, such as hunger, social rank and mood can modulate excitability or promote plasticity to non-linearly modulate firing rate, thus promoting non-linear mixed selectivity across states. These two motifs explain potential circuitry for a neuron to show mixed selectivity, however they do not explain how there can be flexible routing of behavioral output. We propose that a postsynaptic gating circuit motif can explain how mixed selectivity in neurons with collateral projections can allow flexible behavioral outputs. Postsynaptic gating via axo-axonal inhibition or context or state-induced changes in intrinsic excitability could gate the specific downstream pathway that is promoted (Fig. 3; motif 3). Finally, neuromodulators can help bias a given behavioral output. We propose a

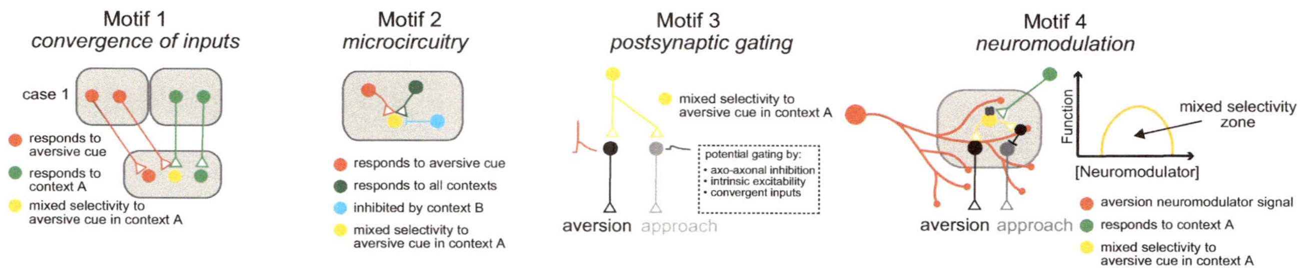

Fig. 3 Circuit motifs as neurobiological mechanisms for mixed selectivity. All motifs represent a mixed selective cell that fires to an aversive cue in a specific context A. Motif 1 shows that the same neuron receives converging input with different synaptic strengths from one input that responds to the aversive cue and another that responds to context A, leading to non-linear mixed selectivity. Motif 2 shows an example of mixed selectivity arising from local microcircuitry. Notice that the mixed selectivity is happening from context and cue information converging locally while inhibition facilitates context selectivity. Motifs 1 and 2 explain how non-linear mixed selectivity can arise from connectivity, while motifs 3 and 4 focus on how circuitry can enable flexible behavioral selection downstream. Motif 3 focuses on gating mechanisms in the downstream regions recruited by mixed selective mPFC cells with collateral projections. Gating mechanisms in the downstream cells, such as axo-axonal inhibition, intrinsic excitability, or converging inputs, can determine which pathway is promoted. Motif 4 -shows a neuron that receives input from a neuromodulator that responds to aversion signals and input that is modulated by context A, which together promote context-specific aversion. By recruiting inhibition of mPFC pathways that promote approach the mixed selective neuron can help route behavioral selection.

motif in which neuromodulator concentration promotes mixed selectivity in mPFC. A recent study showed that dopamine is released to the mPFC and promotes aversion via mPFC output to the brainstem (Vander Weele et al., 2018). mPFC neurons that receive dopaminergic input could promote aversion in context specific manners by integrating contextual inputs and via with disynaptic inhibition decrease approach (Fig. 3; motif 4). With these potential circuit motifs in mind, we review the literature of the mPFC circuits that modulate social behaviors, drug-seeking behavior and anhedonia, as they represent complex behaviors that are mediated by mPFC and could be modulated or encoded via mixed selectivity.

3. mPFC circuits and dynamics flexibly modulate social behaviors

Over the past decade there has been an explosion of research related to how mPFC circuits and pathways modulate social behaviors. Although the hypothalamus and other subcortical circuits are crucial for generating social behaviors, the mPFC is well situated to integrate relevant information (social rank, memories, context) to modulate social behaviors. Individual social interactions are not all the same, they can be positive or negative, depending on the context (previous social history, social rank relationships). mPFC subpopulations that carry valence information can help guide the appropriate behavioral response during a social interaction. For mPFC to flexibly guide social behavior it must have a dynamic and plastic representation of social cues. We review recent studies that support a dynamic representation of social cues in the mPFC. In addition, to flexibly modulate social behaviors the mPFC must have the ability to both promote and inhibit social behaviors. In support of this idea, the mPFC has multiple distinct circuits that either promote or decrease sociability, and some that can be modulated by social experience (Challis, Beck, & Berton, 2014; Franklin et al., 2017; Zhou et al., 2017). Finally, with convergence of shared functions for valence-related and social processing, there is an overlap in the mPFC circuits that regulate sociability, fear and anxiety-like behavior (Felix-Ortiz, Burgos-Robles, Bhagat, Leppla, & Tye, 2016; Huang, Zucca, Levy, & Page, 2020), suggesting that the mPFC integrates negative valence information to decrease social exploration. We review the literature that supports a role of mPFC circuits in flexibly guiding social behavior (Fig. 4) from the lens of a potential interaction of valence encoding and social functions.

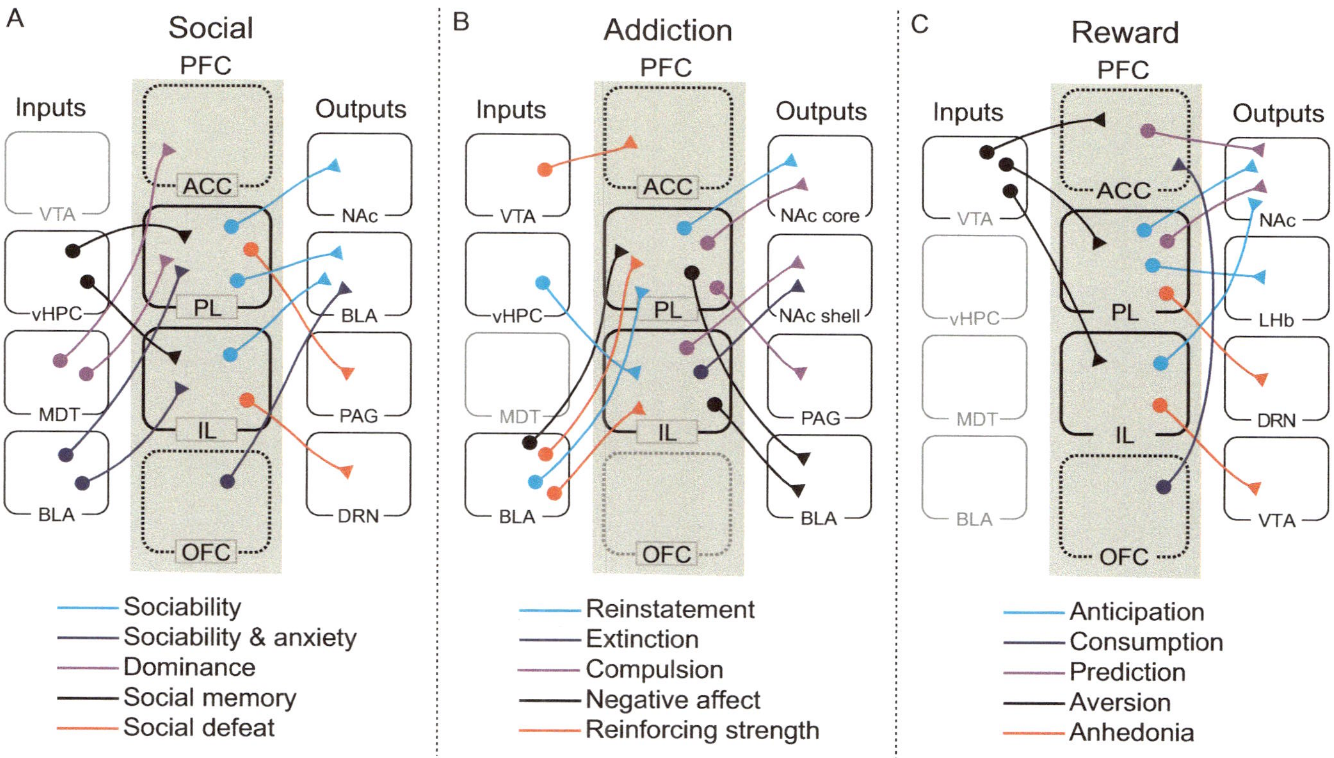

Fig. 4 See figure legend on next page.

3.1 mPFC social representation is dynamic and flexible

For the mPFC to flexibly guide social behavior a static representation of social cues could decrease the ability to flexibly modulate social behaviors. Across species, evidence points to the mPFC having a dynamic representation of social cues. In humans, the mPFC is activated in response to thinking of yourself or others (Seger, Stone, & Keenan, 2004), and the degree of social connections impacts the overlap in the mPFC representation (Courtney & Meyer, 2020). This study suggests that the mPFC social representation is dynamic and depends on social history. In mice, the mPFC represents the conspecifics as well, and mPFC single cell activity can be best predicted when taking into account both the behavior of the self and the other (Kingsbury et al., 2019). In addition, another study showed that mPFC cells can become responsive to social cues with repeated representation (Levy et al., 2019), further demonstrating that mPFC social representation is dynamic. An mPFC circuit that shows a flexible, and mixed selective, social representation is the prelimbic cells that project to the NAc, as they show spatial-social firing patterns (Murugan et al., 2017). Altogether, these studies show that mPFC social representation is not static and could evolve with the needs of the animal and changes in social history.

3.2 Distinct mPFC circuits bidirectionally control sociability

Our ability to socialize is important for survival and health (Cacioppo & Cacioppo, 2018; Matthews & Tye, 2019), and socializing is a complex process influenced by many of the traditional PFC functions such as

Fig. 4 mPFC circuits involved in social behaviors and reward processing. PFC neural circuits involved in social, addition, and reward behaviors. (A) PFC inputs from vHPC, MDT, and BLA implicated in social memory (black), dominance (purple), sociability and anxiety (dark blue). PFC outputs to the NAc, PAG, BLA, and DRN involved in sociability (light blue), social defeat (red), sociability and anxiety (dark blue). (B) PFC inputs from VTA, vHPC, and BLA implicated in reinforcing strength (red), reinstatement (light blue), and negative affect (black). PFC outputs to the NAc core, NAc shell, PAG, and BLA involved in reinstatement (light blue), compulsion (purple), extinction (dark blue), and negative affect (black). (C) PFC inputs from VTA implicated in aversion (black). PFC outputs to the NAc, LHb, DRN, and VTA involved in reward prediction (purple), reward consumption (dark blue) anticipation (light blue), and anhedonia (red). Subdivisions of the PFC: ACC, anterior cingulate cortex; PL, prelimbic; IL, infralimbic; OFC, orbitofrontal cortex. PFC projections/connections to brain regions: vHPC, ventral hippocampus; MDT, mediodorsal thalamus; BLA, basolateral amygdala; NAc, nucleus accumbens; NAc core; NAc shell; PAG, Periaqueductal gray; DRN, dorsal raphe nucleus; VTA, ventral tegmental area; LHb, lateral habenula.

decision-making, cognitive flexibility, memory and compulsive control. Recent optogenetic and chemogenetic studies have revealed that the mPFC has multiple subcortical outputs that modulate sociability and many of these circuits also play a role in valence encoding. Therefore, these mPFC circuits provide a mechanism to flexibly adjust social behavior depending on the potential rewards or dangers in the environment. A recent study showed that the dissociable roles of the infralimbic/prelimbic projections to the BLA in fear and anxiety (Adhikari et al., 2015) extend to sociability (Huang et al., 2020). Activity of PL → BLA cells decreased sociability while IL → BLA activity promoted sociability. Moreover, activation of PL → BLA cells that respond to negative stimuli (shock) was sufficient to see the decrease social preference (Huang et al., 2020), suggesting that negative valence cells are the same that modulate social preference. However, the PL → BLA circuit does not necessarily have a simple and static role in modulating sociability. Another study showed that optogenetic stimulation of PL → BLA neurons did not increase sociability when there was both a novel object and conspecific present (Murugan et al., 2017), supporting a dynamic role for PL → BLA neurons in which the exact context affects their function. Interestingly, the same study showed that PL → NAc stimulation resulted in decreased sociability in this behavioral paradigm (Murugan et al., 2017). Further evidence for a pro-social role of the IL → BLA circuit is that activating it mitigates the social behavioral effects of social defeat in subordinate, but not dominant, syrian hamsters (Dulka et al., 2020). It is of note that the IL → BLA mitigation was dependent on the social rank, as this further supports a non-static role of mPFC circuits in modulating social behaviors.

Common circuitry controls anxiety and social behaviors. Activation of the BLA input to mPFC both increased anxiety-like behavior and decreased sociability in a resident intruder assay (Felix-Ortiz et al., 2016). Importantly, the resident intruder assay is used as a measure of social anxiety in rodents since it is sensitive to anxiolytics (Calhoon & Tye, 2015; File & Animal, 1985). Humans and mice carrying the Val66Met allele (Chen et al., 2006; Li et al., 2019) show increased social anxiety. Moreover, in both humans and mice, social anxiety was linked to altered OFC-amygdala activity (Li et al., 2019). These studies support an interactive relationship between anxiety and social behaviors, and demonstrate that social behaviors can be inhibited due to anxiety signals via the PFC. Overall, mPFC circuits are well situated to modulate sociability bidirectionally to flexibly adjust social behaviors based on valence.

3.3 mPFC circuits change with social defeat and dominance

mPFC social representations are dynamic and change with social experience. What is the mechanism for this change in neural dynamics? One possibility is that there is plasticity in specific mPFC circuits with social experience. Consistent with this hypothesis, recent studies in mice show that mPFC circuits are sensitive to social defeat (Challis et al., 2014; Franklin et al., 2017) and dominance (Zhou et al., 2017).

Two mPFC projections to the brainstem are bidirectionally modulated by social defeat. After social defeat, optogenetic inhibition of the mPFC → DRN pathway rescued social defeat-induced changes in sociability (Challis et al., 2014). This study showed that specifically the ventral mPFC (IL) innervates GABAergic DRN cells, therefore potentially resulting in lower serotonin release which could mediate the social-affective changes induced by social defeat. On the other hand, social defeat decreased mPFC → dPAG connectivity and inhibition of the glutamatergic mPFC → dPAG pathway in non-defeated mice decreased social interaction (Franklin et al., 2017), suggesting that mPFC → dPAG is a pathway that can flexibly modulate sociability depending on social experience. Interestingly, a recent study showed that mPFC → dPAG neurons encode negative valence stimuli (shock) and promote aversion and anxiety-like behavior (Vander Weele et al., 2018). One possibility that could consolidate these seemingly opposing roles of the same circuit is that non-overlapping mPFC → dPAG neurons could be mediating the two distinct roles. Alternatively, the decrease in connectivity in the mPFC → dPAG pathway induced by social defeat could be a plasticity response due to overactivity in this subpopulation as it signals aversion.

An emerging field of the neural circuits underlying social dominance has shined light into the role of mPFC and the medial dorsal thalamus (MDT). Traditionally, the MDT → mPFC pathway has been linked to cognitive flexibility and spatial working memory (Bolkan et al., 2017; Hunt & Aggleton, 1991; Parnaudeau et al., 2013). A recent study showed that social dominance increases plasticity in the MDT-mPFC pathway and LTP induction in this synapse causes instant winning during social competition (Zhou et al., 2017). An additional study showed that MDT was required for the formation of the social hierarchy (Nelson et al., 2019) providing further support that plasticity in the thalamocortical pathway is important for social dominance. Altogether, these studies demonstrate that social experience induces synaptic plasticity in distinct mPFC circuits, thereby highlighting the dynamic role of mPFC circuits in modulating social behaviors.

3.4 Mixed selectivity in social behaviors

Social information is multisensorial and complex. Even in the simplest rodent behavioral assay the presence of another mouse is an olfactory, visual, and auditory cue. In addition to being multisensorial, social behavior is influenced by social experience, context, and internal state. Therefore, any social interaction has multiple parameters and the prefrontal cortex could potentially encode more information by having mixed selectivity to multiple cues and parameters of the social interaction. We see an example of this mixed selectivity in how PL → NAc neurons only encode spatial location when another animal is in the environment, thus demonstrating that mixed selectivity can be observed in a specific subpopulation of neurons (Murugan et al., 2017). In addition, we have reviewed that there is an overlap of the mPFC circuits that mediate valence and social behaviors. An example of this scenario is seen in the recent study by Huang et al., where the same PL → BLA neurons that encode shock modulate sociability. This overlap of mPFC circuits that mediate valence with mPFC circuits that modulate social behavior provides the possibility that the same mPFC neurons may be responding to a combination of threat level, reward assessment, and social stimuli. Mixed selectivity could provide a computational way to encode these parameters more efficiently and to allow the mPFC to flexibly modulate social behaviors.

4. The mPFC is involved in addictive and compulsive behavior

Drug addiction is a chronic, relapsing disorder characterized by the compulsion to seek and take drugs despite associated negative consequences. Addictive drugs are intrinsically rewarding and reinforcing and thus represent positive valence that confers incentive salience (Berridge & Robinson, 2016). Following excessive drug use, drug-associated positive valence has been observed to be both heightened and blunted (Ahmed & Koob, 1998; Grigson & Twining, 2002; Volkow et al., 1997). Heightened drug-associated positive valence can lead to anhedonia, or the loss of pleasure towards non-drug related rewarding stimuli, driving continual drug seeking and abandonment of other rewarding behaviors (e.g., social interaction) (Grigson & Twining, 2002). In other cases, heightened drug-associated positive valence can lead to sensitization or increased positive valence towards other rewarding stimuli (e.g., sucrose) (Bechara, Dolan,

& Hindes, 2002; Fiorino & Phillips, 1999; Goldstein et al., 2007; Nocjar & Panksepp, 2002; Taylor & Horger, 1999; Wyvell & Berridge, 2001). Overactivation of reward pathways in drug addiction leads to concurrent, homeostatic activation of antireward or stress pathways mediating negative reinforcement driving compulsive drug use to alleviate negative internal states created by drug abstinence (Koob & Le Moal, 2008; Solomon, 1980), suggesting that negative valence processing may be particularly important in the development and maintenance of drug addiction. Therefore, both positive and negative valence-associated circuits are likely to be critical for distinct processes in the transition to drug addiction. We review the literature on mPFC circuits involved in drug addiction from the perspective of valence encoding (Fig. 4) and finally integrate the computational framework of mixed selectivity to explain how the mPFC flexibly regulates drug related decisions.

4.1 mPFC-NAc projectors mediate drug reinstatement and extinction

Evidence from clinical and preclinical studies point to a prominent role of the mPFC in addiction (Franklin et al., 2002; George & Koob, 2013; Jasinska, Stein, Kaiser, Naumer, & Yalachkov, 2014; Van den Oever, Spijker, Smit, & De Vries, 2010; Volkow & Fowler, 2000; Volkow, Fowler, & Wang, 2003). Several projection-specific mPFC neurons regulate drug seeking and extinction. The PL and IL subregions of the mPFC predominantly project to the NAc core and shell, respectively (Sesack, Deutch, Roth, & Bunney, 1989), and largely play distinct functional roles in addiction-related behaviors (Moorman, James, McGlinchey, & Aston-Jones, 2015; Peters, Kalivas, & Quirk, 2009). The PL → NAc core is necessary for cue-induced reinstatement whereas the IL → NAc shell is critical for drug extinction. PL neurons projecting to the NAc core are activated by cocaine cues during reinstatement, which positively correlate to cocaine seeking action (McGlinchey, James, Mahler, Pantazis, & Aston-Jones, 2016), suggesting the recruitment of this pathway in drug seeking. Indeed, ablation and silencing of PL → NAc core decreases drug primed- and cue-induced reinstatement of cocaine seeking (Kerstetter et al., 2016; Stefanik et al., 2013; Stefanik, Kupchik, & Kalivas, 2016), consistent with the roles of PL and NAc in drug reinstatement (Kalivas, Volkow, & Seamans, 2005), and sensitivity to an alcohol and nicotine complex interoceptive cues (Randall, McElligott, & Besheer, 2020). In contrast to the PL → NAc core pathway, chemogenetic activation of IL → NAc shell circuit decreases cue-induced reinstatement of cocaine seeking after extinction training (Augur, Wyckoff, Aston-Jones, Kalivas, & Peters, 2016), suggesting a role of the IL → NAc shell in extinction memory. Altogether, these studies suggest that the mPFC → NAc pathway may represent the learned positive valence associated with drug cues.

Notably, with increasing time in abstinence following prolonged drug self-administration, there is a parallel increase in drug cravings, a phenomenon termed *incubation* (Gawin & Kleber, *1988;* Grimm, Hope, Wise, & Shaham, 2001). How drug abstinence alters the representation of drug-associated cues/contexts was recently uncovered, providing insight into cocaine incubation. Cameron, Murugan, Choi, Engel, & Witten (2019) found that IL → NAc shell neurons encode spatial location in a drug context, and the percent of spatial encoding neurons is decreased following prolonged abstinence, suggesting a reduced representation of the drug context. IL → NAc shell projectors exhibit an overall inhibition in activity prior to drug seeking action, which is also reduced following a prolonged period of abstinence. Optogenetic stimulation of IL terminals in the NAc shell decreases drug seeking action following both acute and prolonged abstinence periods (Cameron, et al., 2019). These findings suggest that decreased representation of drug-related context during abstinence in the IL → NAc shell, an extinction-related circuit, may underlie incubation of cocaine craving. The involvement of mPFC projectors in cue-induced, context-induced, and incubation requires the continual, dynamic representation of drug associated cues. We speculate that mixed selectively properties of the mPFC enable this representation of drug associated cues across several timescales.

4.2 Reprogramming corticostriatal circuits for therapeutic plasticity

Drug exposure induced synaptic plasticity in corticostriatal circuits presents a potential avenue for neural circuit reprogramming for therapeutic intervention. In line with the dichotomous corticostriatal pathways, Ma et al. (2014) found that prolonged cocaine abstinence activates drug-induced silent synapses in PL and IL terminals in the NAc core and shell, respectively, through distinct mechanisms. Using an optogenetic stimulation protocol to induce long-term depression (LTD), essentially reversing the activation of cocaine-induced silent synapses in the PL → NAc core and IL → NAc shell led to decreases and increases in cocaine incubation, respectively, which similarly regulates alcohol-seeking behavior as well (Ma et al., 2014; Ma et al., 2018). In addition, using a similar optogenetic LTD stimulation approach, Pascoli et al. (2014) found that reversing the observed cocaine-induced plasticity simultaneously at both the mPFC and vHPC terminals in the NAc shell abolished cue-induced reinstatement of drug-seeking, which persisted for a week (Pascoli et al., 2014). The enduring inhibition of drug-seeking by LTD-induction highlights the potential for reprogramming mPFC circuits for addiction treatment. Interestly, selective manipulation of mPFC terminals in the NAc impairs discrimmination between the drug-associated active lever and inconsequential inactive level without altering drug seeking (Pascoli et al., 2014), suggesting that cocaine-evoked plasticity in mPFC → NAc supports the learned association between action and drug delivery. Beyond corticostriatal circuits, mPFC neurons projecting to the laterodorsal tegmental nucleus (LDT) have been implicated in reward-associated behaviors (Coimbra et al., 2019; Kamii et al., 2015; Lammel et al., 2012; Xiao et al., 2016). mPFC inactivation prevents cocaine-induced plasticity in the LDT (Kurosawa, Taoka, Shinohara, Minami, & Kaneda, 2013), suggesting the importance of mPFC → LDT in cocaine-related positive valence—begging a causal investigation for mPFC → LDT involvement in the drug-associated behavior. Together, drug-induced plasticity is an important mechanism by which mPFC circuits regulate addiction-related behavior, and remodeling of drug-induced plasticity in discrete mPFC circuits holds therapeutic promise for drug relapse.

4.3 Inputs to the mPFC provide valence and context to modulate addiction-related behaviors

Subcortical inputs carrying diverse stimuli-related functional signals are integrated within the mPFC influencing behavior. In particular, valence and context related information is sent to the mPFC and modulates drug reinforcement and drug-seeking behavior.

The BLA sends dense projections to the mPFC that regulate anxiety-like behavior, social interaction, fear learning, and has been proposed to rapidly process and route negative valence information to the mPFC (Burgos-Robles et al., 2017; Felix-Ortizet al., 2016; Klavir, Genud-Gabai, & Paz, 2013). In line with this, Dong, Taylor, Wolf, & Shaham (2017) found that ablation of BLA neurons projecting to the mPFC (BLA → mPFC) increases motivation for alcohol consumption (Dong, et al., 2017), suggesting that BLA → mPFC may limit the reinforcing strength of alcohol consistent with its role in negative valence. It is important to note that BLA → mPFC neurons innervate neighboring BLA neurons as well as send collaterals to other brain regions (Beyeler et al., 2016, 2018), thus it is yet to be determined if the inputs to the mPFC directly mediate alcohol seeking behavior. Contrasting findings from Stefanik & Kalivas (2013) show that optogenetic inhibition of BLA terminals in the PL subdivision of the mPFC inhibits cue-induced reinstatement of cocaine seeking (Stefanik & Kalivas, 2013). Although these studies used different behavioral paradigms, the findings suggest that BLA → mPFC may differentially regulate drug seeking behavior depending on the drug and context. The demonstrated role of the BLA → mPFC pathway in regulating drug seeking and negative valence suggests a potential role of this circuit in compulsive drug use, although this remains to be explored.

The HPC has been implicated in drug reinstatement, and its connectivity with the mPFC regulates relapse to extinguished fear, context-guided memory, and drug-associated cognitive deficits (Fuchs et al., 2005; Kutlu & Gould, 2016; Marek et al., 2018; Place, Farovik, Brockmann, & Eichenbaum, 2016; Rogers & See, 2007; Wang, Jin, & Maren, 2016). Wang et al. found that vHPC-IL projections are activated by context-induced reinstatement of heroin seeking, and chemogenetic inhibition of vHPC-IL decreased heroin seeking (Wang et al., 2018). Moreover, pharmacological manipulation studies show that D1R and D2R signaling in the vHPC → mPFC are important for the acquisition and retrieval of morphine contextual memory depending on opiate exposure history (Wang et al., 2019). However, causal implication of the vHPC → mPFC circuit in drug-associated contextual memory remains to be tested.

The mPFC also receives afferent dopamine projections from the VTA. Interestingly, Narita et al. (2010) found that electrical or μ-opioid receptor agonist (DAMGO) stimulation of the VTA increases dopamine release in the ACC. VTA infusion of DAMGO induces a conditioned place preference that is significantly reduced by chemical lesion of dopamine terminals in the ACC (Hitora-Imamura et al., 2015). These findings suggest that VTA → ACC signals positive valence information related to the reinforcing effects of opioids, which is in opposition to the identified role of VTA → mPFC in aversion (Hitora-Imamura et al., 2015; Lammel et al., 2012; Mantz, Thierry, & Glowinski, 1989; Vander Weele et al., 2018). Alternatively, it is possible that VTA dopamine is necessary for the learned association between drug and context, consistent with the proposed role of mPFC dopamine transients in learning rather than valence (Popescu, Zhou, & Poo, 2016).

While limited studies have directly examined the impact of inputs to the mPFC in addiction-related behaviors, these studies point to a role of BLA, HPC, and VTA inputs to the mPFC in both negative and positive valence aspects of drug reinforcement and context influencing drug seeking behavior. Convergence of these diverse inputs could provide a circuit motif that promotes mixed selectivity in mPFC during drug addiction (Fig. 3).

4.4 Corticoamygdalar circuits mediate abstinence-induced negative affect

Negative affective states created by drug abstinence provide a strong motivational drive to use drugs, contributing to negative reinforcement processes underlying the transition and maintenance of drug dependence (Ahmed & Koob, 1998; Koob & Le Moal, 2008). Multiple studies have now implicated corticoamygdalar circuits in regulating drug abstinence-induced negative affect, consistent with the role of corticoamygdalar circuits in negative valence processing (Burgos-Robles et al., 2017; Felix-Ortiz et al., 2016). McGinnis, Parrish, Chappell, Alexander, & McCool (2020) found that withdrawal from chronic intermittent alcohol exposure strengthens PL→BLA while weakening IL→BLA connectivity. Chemogenetic inhibition of PL terminals in the BLA decreased withdrawal-induced anxiety-like behavior (McGinnis, et al., 2020), highlighting the therapeutic potential of synaptic remodeling in PL→BLA circuits. In line with this, Zhao et al. (2017) found that morphine withdrawal-associated contextual cues activate the mPFC→BLA projectors, suggesting that this circuit may receive contextual information. Chronic morphine exposure increased D1R expression in mPFC→BLA terminals and reversing this D1R overexpression decreased conditioned morphine withdrawal place aversion (Zhao et al., 2017). In addition, amygdala inputs to the mPFC also regulate morphine withdrawal-associated negative affect, as optogenetic inhibition of BLA terminals in the PL inhibits conditioned context-induced place aversion by morphine withdrawal (Song et al., 2019). To date, corticoamygdalar circuits remain the only mPFC circuit directly implicated in regulating drug-associated negative affect. These findings highlight the importance of corticoamygdalar circuitry for representation of negative valence during drug withdrawal states contributing to negative reinforcement driving continual, excessive drug use.

4.5 Cortical regulation of compulsive drug use via striatal and brainstem projections

A defining feature of drug addiction is the compulsion to seek and take drugs despite associated negative consequences. Punishment-resistant drug use is assessed by the introduction of footshock or adulteration of the drug (e.g., alcohol) with the bitter-tastant quinine during drug seeking or taking. In the face of competing rewarding and aversive stimuli, the mPFC is recruited (Burgos-Robles et al., 2017). Indeed, IL neurons, and to a lesser degree PL neurons, encode alcohol seeking action, which is significantly diminished during sessions with punishment (Halladay et al., 2020). In addition, mPFC neurons encode aborts (i.e., when lever approaches are quickly aborted) during punished alcohol seeking and in subsequent unpunished alcohol seeking (Halladay et al., 2020). Of note, encoding of alcohol seeking action and aborts are represented by distinct IL ensembles (Halladay et al., 2020), potentially delineated by anatomical projection although remains to be determined. Causal optogenetic inhibition of IL increases compulsive alcohol seeking, and manipulation of PL bidirectionally regulates compulsive cocaine seeking (Chen et al., 2013). Punishment induced neuroplastic adaptations in IL inputs to NAc shell D1R-expressing neurons, and optogenetic silencing of IL→NAc shell, but not IL→BLA, decreases the effects of punishment on alcohol seeking, suggesting that IL→NAc may encode aborts (Halladay et al., 2020). Consistent with the role of the PL in compulsive cocaine seeking, optogenetic inhibition of PL → NAc core decreases compulsive alcohol intake, but not unpunished alcohol intake, which is dependent on hyperpolarization-activated NMDA receptors (Seif et al., 2013). Interestingly, activation of shock-encoding PL → NAc suppressed reward seeking, suggesting that distinct subpopulations within PL → NAc projectors may differentially regulate reward seeking (Kim et al., 2017). Consistent with the role of the mPFC in processing both reward and aversion, discrete mPFC circuits integrate punishment and drug-associated positive valence to flexibly regulate compulsive drug use.

In addition, mPFC→PAG neurons selectively route aversive-stimuli information (Vander Weele et al., 2018), and Siciliano et al. found that activity patterns in mPFC neurons projecting to the PAG during initial alcohol exposure is predictive of later development of compulsive alcohol drinking behavior (Siciliano et al., 2019). Using machine learning, the authors show that mPFC→PAG neural activity during compulsive alcohol drinking was predictive of whether mice would drink or not drink during the subsequent opportunity to drink (Siciliano et al., 2019), consistent with the mPFC→PAG neurons routing aversive-stimuli information. Moreover, optogenetic manipulation of mPFC terminals in the PAG bidirectionally compulsive alcohol drinking, respectively, but not unpunished alcohol drinking (Siciliano et al., 2019). This study highlights the importance of negative valence processing by the mPFC in regulating compulsive drug taking behavior. Moreover, it suggests that negative valence-related mPFC circuits are important for negative reinforcement and compulsive aspects of drug addiction.

These data highlight a specific role of mPFC→NAc and mPFC→PAG circuits in compulsive drug seeking and taking, and the potential of these circuits to encode both drug-associated positive and punishment-associated negative valence. Mixed selective mPFC neurons in these pathways may be critical in simultaneously encoding reward and punishment associated contingencies for decision making in seeking drugs despite associated consequences.

4.6 Integration of circuits and mixed selectivity in drug addiction

How do these discrete, seemingly static circuits regulate the decision to seek and take drugs? Mixed selectivity properties of the mPFC allow for the complex integration of a multitude of parameters enabling the consequent orchestration of behavior. Several factors influence the decision to use drugs including past rewarding or aversive experiences, drug cravings triggered by drug-associated contexts/cues, internal emotional states, and predictions of future consequences. Specific contexts (i.e., environmental, emotional, internal state, recent prior experience) can elicit cravings and trigger relapse in rodents, primates, and humans (Chaudhri, Sahuque, & Janak, 2008; Fox, Bergquist, Hong, & Sinha, 2007; Litt & Cooney, 1999; Perry, Zbukvic, Kim, & Lawrence, 2014; Reid, Flammino, Starosta, Palamar, & Franck, 2006; Seo et al., 2013; Spealman et al., 2004; Zironi, Burattini, Aicardi, & Janak, 2006), highlighting a conserved behavioral adaptation contributing to addiction. The context in which drug-associated cues are encountered also influence the degree to which drug cues elicit cravings and relapse (Hyman, 2005). Context may modulate drug-associated valence altering drug cue reactivity. For example, during hunger states food-associated positive valence may exceed that of drugs consequently transiently diminishing drug-associated positive valence. The brain continuously assesses this rich multidimensional information to flexibly regulate the decision to use drugs.

As mentioned, the mPFC has mixed selectivity properties enabling simultaneous representation of diverse parameters within and across contexts (Fig. 2). This provides a new conceptual framework in which to understand addiction-related behavioral adaptations such as context-dependent drug craving. Exposure to drug-associated cues under various contexts may trigger activity in mixed selective mPFC neurons encoding salient drug cues and contexts, dynamically shifting population level encoding of the decision to use drugs. For example, mixed selective neurons receiving contextual information from HPC inputs and negative valence information from VTA inputs (Hitora-Imamura et al., 2015; Lammel et al., 2012; Mantz et al., 1989; Vander Weele et al., 2018), may flexibly shift the representation of negative internal states or punishment under various contexts in response to fluctuating dopamine levels to alter motivation for drugs. In line with this idea, mPFC dopamine is necessary for drug relapse under certain conditions (Capriles, Rodaros, Sorge, & Stewart, 2003; James, McGlinchey, Vattikonda, Mahler, & Aston-Jones, 2018; McFarland, Davidge, Lapish, & Kalivas, 2004; Wang et al., 2019). Simultaneous representation of drug reward-associated context and drug abstinence-associated negative internal states may bias population encoding to increase the decision to seek and take drugs despite associated negative consequences underlying drug incentive salience.

Diverse input (e.g., valence- and context-related) to mixed selective mPFC neurons that project to several downstream brain regions may be a mechanism by which hard-wired circuits flexibly direct behavior. Indeed, overlapping mPFC circuits regulating several addiction-related behaviors allow for multiple avenues by which mPFC mixed selective neurons may direct the decision to use drugs. Conceptual integration of circuits and the computational model of mixed selectively provides a new framework for understanding drug addiction.

5. The mPFC as a major component in reward processing

5.1 mPFC circuits involved in "wanting" and "liking"

In reward processing, pleasure is categorized into a anticipatory phase and consummatory phase (Berridge, 2004). During the anticipatory phase, referred to as "wanting," the motivation to receive a reward elicits pleasure (Berridge, 2004); thereby promoting appetitive behaviors that includes foraging for food, sex, and social interactions. Whereas in the consummatory phase, also described as "liking," pleasure is induced by hedonic impact of a reward via initial sensory stimuli (Berridge, 2004). Here we review recent literature describing the role of PFC circuits in motivation and anticipation during reward seeking and consumption (Fig. 4).

Specific mPFC projections participate in anticipatory pleasure. For instance, mPFC layer 5 projection neurons to the NAc (Gabbott et al., 2005)—a major site for reward detection and reward learning (Castro & Berridge, 2014; Day & Carelli, 2007; Day, Roitman, Wightman, & Carelli, 2007)—inhibit reward-seeking behavior in response to aversive stimuli (Kim et al., 2017). For instance, Kim et al identified a specific subset of mPFC → NAc projecting neurons that are recruited in response to electrical shock and are involved in suppressing reward-seeking behavior (Kim et al., 2017). Another downstream target involved in anticipation and reward-seeking behavior is the lateral habenula (LHb) (Benekareddy et al., 2018; Matsumoto & Hikosaka, 2007, 2009; Warden et al., 2012). In rhesus monkeys, the LHb is activated upon presentation of neutral stimuli and punishment, but is inhibited upon reward presentation (Matsumoto & Hikosaka, 2007, 2009). Interestingly, photostimulation of mPFC → LHb in rats significantly reduced mobility in the forced swim task (FST), suggesting a lack of motivated-behavior or a passive coping strategy, in contrast to mPFC → DRN activation that increased kicking and promoted an active coping strategy(Warden et al., 2012). The FST is a commonly used assay to examine anhedonia in rodents, as immobility is a measurement of behavioral despair and mimics depressive-like phenotypes. Nevertheless, these neural circuit studies highlight the diverse pathways that control reward seeking-behavior and motivated behavior.

The OFC, a subregion of the PFC, acts as a primary mediator of consummatory pleasure. It integrates multiple sensory inputs, conveying pleasure sensation in response to gustatory, tactile, visual, olfactory and social stimuli (Grabenhorst & Rolls, 2011), and encodes reward reinforcement and reward prediction via midbrain dopaminergic neurons (Berridge & Kringelbach, 2008; Sul, Kim, Huh, Lee, & Jung, 2010; Takahashi et al., 2009). In a recent study, Jennings et al. identified distinct OFC neuronal ensemble populations which responded to caloric consumption or social interaction, indicative of consummatory pleasure (Jennings et al., 2019). In this study, single-cell activation of social cells suppressed feeding behavior, indicating microcircuit level changes within the OFC that are recruited in response to select reward stimuli. However, whether these OFC neuronal ensembles are affected in specific contexts remains to be investigated.

5.2 mPFC circuits involved in reward-prediction and reward learning

Reward is directly implicated in hedonic pleasure and plays a major role in everyday decision-making processes that can influence survival, lifestyle, or wellbeing. The value of a reward is often described as the neuroeconomic benefit received from a stimulus and can be transient or sustained over long periods of time. Therefore, the reward is evaluated, and the value is determined depending on the salience and/or retrieved memories from prior experiences.

Dopamine (DA) is historically best known for being the principal neurological substrate involved in hedonic pleasure and it is responsible for regulating reward value, reward outcome and motivation (Beyeler et al., 2016; Dong et al., 2017; Klavir et al., 2013). This is predominantly mediated via DA neurons in the VTA of the midbrain, however, the role of VTA DA is heterogenous and depends on the specific downstream circuits (Lammel et al., 2012; Lammel, Ion, Roeper, & Malenka, 2011; Lammel, Tye, & Warden, 2013; Vander Weele, Siciliano, & Tye, 2018). Dopamine in the PFC has been implicated in processing negatively valenced stimuli (Vander Weele et al., 2018), and can produce avoidance (Gunaydin et al., 2014), but is suggested to be a major component in reward prediction (Schultz, Dayan, & Montague, 1997). This suggests that DA modulates mPFC encoding of both positive and negative valence in behavior. Many questions remain unanswered; are there mPFC mixed selectivity neurons that encode for opposite valence? And does DA alter their activity?

The ACC, along with the OFC, encodes reward predictive cues and subjective pleasure (Behrens, Hunt, Woolrich, & Rushworth, 2008; Grabenhorst & Rolls, 2011; Kennerley, Dahmubed, Lara, & Wallis, 2009; Kennerley & Wallis, 2009; Kim, Hwang, & Lee, 2008; Rolls & Grabenhorst, 2008), and is highly active when reward outcome influences decision-making processes (Walton, Devlin, & Rushworth, 2004). Specifically, optogenetic inhibition of ACC or prelimbic projections to the dorsomedial striatum increases reward-seeking behavior during conflict (Friedman et al., 2015). Other studies linked the mPFC → NAc pathway to reward prediction and reward learning. For instance, studies in rodents showed that mPFC → NAc projection neuron activity is amplified in response to reward-predictive cues following reward learning and can induce reward-seeking behavior (Otis et al., 2017). In contrast, activating mPFC → PVT (paraventricular nucleus of the thalamus) neurons suppresses conditioned reward seeking behavior (Otis et al., 2017), and is time-dependent in fear learning, indicating the dynamic role within this pathway. However, it remains to be seen whether the mPFC neuronal projecting populations involved in reward prediction are altered in response to context-specific changes.

5.3 mPFC activity disrupted during anhedonia

Anomalous mPFC activity has been implicated in neuropsychiatric disorders such as major depressive disorders (MDD) and schizophrenia (SCZ). Anhedonia is described as the inability to experience pleasure or hedonic feeling, affecting both consumption and anticipatory reward values, and is observed in both MDD and SCZ (Lee, Jung, Park, & Kim, 2015; Pizzagalli & Depression, 2014). The etiology of anhedonia is linked to disruptions within the brain reward pathways due to dysregulation of dopamine transmission (Chaudhury et al., 2013; Harvey, Pruessner, Czechowska, & Lepage, 2007; Keedwell, Andrew, Williams, Brammer, & Phillips, 2005; Tye et al., 2013) and anomalous mPFC neural activity (Ferenczi et al., 2016; Kim et al., 2017; Warden et al., 2012). For instance, optogenetic studies in rodents revealed that increasing mPFC activity in glutamatergic projection neurons causes a significant decrease in sucrose consumption and social interaction, indicative of a depressive-like phenotype (Ferenczi et al., 2016). Accordingly, in human imaging studies, depressed patients revealed elevated activity in the subgenual ACC of the PFC, which was alleviated following deep brain stimulation (Mayberg et al., 2005). Similarly, deep brain stimulation of the ACC in depressed patients improved depressive-like symptoms in the refractory treatment period (Johansen-Berg et al., 2008; Schlaepfer et al., 2008). Contrastingly, other imaging studies in depressed patients revealed a reduction in PFC activity (Galynker et al., 1998), which may be attributable to a reduction in glutamatergic transmission due to a decrease in the number of synapses (Duman, Sanacora, & Krystal, 2019; Kang et al., 2012). These studies highlight a discrepancy in the literature describing PFC neural activity regarding anhedonia. What could underlie these discrepancies? It is possible that projection-specific neuronal populations are differentially altered during anhedonia, thus resulting in opposing distinct changes in mPFC neural activity.

5.4 mPFC projecting neurons involved in anhedonia

mPFC projection target regions were identified to be involved in anhedonia. For instance, acute optogenetic stimulation of mPFC-projecting neurons in the dorsal raphe nucleus (DRN) in rats increased mobility in the forced swim task, promoting motivation and active coping strategy (Warden et al., 2012). This study highlighted the importance of circuit-specific studies, as acute optogenetic stimulation of the mPFC caused no significant changes in mobility. The mPFC → DRN pathway is also involved in social interaction, as photoinhibition of this circuit prevents social withdrawal following a social defeat paradigm (Challis, et al., 2014). The DRN, a major target in depression (Maier, 2015; Sf & Lr, 2005), consists of both serotonergic and GABAergic neurons that primarily receive inputs from PL projections onto GABAergic interneurons (Peyron, Petit, Rampon, Jouvet, & Luppi, 1997; Vertes, 2004), which indirectly inhibits serotonergic neurons (Jankowski & Sesack, 2004). Interestingly, it was recently shown that both serotonergic and GABAergic neurons in the DRN exhibit context-specific changes in negative environments (Seo et al., 2019). It is likely mPFC-DRN projections change depending on mood or during disease states; however, these hypotheses have yet to be investigated.

The VTA is another major region implicated in anhedonia, and can be regulated via top-down connections from cortical regions that modulate DA activity (Chaudhury et al., 2013; Ferenczi et al., 2016; Tye et al., 2013). Accordingly, the VTA receives sparse projections from the mPFC that innervate both GABA and DA neurons and have no preferential input (Beier et al., 2015; Carr & Sesack, 2000a). Interestingly, activating the IL-PFC reduced VTA DA excitability in rats, specifically in medial VTA neurons that were most sensitive to chronic mild stress (Moreines, Owrutsky, & Grace, 2017). It is likely mPFC projections strongly influence VTA DA activity attributable to anhedonia. However, whether there are differential effects of mPFC projections onto VTA DA and GABAergic neurons remains to be understood.

5.5 Mixed selectivity neurons during anhedonia

The PFC, as a higher-order processing center, is highly adaptable and dynamic. We propose that PFC function is compromised during anhedonia which may limit flexibility. For instance, human fMRI studies of MDD patients showed a marked decrease in PFC activity that is likely linked to glutamatergic dysconnectivity (Abdallah et al., 2017; Murrough, Abdallah, & Mathew, 2017). This corroborated previous findings that revealed PFC hypofrontality within MDD patients contribute to the negative symptoms of depression (Galynker et al., 1998). We suggest that during anhedonia, mixed selectivity neurons have a decreased capacity to encode reward due to changes in internal homeostasis. Accordingly, we expect an overall reduction in the activity of PFC mixed selectivity projections neurons, consistent with the observed decrease in PFC neural activity in MDD patients that lead to impairments in anticipation and reward consumption.

Additionally, mixed selectivity neurons may leverage axonal collateralization and bias preferential outputs onto specific neuronal populations. For instance, mPFC neurons innervate both GABA and DA neurons in the VTA, and receive reciprocal connections from both cell types (Carr & Sesack, 2000a, 2000b). The mPFC mixed selectivity neurons could provide preferential inputs that influence the postsynaptic response on specific cell types of the VTA that lead to differential behavioral outcomes. Indeed, we suggest postsynaptic gating mechanisms via axo-axonal inhibition, intrinsic excitability, and convergent inputs modulate the mixed selectivity outputs (Fig. 3). Accordingly, we propose that within an anhedonic state, changes in postsynaptic gating mechanisms decrease the ability to promote reward in a cell-type specific and brain region-specific manner. For example, changes in postsynaptic gating could shift mixed selectivity mPFC-VTA neurons to preferentially modulate VTA GABAergic neurons leading to an increase in the postsynaptic response. The increase in VTA GABAergic neuron activity would cause an inhibition in VTA DA release leading to depressive-like phenotypes. Altogether, this theoretical model would imply mPFC mixed selectivity neuronal output activity is tuned during anhedonia that would lead to deficits in anticipatory and consummatory pleasure.

6. Summary

In summary, we provide this review of the current state of PFC research to support the widely-appreciated role of the PFC in higher-order cognitive functions by highlighting the role of PFC ensembles and circuitry in mediating emotional processing and the cognitive appraisal of subcortically-mediated emotional processes in the context of mood disorders, addiction, and social behaviors. In addition to diversification of scope in function, we confront the growing chasm between the systems-level focus on the ensemble dynamics with flexible functions of individual "mixed selectivity" neurons (which point to the *ensemble* as the functional unit), and the circuit-level focus on the on the circuit components of input-output architectures and wiring diagrams (that point to the "hard-wired" *circuit component* as the functional unit). Moving forward, we anticipate the development of novel computational models for PFC function that may unite these perspectives (Tsuda, Tye, Siegelmann, & Sejnowski, 2020). At this point in the field with these rapidly diverging subfields, we are in need of a cohesive, integrative model that reconciles these contrasting viewpoints and ushers us back into a converging trajectory.

While perspectives will continue to diversify, cross-pollination across subfields of research will only enhance our understanding of the most uniquely-evolved structure in the human brain, and therefore ourselves.

Acknowledgments

We thank Amy Cao for her illustration of Fig. 1. We thank Romy Wichmann and Gillian Matthews for comments on the manuscript, and Earl Miller for helpful discussion. We acknowledge funding sources K99 MH124435-01 (N.P.-C.), 8 K00 MH124182-03 (A.A.C.) and R01-MH115920 (NIMH; K.M.T.) Pioneer Award DP1-AT009925 (NCCIH; K.M.T.).

Competing interest

The authors report no competing interest or biomedical financial interests.

Author contributions

All authors contribute to conception, writing, figure making and editing/revising.

References

Abdallah, C. G., et al. (2017). Prefrontal connectivity and glutamate transmission: relevance to depression pathophysiology and ketamine treatment. *Biological psychiatry. Cognitive Neuroscience and Neuroimaging*, *2*, 566–574.

Abimbola, S., & Awolowo, O. (2006). The white cut: Egas Moniz, lobotomy, and the Nobel prize. *BMJ*, *332*.

Adhikari, A., et al. (2015). Basomedial amygdala mediates top-down control of anxiety and fear. *Nature*, *527*, 179–185.

Adolphs, R., Tranel, D., Damasio, H., & Damasio, A. (1994). Impaired recognition of emotion in facial expressions following bilateral damage to the human amygdala. *Nature*, *372*, 669–672.

Ahmed, S. H., & Koob, G. F. (1998). Transition from moderate to excessive drug intake: Change in hedonic set point. *Science*, *282*, 298–300.

Aleyasin, H., Flanigan, M., & Russo, S. J. (2018). Neurocircuitry of aggression and aggression seeking behavior: Nose poking into brain circuitry controlling aggression. *Current Opinion in Neurobiology*, *49*, 184–191.

Allsop, S. A., Vander Weele, C. M., Wichmann, R., & Tye, K. M. (2014). Optogenetic insights on the relationship between anxiety-related behaviors and social deficits. *Frontiers in Behavioral Neuroscience*, *8*, 241.

Allsop, S. A., et al. (2018). Corticoamygdala transfer of socially derived information gates observational learning. *Cell*, *173*, 1329. 1342.e18.

Anderson, S. W., Barrash, J., Bechara, A., & Tranel, D. (2006). Impairments of emotion and real-world complex behavior following childhood- or adult-onset damage to ventromedial prefrontal cortex. *Journal of the International Neuropsychological Society*, *12*, 224–235.

Anderson, S. W., et al. (2007). The earliest behavioral expression of focal damage to human prefrontal cortex. *Cortex*, *43*, 806–816.

Augur, I. F., Wyckoff, A. R., Aston-Jones, G., Kalivas, P. W., & Peters, J. (2016). Chemogenetic activation of an extinction neural circuit reduces cue-induced reinstatement of cocaine seeking. *Journal of Neuroscience: The Official Journal of the Society for Neuroscience*, *36*, 10174–10180.

Bechara, A., Damasio, H., & Damasio, A. R. (2000). Emotion, decision making and the orbitofrontal cortex. *Cereb. Cortex N. Y. N, 1991*(10), 295–307.

Bechara, A., Damasio, A. R., Damasio, H., & Anderson, S. W. (1994). Insensitivity to future consequences following damage to human prefrontal cortex. *Cognition, 50*, 7–15.

Bechara, A., Dolan, S., & Hindes, A. (2002). Decision-making and addiction (part II): Myopia for the future or hypersensitivity to reward? *Neuropsychologia, 40*, 1690–1705.

Behrens, T. E. J., Hunt, L. T., Woolrich, M. W., & Rushworth, M. F. S. (2008). Associative learning of social value. *Nature, 456*, 245–249.

Beier, K. T., et al. (2015). Circuit architecture of VTA dopamine neurons revealed by systematic Input-output mapping. *Cell, 162*, 622–634.

Benekareddy, M., et al. (2018). Identification of a corticohabenular circuit regulating socially directed behavior. *Biological psychiatry, 83*. https://pubmed.ncbi.nlm.nih.gov/29336819/.

Berridge, K. C. (2004). Motivation concepts in behavioral neuroscience. *Physiology & Behavior, 81*, 179–209.

Berridge, K. C., & Kringelbach, M. L. (2008). Affective neuroscience of pleasure: Reward in humans and animals. *Psychopharmacology, 199*(457–480).

Berridge, K. C., & Robinson, T. E. (2016). Liking, Wanting and the Incentive-Sensitization Theory of Addiction. *The American Psychologist, 71*, 670–679.

Beyeler, A., et al. (2016). Divergent routing of positive and negative information from the amygdala during memory retrieval. *Neuron, 90*, 348–361.

Beyeler, A., et al. (2018). Organization of valence-encoding and projection-defined neurons in the basolateral amygdala. *Cell Reports, 22*, 905–918.

Blair, R. J. R., & Cipolotti, L. (2000). Impaired social response reversal. A case of 'acquired sociopathy'. *Brain: A Journal of Neurology, 123*, 1122–1141.

Bolkan, S. S., et al. (2017). Thalamic projections sustain prefrontal activity during working memory maintenance. *Nature Neuroscience, 20*, 987.

Bossert, J. M., et al. (2011). Ventral medial prefrontal cortex neuronal ensembles mediate context-induced relapse to heroin. *Nature Neuroscience, 14*, 420–422.

Bossert, J. M., et al. (2012). Role of projections from ventral medial prefrontal cortex to nucleus accumbens shell in context-induced reinstatement of heroin seeking. *The Journal of Neuroscience, 32*, 4982–4991.

Brodmann, K. (1909). *Vergleichende Lokalisationslehre der Grosshirnrinde in ihren Prinzipien dargestellt auf Grund des Zellenbaues*. Barth.

Brown, S., & Sharpey-Schafer, E. A., III. (1888). An investigation into the function of the occipital and temporal lobes of the monkey's brain. *Proceedings of the Royal Society of London, 43*, 276.

Brozoski, T. J., Brown, R. M., Rosvold, H. E., & Goldman, P. S. (1979). Cognitive deficit caused by regional depletion of dopamine in prefrontal cortex of rhesus monkey. *Science, 205*, 929–932.

Burgos-Robles, A., et al. (2017). Amygdala inputs to prefrontal cortex guide behavior amid conflicting cues of reward and punishment. *Nature Neuroscience, 20*, 824–835.

Cacioppo, J. T., & Cacioppo, S. (2018). The growing problem of loneliness. *The Lancet, 391*, 426.

Calhoon, G. G., & Tye, K. M. (2015). Resolving the neural circuits of anxiety. *Nature Neuroscience, 18*, 1394–1404.

Calhoon, G. G., et al. (2018). Acute food deprivation rapidly modifies valence-coding microcircuits in the amygdala. *bioRxiv*, 285189. https://doi.org/10.1101/285189.

Callicott, J. H., et al. (2003). Abnormal fMRI response of the dorsolateral prefrontal cortex in cognitively intact siblings of patients with schizophrenia. *The American Journal of Psychiatry, 160*, 709–719.

Cameron, C. M., Murugan, M., Choi, J. Y., Engel, E. A., & Witten, I. B. (2019). Increased cocaine motivation is associated with degraded spatial and temporal representations in IL-NAc neurons. *Neuron, 103*, 80. 91.e7.

Capriles, N., Rodaros, D., Sorge, R. E., & Stewart, J. (2003). A role for the prefrontal cortex in stress- and cocaine-induced reinstatement of cocaine seeking in rats. *Psychopharmacology, 168*(66–74).
Carmichael, S. T., & Price, J. L. (1994). Architectonic subdivision of the orbital and medial prefrontal cortex in the macaque monkey. *Journal of Comparative Neurology, 346*, 366–402.
Carr, D. B., & Sesack, S. R. (2000a). Projections from the rat prefrontal cortex to the ventral tegmental area: Target specificity in the synaptic associations with mesoaccumbens and mesocortical neurons. *The Journal of Neuroscience, 20*, 3864–3873.
Carr, D. B., & Sesack, S. R. (2000b). GABA-containing neurons in the rat ventral tegmental area project to the prefrontal cortex. *Synapse, 38*, 114–123.
Castro, D. C., & Berridge, K. C. (2014). Opioid hedonic hotspot in nucleus accumbens shell: Mu, delta, and kappa maps for enhancement of sweetness "liking" and "wanting". *The Journal of Neuroscience, 34*, 4239–4250.
Challis, C., Beck, S. G., & Berton, O. (2014). Optogenetic modulation of descending prefrontocortical inputs to the dorsal raphe bidirectionally bias socioaffective choices after social defeat. *Frontiers in Behavioral Neuroscience, 8*, 43.
Chaudhri, N., Sahuque, L. L., & Janak, P. H. (2008). Context-induced relapse of conditioned behavioral responding to ethanol cues in rats. *Biological Psychiatry, 64*, 203–210.
Chaudhury, D., et al. (2013). Rapid regulation of depression-related behaviours by control of midbrain dopamine neurons. *Nature, 493*, 532–536.
Chen, Z.-Y., et al. (2006). Genetic variant BDNF (Val66Met) polymorphism alters anxiety-related behavior. *Science, 314*, 140–143.
Chen, B. T., et al. (2013). Rescuing cocaine-induced prefrontal cortex hypoactivity prevents compulsive cocaine seeking. *Nature, 496*, 359–362.
Ciocchi, S., et al. (2010). Encoding of conditioned fear in central amygdala inhibitory circuits. *Nature, 468*, 277–282.
Clark, L., et al. (2008). Differential effects of insular and ventromedial prefrontal cortex lesions on risky decision-making. *Brain, 131*, 1311–1322.
Clem, R. L., & Huganir, R. L. (2010). Calcium-permeable AMPA receptor dynamics mediate fear memory erasure. *Science, 330*, 1108–1112.
Coimbra, B., et al. (2019). Role of laterodorsal tegmentum projections to nucleus accumbens in reward-related behaviors. *Nature Communications, 10*, 4138.
Courtney, A. L., & Meyer, M. L. (2020). Self-other representation in the social brain reflects social connection. *The Journal of Neuroscience, 40*, 5616–5627.
Davis, M., & Whalen, P. J. (2001). The amygdala: Vigilance and emotion. *Molecular Psychiatry, 6*, 13–34.
Day, J. J., & Carelli, R. M. (2007). The nucleus accumbens and pavlovian reward learning. *The Neuroscientist: A Review Journal Bringing Neurobiology, Neurology and Psychiatry, 13*, 148–159.
Day, J. J., Roitman, M. F., Wightman, R. M., & Carelli, R. M. (2007). Associative learning mediates dynamic shifts in dopamine signaling in the nucleus accumbens. *Nature Neuroscience, 10*, 1020–1028.
Dias, R., Robbins, T. W., & Roberts, A. C. (1996). Dissociation in prefrontal cortex of affective and attentional shifts. *Nature, 380*, 69–72.
Dong, Y., Taylor, J. R., Wolf, M. E., & Shaham, Y. (2017). Circuit and synaptic plasticity mechanisms of drug relapse. *The Journal of Neuroscience, 37*, 10867–10876.
Donoghue, J. P., & Wise, S. P. (1982). The motor cortex of the rat: Cytoarchitecture and microstimulation mapping. *Journal of Comparative Neurology, 212*, 76–88.
Dulka, B. N., et al. (2020). Chemogenetic activation of an infralimbic cortex to basolateral amygdala projection promotes resistance to acute social defeat stress. *Scientific Reports, 10*, 6884.

Duman, R. S., Sanacora, G., & Krystal, J. H. (2019). Altered connectivity in depression: Gaba and glutamate neurotransmitter deficits and reversal by novel treatments. *Neuron, 102*, 75–90.
Etkin, A., Egner, T., & Kalisch, R. (2011). Emotional processing in anterior cingulate and medial prefrontal cortex. *Trends in Cognitive Sciences, 15*, 85–93.
Euston, D. R., Gruber, A. J., & McNaughton, B. L. (2012). The role of medial prefrontal cortex in memory and decision making. *Neuron, 76*, 1057–1070.
Fanselow, M. S., & Dong, H.-W. (2010). Are the dorsal and ventral hippocampus functionally distinct structures? *Neuron, 65*, 7–19.
Fanselow, M. S., & LeDoux, J. E. (1999). Why we think plasticity underlying Pavlovian fear conditioning occurs in the basolateral amygdala. *Neuron, 23*, 229–232.
Felix-Ortiz, A. C., Burgos-Robles, A., Bhagat, N. D., Leppla, C. A., & Tye, K. M. (2016). Bidirectional modulation of anxiety-related and social behaviors by amygdala projections to the medial prefrontal cortex. *Neuroscience, 321*, 197–209.
Ferenczi, E. A., et al. (2016). Prefrontal cortical regulation of brainwide circuit dynamics and reward-related behavior. *Science, 351*, aac9698.
File, S., & Animal, E. (1985). Models for predicting clinical efficacy of anxiolytic drugs: Social behaviour. *Neuropsychobiology, 13*, 55–62.
Fiorino, D. F., & Phillips, A. G. (1999). Facilitation of sexual behavior and enhanced dopamine efflux in the nucleus accumbens of male rats afterd-amphetamine-induced behavioral sensitization. *The Journal of Neuroscience, 19*, 456–463.
Floden, D., Alexander, M. P., Kubu, C. S., Katz, D., & Stuss, D. T. (2008). Impulsivity and risk-taking behavior in focal frontal lobe lesions. *Neuropsychologia, 46*, 213–223.
Forbes, C. E., & Grafman, J. (2010). The role of the human prefrontal cortex in social cognition and moral judgment. *Annual Review of Neuroscience, 33*, 299–324.
Fox, H. C., Bergquist, K. L., Hong, K.-I., & Sinha, R. (2007). Stress-induced and alcohol cue-induced craving in recently abstinent alcohol-dependent individuals. *Alcoholism, Clinical and Experimental Research, 31*, 395–403.
Franklin, T. R., et al. (2002). Decreased gray matter concentration in the insular, orbitofrontal, cingulate, and temporal cortices of cocaine patients. *Biological Psychiatry, 51*, 134–142.
Franklin, T. B., et al. (2017). Prefrontal cortical control of a brainstem social behavior circuit. *Nature Neuroscience, 20*, 260–270.
Friedman, A., et al. (2015). A corticostriatal path targeting striosomes controls decision-making under conflict. *Cell, 161*, 1320–1333.
Fuchs, R. A., et al. (2005). The role of the dorsomedial prefrontal cortex, basolateral amygdala, and dorsal hippocampus in contextual reinstatement of cocaine seeking in rats. *Neuropsychopharmacology: Official Publication of the American College of Neuropsychopharmacology, 30*, 296–309.
Fusi, S., Miller, E. K., & Rigotti, M. (2016). Why neurons mix: High dimensionality for higher cognition. *Current Opinion in Neurobiology, 37*, 66–74.
Gabbott, P. L. A., Warner, T. A., Jays, P. R. L., Salway, P., & Busby, S. J. (2005). Prefrontal cortex in the rat: Projections to subcortical autonomic, motor, and limbic centers. *Journal of Comparative Neurology, 492*, 145–177.
Galynker, I. I., et al. (1998). Hypofrontality and negative symptoms in major depressive disorder. *The Journal of Nuclear Medicine: The Official Journal of the Society of Nuclear Medicine, 39*, 608–612.
Gawin, F. H., & Kleber, H. D. (1988). Evolving conceptualizations of cocaine dependence. *The Yale Journal of Biology and Medicine, 61*, 123–136.
George, O., & Koob, G. F. (2013). Control of craving by the prefrontal cortex. *Proceedings of the National Academy of Sciences of the United States of America, 110*, 4165–4166.
Goldman-Rakic, P. S. (1996). Regional and cellular fractionation of working memory. *Proceedings of the National Academy of Sciences, 93*, 13473–13480.

Goldstein, R. Z., et al. (2007). Is decreased prefrontal cortical sensitivity to monetary reward associated with impaired motivation and self-control in cocaine addiction? *The American Journal of Psychiatry*, *164*, 43–51.

Grabenhorst, F., & Rolls, E. T. (2011). Value, pleasure and choice in the ventral prefrontal cortex. *Trends in Cognitive Sciences*, *15*, 56–67.

Grace, A. A., Floresco, S. B., Goto, Y., & Lodge, D. J. (2007). Regulation of firing of dopaminergic neurons and control of goal-directed behaviors. *Trends in Neurosciences*, *30*, 220–227.

Grafman, J., et al. (1996). Frontal lobe injuries, violence, and aggression: A report of the Vietnam Head Injury Study. *Neurology*, *46*, 1231–1238.

Grigson, P. S., & Twining, R. C. (2002). Cocaine-induced suppression of saccharin intake: A model of drug-induced devaluation of natural rewards. *Behavioral Neuroscience*, *116*, 321–333.

Grimm, J. W., Hope, B. T., Wise, R. A., & Shaham, Y. (2001). Incubation of cocaine craving after withdrawal. *Nature*, *412*, 141–142.

Grunfeld, I. S., & Likhtik, E. (2018). Mixed selectivity encoding and action selection in the prefrontal cortex during threat assessment. *Current Opinion in Neurobiology*, *49*, 108–115.

Gunaydin, L. A., et al. (2014). Natural neural projection dynamics underlying social behavior. *Cell*, *157*, 1535–1551.

Halladay, L. R., et al. (2020). Prefrontal regulation of punished ethanol self-administration. *Biological Psychiatry*, *87*, 967–978.

Hamani, C., et al. (2011). The subcallosal cingulate gyrus in the context of major depression. *Biological Psychiatry*, *69*, 301–308.

Han, J.-H., et al. (2007). Neuronal competition and selection during memory formation. *Science*, *316*, 457–460.

Han, J.-H., et al. (2009). Selective erasure of a fear memory. *Science*, *323*, 1492–1496.

Harlow, J. M. (1868). Recovery from passage of an iron bar through the head. B. Mass. *Medicine and Society*, *2*, 3–20.

Harlow, J. M. (1993). Recovery from the passage of an iron bar through the head. *History of Psychiatry*, *4*, 274–281.

Harvey, P.-O., Pruessner, J., Czechowska, Y., & Lepage, M. (2007). Individual differences in trait anhedonia: A structural and functional magnetic resonance imaging study in non-clinical subjects. *Molecular Psychiatry*, *12*(703), 767–775.

Heidbreder, C. A., & Groenewegen, H. J. (2003). The medial prefrontal cortex in the rat: Evidence for a dorso-ventral distinction based upon functional and anatomical characteristics. *Neuroscience and Biobehavioral Reviews*, *27*, 555–579.

Hitora-Imamura, N., et al. (2015). Prefrontal dopamine regulates fear reinstatement through the downregulation of extinction circuits. *eLife*, *4*, e08274.

Hornak, J., Rolls, E. T., & Wade, D. (1996). Face and voice expression identification in patients with emotional and behavioural changes following ventral frontal lobe damage. *Neuropsychologia*, *34*, 247–261.

Hornak, J., et al. (2003). Changes in emotion after circumscribed surgical lesions of the orbitofrontal and cingulate cortices. *Brain: A Journal of Neurology*, *126*, 1691–1712.

Huang, W.-C., Zucca, A., Levy, J., & Page, D. T. (2020). Social behavior is modulated by valence-encoding mPFC-amygdala sub-circuitry. *Cell Reports*, *32*, 107899.

Hunt, P. R., & Aggleton, J. P. (1991). Medial dorsal thalamic lesions and working memory in the rat. *Behavioral and Neural Biology*, *55*, 227–246.

Hyman, S. E. (2005). Addiction: A disease of learning and memory. *The American Journal of Psychiatry*, *162*, 1414–1422.

James, M. H., McGlinchey, E. M., Vattikonda, A., Mahler, S. V., & Aston-Jones, G. (2018). Cued reinstatement of cocaine but not sucrose seeking is dependent on dopamine signaling in prelimbic cortex and is associated with recruitment of prelimbic neurons that project to contralateral nucleus accumbens core. *The International Journal of Neuropsychopharmacology*, *21*, 89–94.

Janak, P. H., & Tye, K. M. (2015). From circuits to behaviour in the amygdala. *Nature*, *517*, 284–292.

Jankowski, M. P., & Sesack, S. R. (2004). Prefrontal cortical projections to the rat dorsal raphe nucleus: Ultrastructural features and associations with serotonin and gamma-aminobutyric acid neurons. *Journal of Comparative Neurology*, *468*, 518–529.

Jasinska, A. J., Stein, E. A., Kaiser, J., Naumer, M. J., & Yalachkov, Y. (2014). Factors modulating neural reactivity to drug cues in addiction: A survey of human neuroimaging studies. *Neuroscience and Biobehavioral Reviews*, *38*, 1–16.

Jennings, J. H., et al. (2019). Interacting neural ensembles in orbitofrontal cortex for social and feeding behaviour. *Nature*, *565*, 645–649.

Johansen-Berg, H., et al. (2008). Anatomical connectivity of the subgenual cingulate region targeted with deep brain stimulation for treatment-resistant depression. *Cerebral Cortex (New York, N.Y.)*, *1991*(18), 1374–1383.

Johnston, W. J., Palmer, S. E., & Freedman, D. J. (2020). Nonlinear mixed selectivity supports reliable neural computation. *PLoS Computational Biology*, *16*, e1007544.

Kalivas, P. W., Volkow, N., & Seamans, J. (2005). Unmanageable motivation in addiction: A pathology in prefrontal-accumbens glutamate transmission. *Neuron*, *45*, 647–650.

Kamii, H., et al. (2015). Intrinsic membrane plasticity via increased persistent sodium conductance of cholinergic neurons in the rat laterodorsal tegmental nucleus contributes to cocaine-induced addictive behavior. *The European Journal of Neuroscience*, *41*, 1126–1138.

Kang, H. J., et al. (2012). Decreased expression of synapse-related genes and loss of synapses in major depressive disorder. *Nature Medicine*, *18*, 1413–1417.

Keedwell, P. A., Andrew, C., Williams, S. C. R., Brammer, M. J., & Phillips, M. L. (2005). The neural correlates of anhedonia in major depressive disorder. *Biological Psychiatry*, *58*, 843–853.

Kennerley, S. W., Dahmubed, A. F., Lara, A. H., & Wallis, J. D. (2009). Neurons in the frontal lobe encode the value of multiple decision variables. *Journal of Cognitive Neuroscience*, *21*, 1162–1178.

Kennerley, S. W., & Wallis, J. D. (2009). Encoding of reward and space during a working memory task in the orbitofrontal cortex and anterior cingulate sulcus. *Journal of Neurophysiology*, *102*, 3352–3364.

Kennerley, S. W., & Walton, M. E. (2011). Decision making and reward in frontal cortex. *Behavioral Neuroscience*, *125*, 297–317.

Kerstetter, K. A., et al. (2016). Corticostriatal afferents modulate responsiveness to psychostimulant drugs and drug-associated stimuli. *Neuropsychopharmacology*, *41*, 1128–1137.

Kim, S., Hwang, J., & Lee, D. (2008). Prefrontal coding of temporally discounted values during inter-temporal choice. *Neuron*, *59*, 161–172.

Kim, C. K., et al. (2017). Molecular and circuit-dynamical identification of top-down neural mechanisms for restraint of reward seeking. *Cell*, *170*, 1013. 1027.e14.

Kingsbury, L., et al. (2019). Correlated neural activity and encoding of behavior across brains of socially interacting animals. *Cell*, *178*, 429. 446.e16.

Klavir, O., Genud-Gabai, R., & Paz, R. (2013). Functional connectivity between amygdala and cingulate cortex for adaptive aversive learning. *Neuron*, *80*, 1290–1300.

Klüver, H., & Bucy, P. C. (1937). 'Psychic blindness' and other symptoms following bilateral temporal lobectomy in Rhesus monkeys. *The American Journal of Physiology*, *119*, 352–353.

Koob, G. F., & Le Moal, M. (2008). Neurobiological mechanisms for opponent motivational processes in addiction. *Philosophical Transactions of the Royal Society B: Biological Sciences*, *363*, 3113–3123.

Kunwar, P. S., et al. (2015). Ventromedial hypothalamic neurons control a defensive emotion state. *eLife*, *4*, e06633.

Kurosawa, R., Taoka, N., Shinohara, F., Minami, M., & Kaneda, K. (2013). Cocaine exposure enhances excitatory synaptic drive to cholinergic neurons in the laterodorsal tegmental nucleus. *The European Journal of Neuroscience*, *38*, 3027–3035.

Kutlu, M. G., & Gould, T. J. (2016). Effects of drugs of abuse on hippocampal plasticity and hippocampus-dependent learning and memory: Contributions to development and maintenance of addiction. *Learning & Memory (Cold Spring Harbor, N.Y.)*, *23*, 515–533.

Kyriazi, P., Headley, D. B., & Pare, D. (2018). Multi-dimensional coding by basolateral amygdala neurons. *Neuron*. https://doi.org/10.1016/j.neuron.2018.07.036.

Lammel, S., Ion, D. I., Roeper, J., & Malenka, R. C. (2011). Projection-specific modulation of dopamine neuron synapses by aversive and rewarding stimuli. *Neuron*, *70*, 855–862.

Lammel, S., Tye, K. M., & Warden, M. R. (2013). Progress in understanding mood disorders: Optogenetic dissection of neural circuits. *Genes, Brain, and Behavior*. https://doi.org/10.1111/gbb.12049.

Lammel, S., et al. (2012). Input-specific control of reward and aversion in the ventral tegmental area. *Nature*, *491*, 212–217.

Lang, P. J. (1995). The emotion probe: Studies of motivation and attention. *The American Psychologist*, *50*, 372–385.

Lang, P. J., Bradley, M. M., & Cuthbert, B. N. (1990). Emotion, attention, and the startle reflex. *Psychological Review*, *97*, 377–395.

Laubach, M., Amarante, L. M., Swanson, K., & White, S. R. (2018). What, if anything, is rodent prefrontal cortex? *eNeuro*, *5*.

LeDoux, J. E. (1996). *The emotional brain: The mysterious underpinnings of emotional life* (p. 384). Simon & Schuster.

LeDoux, J. E. (2000). Emotion circuits in the brain. *Annual Review of Neuroscience*, *23*, 155–184.

Lee, T. G., & D'Esposito, M. (2012). The dynamic nature of top-down signals originating from prefrontal cortex: A combined fMRI–TMS study. *The Journal of Neuroscience*, *32*, 15458–15466.

Lee, J. S., Jung, S., Park, I. H., & Kim, J.-J. (2015). Neural basis of anhedonia and amotivation in patients with schizophrenia: The role of reward system. *Current Neuropharmacology*, *13*, 750–759.

Levy, D. R., et al. (2019). Dynamics of social representation in the mouse prefrontal cortex. *Nature Neuroscience*, *22*, 2013–2022.

Li, A., et al. (2019). Role of BDNF in the development of an OFC-amygdala circuit regulating sociability in mouse and human. *Molecular Psychiatry*. https://doi.org/10.1038/s41380-019-0422-4.

Lin, D., et al. (2011). Functional identification of an aggression locus in the mouse hypothalamus. *Nature*, *470*, 221–226.

Litt, M. D., & Cooney, N. L. (1999). Inducing craving for alcohol in the laboratory. *Alcohol Research & Health: The Journal of the National Institute on Alcohol Abuse and Alcoholism*, *23*, 174–178.

Lo, L., et al. (2019). Connectional architecture of a mouse hypothalamic circuit node controlling social behavior. *Proceedings of the National Academy of Sciences*, *116*, 7503–7512.

Ma, T., Cheng, Y., Hellard, E., Wang, X., Lu, J., Gao, X., et al. (2018). Bidirectional and long-lasting control of alcohol-seeking behavior by corticostriatal LTP and LTD. *Nature Neuroscience*, *21*(3), 373–383. https://doi.org/10.1038/s41593-018-0081-9.

Ma, Y.-Y., et al. (2014). Bi-directional modulation of incubation of cocaine craving by silent synapse-based remodeling of prefrontal cortex to accumbens projections. *Neuron*, *83*, 1453–1467.

Machens, C. K., Romo, R., & Brody, C. D. (2010). Functional, but not anatomical, separation of "what" and "when" in prefrontal cortex. *The Journal of Neuroscience*, *30*, 350–360.

Maier, S. (2015). Behavioral control blunts reactions to contemporaneous and future adverse events: Medial prefrontal cortex plasticity and a corticostriatal network. *Neurobiol. Stress*. https://doi.org/10.1016/j.ynstr.2014.09.003.
Mante, V., Sussillo, D., Shenoy, K. V., & Newsome, W. T. (2013). Context-dependent computation by recurrent dynamics in prefrontal cortex. *Nature*, *503*, 78–84.
Mantz, J., Thierry, A. M., & Glowinski, J. (1989). Effect of noxious tail pinch on the discharge rate of mesocortical and mesolimbic dopamine neurons: Selective activation of the mesocortical system. *Brain Research*, *476*, 377–381.
Marek, R., et al. (2018). Hippocampus-driven feed-forward inhibition of the prefrontal cortex mediates relapse of extinguished fear. *Nature Neuroscience*, *21*, 384–392.
Maren, S. (2001). Neurobiology of pavlovian fear conditioning. *Annual Review of Neuroscience*, *24*, 897–931.
Maren, S., & Quirk, G. J. (2004). Neuronal signalling of fear memory. *Nature Reviews. Neuroscience*, *5*, 844–852.
Matsumoto, M., & Hikosaka, O. (2007). Lateral habenula as a source of negative reward signals in dopamine neurons. *Nature*, *447*, 1111–1115.
Matsumoto, M., & Hikosaka, O. (2009). Representation of negative motivational value in the primate lateral habenula. *Nature Neuroscience*, *12*, 77–84.
Matthews, G. A., & Tye, K. M. (2019). Neural mechanisms of social homeostasis. *Annals. New York Academy of Sciences*, *1457*, 5–25.
Mayberg, H. S., et al. (2005). Deep brain stimulation for treatment-resistant depression. *Neuron*, *45*, 651–660.
McFarland, K., Davidge, S. B., Lapish, C. C., & Kalivas, P. W. (2004). Limbic and motor circuitry underlying footshock-induced reinstatement of cocaine-seeking behavior. *The Journal of Neuroscience*, *24*, 1551–1560.
McGinnis, M. M., Parrish, B. C., Chappell, A. M., Alexander, N. J., & McCool, B. A. (2020). Chronic ethanol differentially modulates glutamate release from dorsal and ventral prefrontal cortical inputs onto rat basolateral amygdala principal neurons. *eNeuro*, *7*.
McGlinchey, E. M., James, M. H., Mahler, S. V., Pantazis, C., & Aston-Jones, G. (2016). Prelimbic to accumbens core pathway is recruited in a dopamine-dependent manner to drive cued reinstatement of cocaine seeking. *The Journal of Neuroscience*, *36*, 8700–8711.
Méndez-Bértolo, C., et al. (2016). A fast pathway for fear in human amygdala. *Nature Neuroscience*, *19*, 1041–1049.
Miller, E. K. (1999). The prefrontal cortex: Complex neural properties for complex behavior. *Neuron*, *22*, 15–17.
Miller, E. K., & Buschman, T. J. (2013). Cortical circuits for the control of attention. *Current Opinion in Neurobiology*, *23*, 216–222.
Miller, E. K., & Cohen, J. D. (2001). An integrative theory of prefrontal cortex function. *Annual Review of Neuroscience*, *24*, 167–202.
Moniz, E. (1937). Prefrontal leucotomy in the treatment of mental disorders. *The American Journal of Psychiatry*, *93*, 1379–1385.
Moorman, D. E., James, M. H., McGlinchey, E. M., & Aston-Jones, G. (2015). Differential roles of medial prefrontal subregions in the regulation of drug seeking. *Brain Research*, *1628*, 130–146.
Moreines, J. L., Owrutsky, Z. L., & Grace, A. A. (2017). Involvement of infralimbic prefrontal cortex but not lateral habenula in dopamine attenuation after chronic mild stress. *Neuropsychopharmacology*, *42*, 904–913.
Muir, J., Lopez, J., & Bagot, R. C. (2019). Wiring the depressed brain: Optogenetic and chemogenetic circuit interrogation in animal models of depression. *Neuropsychopharmacology*, *44*, 1013–1026.

Murrough, J. W., Abdallah, C. G., & Mathew, S. J. (2017). Targeting glutamate signalling in depression: Progress and prospects. *Nature Reviews. Drug Discovery*, *16*, 472–486.
Murugan, M., et al. (2017). Combined social and spatial coding in a descending projection from the prefrontal cortex. *Cell*, *171*, 1663. 1677.e16.
Namburi, P., et al. (2015). A circuit mechanism for differentiating positive and negative associations. *Nature*, *520*, 675–678.
Naya, Y., Chen, H., Yang, C., & Suzuki, W. A. (2017). Contributions of primate prefrontal cortex and medial temporal lobe to temporal-order memory. *Proceedings of the National Academy of Sciences*, *114*, 13555–13560.
Narita, M., et al. (2010). Implication of dopaminergic projection from the ventral tegmental area to the anterior cingulate cortex in μ-opioid-induced place preference. *Addiction Biology*, *14*, 434–447.
Nelson, A. C., et al. (2019). Molecular and circuit architecture of social hierarchy. *bioRxiv*, 838664.
Nocjar, C., & Panksepp, J. (2002). Chronic intermittent amphetamine pretreatment enhances future appetitive behavior for drug- and natural-reward: Interaction with environmental variables. *Behavioural Brain Research*, *128*, 189–203.
Ochsner, K. N., Bunge, S. A., Gross, J. J., & Gabrieli, J. D. E. (2002). Rethinking feelings: An FMRI study of the cognitive regulation of emotion. *Journal of Cognitive Neuroscience*, *14*, 1215–1229.
Ochsner, K. N., et al. (2009). Bottom-up and top-down processes in emotion generation. *Psychological Science*, *20*, 1322–1331.
Ongür, D., Ferry, A. T., & Price, J. L. (2003). Architectonic subdivision of the human orbital and medial prefrontal cortex. *Journal of Comparative Neurology*, *460*, 425–449.
Ongür, D., & Price, J. L. (2000). The organization of networks within the orbital and medial prefrontal cortex of rats, monkeys and humans. *Cereb. Cortex N. Y. N*, *1991*(10), 206–219.
Otis, J. M., et al. (2017). Prefrontal cortex output circuits guide reward seeking through divergent cue encoding. *Nature*, *543*, 103–107.
Papez, J. W. (1937). A proposed mechanism of emotion. *Archives of Neurology and Psychiatry*, *38*, 725–743.
Parnaudeau, S., et al. (2013). Inhibition of mediodorsal thalamus disrupts thalamofrontal connectivity and cognition. *Neuron*, *77*, 1151–1162.
Parthasarathy, A., et al. (2017). Mixed selectivity morphs population codes in prefrontal cortex. *Nature Neuroscience*, *20*, 1770–1779.
Pascoli, V., et al. (2014). Contrasting forms of cocaine-evoked plasticity control components of relapse. *Nature*, *509*, 459–464.
Passingham, R. E., & Wise, S. P. (2014). *The neurobiology of the prefrontal cortex: Anatomy, evolution, and the origin of insight*. xxii, 399 Oxford University Press.
Paton, J. J., Belova, M. A., Morrison, S. E., & Salzman, C. D. (2006). The primate amygdala represents the positive and negative value of visual stimuli during learning. *Nature*, *439*, 865–870.
Pennington, B. F., & Bennetto, L. (1993). Main effects or transactions in the neuropsychology of conduct disorder? Commentary on "The neuropsychology of conduct disorder". *Development and Psychopathology*, *5*, 153–164.
Perry, C. J., Zbukvic, I., Kim, J. H., & Lawrence, A. J. (2014). Role of cues and contexts on drug-seeking behaviour. *British Journal of Pharmacology*, *171*, 4636–4672.
Peters, J., Kalivas, P. W., & Quirk, G. J. (2009). Extinction circuits for fear and addiction overlap in prefrontal cortex. *Learning & Memory*, *16*, 279–288.
Petrides, M. (1995). Functional organization of the human frontal cortex for mnemonic processing. Evidence from neuroimaging studies. *Annals of the New York Academy of Sciences*, *769*, 85–96.

Peyron, C., Petit, J.-M., Rampon, C., Jouvet, M., & Luppi, P.-H. (1997). Forebrain afferents to the rat dorsal raphe nucleus demonstrated by retrograde and anterograde tracing methods. *Neuroscience, 82*, 443–468.

Phelps, E. A., & LeDoux, J. E. (2005). Contributions of the amygdala to emotion processing: From animal models to human behavior. *Neuron, 48*, 175–187.

Pizzagalli, D., & Depression, A. (2014). Stress, and anhedonia: Toward a synthesis and integrated model. *Annual Review of Clinical Psychology, 10*, 393–423.

Place, R., Farovik, A., Brockmann, M., & Eichenbaum, H. (2016). Bidirectional prefrontal-hippocampal interactions support context-guided memory. *Nature Neuroscience, 19*, 992–994.

Popescu, A. T., Zhou, M. R., & Poo, M.-M. (2016). Phasic dopamine release in the medial prefrontal cortex enhances stimulus discrimination. *Proceedings of the National Academy of Sciences of the United States of America, 113*, E3169–E3176.

Powell, N. J., & Redish, A. D. (2014). Complex neural codes in rat prelimbic cortex are stable across days on a spatial decision task. *Frontiers in Behavioral Neuroscience, 8*.

Quirk, G. J., & Beer, J. S. (2006). Prefrontal involvement in the regulation of emotion: Convergence of rat and human studies. *Current Opinion in Neurobiology, 16*, 723–727.

Ramirez-Cardenas, A., & Viswanathan, P. (2016). The role of prefrontal mixed selectivity in cognitive control. *The Journal of Neuroscience, 36*, 9013–9015.

Randall, P. A., McElligott, Z. A., & Besheer, J. (2020). Role of mPFC and nucleus accumbens circuitry in modulation of a nicotine plus alcohol compound drug state. *Addiction Biology, 25*, e12782.

Reid, M. S., Flammino, F., Starosta, A., Palamar, J., & Franck, J. (2006). Physiological and subjective responding to alcohol cue exposure in alcoholics and control subjects: Evidence for appetitive responding. *Journal of Neural Transmission (Vienna, Austria), 1996*(113), 1519–1535.

Reijmers, L. G., Perkins, B. L., Matsuo, N., & Mayford, M. (2007). Localization of a stable neural correlate of associative memory. *Science, 317*, 1230–1233.

Ridderinkhof, K. R., Ullsperger, M., Crone, E. A., & Nieuwenhuis, S. (2004). The role of the medial frontal cortex in cognitive control. *Science, 306*, 443–447.

Riga, D., et al. (2014). Optogenetic dissection of medial prefrontal cortex circuitry. *Frontiers in Systems Neuroscience, 8*.

Rigotti, M., et al. (2013). The importance of mixed selectivity in complex cognitive tasks. *Nature, 497*, 585–590.

Rogers, J. L., & See, R. E. (2007). Selective inactivation of the ventral hippocampus attenuates cue-induced and cocaine-primed reinstatement of drug-seeking in rats. *Neurobiology of Learning and Memory, 87*, 688–692.

Rolls, E. T., & Grabenhorst, F. (2008). The orbitofrontal cortex and beyond: From affect to decision-making. *Progress in Neurobiology, 86*, 216–244.

Rumpel, S., LeDoux, J., Zador, A., & Malinow, R. (2005). Postsynaptic receptor trafficking underlying a form of associative learning. *Science, 308*, 83–88.

Russell, J. A. (1980). A circumplex model of affect. *Journal of Personality and Social Psychology, 39*, 1161–1178.

Sawaguchi, T., & Goldman-Rakic, P. S. (1991). D1 dopamine receptors in prefrontal cortex: Involvement in working memory. *Science, 251*, 947–950.

Schlaepfer, T. E., et al. (2008). Deep brain stimulation to reward circuitry alleviates anhedonia in refractory major depression. *Neuropsychopharmacology, 33*, 368–377.

Schultz, W., Dayan, P., & Montague, P. R. (1997). A neural substrate of prediction and reward. *Science, 275*, 1593–1599.

Seger, C. A., Stone, M., & Keenan, J. P. (2004). Cortical Activations during judgments about the self and an other person. *Neuropsychologia, 42*, 1168–1177.

Seif, T., et al. (2013). Cortical activation of accumbens hyperpolarization-active NMDARs mediates aversion-resistant alcohol intake. *Nature Neuroscience, 16*, 1094–1100.

Semendeferi, K., Damasio, H., Frank, R., & Van Hoesen, G. W. (1997). The evolution of the frontal lobes: A volumetric analysis based on three-dimensional reconstructions of magnetic resonance scans of human and ape brains. *Journal of Human Evolution, 32*, 375–388.

Semendeferi, K., Lu, A., Schenker, N., & Damasio, H. (2002). Humans and great apes share a large frontal cortex. *Nature Neuroscience*, *5*, 272–276.
Seo, D., et al. (2013). Disrupted ventromedial prefrontal function, alcohol craving, and subsequent relapse risk. *JAMA Psychiatry Chic. Ill*, *70*, 727–739.
Seo, C., et al. (2019). Intense threat switches dorsal raphe serotonin neurons to a paradoxical operational mode. *Science*, *363*, 538–542.
Sesack, S. R., Deutch, A. Y., Roth, R. H., & Bunney, B. S. (1989). Topographical organization of the efferent projections of the medial prefrontal cortex in the rat: An anterograde tract-tracing study with Phaseolus vulgaris leucoagglutinin. *Journal of Comparative Neurology*, *290*, 213–242.
Sf, M., & Lr, W. (2005). Stressor controllability and learned helplessness: The roles of the dorsal raphe nucleus, serotonin, and corticotropin-releasing factor. *Neuroscience and Biobehavioral Reviews*, *29*. https://pubmed.ncbi.nlm.nih.gov/15893820/.
Shabel, S. J., & Janak, P. H. (2009). Substantial similarity in amygdala neuronal activity during conditioned appetitive and aversive emotional arousal. *Proceedings of the National Academy of Sciences*, *106*, 15031–15036.
Siciliano, C. A., et al. (2019). A cortical-brainstem circuit predicts and governs compulsive alcohol drinking. *Science*, *366*, 1008–1012.
Solomon, R. L. (1980). The opponent-process theory of acquired motivation: The costs of pleasure and the benefits of pain. *The American Psychologist*, *35*, 691–712.
Song, J., et al. (2019). Crucial role of feedback signals from prelimbic cortex to basolateral amygdala in the retrieval of morphine withdrawal memory. *Science Advances*, *5*, eaat3210.
Sotres-Bayon, F., & Quirk, G. J. (2010). Prefrontal control of fear: More than just extinction. *Current Opinion in Neurobiology*, *20*, 231–235.
Spealman, R. D., et al. (2004). Triggers of relapse: Nonhuman primate models of reinstated cocaine seeking. *Nebraska Symposium on Motivation*, *50*, 57–84.
Stefanik, M. T., & Kalivas, P. W. (2013). Optogenetic dissection of basolateral amygdala projections during cue-induced reinstatement of cocaine seeking. *Frontiers in Behavioral Neuroscience*, *7*.
Stefanik, M. T., Kupchik, Y. M., & Kalivas, P. W. (2016). Optogenetic inhibition of cortical afferents in the nucleus accumbens simultaneously prevents cue-induced transient synaptic potentiation and cocaine-seeking behavior. *Brain Structure & Function*, *221*, 1681–1689.
Stefanik, M. T., et al. (2013). Optogenetic inhibition of cocaine seeking in rats. *Addiction Biology*, *18*, 50–53.
Sul, J. H., Kim, H., Huh, N., Lee, D., & Jung, M. W. (2010). Distinct roles of rodent orbitofrontal and medial prefrontal cortex in decision making. *Neuron*, *66*, 449–460.
Takahashi, Y. K., et al. (2009). The orbitofrontal cortex and ventral tegmental area are necessary for learning from unexpected outcomes. *Neuron*, *62*, 269–280.
Tamietto, M., & de Gelder, B. (2010). Neural bases of the non-conscious perception of emotional signals. *Nature Reviews. Neuroscience*, *11*, 697–709.
Taylor, J. R., & Horger, B. A. (1999). Enhanced responding for conditioned reward produced by intra-accumbens amphetamine is potentiated after cocaine sensitization. *Psychopharmacology*, *142*(31–40).
Tierney, A., & Egas, J. (2000). Moniz and the origins of psychosurgery: A review commemorating the 50th anniversary of Moniz's Nobel Prize. *Journal of the History of the Neurosciences*, *9*, 22–36.
Tsuchida, A., & Fellows, L. K. (2012). Are You Upset? Distinct roles for orbitofrontal and lateral prefrontal cortex in detecting and distinguishing facial expressions of emotion. *Cerebral Cortex*, *22*, 2904–2912.
Tsuda, B., Tye, K. M., Siegelmann, H. T., & Sejnowski, T. J. (2020). A modeling framework for adaptive lifelong learning with transfer and savings through gating in the prefrontal cortex. *bioRxiv*. https://doi.org/10.1101/2020.03.11.984757. 2020.03.11.984757.
Tye, K. M. (2018). Neural circuit motifs in valence processing. *Neuron*, *100*, 436–452.
Tye, K. M., et al. (2011). Amygdala circuitry mediating reversible and bidirectional control of anxiety. *Nature*, *471*, 358–362.

Tye, K. M., et al. (2013). Dopamine neurons modulate neural encoding and expression of depression-related behaviour. *Nature, 493*, 537–541.

Van den Oever, M. C., Spijker, S., Smit, A. B., & De Vries, T. J. (2010). Prefrontal cortex plasticity mechanisms in drug seeking and relapse. *Neuroscience and Biobehavioral Reviews, 35*, 276–284.

Vander Weele, C. M., Siciliano, C. A., & Tye, K. M. (2018). Dopamine tunes prefrontal outputs to orchestrate aversive processing. *Brain Research*. https://doi.org/10.1016/j.brainres.2018.11.044.

Vander Weele, C. M., et al. (2018). Dopamine enhances signal-to-noise ratio in cortical-brainstem encoding of aversive stimuli. *Nature, 563*, 397–401.

Vertes, R. P. (2004). Differential projections of the infralimbic and prelimbic cortex in the rat. *Synapse, 51*, 32–58.

Volkow, N. D., & Fowler, J. S. (2000). Addiction, a disease of compulsion and drive: Involvement of the orbitofrontal cortex. *Cerebral Cortex (New York, N.Y.), 1991*(10), 318–325.

Volkow, N. D., Fowler, J. S., & Wang, G.-J. (2003). The addicted human brain: Insights from imaging studies. *The Journal of Clinical Investigation, 111*, 1444–1451.

Volkow, N. D., et al. (1997). Relationship between subjective effects of cocaine and dopamine transporter occupancy. *Nature, 386*, 827–830.

Wallis, J., & Orbitofrontal, D. (2007). Cortex and its contribution to decision-making. *Annual Review of Neuroscience, 30*, 31–56.

Walton, M. E., Devlin, J. T., & Rushworth, M. F. S. (2004). Interactions between decision making and performance monitoring within prefrontal cortex. *Nature Neuroscience, 7*, 1259–1265.

Wang, Q., Jin, J., & Maren, S. (2016). Renewal of extinguished fear activates ventral hippocampal neurons projecting to the prelimbic and infralimbic cortices in rats. *Neurobiology of Learning and Memory, 134*(Pt A), 38–43.

Wang, N., et al. (2018). Role of Glutamatergic Projections from the Ventral CA1 to Infralimbic Cortex in Context-Induced Reinstatement of Heroin Seeking. *Neuropsychopharmacology, 43*, 1373–1384.

Wang, Y., et al. (2019). Opiate-associated contextual memory formation and retrieval are differentially modulated by dopamine D1 and D2 signaling in hippocampal-prefrontal connectivity. *Neuropsychopharmacol. Off. Publ. Am. Coll. Neuropsychopharmacol., 44*, 334–343.

Warden, M. R., et al. (2012). A prefrontal cortex–brainstem neuronal projection that controls response to behavioural challenge. *Nature, 492*, 428–432.

Weiskrantz, L. (1956). Behavioral changes associated with ablation of the amygdaloid complex in monkeys. *Journal of Comparative and Physiological Psychology, 49*, 381–391.

Wyvell, C. L., & Berridge, K. C. (2001). Incentive sensitization by previous amphetamine exposure: Increased cue-triggered 'wanting' for sucrose reward. *Journal of Neuroscience: The Official Journal of the Society for Neuroscience, 21*, 7831–7840.

Xiao, C., et al. (2016). Cholinergic mesopontine signals govern locomotion and reward through dissociable midbrain pathways. *Neuron, 90*, 333–347.

Yizhar, O., & Klavir, O. (2018). Reciprocal amygdala–prefrontal interactions in learning. *Current Opinion in Neurobiology, 52*, 149–155.

Zhao, Y., et al. (2017). Memory retrieval in addiction: A role for miR-105-mediated regulation of D1 receptors in mPFC neurons projecting to the basolateral amygdala. *BMC Biology, 15*.

Zhou, T., et al. (2017). History of winning remodels thalamo-PFC circuit to reinforce social dominance. *Science, 357*, 162–168.

Zironi, I., Burattini, C., Aicardi, G., & Janak, P. H. (2006). Context is a trigger for relapse to alcohol. *Behavioural Brain Research, 167*, 150–155.

CHAPTER EIGHT

Social processing by the primate medial frontal cortex

Philip T. Putnam[a,*] and Steve W.C. Chang[a,b,c]
[a]Department of Psychology, Yale University, New Haven, CT, United States
[b]Department of Neuroscience, Yale University School of Medicine, New Haven, CT, United States
[c]Kavli Institute for Neuroscience, Yale University School of Medicine, New Haven, CT, United States
*Corresponding author: e-mail address: philip.putnam@yale.edu

Contents

Abstract

The primate medial frontal cortex is comprised of several brain regions that are consistently implicated in regulating complex social behaviors. The medial frontal cortex is also critically involved in many non-social behaviors, such as those involved in reward, affective, and decision-making processes, broadly implicating the fundamental role of the medial frontal cortex in internally guided cognition. An essential question therefore is what unique contributions, if any, does the medial frontal cortex make to social behaviors? In this chapter, we outline several neural algorithms necessary for mediating adaptive social interactions and discuss selected evidence from behavioral neurophysiology experiments supporting the role of the medial frontal cortex

International Review of Neurobiology, Volume 158
ISSN 0074-7742
https://doi.org/10.1016/bs.irn.2020.12.003

in implementing these algorithms. By doing so, we primarily focus on research in nonhuman primates and examine several key attributes of the medial frontal cortex. Specifically, we review neuronal substrates in the medial frontal cortex uniquely suitable for enabling social monitoring, observational and vicarious learning, as well as predicting the behaviors of social partners. Moreover, by utilizing the three levels of organization in information processing systems proposed by Marr (1982) and recently adapted by Lockwood, Apps, and Chang (2020) for social information processing, we survey selected social functions of the medial frontal cortex through the lens of socially relevant algorithms and implementations. Overall, this chapter provides a broad overview of the behavioral neurophysiology literature endorsing the importance of socially relevant neural algorithms implemented by the primate medial frontal cortex for regulating social interactions.

1. Introduction

Social interactions cannot be understood simply through the framework of classical learning processes (Bandura & Walters, 1977; Frith & Frith, 2010; Redcay & Schilbach, 2019). Rather, highly adaptative and contingent social behaviors require specialized neural processes to observe and predict the actions of others, to learn from these actions, and make inferences about a social partner's motivations and intentions. Here we will present emerging evidence from behavioral neurophysiology research that suggests that the medial frontal cortex (MFC) performs a unique role in shaping social behaviors by performing many of these processes. Additionally, by applying the three levels of organization proposed by David Marr in his seminal 1982 book *Vision* (Marr, 1982) to social information processing (Lockwood, Apps, & Chang, 2020), we will examine in detail why certain regions within the MFC are particularly suited to support socially relevant computations with specific algorithms and implementations.

First, it is important to recognize that the MFC is just one collection of brain regions that is critically implicated in social functions out of many other collections of cortical and subcortical regions implicated in social behaviors (Adolphs, 2009; Behrens, Hunt, & Rushworth, 2009). Accumulating evidence strongly suggests that social behaviors are accomplished through the coordination of many different brain regions working in tandem (Chen & Hong, 2018; Dal Monte, Chu, Fagan, & Chang, 2020; Dölen, Darvishzadeh, Huang, & Malenka, 2013; Sliwa & Freiwald, 2017; Zhou et al., 2017). Based on existing evidence it is plausible to argue that the regions in the MFC serve as central nodes of social processing networks in the brain across multiple animal species. These MFC brain regions include

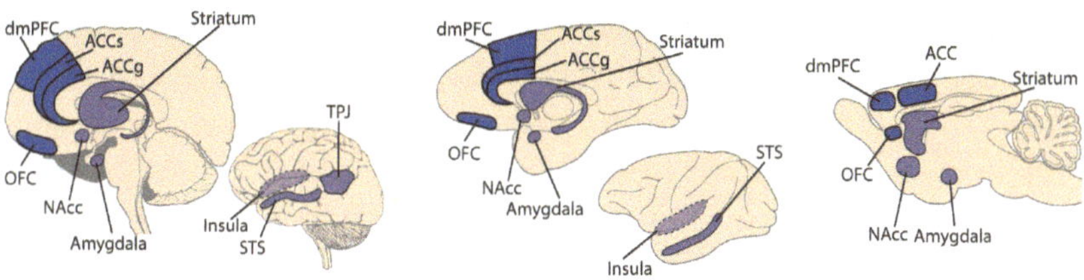

Fig. 1 Sagittal brain illustrations with selected brain structures implicated in social functions in humans (A), rhesus macaques (B), and mice (C). Highlighted are the key components of the MFC (darker shading) and other regions (lighter shading) that are implicated in various social behaviors in each species. ACCg, anterior cingulate gyrus; ACCs, anterior cingulate sulcus; dmPFC, dorsomedial prefrontal cortex; OFC, orbitofrontal cortex; NAcc, nucleus accumbens; STS, superior temporal sulcus; TPJ, temporal parietal junction.

the anterior cingulate cortex (ACC), dorsal, medial, and ventral sub-divisions of the medial prefrontal cortex (mPFC), supplemental motor area (SMA), pre-SMA, as well as the medial aspects of the orbitofrontal cortex (OFC), among others (Wise, 2008). The MFC is centrally positioned as a key node in a network of brain regions involved in social behaviors (Fig. 1), which some scholars have referred to as the "social brain" (Adolphs, 2009; Lockwood et al., 2020), and is a major cortical node of the broader limbic network (Morecraft & Van Hoesen, 1998; Porrino, Crane, & Goldman-Rakic, 1981), allowing integrations of reward and affective information critical for social functions. As part of this network (Fig. 1), the MFC is highly interconnected with other frontal regions (Carmichael & Price, 1996), other cortical areas such as the superior temporal sulcus (STS) and temporoparietal junction (TPJ) (Bachevalier, Meunier, Lu, & Ungerleider, 1997; Carmichael & Price, 1995b), and subcortical structures such as the amygdala (Amaral & Insausti, 1992; Carmichael & Price, 1995a; Morecraft et al., 2012) and nucleus accumbens (NAcc).

Experimental evidence supports a causal role for the MFC in social behaviors (Amodio & Frith, 2006). In both humans (Bechara, Damasio, & Damasio, 2000; Saver & Damasio, 1991) and rhesus monkeys (Basile, Schafroth, Karaskiewicz, Chang, & Murray, 2020; Rudebeck, Buckley, Walton, & Rushworth, 2006), lesions within the MFC regions alter social behaviors, such as social valuation and forming prosocial preferences from vicarious social reward. However, lesion-induced social behavioral deficits often do not truly distinguish the MFC from other interconnected brain areas, such as the amygdala, that are also implicated in social behaviors (Adolphs, 2010). On the other hand, evidence from behavioral neurophysiology has revealed that the MFC may be uniquely positioned to perform several key algorithms necessary for social interactions. In this chapter, we first briefly outline selected key algorithms necessary for social interactions in humans and non-human primates, and then provide experimental evidence suggesting that the neurons in the primate MFC are performing these algorithms. Finally, we discuss why the MFC is structurally and evolutionarily suited for social functions. In doing so, we will frequently reference back to the three levels of information processing—computational, algorithmic, and implementational—in discussing how the MFC regions are relevant for social functions.

2. Key algorithms in social interactions

As a framework to analyze information processing systems, David Marr proposed three levels of organization that should be considered (Fig. 2) (Marr, 1982). The most macroscopic level is *computational*, or understanding the ultimate goal of the process, and the logic of the strategy by which it will be accomplished. Supporting this goal is the *algorithm*, the logical operation or set of rules that is applied to accomplish the goals of the system. Finally, there is the *implementation*, or how these algorithms are realized physically. As an illustration, Marr applied this framework to a cash register, where the *computational* goal is to accomplish arithmetic and thus it must "master the theory of addition." The specific *algorithms* used to accomplish this can take various forms (e.g. Arabic numerals require "adding the least significant digits first and 'carrying' if the sum exceeds 9" while other numeric representations would require different parallel algorithms) but must ultimately serve the exactly same computational goal. These algorithms must be instantiated in a physical form at the *implementational* level, which for a mechanical cash register may be tracking numerical digits by a position on a notched metal wheel or the logic gates on a computer chip inside a digital cash register. This three-level framework is particularly advantageous for studying neuroscience when considering many varied *computational goals* the brain must accomplish. For example, an individual may be engaging in cooperative social behaviors in one context, and then must quickly transition to competitive behaviors in the next. Similarly, an individual may learn a choice-outcome association in an isolated context, and then later needs to apply that knowledge in a social context. To support these varied social computations, many different algorithmic processes have been proposed, such as monitoring the actions of others in an agent-specific reference

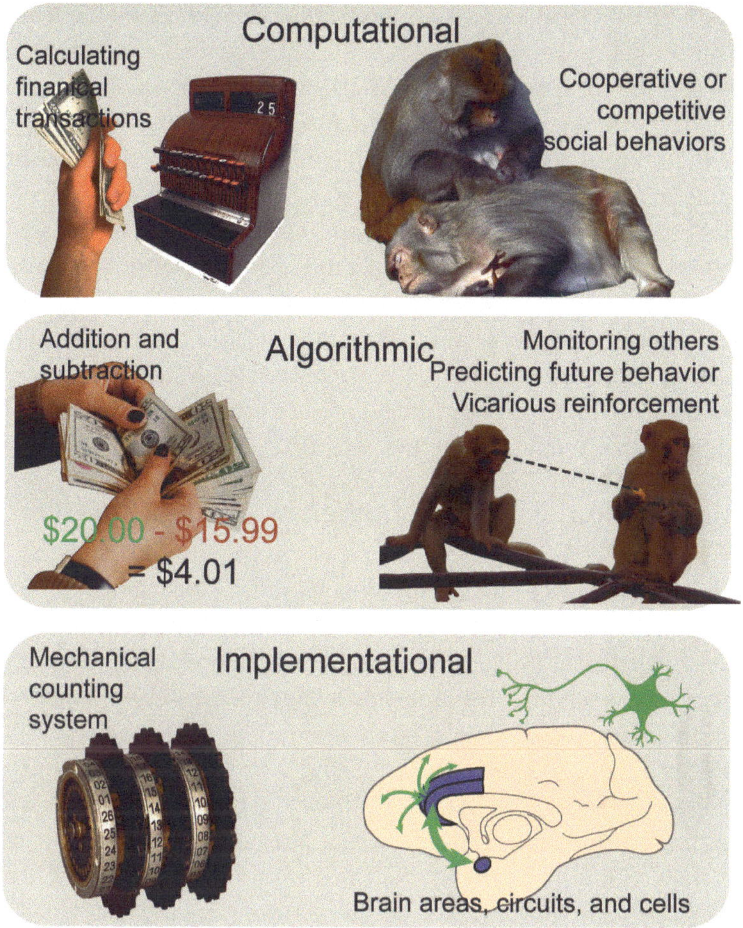

Fig. 2 Levels of organization proposed by David Marr to understand information processing systems. The computational level (top) encompasses the ultimate goal of the process, for example calculating financial transactions is the goal of a cash register (left) or engaging in cooperative social behaviors is the goal of certain brains areas. The algorithmic level (middle) is the logical processes by which this goal is accomplished. For a cash register to tally purchases it must perform arithmetic (left), while social brain regions must perform certain processes such as monitoring others or predicting their future behaviors (right). Finally, these processes must be realized at the implementational level (bottom). In a cash register the actual arithmetic is performed by a mechanical counting system (left), while in the brain different areas, circuits, and cells work together to process information (right).

frame or predicting the actions of others. These algorithms are implemented in circuits of neurons, linked by synaptic connections and oftentimes clustered in specific brain areas. Applying this framework to social functions, it has been recently suggested that some neural processes are socially specialized at the level of algorithms and implementations, and that altering the computational goals can change social specialization at these other levels (Lockwood et al., 2020). Given the enormous diversity of behaviors which the MFC is implicated in, it is likely that many of these same algorithms and implementations are shared between social and non-social behaviors and can be flexibly recruited to serve different computational goals. Below we outline a subset of the computational goals required for adaptative social behaviors, before providing experimental evidence supporting the role of the MFC at the algorithmic and implementational levels.

2.1 Monitoring the actions of others with agent-specific computation

Fundamental to any social interaction is an individual's ability to observe the actions of another, and to understand these actions from the perspective of the social partner. These actions may result in rewarding outcomes for the social partner, which the individual must process in reference to the partner in order to accurately guide their own social behaviors. This agent-specific computation, i.e. processing a reward in a "self-oriented" or "other-oriented" reference frame is theorized to be supported by a coordinate transformation framework (Chang, 2013), analogous to how distinctly referenced information is transformed from one frame of reference to another to perform sensorimotor transformation (Snyder, 2000). Even the most basic social interactions require agent-specific referencing of social information, as otherwise representing others separately from self in the brain would be impossible. These other-oriented signals may represent various domains of information; such as the overt actions of a social partner, errors in those actions, or information about rewarding or aversive outcomes received by others. At an implementational level, it would be predicted that populations of individual neurons would separately represent other-referenced information, self-referenced information, or commonly (both-referenced) information in order to perform both self- and other-regarding behaviors. Importantly, monitoring the actions of others in an agent-specific manner is also crucial to other algorithms used to shape social behaviors. For example, monitoring reward received by another is crucial for observational learning as well as deriving vicarious reinforcement.

2.2 Predicting the actions of others

Building upon the ability to monitor the actions of others, being able to predict the actions of others is a crucial computation that grants an individual the ability to engage in strategic, cooperative, or competitive social behaviors, or to learn through observing another's actions and outcomes. This prediction extends beyond merely learning the sensorimotor relationships between another's actions and associated outcomes but requires the observer to model the intentions that motivate these actions to discern what outcomes the partner values and expects (Apps, Balsters, & Ramnani, 2012; Apps, Rushworth, & Chang, 2016; Haroush & Williams, 2015; Joiner, Piva, Turrin, & Chang, 2017). An individual may choose to use these predictions in turn to guide their own behaviors, such as choosing to cooperate with others, or to compete for those same valued outcomes (Haroush & Williams, 2015). Predictions like these allow an individual to prospectively predict how their own actions will affect the behaviors of others, where errors in these predictions permit generalized algorithms, such as reinforcement learning, to guide learning about the internal states and motivations of others to support more complex forms of social learning and mentalizing (Joiner et al., 2017).

2.3 Observational learning

Observing the actions of others is a key component of how many species learn about the world. Crucially, observational learning enables the observer to learn about potential dangers or rewards from conspecifics while circumventing risks associated with direct experience. Additionally, observers can also learn about more abstract information such as the motivations or intentions of others and even specific personal characteristics of other individuals. The process of observational learning builds upon both monitoring and predicting the behaviors of another individual, necessitating both tracking the partner's actions and their associated outcomes in a self-referenced or other-referenced processing framework, and using this perspective to discern what outcomes the partner values and thus to predict their future actions. Observational learning begins early in development, with young children imitating the actions of older individuals (Bandura, Ross, & Ross, 1961), and continues into adulthood in multiple animal species, including rats (Heyes, Jaldow, Nokes, & Dawson, 1994), birds (Akins, Klein, & Zentall, 2002), monkeys (Grabenhorst, Hernádi, & Schultz, 2012;

Tomasello, Davis-Dasilva, Camak, & Bard, 1987) and humans (Bandura, 2008; Debiec & Olsson, 2017; Hill, Boorman, & Fried, 2016; Olsson, Knapska, & Lindström, 2020). Moreover, a higher degree of both attentional (Pagnotta, Laland, & Coco, 2020), autonomic (Pärnamets, Espinosa, & Olsson, 2020), or brain-to-brain (Bevilacqua et al., 2019; Dikker et al., 2017; Kostorz, Flanagin, & Glasauer, 2020) synchrony has been shown to enhance learning rates during observational learning. Such synchrony between a demonstrator and an observer may, in turn, facilitate socially specific algorithms such as theory of mind.

2.4 Theory of mind

Theory of mind is the process by which one attributes mental states, beliefs, and intentions to others. While the presence of a theory of mind in non-human primates is an ongoing and controversial question (Horschler, MacLean, & Santos, 2020), some components of theory of mind may be shared across primates (Drayton & Santos, 2016; Horschler et al., 2020). The most common test for theory of mind is the *false belief task*, where participants are tested to see if they can understand whether another social agent has a false belief about an object's location. In the false belief task, the subject observes an agent placing an object in set location, but another agent subsequently moves that object to a different location (unknown to the first agent). When the first agent returns to look for the object, the subject may anticipate that the first agent will search for the original but wrong location or the new and correct location depending on whether the subject possesses the capacity to represent a false belief. In brief, it is theorized that having a false belief representation would lead for the subject to anticipate that the first agent would search for the original yet wrong location. This specialized task thus attempts to examine if an individual can model the beliefs of another social agent distinctly from one's own, which can be argued to be the most complex form of agent-specific referencing. Human infants, even from a very young age, are generally able to use the mental states of others to explain their behaviors (Onishi & Baillargeon, 2005). However, whether non-human primates can "pass" the false belief task remains controversial (Hayashi et al., 2020; Horschler et al., 2020; Krupenye, Kano, Hirata, Call, & Tomasello, 2016). Overall, a theory of mind is a highly adaptive social algorithm, allowing an individual to model another's psychological states and can be used to predict, understand, and influence the behavior of others. Given the highly specific nature of theory of mind, it is exceedingly likely to be a socially specific algorithm that critically builds upon monitoring others in an agent-specific fashion, predicting other's behaviors, and learning about these mental states through observation.

3. Evidence from behavioral neurophysiology during social interactions

Examining the neuronal activity of MFC neurons while individuals participate in social behaviors grants a direct window into how various socially relevant algorithms are implemented in the MFC subregions. Uniquely suited to addressing these question is the emerging field of primate social neurophysiology; the investigation of the neurophysiological mechanisms underlying social interactions using a nonhuman primate model system (Chang, 2017; Gothard et al., 2017). Here we present studies from this subfield of neuroscience that provide insight into how the MFC regulate social behaviors by implementing socially relevant algorithms.

3.1 Neurons in the MFC monitor others in an agent-specific manner

Representations of value are of primary importance for the survival of a species, and thus are computed across many different brain regions including the medial and orbital frontal areas (Azab & Hayden, 2017; Hayden & Platt, 2010; Kennerley, Behrens, & Wallis, 2011; Padoa-Schioppa & Assad, 2006; Paton, Belova, Morrison, & Salzman, 2006; Rogers et al., 2004). However, the computations of value in a social context, especially those involving the process of deriving reward from others such as vicarious reward appears to be more selectively mediated by specific brain regions. Monitoring rewarding outcomes of others and estimating internal states of others is arguably the most fundamental computation that is necessary for normal social interactions. It goes without saying that such other-referenced processes are central to observational learning, empathy, and theory of mind.

To examine single neuron correlates of other-referenced reward processing in the MFC, researchers have recorded single unit activity from the anterior cingulate cortex (ACC) while monkeys were engaged in various social reward paradigms. In one of such paradigms known as the social reward allocation task (Fig. 3A), an actor monkey expressed either prosocial or antisocial preference by deciding to share with or withhold juice rewards from a conspecific recipient monkey in the same setup (Chang, Winecoff, & Platt, 2011). In one context, the actor monkey chose between donating

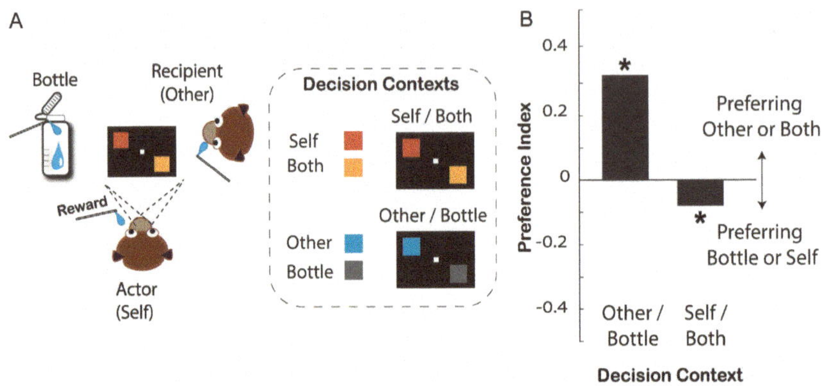

Fig. 3 Social reward allocation paradigm. (A) Task design. An actor monkey makes decisions to share with, or withhold juice rewards from, a recipient monkey seated adjacent to the actor. In the Self/Both context, the actor chooses between delivering a juice reward to himself and to both himself and the recipient. In the Other/Bottle context, the actor chooses between delivering a juice rewards to the recipient and wasting it in an empty bottle. (B) Social decision preference. Actor monkeys prefer to deliver juice rewards to the recipient over wasting them in the Other/Bottle context (exhibiting prosocial preference), whereas they prefer to deliver juice rewards just to themselves over both monkeys (exhibiting antisocial preference). *Figures adapted from Dal Monte, O., Chu, C. C. J., Fagan, N. A., & Chang, S. W. C. (2020). Specialized medial prefrontal–amygdala coordination in other-regarding decision preference.* Nature Neuroscience, 23*(4), 565–574. doi:10.1038/s41593-020-0593-y.*

juice rewards to the recipient and no one (Other/Bottle or Other/Neither context), whereas in the other context the actor chose between drinking juice rewards alone and together with the recipient (Self/Both context). Importantly, the choices in each context were paired such that choosing one or the other option did not have any direct reward impact to the actor—that is, the actor always received juice rewards in the Self/Both context, but never received juice rewards in the Other/Bottle context. Behaviorally, actor monkeys have been consistently found to prefer donating juice to the recipient over wasting, exhibiting a prosocial preference (Basile et al., 2020; Chang, Barter, Ebitz, Watson, & Platt, 2012; Chang et al., 2011, 2015; Chang, Gariépy, & Platt, 2013; Dal Monte et al., 2020) (Fig. 3B). By contrast, when the options for delivering juice rewards to the actors and both the actor and recipient monkeys were available, the actors consistently preferred drinking alone, exhibiting an antisocial preference (Chang et al., 2011, 2015; Dal Monte et al., 2020) (Fig. 3B).

Using the social reward allocation task, Chang and colleagues found that distinct neurons in the anterior cingulate gyrus (ACCg) exclusively encoded

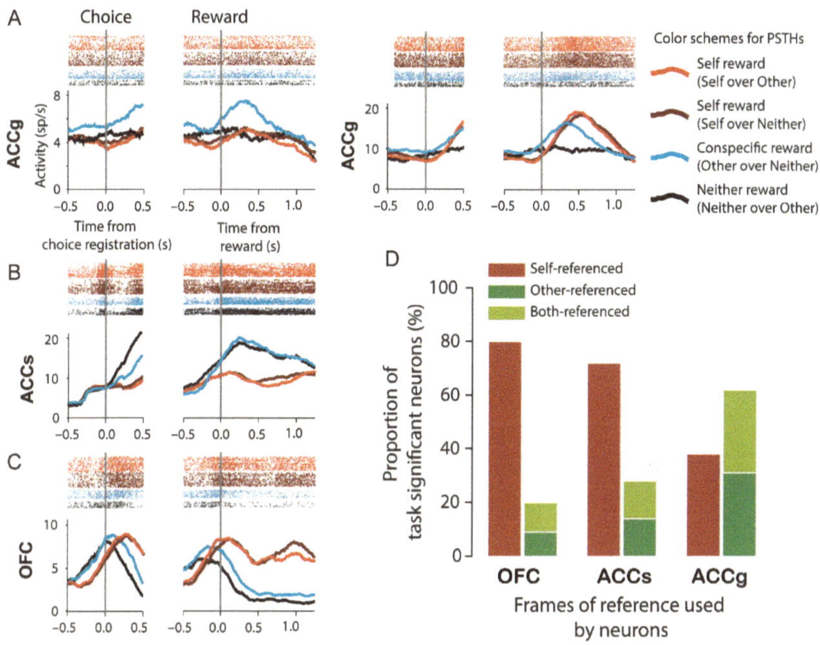

Fig. 4 Single unit activity of ACCg, ACCs, and OFC neurons collected during the social reward allocation task. (A) Single unit activity of two example ACCg cells depicted as a raster (above) and peri-stimulus time histogram (below) aligned to the time when the actor monkey made the choice (left) and when the reward was delivered (right). In the "other-referenced" ACCg cell shown on the left, note the specific increase of firing rate for the decisions to deliver juice rewards to the recipient and the reward outcome of the recipient. In the "both-referenced" ACCg cell shown on the right, note the increase in firing rate for the decisions to deliver juice rewards to the actor, the recipient, as well as both monkeys, but not for delivering juice to no one, and the associated reward outcomes. (B) Single unit activity of an example ACCs cell. In this "self-referenced" cell, note the increase in firing rate for decisions and reward outcomes that resulted in the foregoing of actor's rewards. (C) Single unit activity of an example OFC cell. In this "self-referenced" cell, note the increase in firing rate for all decisions but only for reward outcomes that resulted in the receipt of actor's rewards. (D) Plots summarizing the proportion of neurons encoding reward outcomes in self-referenced, other-referenced, or both-referenced frames. *Figures adapted from Chang, S. W. C., Gariépy, J.-F., & Platt, M. L. (2013). Neuronal reference frames for social decisions in primate frontal cortex.* Nature Neuroscience, 16*(2), 243–250. doi:10.1038/nn.3287.*

reward allocations to the other monkey (other-referenced), reward allocations to oneself (self-referenced), or both monkeys (both-referenced) (Chang et al., 2013) (Figs. 4A and 5A). Notably, the vast majority of neurons in the neighboring anterior cingulate sulcus (ACCs) signaled actor's foregone rewards by increasing activity similarly for the other monkey's rewards

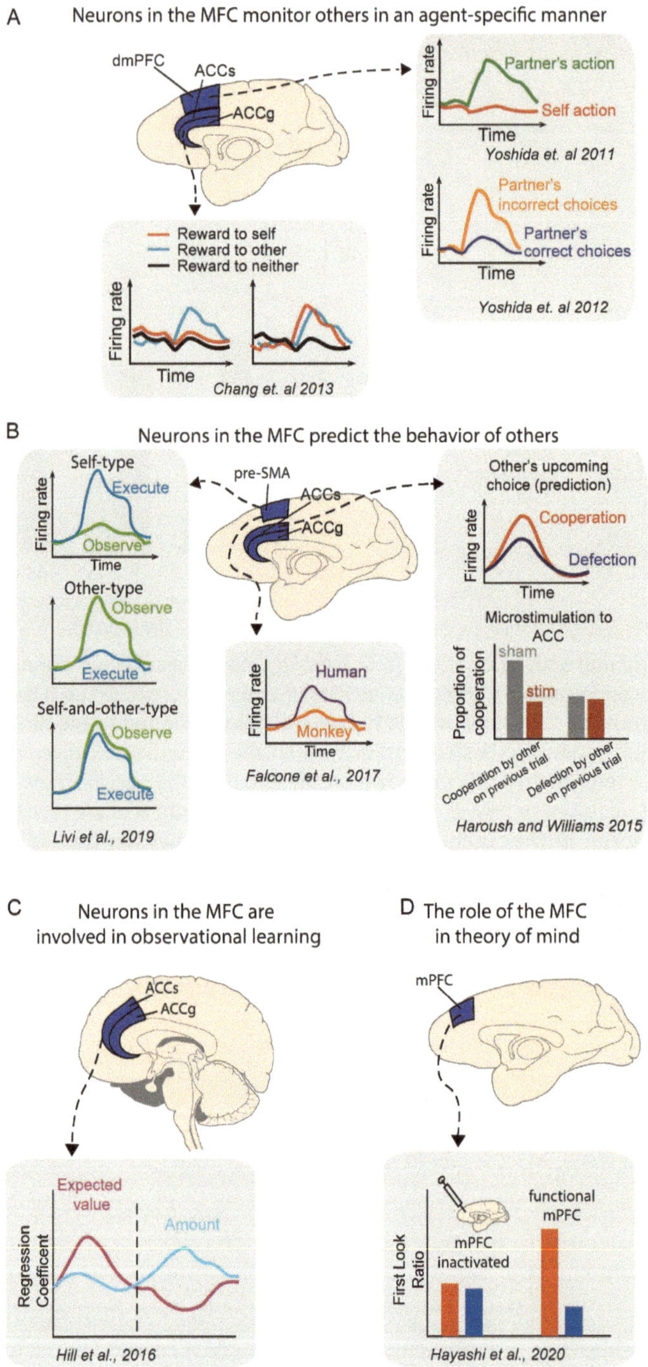

Fig. 5 Illustrations of selected findings from MFC during social interactions. *These illustrations are adapted from Gangopadhyay, P., Chawla, M., Dal Monte, O., & Chang, S. W. C. (2020). Neural mechanisms guiding social decision-making in the prefrontal-amygdala circuits.* Nature Neuroscience, *(in press).*

and wasted rewards (self-referenced) (Fig. 4B). Neurons in the orbitofrontal cortex (OFC), on the other hand, principally encoded rewards delivered to oneself (self-referenced) (Fig. 4C). Across the three neural populations, ACCg contained the highest number of neurons that were classified as either other-referenced or both-referenced, whereas the majority of ACCs and OFC neurons were classified as self-referenced (Fig. 4D). This evidence shows the ACCg is a specialized MFC region tracking rewarding outcomes of others, a key computation for monitoring others in an agent-specific manner (Apps et al., 2016; Chang et al., 2015). However, agent-specific signals are not confined to ACCg.

In the domains of performing and observing reward-guided actions, neurons in the dorsal bank of ACCs and pre-SMA have been found to exhibit robust correlates of social monitoring—that is, monitoring actions and errors of another individual. Yoshida, Saito, Iriki, and Isoda (2011) first devised an experimental paradigm in which an actor monkey chose between a pair of targets, and the partner monkey observed the actor to gain helpful information to guide their subsequent actions to obtain rewards. By recording single unit activity in the MFC, they identified a selective population of cells that encoded the actions of the other monkey in an agent-specific manner (Yoshida et al., 2011) (Fig. 5A). Notably, subsequent analyses of this same task found a special population of MFC neurons (dorsal bank of ACCs and pre-SMA) that substantially modulated their activity exclusively in response to another's errors (Yoshida, Saito, Iriki, & Isoda, 2012) (Fig. 5A). This agent-specific encoding of others' actions and errors is a core algorithmic component required for monitoring social partners. However, the social monitoring functions of MFC are not limited to the action and error monitoring domain. A recent study by Noritake, Ninomiya, and Isoda (2018) employed a social conditioning procedure where different amounts of juice rewards were delivered between a pair of monkeys. In this study, they found that pre-SMA and its rostrally adjacent area 9 of the MFC encoded reward values in either self-referenced or other-reference frames in an agent-specific manner, whereas midbrain dopaminergic neurons encoded an integrated subjective value (as modulated by the likelihood that another individual would receive a larger or smaller reward) (Noritake et al., 2018). Furthermore, single neurons in the MFC (pre-SMA) have been shown to encode representations of objects in an agent-specific manner. In a study by Livi et al. (2019), the activity of pre-SMA neurons was compared between trials where monkeys were instructed to grasp (or not grasp) one of three possible objects and trials where subject monkeys observed a human doing the same action to the objects. In this experimental setting, pre-SMA neurons either selectively encoded monkey's own actions, the actions of another agent, or both the monkey and other agent's actions (Livi et al., 2019) (Fig. 5B), again supporting the role of MFC in agent-specific representation of behaviors involving self-, other-, and both-reference frames. Therefore, social monitoring functions by the MFC seems to be more generalized to include social value monitoring, positioning MFC as a core brain region implicated in other-referenced computations that might be central to mentalizing and empathetic operations in the primate brain.

Taken together, these behavioral neurophysiology studies parallel functional neuroimaging and electroencephalographic data from human subjects that has identified neural signals of reward outcome or value across self and other in the MFC (Apps et al., 2012; Apps, Lesage, & Ramnani, 2015; Janowski, Camerer, & Rangel, 2013; Jenkins, Macrae, & Mitchell, 2008; Lockwood, Apps, Roiser, & Viding, 2015; Piva et al., 2019). However, there seems to be differently specialized functions across distinct subregions in the primate MFC with respect to their agent-specific processing. For example, in the case of ACCg, converging evidence from behavioral neurophysiology in monkeys and functional neuroimaging in humans suggests a specialized role of ACCg in computing other-referenced reward information used for tracking the motivations of others to possibly enable empathy and prosocial decision-making (Apps et al., 2016). By contrast, again based on both behavioral neurophysiology in monkeys and functional neuroimaging in humans, more dorsal MFC regions relative to ACCg, such as pre-SMA, seems to be more dedicated to understanding other's mental states through observing and predicting other's actions and intentions (Isoda & Noritake, 2013).

3.2 Neurons in the MFC predict the behavior of others

Although monitoring the actions of others is a supremely useful function, using past knowledge gleaned from doing so to predict their *future* behaviors has even greater behavioral utility. Neurons in the MFC have been shown to do exactly this; predict the future behaviors of social partners. A study by Haroush and Williams (2015) revealed that primate ACC neurons predict the behavior of an opponent during a cooperative social exchange. The authors trained pairs of rhesus monkeys to play an interactive prisoner's dilemma game, and recorded the activity of single neurons in the ACC. By analyzing the activity of these ACC neurons, they found two non-overlapping neuronal populations that encoded either the subject's (the monkey from whom the neurons were recorded) own choice, or the other monkey's upcoming choice, predicted by the previous pattern of cooperative behaviors (Fig. 5B). Importantly, a relatively high percentage of neurons (about 28%) in the ACC encoded the other's upcoming choice (Haroush & Williams, 2015). The activity of these ACC neurons found in this study can be interpreted as a prediction of the other's choice during social interactions, as it was yet unknown to the subject. Moreover, when decoding the other's choice using the simultaneously recorded population of neurons, it was possible to classify the opponent's choice with high accuracy (about 80%). Causally, microstimulations to the ACC reduced the frequency of cooperative decisions following past cooperative interactions by others, likely by disrupting the predictability necessary to guide cooperative decisions in the iterative prisoner's dilemma task. These results support the view that ACC neurons encode information about other social agents and make predictions about their behavior during social interaction.

Another direct neurophysiological evidence supporting the role of MFC in social predictions was reported during a monkey-human interaction paradigm used by Falcone, Cirillo, Ferraina, and Genovesio (2017). In this study, a human experimenter took turns alternating with a rhesus monkey in selecting between pairs of targets (from a potential pool of four), where choosing the target that did not match previously chosen target led to rewards, thus requiring the observer to monitor the other actor's choice when it was not their turn. Recordings of single unit activity in the MFC (pre-SMA, SMA, and a region rostral to them) of the participating monkey revealed that selective neurons encoded the future choice with in an agent-specific and often target-direction selective manner (Falcone et al., 2017) (Fig. 5B). This encoding of a prediction for a partner's future behaviors by MFC neurons represents a necessary algorithmic process to regulate social learning and adaptative social behaviors and supports the notion that a key implantation of these algorithm occurs in the MFC.

3.3 Neurons in the MFC are involved in observational and learning from vicarious reinforcement

The ability to predict other's future behaviors is, in turn, a necessary component of social learning that requires learning derived from other-referenced predictions and errors (Joiner et al., 2017). The role of MFC in observational learning has been documented in neurons in the brains of epileptic patients by Hill et al. (2016). In this study, human subjects performed a card game where they attempted to differentiate between two decks of cards as high- or low-probability of winning. Subjects learned through both self-experienced trials (i.e. they picked for themselves) and observational trials (i.e. they observed two other players pick between the decks of cards). Although neurons in the amygdala and prefrontal regions encoded the outcome (winning or losing) of trials, only neurons in rostral ACC neurons encoded observational reward prediction errors (Fig. 5C). These observational prediction errors, or the difference between the expected and actual reward, likely serve as an important algorithm supporting observational learning (Apps et al., 2012, 2015, 2016; Burke, Tobler, Baddeley, & Schultz, 2010; Joiner et al., 2017; Lockwood et al., 2015).

Learning from what happens to others is also important for developing and maintaining social preference. In the previously mentioned social reward allocation task, monkeys will form a prosocial preference (i.e. choosing to donate juice to the other monkey instead of no one) and antisocial preference (delivering juice to themselves over both monkeys) in a context-dependent manner (Chang et al., 2011). The process by which these preferences develop requires learning from vicarious reinforcement, where the actor monkey must monitor their social partner and understand the rewarding impact of their choices to experience vicarious reinforcement in order to develop and maintain a prosocial or an antisocial preference. Does the MFC contribute causally to learning to develop a social preference derived from vicarious reward? Recently, Basile and colleagues addressed this precise question using a selective excitotoxic lesion approach to the ACC. In this study, all six actors in a modified version of the social allocation task showed a prosocial preference (preferring other's reward over wasted reward) before excitotoxic lesion to ACC. However, postoperatively, the ACC-lesioned actors ($n=3$), but not control actors ($n=3$), could no longer learn to acquire a prosocial preference based on vicarious reinforcement derived from the reward delivered to the recipient monkeys (Basile et al., 2020) (Fig. 6). Importantly, the ACC-lesioned actors had no problem on simple learning to prefer one's own rewards over other's rewards or wasted rewards to an empty bottle (Basile et al., 2020). This deficit in forming a prosocial preference is possibly due to the lack of a social prediction error signal originating from the ACC. Already formed prosocial preferences (those with preoperatively learned cues) were reduced compared to the control animals but were still present, suggesting that the ACC is necessary for learning from vicarious reinforcement but perhaps only partially involved in the process of maintaining social preferences.

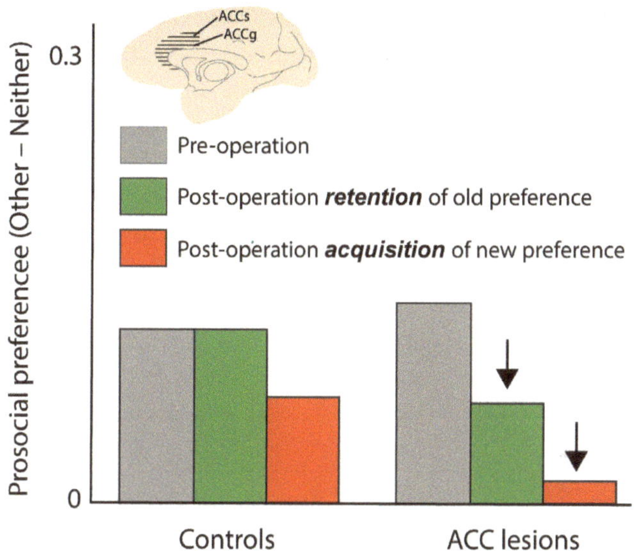

Fig. 6 ACC lesion (indicated in the inset brain diagram) reduced preoperatively learned prosocial preference and eliminated formation of postoperative prosocial preference in a vicarious social reward task (i.e., a preference for a juice reward delivered to a conspecific monkey ["Other"] over an empty bottle ["Neither"]). There was no effect from the ACC lesion on learning to prefer a juice reward to self over the conspecific monkey or over the empty bottle. Further, there was no overall changes to pupillary responses from the lesion. Data from three ACC lesion animals and three control animals. *Illustration based on Basile, B. M., Schafroth, J. L., Karaskiewicz, C. L., Chang, S. W. C., & Murray, E. A. (2020). The anterior cingulate cortex is necessary for forming prosocial preferences from vicarious reinforcement in monkeys.* PLoS Biology, 18*(6), e3000677. doi:10.1371/journal.pbio.3000677.*

3.4 The MFC and theory of mind

It remains unclear if any primate species aside from humans truly exhibit theory of mind (Drayton & Santos, 2016; Horschler et al., 2020; Krupenye et al., 2016), making laboratory experiments aimed at studying the neural mechanisms underlying theory of mind extremely challenging. However, a recent study Hayashi et al. (2020) implicates the MFC as being causally involved in theory of mind based on the false belief task (Wimmer & Perner, 1983) in Japanese macaques. In this experiment employing a dramatized and highly salient version of the false belief task similar to the one used recently by Krupenye and colleagues in apes (Krupenye et al., 2016), the eye movements of monkeys were recorded while they watched videos showing the aforementioned scenarios. During each video, the agent pursued a target that was placed in one of two different locations. By comparing anticipatory looking between the false-belief location (where the first agent last saw the object) and the new location truly containing the object, the authors showed that macaques looked first and more often at the false-belief location (Hayashi et al., 2020). The authors then tested the role of the MFC in this false belief mentalizing by disrupting MFC function using designer receptor exclusively activated by designer drugs (DREADDs). They found that MFC inactivation abolished the anticipatory looking toward the false-belief location (Fig. 5D), suggesting that the MFC plays a casual role in this mentalizing behavior. However, it is worthwhile to note that the behavioral demonstration of such a false belief representation in monkeys still remains controversial (see Horschler et al., 2020). Nevertheless, this study still lends a causal support for other-referenced information processing in the MFC.

4. Attributes of the MFC for social processing

The role of the MFC in performing specific algorithms that support necessary social computations is further supported by anatomical and comparative evidence. First, the MFC regions have bidirectional anatomical connections with the limbic system, allowing integrations across affective and reward-related information (Barbas, 2000; Carmichael & Price, 1995a). This is in contrast to the lateral aspects of the frontal cortex which interfaces more strongly with the sensorimotor processing regions in the brain, including the temporal and parietal cortices (Carmichael & Price, 1995b) for predominantly guiding externally guided cognition. Therefore, the MFC is anatomically equipped to synergistically work with the limbic systems to enable internally guided computations involving affective and reward-related variables suitable for social interactions. Second, possibly as a result of the anatomical connections, functional characteristics of the MFC in relation to executing socially relevant algorithms seem to differ from many other key brain regions implicated in social behaviors.

4.1 Bidirectional anatomical connections with the limbic networks

The interconnectedness of the MFC with limbic structures (Amaral & Insausti, 1992; Carmichael & Price, 1995a; Morecraft et al., 2012), particularly the amygdala, is a key feature that enables the MFC to perform these social computations. The amygdala contains many neural specializations for social perception, such as neurons that respond selectively to faces

(Gothard, Battaglia, Erickson, Spitler, & Amaral, 2007; Minxha et al., 2017; Rutishauser et al., 2011), emotions (Kuraoka & Nakamura, 2007; Wang et al., 2017), and eye contact (Mosher, Zimmerman, & Gothard, 2014; Putnam & Gothard, 2019). Amygdala damage in humans (Adolphs, Tranel, Damasio, & Damasio, 1994; Spezio, Huang, Castelli, & Adolphs, 2007) and monkeys (Emery et al., 2001; Taubert et al., 2018) impairs social perception and social interactions. Given the anatomical connections between MFC and the amygdala (Amaral & Insausti, 1992; Carmichael & Price, 1995a; Morecraft et al., 2012), it is not surprising that recent studies suggest functional relationships between MFC and the amygdala in the production of social behaviors. A recent study by Dal Monte et al. (2020) revealed that prosocial decisions were associated with enhanced synchronization of spikes and local field potential (LFP) activity between the MFC (ACCg) and the amygdala, whereas antisocial decisions measured from a different context were associated with suppressed synchronization of the same processes (Fig. 7) (Dal Monte et al., 2020). Moreover, this spike-LFP synchrony between the two areas was frequency-specific, where the frequency channel used was determined by the area contributing the spikes (i.e., spikes from the amygdala was synchronized to the beta frequency of ACCg LFP, whereas spikes from ACCg was synchronized to the gamma frequency of the amygdala) (Dal Monte et al., 2020). Interestingly, the researchers also found that prosocial decisions were associated with an increase in directional information transfer from the amygdala to ACCg in the same beta frequency band, whereas antisocial decisions were associated with an increased in information flow from ACCg to the amygdala more broadly across different frequency bands (Dal Monte et al., 2020). Other studies examining non-social behaviors collaborate the importance of the interactions between the ACC and the amygdala in guiding behaviors. For example, these interactions are important in shaping reward learning (Fiuzat, Rhodes, & Murray, 2017; Klavir, Genud-Gabai, & Paz, 2013; Livneh & Paz, 2012; Rudebeck, Mitz, Chacko, & Murray, 2013; Rudebeck, Ripple, Mitz, Averbeck, & Murray, 2017) and observational learning (Burgos-Robles, Gothard, Monfils, Morozov, & Vicentic, 2019). The amygdalo-cingulate connectivity is excellently reviewed in depth by Burgos-Robles and colleagues, where the authors discuss how these cytoarchitectonic features and functional connections support the role of the ACC in observational learning (Burgos-Robles et al., 2019).

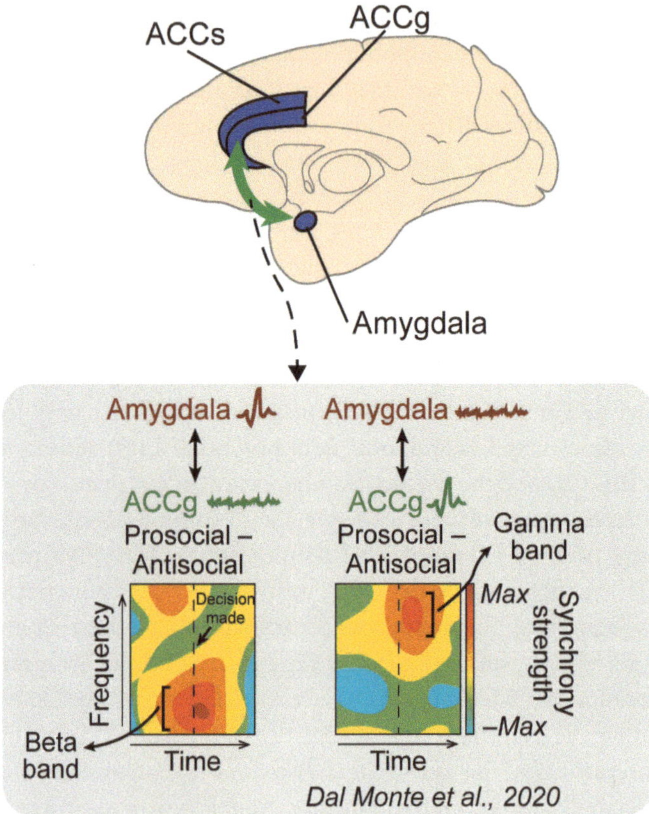

Fig. 7 Coherence between spiking activity and local field potential (LFP) signals between ACCg and the basolateral amygdala was found to be enhanced for making prosocial decisions but suppressed for making antisocial decisions in monkeys. This coherence occurred in dedicated frequency channels critically depending on the brain region contributing the spikes to the coherence. *Illustrations based on the data from Dal Monte, O., Chu, C. C. J., Fagan, N. A., & Chang, S. W. C. (2020). Specialized medial prefrontal–amygdala coordination in other-regarding decision preference.* Nature Neuroscience, *23(4), 565–574. doi:10.1038/s41593-020-0593-y and the summary figure from Gangopadhyay, P., Chawla, M., Dal Monte, O., & Chang, S. W. C. (2020). Neural mechanisms guiding social decision-making in the prefrontal-amygdala circuits.* Nature Neuroscience, *(in press).*

4.2 Distinctive attributes of the MFC for socially relevant algorithms

Although the MFC represents a collection of brain regions out of many that are implicated in guiding social behaviors, several functional characteristics of the MFC distinguish its contributions to guiding social behaviors from other brain regions. Here we will compare the MFC to a small selection of other brain regions and example studies to explore the potential uniqueness of algorithms employed by the MFC. Our goal in this section is not to be exhaustive but rather to briefly highlight one or two studies from other brain regions in order to inform how neurons in the MFC might exhibit different socially relevant algorithms from other brain regions.

4.2.1 Compared to the lateral frontal cortex

The lateral frontal cortex is also comprised of a collection of brain regions, including the ventral premotor cortex (PMv) and dorsolateral prefrontal cortex (dlPFC). These areas are strongly implicated in cognitive operations such as working memory (Barbey, Koenigs, & Grafman, 2013; Curtis & D'Esposito, 2003; Funahashi, Bruce, & Goldman-Rakic, 1989; Fuster & Alexander, 1971; Miller, Erickson, & Desimone, 1996), outcome valuation (Cai & Padoa-Schioppa, 2014; Yim, Cai, & Wang, 2019), and strategy implementation (Baxter, Gaffan, Kyriazis, & Mitchell, 2009; Seo, Cai, Donahue, & Lee, 2014). Most pertinently to social processing, the lateral frontal cortex contains a network of neurons, known as "mirror neurons," that respond to both the observation of an action by another, and the performance of the same action by oneself (Fabbri-Destro & Rizzolatti, 2008; Rizzolatti & Craighero, 2004; Rizzolatti, Fadiga, Gallese, & Fogassi, 1996; Rizzolatti, Fogassi, & Gallese, 2001). Mirror neurons were first identified in PMv but also since found in many different brain regions such as the inferior parietal lobule, the primary motor cortex, and the superior temporal sulcus (Dushanova & Donoghue, 2010; Ferrari, Bonini, & Fogassi, 2009; Hamilton, 2013; Rizzolatti et al., 2001; Suzuki et al., 2015), forming a mirror neuron network in the brain (Bonini, 2017). Mirror neurons have many parallels to neurons found in the MFC; mirror neurons seem to represent the actions of self and other and necessitate monitoring another individual's actions with a theorized role of understanding other's intentions. However, one key differentiation exists; mirror neurons can be algorithmically described as a stimulation of other's behavior based on externally guided information. A key feature of mirror neurons is that observing and executing the same actions are encoded in the same manner, hence the term mirroring. By contrast, a consistent feature of agent-specific processing in the MFC is the independent and coexistent representations of self-referenced, other-referenced, and a mirror-like combined self-and-other-referenced representations. Having separate representations of self-referenced and other-referenced signals within different population of neurons in the MFC is critical for many of the algorithms implemented for social behaviors such that the brain can differentiate self from other or represent information in a relative fashion between self and other. Mirroring by itself does not provide such separate representations and would not be possible to predict other's behaviors during social interactions (except when the interactions consists of simply mimicking one another). While the mirror neuron network is implicated as one mechanism for transforming of visual information into a motor format (Fabbri-Destro & Rizzolatti, 2008), it remains undecided if this network is required for visuomotor transformations (Hickok, 2009), and if this system causally impacts social behaviors (Hamilton, 2013).

Multiple studies also suggest that the dlPFC, another subdivision of the lateral frontal cortex, is centrally involved in strategic social interactions. In experiments where a monkey was strategically pitted against a computer opponent, neurons in the dlPFC tracked not only the subject's previous choices but also previous choices made by the opponent (Seo, Barraclough, & Lee, 2007; Seo et al., 2014). Moreover, many neurons in this region have been shown to encode winning and losing resulting from competition, but not from non-competition, when two monkeys played a missile commando game (Hosokawa & Watanabe, 2012). In fact, tracking the actions of opponents and the respective reward outcomes emerging from competition are both important algorithmic processes required for strategic behaviors. The involvement of the dlPFC in strategic decision-making is consistent with the broader role of the dlPFC in executive functions (Lee & Seo, 2007). One important contrast is that, compared to MFC neurons with stronger associations with the limbic networks and possibly under a greater influence of internal states, the information represented by dlPFC neurons might be more tightly linked with sensorimotor or externally guided processing of actions linked to rewarding outcomes (Donahue & Lee, 2015; Lee & Seo, 2007).

4.2.2 Compared to the amygdala

The amygdala is a subcortical region strongly implicated in social behaviors across many species (Adolphs, 2010). Like the MFC, the amygdala plays a role non-social learning such as reward valuation (Baxter & Murray, 2002; Bermudez, Göbel, & Schultz, 2012; Bermudez & Schultz, 2010; Burgos-Robles et al., 2017; Fiuzat et al., 2017; Paton et al., 2006; Putnam & Gothard, 2019; Rudebeck et al., 2013, 2017). A recent study by

Grabenhorst et al. (2012) provides strong evidence of functional overlap between the amygdala and MFC. In this study, two monkeys alternated between making reward-based choices, choosing between one of two fractals unique to each monkey that were associated with different reward probabilities. Halfway through each recording session, the pairs of fractals switched between the monkeys so that the observational learning could be tested. By recording single unit activity from the amygdala in one of the two monkeys, it was found that amygdala neurons encoded the reward values of the fractals, regardless of if this value was extract from personal experience or observation (Grabenhorst et al., 2012). The neuronal selectivity to the fractal stimuli was not merely for certain visual patterns, as this reward-value encoding tracked the unannounced probability reversal during each of the two halves. This study provides evidence that amygdala neurons, like neurons in the ACC (Haroush & Williams, 2015), are capable of predicting the actions of others by simulating their decisions. Indeed, it would be interesting to understand if these predictive signals of other's behaviors observed in the ACC (Haroush & Williams, 2015) originate from the amygdala, or vice versa, given the rich connections between the basolateral amygdala (the amygdala subdivision with the neural population showing the strongest shared value representation between self and another individual based on cross-decoding of reward value from self to other; Grabenhorst et al., 2012) and the ACC (Morecraft & Van Hoesen, 1998). While this seemingly shared algorithm in both the ACC and the amygdala provides evidence for functional overlaps between the two regions, there are still notable differences between them. It is possible that this algorithm is shared between the two different regions, however the overall computational goal might be different. One hypothesis is that the computational goal of the amygdala in observational learning is to support one own's future goal-directed behaviors, i.e. the subjects in the task were implicitly aware of an impending stimulus switch and thus tracking the stimulus-reward associations of the other's monkey was beneficial. However, this can be contrasted with neurons in the ACCg of the MFC that signal other's reward outcome even though representing this reward outcome had no behavioral contingencies for one's own reward (Chang et al., 2013), an independent representation of other's outcome that may be directly relevant for enabling vicarious reinforcement. This hypothesis, however, is mostly speculative and future studies will be needed to dissect the similarities and differences between the MFC and amygdala in representing social variables with respect to self and others.

4.2.3 Compared to the lateral intraparietal cortex

The lateral intraparietal cortex (LIP) is a region broadly associated with spatial attention (Bisley & Goldberg, 2003) and planning eye movements (Snyder, Batista, & Andersen, 2000). There exists evidence that, additionally, LIP neurons track the gaze of others. Shepherd, Klein, Deaner, and Platt (2009) recorded single unit activity from the LIP in a monkey while the subject participated in a simple visual orienting task. In this task, the monkey fixated on a center target, and once fixation was initiated, an image of a conspecific was presented at the center serving as a cue for the subject to shift their gaze toward a peripheral target. The direction of this saccade was either congruent or incongruent with the conspecific's gaze direction in the image. The results showed that LIP neurons are modulated by social gaze cues by exhibiting enhanced or suppressed firing rates when the social gaze was directed toward the neuron's receptive field (Shepherd et al., 2009). These results may capture a social gaze-related sensorimotor transformation by LIP neurons, which may underlie the transformation required for gaze-following behaviors. As of present, no data suggests that LIP neurons are involved in predicting other's behaviors in social interactions, nor do LIP neurons represent the gaze of others using a dissociated representation from one's own gaze direction. One plausible explanation is that LIP neurons may perform an attentional allocation or motor planning algorithm based on other's gaze direction to accomplish the sensorimotor transformation according to the information available from others. By contrast, social monitoring algorithms performed by the neurons in the MFC might be purposed for understanding other's preferences and other's internal (e.g., motivational) states (Haroush & Williams, 2015; Joiner et al., 2017; Lockwood et al., 2015) to support social learning and social decision-making (Apps et al., 2012, 2015, 2016; Joiner et al., 2017).

4.2.4 Compared to the striatum

The striatum is key node of decision-making and reward circuity in the brain (Burton, Nakamura, & Roesch, 2015). Unsurprisingly, as rewards powerfully shape social interactions, the striatum is critically implicated in social behaviors (Báez-Mendoza & Schultz, 2013). Báez-Mendoza and colleagues tested how neurons in the monkey striatum represented social action and reward by implementing a task where the subject monkey and another partner (either a second money or a computer actor) took turns making operant responses for rewards. They found that striatal neurons encoded signals for the actions of both self and other, but only encoded reward signals for self

(Báez-Mendoza, Harris, & Schultz, 2013). This coding of social action in the striatum parallels the findings from the MFC (Yoshida et al., 2011, 2012), and may suggest that striatal neurons work in close communication with the MFC to associate the information about other's actions and rewards given the well-known role of the striatum in linking action to reward (Balleine, Delgado, & Hikosaka, 2007; Monosov, Leopold, & Hikosaka, 2015). However, the absence of a localized other-referenced reward signal in that study also highlights the difference of the striatum from the MFC (Chang et al., 2013; Dal Monte et al., 2020). There is another excellent study from the ventral striatum for comparing representations of other's reward by the ventral striatum and the MFC. A study by Kashtelyan, Lichtenberg, Chen, Cheer, and Roesch (2014) measured dopamine release in rats when they observed rewards being delivered to conspecifics. When rats *first* observed conspecific receive rewards, there was an increase of dopamine release in the ventral striatum, along with appetitive ultrasonic vocalizations (Kashtelyan et al., 2014). Critically, later trials did not result in increased dopamine but rather a reduction in dopamine along with an increased rate of aversive ultrasonic vocalizations. These results suggest that representations of other's reward in the ventral striatum is highly dependent on affective states. In the MFC, representations of other's reward were found to be stable over the course of entire sessions (Chang et al., 2013), although existing studies have not yet tested how these socially relevant signals are gated by fluctuating internal states.

5. Social specificity of the MFC: Algorithms and implementations

An outstanding and critical question remains on understanding the role of the MFC in guiding social behaviors. Are these neuronal processes and algorithms in the MFC specific to social behaviors or are they shared across both social and non-social functions? Generally, the question of the existence of dedicated regions, circuits, or processes in the brain for social behaviors (compared to an alternative hypothesis that all of these are all shared between social and non-social faculties) is still under debate. Some researchers have put forward the concept of a "social brain," of which the MFC is a major node, that implies specialized systems for social cognition (Adolphs, 2009). Despite the presence of functionality that is inherently social, such as the perception of faces and emotions, there is increasing evidence of overlap between social and non-social functions at the neuronal level. For example, although the evaluation of social stimuli, particularly faces, is often cited as a key specialized module of the social brain, it was recently shown in rhesus monkeys that the same amygdala neurons encode faces (Putnam & Gothard, 2019) or social hierarchies (Munuera, Rigotti, & Salzman, 2018) also track non-social features such as reward and task dynamics. This outstanding question is where the framework of Marr is most useful as we can examine social specificity at different organizational levels, and this approach has been recently applied in order to redefine the concept of social specificity, or lack thereof, through the lens of computational, algorithmic, and implementational levels (Lockwood et al., 2020). In this last section, we examine the MFC in social functions specifically at the level of algorithms and implementations.

At the most macroscopic level, the computational goal of social interactions is compelled by the fundamental drives (e.g. needing to compete or cooperate with others to obtain food). While these goals can be innately social, the algorithms that support them are unlikely to be socially specific. Algorithmic process, such as reinforcement learning, have been used to model both social and non-social behaviors with high accuracy (Behrens, Hunt, Woolrich, & Rushworth, 2008; Charpentier, Iigaya, & O'Doherty, 2020; Ereira, Dolan, & Kurth-Nelson, 2018; Olsson et al., 2020). It has been argued that the ACCg, an area with a high degree of social specificity at the implementational level, may be utilizing reinforcement learning as an common underlying algorithm (Apps et al., 2015, 2016; Behrens et al., 2008; Hill et al., 2016; Joiner et al., 2017; Lockwood et al., 2015, 2018). However, there also exist algorithmic processes that appear to be exclusive to social functions. One example of this is the process of metalizing or theory of mind. In theory of mind tasks, such as the one used by Hayashi et al. (2020), general algorithms such as reinforcement learning cannot fully explain the experimental results (Devaine et al., 2017). In general, while it is clearer that the MFC implements socially non-specific algorithms in the production of social behaviors such as reinforcement learning algorithms, it remains to be further tested if there are also socially specific algorithms being implemented.

There is some evidence that there are socially specific implementations in the MFC. As referenced previously, a study by Chang et al. (2013) found that different proportions of neurons in the ACCg and ACCs encode other-referenced and self-referenced reward, respectively, with varying degrees of specialization. Neurons in the ACCg principally encoded rewards delivered to the *other* monkey, either through coding other- or both-referenced rewards. In contrast, neurons in the ACCs principally encoded rewards delivered to self, although other- or both-referenced reward coding neurons constituted a sizeable minority. In contrast still, nearly 80% of reward-encoding neurons recorded from the OFC were self-referenced. These findings support a division of socially specific implementations between the gyrus and the sulcus of the ACC at the neuronal level. A potentially specialized implementation in ACCg has been supported from human neuroimaging literature as well (Apps et al., 2015, 2016; Behrens et al., 2008; Hill et al., 2016; Joiner et al., 2017; Lockwood et al., 2015, 2018). Interestingly, in a recent study examining shared coding for reward value and hierarchical rank in monkeys, Munuera et al. (2018) found that the OFC and ACC encoded reward values but failed to robustly encode representations of hierarchical rank in a shared manner. This was in contrast to the neurons recorded from the amygdala that used the same neuronal ensembles to encode both hierarchical rank and reward values associated with non-social stimuli (Munuera et al., 2018). These findings suggest, accepting the premise that a single hierarchical algorithm can be used to sort information such as reward value or rank, the presence of potential socially specific implementations in the ACC and OFC but not in the amygdala. Given the range of functions associated with the MFC, it seems nearly certain that some MFC neurons implement generalized algorithms spanning both social and non-social behaviors during social interactions. Future studies may identify possibly differentiated neuronal ensembles in the MFC regions that implement socially specific and socially non-specific algorithms for guiding social behaviors.

6. Conclusions

In this chapter, we outline how the MFC contributes to shaping adaptive social behaviors through the use of several key algorithms. Experimental evidence, from behavioral neurophysiology, supports the involvement of MFC neurons in each of these algorithms. Single neuron recording studies from the MFC provide evidence for the function of the MFC in monitoring others in an agent-specific manner (Chang et al., 2013; Dal Monte et al., 2020; Noritake et al., 2018; Yoshida et al., 2011, 2012) and in predicting the behavior of social partners (Falcone, Brunamonti, Ferraina, & Genovesio, 2016; Falcone et al., 2017; Haroush & Williams, 2015; Livi et al., 2019). Moreover, single unit recording from human patients in the rostral ACC as well lesion studies in non-human primates implicate the

MFC in observational learning and learning from vicarious reinforcement (Basile et al., 2020; Hill et al., 2016). Finally, a recent study utilizing chemogenetic inactivation of the MFC suggests a role for the MFC in theory of mind like processes (Hayashi et al., 2020). Together, socially relevant algorithms found in MFC neurons support necessary social computations required for effective social interactions. Although the brain regions within the MFC are not the only brain areas implicated in shaping social functions, the MFC has various unique attributes that allow the MFC to play a central role in shaping interactive social behaviors. For example, strong connectivity with the limbic network supports the implantation of algorithmic processes such as observational learning (Burgos-Robles et al., 2019) and prosocial decision-making (Dal Monte et al., 2020). When compared to some of other brain areas, there seems to be some unique algorithms performed by MFC neurons. To continue to investigate and understand social functions in the MFC, applying the three levels of organization proposed by Marr (1982) to social behaviors (Lockwood et al., 2020) may help guide and refine research questions, designs, and interpretations. Fruitful research directions include examining socially relevant algorithms and implementations in the MFC by holding one level constant through innovative experimental designs (Lockwood et al., 2020). Continued technological innovations will in turn allow new discoveries of how socially specific and socially non-specific processes in different MFC ensembles interact to guide social interactions.

Acknowledgment

This work was supported by the National Institute of Mental Health (R01 MH120081, R01MH110750). We thank Colin Stanton for help with some of the figure illustrations.

References

Adolphs, R. (2009). The social brain: Neural basis of social knowledge. *Annual Review of Psychology*, *60*, 693–716. https://doi.org/10.1146/annurev.psych.60.110707.163514.

Adolphs, R. (2010). What does the amygdala contribute to social cognition? *Annals of the New York Academy of Sciences*, *1191*(1), 42–61. https://doi.org/10.1111/j.1749-6632.2010.05445.x.

Adolphs, R., Tranel, D., Damasio, H., & Damasio, A. (1994). Impaired recognition of emotion in facial expressions following bilateral damage to the human amygdala. *Nature*, *372*(6507), 669–672. https://doi.org/10.1038/372669a0.

Akins, C. K., Klein, E. D., & Zentall, T. R. (2002). Imitative learning in Japanese quail (Coturnix japonica) using the bidirectional control procedure. *Animal Learning & Behavior*, *30*(3), 275–281. https://doi.org/10.3758/BF03192836.

Amaral, D. G., & Insausti, R. (1992). Retrograde transport of D-[3H]-aspartate injected into the monkey amygdaloid complex. *Experimental Brain Research*, *88*(2), 375–388. https://doi.org/10.1007/BF02259113.

Amodio, D. M., & Frith, C. D. (2006). Meeting of minds: The medial frontal cortex and social cognition. *Nature Reviews. Neuroscience*, 7(4), 268–277. https://doi.org/10.1038/nrn1884.

Apps, M. A. J., Balsters, J. H., & Ramnani, N. (2012). The anterior cingulate cortex: Monitoring the outcomes of others' decisions. *Social Neuroscience*, 7(4), 424–435. https://doi.org/10.1080/17470919.2011.638799.

Apps, M. A. J., Lesage, E., & Ramnani, N. (2015). Vicarious reinforcement learning signals when instructing others. *Journal of Neuroscience*, *35*(7), 2904–2913. https://doi.org/10.1523/JNEUROSCI.3669-14.2015.

Apps, M. A. J., Rushworth, M. F. S., & Chang, S. W. C. (2016). The anterior cingulate gyrus and social cognition: Tracking the motivation of others. *Neuron*, *90*(4), 692–707. https://doi.org/10.1016/j.neuron.2016.04.018.

Azab, H., & Hayden, B. Y. (2017). Correlates of decisional dynamics in the dorsal anterior cingulate cortex. *PLoS Biology*, *15*(11), e2003091. https://doi.org/10.1371/journal.pbio.2003091.

Bachevalier, J., Meunier, M., Lu, M. X., & Ungerleider, L. G. (1997). Thalamic and temporal cortex input to medial prefrontal cortex in rhesus monkeys. *Experimental Brain Research*, *115*(3), 430–444.

Báez-Mendoza, R., Harris, C. J., & Schultz, W. (2013). Activity of striatal neurons reflects social action and own reward. *Proceedings of the National Academy of Sciences of the United States of America*, *110*(41), 16634–16639. https://doi.org/10.1073/pnas.1211342110.

Báez-Mendoza, R., & Schultz, W. (2013). The role of the striatum in social behavior. *Frontiers in Neuroscience*, 7, 233. https://doi.org/10.3389/fnins.2013.00233.

Balleine, B. W., Delgado, M. R., & Hikosaka, O. (2007). The role of the dorsal striatum in reward and decision-making. *Journal of Neuroscience*, *27*(31), 8161–8165. https://doi.org/10.1523/JNEUROSCI.1554-07.2007.

Bandura, A. (2008). Observational learning. In *The International encyclopedia of communication* American Cancer Society. https://doi.org/10.1002/9781405186407.wbieco004.

Bandura, A., Ross, D., & Ross, S. A. (1961). Transmission of aggression through imitation of aggressive models. *The Journal of Abnormal and Social Psychology*, *63*(3), 575–582. https://doi.org/10.1037/h0045925.

Bandura, A., & Walters, R. H. (1977). *Social learning theory. Vol. 1.* NJ: Prentice-Hall Englewood Cliffs.

Barbas, H. (2000). Connections underlying the synthesis of cognition, memory, and emotion in primate prefrontal cortices. *Brain Research Bulletin*, *52*(5), 319–330. https://doi.org/10.1016/s0361-9230(99)00245-2.

Barbey, A. K., Koenigs, M., & Grafman, J. (2013). Dorsolateral prefrontal contributions to human working memory. *Cortex: A Journal Devoted to the Study of the Nervous System and Behavior*, *49*(5), 1195–1205. https://doi.org/10.1016/j.cortex.2012.05.022.

Basile, B. M., Schafroth, J. L., Karaskiewicz, C. L., Chang, S. W. C., & Murray, E. A. (2020). The anterior cingulate cortex is necessary for forming prosocial preferences from vicarious reinforcement in monkeys. *PLoS Biology*, *18*(6). https://doi.org/10.1371/journal.pbio.3000677, e3000677.

Baxter, M. G., Gaffan, D., Kyriazis, D. A., & Mitchell, A. S. (2009). Ventrolateral prefrontal cortex is required for performance of a strategy implementation task but not reinforcer devaluation effects in rhesus monkeys. *The European Journal of Neuroscience*, *29*(10), 2049–2059. https://doi.org/10.1111/j.1460-9568.2009.06740.x.

Baxter, M. G., & Murray, E. A. (2002). The amygdala and reward. *Nature Reviews. Neuroscience*, *3*(7), 563–573. https://doi.org/10.1038/nrn875.

Bechara, A., Damasio, H., & Damasio, A. R. (2000). Emotion, decision making and the orbitofrontal cortex. *Cerebral Cortex*, *10*(3), 295–307. https://doi.org/10.1093/cercor/10.3.295.

Behrens, T. E. J., Hunt, L. T., & Rushworth, M. F. S. (2009). The computation of social behavior. *Science*, *324*(5931), 1160–1164. https://doi.org/10.1126/science.1169694.
Behrens, T. E. J., Hunt, L. T., Woolrich, M. W., & Rushworth, M. F. S. (2008). Associative learning of social value. *Nature*, *456*(7219), 245–249. https://doi.org/10.1038/nature07538.
Bermudez, M. A., Göbel, C., & Schultz, W. (2012). Sensitivity to temporal reward structure in amygdala neurons. *Current Biology: CB*, *22*(19), 1839–1844. https://doi.org/10.1016/j.cub.2012.07.062.
Bermudez, M. A., & Schultz, W. (2010). Reward magnitude coding in primate amygdala neurons. *Journal of Neurophysiology*, *104*(6), 3424–3432. https://doi.org/10.1152/jn.00540.2010.
Bevilacqua, D., Davidesco, I., Wan, L., Chaloner, K., Rowland, J., Ding, M., et al. (2019). Brain-to-brain synchrony and learning outcomes vary by student-teacher dynamics: Evidence from a real-world classroom electroencephalography study. *Journal of Cognitive Neuroscience*, *31*(3), 401–411. https://doi.org/10.1162/jocn_a_01274.
Bisley, J. W., & Goldberg, M. E. (2003). Neuronal activity in the lateral intraparietal area and spatial attention. *Science (New York, N.Y.)*, *299*(5603), 81–86. https://doi.org/10.1126/science.1077395.
Bonini, L. (2017). The extended mirror neuron network: Anatomy, origin, and functions. *The Neuroscientist*, *23*(1), 56–67. https://doi.org/10.1177/1073858415626400.
Burgos-Robles, A., Gothard, K. M., Monfils, M. H., Morozov, A., & Vicentic, A. (2019). Conserved features of anterior cingulate networks support observational learning across species. *Neuroscience and Biobehavioral Reviews*, *107*, 215–228. https://doi.org/10.1016/j.neubiorev.2019.09.009.
Burgos-Robles, A., Kimchi, E. Y., Izadmehr, E. M., Porzenheim, M. J., Ramos-Guasp, W. A., Nieh, E. H., et al. (2017). Amygdala inputs to prefrontal cortex guide behavior amid conflicting cues of reward and punishment. *Nature Neuroscience*, *20*(6), 824–835. https://doi.org/10.1038/nn.4553.
Burke, C. J., Tobler, P. N., Baddeley, M., & Schultz, W. (2010). Neural mechanisms of observational learning. *Proceedings of the National Academy of Sciences of the United States of America*, *107*(32), 14431–14436. https://doi.org/10.1073/pnas.1003111107.
Burton, A. C., Nakamura, K., & Roesch, M. R. (2015). From ventral-medial to dorsal-lateral striatum: Neural correlates of reward-guided decision-making. *Neurobiology of Learning and Memory*, *117*, 51–59. https://doi.org/10.1016/j.nlm.2014.05.003.
Cai, X., & Padoa-Schioppa, C. (2014). Contributions of orbitofrontal and lateral prefrontal cortices to economic choice and the good-to-action transformation. *Neuron*, *81*(5), 1140–1151. https://doi.org/10.1016/j.neuron.2014.01.008.
Carmichael, S. T., & Price, J. L. (1995a). Limbic connections of the orbital and medial prefrontal cortex in macaque monkeys. *The Journal of Comparative Neurology*, *363*(4), 615–641. https://doi.org/10.1002/cne.903630408.
Carmichael, S. T., & Price, J. L. (1995b). Sensory and premotor connections of the orbital and medial prefrontal cortex of macaque monkeys. *The Journal of Comparative Neurology*, *363*(4), 642–664. https://doi.org/10.1002/cne.903630409.
Carmichael, S. T., & Price, J. L. (1996). Connectional networks within the orbital and medial prefrontal cortex of macaque monkeys. *The Journal of Comparative Neurology*, *371*(2), 179–207. https://doi.org/10.1002/(SICI)1096-9861(19960722)371:2<179::AID-CNE1>3.0.CO;2-#.
Chang, S. W. C. (2013). Coordinate transformation approach to social interactions. *Frontiers in Neuroscience*, 7, 147. https://doi.org/10.3389/fnins.2013.00147.
Chang, S. W. C. (2017). An emerging field of primate social neurophysiology: Current developments. *eNeuro*, *4*(5). https://doi.org/10.1523/ENEURO.0295-17.2017. ENEURO.0295-17.2017.

Chang, S. W. C., Barter, J. W., Ebitz, R. B., Watson, K. K., & Platt, M. L. (2012). Inhaled oxytocin amplifies both vicarious reinforcement and self reinforcement in rhesus macaques (Macaca mulatta). *Proceedings of the National Academy of Sciences of the United States of America*, *109*(3), 959–964. https://doi.org/10.1073/pnas.1114621109.

Chang, S. W. C., Fagan, N. A., Toda, K., Utevsky, A. V., Pearson, J. M., & Platt, M. L. (2015). Neural mechanisms of social decision-making in the primate amygdala. *Proceedings of the National Academy of Sciences of the United States of America*, *112*(52), 16012–16017. https://doi.org/10.1073/pnas.1514761112.

Chang, S. W. C., Gariépy, J.-F., & Platt, M. L. (2013). Neuronal reference frames for social decisions in primate frontal cortex. *Nature Neuroscience*, *16*(2), 243–250. https://doi.org/10.1038/nn.3287.

Chang, S. W. C., Winecoff, A. A., & Platt, M. L. (2011). Vicarious reinforcement in Rhesus Macaques (Macaca Mulatta). *Frontiers in Neuroscience*, *5*, 27. https://doi.org/10.3389/fnins.2011.00027.

Charpentier, C. J., Iigaya, K., & O'Doherty, J. P. (2020). A neuro-computational account of arbitration between choice imitation and goal emulation during human observational learning. *Neuron*, *106*(4), 687–699.e7. https://doi.org/10.1016/j.neuron.2020.02.028.

Chen, P., & Hong, W. (2018). Neural circuit mechanisms of social behavior. *Neuron*, *98*(1), 16–30. https://doi.org/10.1016/j.neuron.2018.02.026.

Curtis, C. E., & D'Esposito, M. (2003). Persistent activity in the prefrontal cortex during working memory. *Trends in Cognitive Sciences*, 7(9), 415–423. https://doi.org/10.1016/S1364-6613(03)00197-9.

Dal Monte, O., Chu, C. C. J., Fagan, N. A., & Chang, S. W. C. (2020). Specialized medial prefrontal–amygdala coordination in other-regarding decision preference. *Nature Neuroscience*, *23*(4), 565–574. https://doi.org/10.1038/s41593-020-0593-y.

Debiec, J., & Olsson, A. (2017). Social fear learning: From animal models to human function. *Trends in Cognitive Sciences*, *21*(7), 546–555. https://doi.org/10.1016/j.tics.2017.04.010.

Devaine, M., San-Galli, A., Trapanese, C., Bardino, G., Hano, C., Jalme, M. S., et al. (2017). Reading wild minds: A computational assay of theory of mind sophistication across seven primate species. *PLoS Computational Biology*, *13*(11), e1005833. https://doi.org/10.1371/journal.pcbi.1005833.

Dikker, S., Wan, L., Davidesco, I., Kaggen, L., Oostrik, M., McClintock, J., et al. (2017). Brain-to-brain synchrony tracks real-world dynamic group interactions in the classroom. *Current Biology*, *27*(9), 1375–1380. https://doi.org/10.1016/j.cub.2017.04.002.

Dölen, G., Darvishzadeh, A., Huang, K. W., & Malenka, R. C. (2013). Social reward requires coordinated activity of nucleus accumbens oxytocin and serotonin. *Nature*, *501*(7466), 179. https://doi.org/10.1038/nature12518.

Donahue, C. H., & Lee, D. (2015). Dynamic routing of task-relevant signals for decision making in dorsolateral prefrontal cortex. *Nature Neuroscience*, *18*(2), 295–301. https://doi.org/10.1038/nn.3918.

Drayton, L. A., & Santos, L. R. (2016). A decade of theory of mind research on Cayo Santiago: Insights into rhesus macaque social cognition. *American Journal of Primatology*, *78*(1), 106–116. https://doi.org/10.1002/ajp.22362.

Dushanova, J., & Donoghue, J. (2010). Neurons in primary motor cortex engaged during action observation. *The European Journal of Neuroscience*, *31*(2), 386–398. https://doi.org/10.1111/j.1460-9568.2009.07067.x.

Emery, N. J., Capitanio, J. P., Mason, W. A., Machado, C. J., Mendoza, S. P., & Amaral, D. G. (2001). The effects of bilateral lesions of the amygdala on dyadic social interactions in rhesus monkeys (Macaca mulatta). *Behavioral Neuroscience*, *115*(3), 515–544.

Ereira, S., Dolan, R. J., & Kurth-Nelson, Z. (2018). Agent-specific learning signals for self–other distinction during mentalising. *PLoS Biology*, *16*(4), e2004752. https://doi.org/10.1371/journal.pbio.2004752.

Fabbri-Destro, M., & Rizzolatti, G. (2008). Mirror neurons and mirror systems in monkeys and humans. *Physiology, 23*(3), 171–179. https://doi.org/10.1152/physiol.00004.2008.

Falcone, R., Brunamonti, E., Ferraina, S., & Genovesio, A. (2016). Neural encoding of self and another agent's goal in the primate prefrontal cortex: Human-monkey interactions. *Cerebral Cortex (New York, N.Y.: 1991), 26*(12), 4613–4622. https://doi.org/10.1093/cercor/bhv224.

Falcone, R., Cirillo, R., Ferraina, S., & Genovesio, A. (2017). Neural activity in macaque medial frontal cortex represents others' choices. *Scientific Reports*, 7, 12663. https://doi.org/10.1038/s41598-017-12822-5.

Ferrari, P. F., Bonini, L., & Fogassi, L. (2009). From monkey mirror neurons to primate behaviours: Possible 'direct' and 'indirect' pathways. *Philosophical Transactions of the Royal Society, B: Biological Sciences, 364*(1528), 2311–2323. https://doi.org/10.1098/rstb.2009.0062.

Fiuzat, E. C., Rhodes, S. E. V., & Murray, E. A. (2017). The role of orbitofrontal-amygdala interactions in updating action-outcome valuations in macaques. *The Journal of Neuroscience: The Official Journal of the Society for Neuroscience, 37*(9), 2463–2470. https://doi.org/10.1523/JNEUROSCI.1839-16.2017.

Frith, C., & Frith, U. (2010). Learning from others: Introduction to the special review series on social neuroscience. *Neuron, 65*(6), 739–743.

Funahashi, S., Bruce, C. J., & Goldman-Rakic, P. S. (1989). Mnemonic coding of visual space in the monkey's dorsolateral prefrontal cortex. *Journal of Neurophysiology, 61*(2), 331–349. https://doi.org/10.1152/jn.1989.61.2.331.

Fuster, J. M., & Alexander, G. E. (1971). Neuron activity related to short-term memory. *Science, 173*(3997), 652–654. https://doi.org/10.1126/science.173.3997.652.

Gothard, K. M., Battaglia, F. P., Erickson, C. A., Spitler, K. M., & Amaral, D. G. (2007). Neural responses to facial expression and face identity in the monkey amygdala. *Journal of Neurophysiology, 97*(2), 1671–1683. https://doi.org/10.1152/jn.00714.2006.

Gothard, K. M., Mosher, C. P., Zimmerman, P. E., Putnam, P. T., Morrow, J. K., & Fuglevand, A. J. (2017). New perspectives on the neurophysiology of primate amygdala emerging from the study of naturalistic social behaviors. *Wiley Interdisciplinary Reviews: Cognitive Science, 9*(1), e1449. https://doi.org/10.1002/wcs.1449.

Grabenhorst, F., Hernádi, I., & Schultz, W. (2012). Prediction of economic choice by primate amygdala neurons. *Proceedings of the National Academy of Sciences of the United States of America, 109*(46), 18950–18955. https://doi.org/10.1073/pnas.1212706109.

Hamilton, A. F.d. C. (2013). Reflecting on the mirror neuron system in autism: A systematic review of current theories. *Developmental Cognitive Neuroscience, 3*, 91–105. https://doi.org/10.1016/j.dcn.2012.09.008.

Haroush, K., & Williams, Z. M. (2015). Neuronal prediction of opponent's behavior during cooperative social interchange in primates. *Cell, 160*(6), 1233–1245. https://doi.org/10.1016/j.cell.2015.01.045.

Hayashi, T., Akikawa, R., Kawasaki, K., Egawa, J., Minamimoto, T., Kobayashi, K., et al. (2020). Macaques exhibit implicit gaze bias anticipating others' false-belief-driven actions via medial prefrontal cortex. *Cell Reports, 30*(13), 4433–4444.e5. https://doi.org/10.1016/j.celrep.2020.03.013.

Hayden, B. Y., & Platt, M. L. (2010). Neurons in anterior cingulate cortex multiplex information about reward and action. *Journal of Neuroscience, 30*(9), 3339–3346. https://doi.org/10.1523/JNEUROSCI.4874-09.2010.

Heyes, C. M., Jaldow, E., Nokes, T., & Dawson, G. R. (1994). Imitation in rats (Rattus norvegicus): The role of demonstrator action. *Behavioural Processes, 32*(2), 173–182. https://doi.org/10.1016/0376-6357(94)90074-4.

Hickok, G. (2009). Eight problems for the mirror neuron theory of action understanding in monkeys and humans. *Journal of Cognitive Neuroscience*, *21*(7), 1229–1243. https://doi.org/10.1162/jocn.2009.21189.

Hill, M. R., Boorman, E. D., & Fried, I. (2016). Observational learning computations in neurons of the human anterior cingulate cortex. *Nature Communications*, 7(1), 12722. https://doi.org/10.1038/ncomms12722.

Horschler, D. J., MacLean, E. L., & Santos, L. R. (2020). Do non-human primates really represent others' beliefs? *Trends in Cognitive Sciences*, *24*(8), 594–605. https://doi.org/10.1016/j.tics.2020.05.009.

Hosokawa, T., & Watanabe, M. (2012). Prefrontal neurons represent winning and losing during competitive video shooting games between monkeys. *The Journal of Neuroscience: The Official Journal of the Society for Neuroscience*, *32*(22), 7662–7671. https://doi.org/10.1523/JNEUROSCI.6479-11.2012.

Isoda, M., & Noritake, A. (2013). What makes the dorsomedial frontal cortex active during reading the mental states of others? *Frontiers in Neuroscience*, 7. https://doi.org/10.3389/fnins.2013.00232.

Janowski, V., Camerer, C., & Rangel, A. (2013). Empathic choice involves vmPFC value signals that are modulated by social processing implemented in IPL. *Social Cognitive and Affective Neuroscience*, *8*(2), 201–208. https://doi.org/10.1093/scan/nsr086.

Jenkins, A. C., Macrae, C. N., & Mitchell, J. P. (2008). Repetition suppression of ventromedial prefrontal activity during judgments of self and others. *Proceedings of the National Academy of Sciences of the United States of America*, *105*(11), 4507–4512. https://doi.org/10.1073/pnas.0708785105.

Joiner, J., Piva, M., Turrin, C., & Chang, S. W. C. (2017). Social learning through prediction error in the brain. *NPJ Science of Learning*, *2*(1), 1–9. https://doi.org/10.1038/s41539-017-0009-2.

Kashtelyan, V., Lichtenberg, N. T., Chen, M. L., Cheer, J. F., & Roesch, M. R. (2014). Observation of reward delivery to a conspecific modulates dopamine release in ventral striatum. *Current Biology*, *24*(21), 2564–2568. https://doi.org/10.1016/j.cub.2014.09.016.

Kennerley, S. W., Behrens, T. E. J., & Wallis, J. D. (2011). Double dissociation of value computations in orbitofrontal and anterior cingulate neurons. *Nature Neuroscience*, *14*(12), 1581–1589. https://doi.org/10.1038/nn.2961.

Klavir, O., Genud-Gabai, R., & Paz, R. (2013). Functional connectivity between amygdala and cingulate cortex for adaptive aversive learning. *Neuron*, *80*(5), 1290–1300. https://doi.org/10.1016/j.neuron.2013.09.035.

Kostorz, K., Flanagin, V. L., & Glasauer, S. (2020). Synchronization between instructor and observer when learning a complex bimanual skill. *NeuroImage*. https://doi.org/10.1016/j.neuroimage.2020.116659, 116659.

Krupenye, C., Kano, F., Hirata, S., Call, J., & Tomasello, M. (2016). Great apes anticipate that other individuals will act according to false beliefs. *Science*, *354*(6308), 110–114. https://doi.org/10.1126/science.aaf8110.

Kuraoka, K., & Nakamura, K. (2007). Responses of single neurons in monkey amygdala to facial and vocal emotions. *Journal of Neurophysiology*, *97*(2), 1379–1387. https://doi.org/10.1152/jn.00464.2006.

Lee, D., & Seo, H. (2007). Mechanisms of reinforcement learning and decision making in the primate dorsolateral prefrontal cortex. *Annals of the New York Academy of Sciences*, *1104*(1), 108–122. https://doi.org/10.1196/annals.1390.007.

Livi, A., Lanzilotto, M., Maranesi, M., Fogassi, L., Rizzolatti, G., & Bonini, L. (2019). Agent-based representations of objects and actions in the monkey pre-supplementary motor area. *Proceedings of the National Academy of Sciences of the United States of America*, *116*(7), 2691–2700. https://doi.org/10.1073/pnas.1810890116.

Livneh, U., & Paz, R. (2012). Amygdala-prefrontal synchronization underlies resistance to extinction of aversive memories. *Neuron*, *75*(1), 133–142. https://doi.org/10.1016/j.neuron.2012.05.016.

Lockwood, P. L., Apps, M. A. J., & Chang, S. W. C. (2020). Is there a "social" brain? Implementations and algorithms. *Trends in Cognitive Sciences*, *24*(10), 802–813. https://doi.org/10.1016/j.tics.2020.06.011.

Lockwood, P. L., Apps, M. A. J., Roiser, J. P., & Viding, E. (2015). Encoding of vicarious reward prediction in anterior cingulate cortex and relationship with trait empathy. *Journal of Neuroscience*, *35*(40), 13720–13727. https://doi.org/10.1523/JNEUROSCI.1703-15.2015.

Lockwood, P. L., Wittmann, M. K., Apps, M. A. J., Klein-Flügge, M. C., Crockett, M. J., Humphreys, G. W., et al. (2018). Neural mechanisms for learning self and other ownership. *Nature Communications*, *9*(1), 4747. https://doi.org/10.1038/s41467-018-07231-9.

Marr, D. (1982). *Vision: A computational investigation into the human representation and processing of visual information*. New York, NY: Henry Holt and Co Inc. 2(4.2).

Miller, E. K., Erickson, C. A., & Desimone, R. (1996). Neural mechanisms of visual working memory in prefrontal cortex of the macaque. *Journal of Neuroscience*, *16*(16), 5154–5167. https://doi.org/10.1523/JNEUROSCI.16-16-05154.1996.

Minxha, J., Mosher, C., Morrow, J. K., Mamelak, A. N., Adolphs, R., Gothard, K. M., et al. (2017). Fixations gate species-specific responses to free viewing of faces in the human and macaque amygdala. *Cell Reports*, *18*(4), 878–891. https://doi.org/10.1016/j.celrep.2016.12.083.

Monosov, I. E., Leopold, D. A., & Hikosaka, O. (2015). Neurons in the primate medial basal forebrain signal combined information about reward uncertainty, value, and punishment anticipation. *Journal of Neuroscience*, *35*(19), 7443–7459. https://doi.org/10.1523/JNEUROSCI.0051-15.2015.

Morecraft, R. J., Stilwell-Morecraft, K. S., Cipolloni, P. B., Ge, J., McNeal, D. W., & Pandya, D. N. (2012). Cytoarchitecture and cortical connections of the anterior cingulate and adjacent somatomotor fields in the rhesus monkey. *Brain Research Bulletin*, *87*(4–5), 457–497. https://doi.org/10.1016/j.brainresbull.2011.12.005.

Morecraft, R. J., & Van Hoesen, G. W. (1998). Convergence of limbic input to the cingulate motor cortex in the rhesus monkey. *Brain Research Bulletin*, *45*(2), 209–232. https://doi.org/10.1016/S0361-9230(97)00344-4.

Mosher, C. P., Zimmerman, P. E., & Gothard, K. M. (2014). Neurons in the monkey amygdala detect eye contact during naturalistic social interactions. *Current Biology: CB*, *24*(20), 2459–2464. https://doi.org/10.1016/j.cub.2014.08.063.

Munuera, J., Rigotti, M., & Salzman, C. D. (2018). Shared neural coding for social hierarchy and reward value in primate amygdala. *Nature Neuroscience*, *21*(3), 415–423. https://doi.org/10.1038/s41593-018-0082-8.

Noritake, A., Ninomiya, T., & Isoda, M. (2018). Social reward monitoring and valuation in the macaque brain. *Nature Neuroscience*, *21*(10), 1452–1462. https://doi.org/10.1038/s41593-018-0229-7.

Olsson, A., Knapska, E., & Lindström, B. (2020). The neural and computational systems of social learning. *Nature Reviews Neuroscience*, *21*(4), 197–212. https://doi.org/10.1038/s41583-020-0276-4.

Onishi, K. H., & Baillargeon, R. (2005). Do 15-month-old infants understand false beliefs? *Science (New York, N.Y.)*, *308*(5719), 255–258. https://doi.org/10.1126/science.1107621.

Padoa-Schioppa, C., & Assad, J. A. (2006). Neurons in the orbitofrontal cortex encode economic value. *Nature*, *441*(7090), 223–226. https://doi.org/10.1038/nature04676.

Pagnotta, M., Laland, K. N., & Coco, M. I. (2020). Attentional coordination in demonstrator-observer dyads facilitates learning and predicts performance in a novel manual task. *Cognition*, *201*. https://doi.org/10.1016/j.cognition.2020.104314, 104314.

Pärnamets, P., Espinosa, L., & Olsson, A. (2020). Physiological synchrony predicts observational threat learning in humans. *Proceedings of the Biological Sciences*, *287*(1927), 20192779. https://doi.org/10.1098/rspb.2019.2779.

Paton, J. J., Belova, M. A., Morrison, S. E., & Salzman, C. D. (2006). The primate amygdala represents the positive and negative value of visual stimuli during learning. *Nature*, *439*(7078), 865–870. https://doi.org/10.1038/nature04490.

Piva, M., Velnoskey, K., Jia, R., Nair, A., Levy, I., & Chang, S. W. (2019). The dorsomedial prefrontal cortex computes task-invariant relative subjective value for self and other. *eLife*, *8*. https://doi.org/10.7554/eLife.44939.

Porrino, L. J., Crane, A. M., & Goldman-Rakic, P. S. (1981). Direct and indirect pathways from the amygdala to the frontal lobe in rhesus monkeys. *Journal of Comparative Neurology*, *198*(1), 121–136. https://doi.org/10.1002/cne.901980111.

Putnam, P. T., & Gothard, K. M. (2019). Multidimensional neural selectivity in the primate amygdala. *ENeuro*, *6*(5). https://doi.org/10.1523/ENEURO.0153-19.2019.

Redcay, E., & Schilbach, L. (2019). Using second-person neuroscience to elucidate the mechanisms of social interaction. *Nature Reviews. Neuroscience*, *20*(8), 495–505. https://doi.org/10.1038/s41583-019-0179-4.

Rizzolatti, G., & Craighero, L. (2004). The mirror-neuron system. *Annual Review of Neuroscience*, *27*(1), 169–192. https://doi.org/10.1146/annurev.neuro.27.070203.144230.

Rizzolatti, G., Fadiga, L., Gallese, V., & Fogassi, L. (1996). Premotor cortex and the recognition of motor actions. *Cognitive Brain Research*, *3*(2), 131–141. https://doi.org/10.1016/0926-6410(95)00038-0.

Rizzolatti, G., Fogassi, L., & Gallese, V. (2001). Neurophysiological mechanisms underlying the understanding and imitation of action. *Nature Reviews. Neuroscience*, *2*(9), 661–670. https://doi.org/10.1038/35090060.

Rogers, R. D., Ramnani, N., Mackay, C., Wilson, J. L., Jezzard, P., Carter, C. S., et al. (2004). Distinct portions of anterior cingulate cortex and medial prefrontal cortex are activated by reward processing in separable phases of decision-making cognition. *Biological Psychiatry*, *55*(6), 594–602. https://doi.org/10.1016/j.biopsych.2003.11.012.

Rudebeck, P. H., Buckley, M. J., Walton, M. E., & Rushworth, M. F. S. (2006). A role for the macaque anterior cingulate gyrus in social valuation. *Science (New York, N.Y.)*, *313*(5791), 1310–1312. https://doi.org/10.1126/science.1128197.

Rudebeck, P. H., Mitz, A. R., Chacko, R. V., & Murray, E. A. (2013). Effects of amygdala lesions on reward-value coding in orbital and medial prefrontal cortex. *Neuron*, *80*(6), 1519–1531. https://doi.org/10.1016/j.neuron.2013.09.036.

Rudebeck, P. H., Ripple, J. A., Mitz, A. R., Averbeck, B. B., & Murray, E. A. (2017). Amygdala contributions to stimulus-reward encoding in the macaque medial and orbital frontal cortex during learning. *The Journal of Neuroscience: The Official Journal of the Society for Neuroscience*, *37*(8), 2186–2202. https://doi.org/10.1523/JNEUROSCI.0933-16.2017.

Rutishauser, U., Tudusciuc, O., Neumann, D., Mamelak, A. N., Heller, A. C., Ross, I. B., et al. (2011). Single-unit responses selective for whole faces in the human amygdala. *Current Biology: CB*, *21*(19), 1654–1660. https://doi.org/10.1016/j.cub.2011.08.035.

Saver, J. L., & Damasio, A. R. (1991). Preserved access and processing of social knowledge in a patient with acquired sociopathy due to ventromedial frontal damage. *Neuropsychologia*, *29*(12), 1241–1249. https://doi.org/10.1016/0028-3932(91)90037-9.

Seo, H., Barraclough, D. J., & Lee, D. (2007). Dynamic signals related to choices and outcomes in the dorsolateral prefrontal cortex. *Cerebral Cortex (New York, N.Y.: 1991)*, *17*(Suppl. 1), i110–i117. https://doi.org/10.1093/cercor/bhm064.

Seo, H., Cai, X., Donahue, C. H., & Lee, D. (2014). Neural correlates of strategic reasoning during competitive games. *Science*, *346*(6207), 340–343. https://doi.org/10.1126/science.1256254.

Shepherd, S. V., Klein, J. T., Deaner, R. O., & Platt, M. L. (2009). Mirroring of attention by neurons in macaque parietal cortex. *Proceedings of the National Academy of Sciences of the United States of America, 106*(23), 9489–9494. https://doi.org/10.1073/pnas.0900419106.

Sliwa, J., & Freiwald, W. A. (2017). A dedicated network for social interaction processing in the primate brain. *Science (New York, N.Y.), 356*(6339), 745–749. https://doi.org/10.1126/science.aam6383.

Snyder, L. H. (2000). Coordinate transformations for eye and arm movements in the brain. *Current Opinion in Neurobiology, 10*(6), 747–754. https://doi.org/10.1016/s0959-4388(00)00152-5.

Snyder, L. H., Batista, A. P., & Andersen, R. A. (2000). Intention-related activity in the posterior parietal cortex: A review. *Vision Research, 40*(10), 1433–1441. https://doi.org/10.1016/S0042-6989(00)00052-3.

Spezio, M. L., Huang, P.-Y. S., Castelli, F., & Adolphs, R. (2007). Amygdala damage impairs eye contact during conversations with real people. *The Journal of Neuroscience, 27*(15), 3994–3997. https://doi.org/10.1523/JNEUROSCI.3789-06.2007.

Suzuki, W., Banno, T., Miyakawa, N., Abe, H., Goda, N., & Ichinohe, N. (2015). Mirror neurons in a new world monkey, common marmoset. *Frontiers in Neuroscience, 9*. https://doi.org/10.3389/fnins.2015.00459.

Taubert, J., Flessert, M., Wardle, S. G., Basile, B. M., Murphy, A. P., Murray, E. A., et al. (2018). Amygdala lesions eliminate viewing preferences for faces in rhesus monkeys. *Proceedings of the National Academy of Sciences of the United States of America, 115*(31), 8043–8048. https://doi.org/10.1073/pnas.1807245115.

Tomasello, M., Davis-Dasilva, M., Camak, L., & Bard, K. (1987). Observational learning of tool-use by young chimpanzees. *Human Evolution, 2*(2), 175–183. https://doi.org/10.1007/BF02436405.

Wang, S., Yu, R., Tyszka, J. M., Zhen, S., Kovach, C., Sun, S., et al. (2017). The human amygdala parametrically encodes the intensity of specific facial emotions and their categorical ambiguity. *Nature Communications, 8*, 14821. https://doi.org/10.1038/ncomms14821.

Wimmer, H., & Perner, J. (1983). Beliefs about beliefs: Representation and constraining function of wrong beliefs in young children's understanding of deception. *Cognition, 13*(1), 103–128. https://doi.org/10.1016/0010-0277(83)90004-5.

Wise, S. P. (2008). Forward frontal fields: Phylogeny and fundamental function. *Trends in Neurosciences, 31*(12), 599–608. https://doi.org/10.1016/j.tins.2008.08.008.

Yim, M. Y., Cai, X., & Wang, X.-J. (2019). Transforming the choice outcome to an action plan in monkey lateral prefrontal cortex: A neural circuit model. *Neuron, 103*(3), 520–532.e5. https://doi.org/10.1016/j.neuron.2019.05.032.

Yoshida, K., Saito, N., Iriki, A., & Isoda, M. (2011). Representation of others' action by neurons in monkey medial frontal cortex. *Current Biology: CB, 21*(3), 249–253. https://doi.org/10.1016/j.cub.2011.01.004.

Yoshida, K., Saito, N., Iriki, A., & Isoda, M. (2012). Social error monitoring in macaque frontal cortex. *Nature Neuroscience, 15*(9), 1307–1312. https://doi.org/10.1038/nn.3180.

Zhou, T., Zhu, H., Fan, Z., Wang, F., Chen, Y., Liang, H., et al. (2017). History of winning remodels thalamo-PFC circuit to reinforce social dominance. *Science, 357*(6347), 162–168. https://doi.org/10.1126/science.aak9726.

CHAPTER NINE

Potential roles of the rodent medial prefrontal cortex in conflict resolution between multiple decision-making systems

Amber E. McLaughlin[†], Geoffrey W. Diehl[†], and A. David Redish*
Department of Neuroscience, University of Minnesota, Minneapolis, MN, United States
*Corresponding author: e-mail address: redish@umn.edu

Contents

Abstract

Mammalian decision-making is mediated by the interaction of multiple, neurally and computationally separable decision systems. Having multiple systems requires a mechanism to manage conflict and converge onto the selection of singular actions. A long history of evidence has pointed to the prefrontal cortex as a central component in processing the interactions between distinct decision systems and resolving conflicts among them. In this chapter we review four theories of how that interaction might occur and identify how the medial prefrontal cortex in the rodent may be involved in each theory. We then present experimental predictions implied by the neurobiological data in the context of each theory as a starting point for future investigation of medial prefrontal cortex and decision-making.

[†] Contributed equally.

International Review of Neurobiology, Volume 158
ISSN 0074-7742
https://doi.org/10.1016/bs.irn.2020.11.009

1. Multiple decision-making systems interact to inform action-selection

Current theories suggest multiple decision systems underlie action-selection in mammalian brains (Daw, Niv, & Dayan, 2005; Kahneman, 2011; O'Keefe & Nadel, 1978; Redish, 1999, 2013), identifying three action-selection systems: *Pavlovian*, *Deliberative*, and *Procedural* (see van der Meer, Kurth-Nelson, & Redish, 2012, for review). These different decision-making systems entail different computational algorithms that are instantiated through distinct neural circuits. Yet common among them is evidence that the medial prefrontal cortex (mPFC) plays an important role. Interestingly, newer work has pointed to subregional specificity of mPFC's involvement in these action-selection processes and their interactions (Balaguer-Ballester, Lapish, Seamans, & Durstewitz, 2011; Chudasama & Robbins, 2003; Durstewitz, Vittoz, Floresco, & Seamans, 2010; Euston, Gruber, & McNaughton, 2012; Horst & Laubach, 2012; Kesner & Churchwell, 2011; Milad & Quirk, 2002; Narayanan, Horst, & Laubach, 2006; Padilla-Coreano et al., 2016; Powell & Redish, 2016; Schmidt, Duin, & Redish, 2019; Sharpe & Killcross, 2015; Smith & Graybiel, 2013).

A multiple decision-system model necessitates a mechanism of interaction, particularly when systems suggest conflicting actions. However, while the identity of this mechanism remains unknown, there is reason to believe mPFC is fundamentally involved. Building from computational definitions of each system and points of interaction between them, we derive experimental predictions that will shed light on the underlying neurobiological mechanisms. This approach reveals potential explanations of how the multiple decision systems interact and the role of mPFC in decision-making.

2. Computation and neurobiology of decision systems

A predominant taxonomy categorizes action-selection algorithms based on their fundamentally different computations—*Pavlovian* responses, goal-directed *Deliberation*, and *Procedural* action-selection. The Pavlovian system builds associations between environmental stimuli allowing the release of species-important responses to those learned stimuli (Bouton, 2007; Dayan, Niv, Seymour, & Daw, 2006; Pavlov, 1927). Deliberative decisions depend on a constructed representation, evaluation, and selection of potential outcomes. As outcome prediction depends on a model of the transition structure of the world, Deliberative processes are often referred to as "model-based" (Gilbert & Wilson, 2007; Johnson, van der Meer, & Redish, 2007; Niv, Joel, & Dayan, 2006; Redish, 2016). Procedural decision-making relies on recognizing and categorizing situations based on experience and responding with appropriate, well-learned action-chains (Barnes, Kubota, Hu, Jin, & Graybiel, 2005; Dezfouli & Balleine, 2012; Graybiel, 1998; Mishkin & Appenzeller, 1987) (see Redish, 2013, for review). As these processes do not require modeling explicit transition structures, Procedural responding is often referred to as "model-free"(Daw et al., 2005; Niv et al., 2006), but this is a misnomer as the situation-categorization process does contain a model of what aspects in the world are relevant (Cochran & Cisler, 2019; Gershman & Niv, 2010; Redish, Jensen, Johnson, & Kurth-Nelson, 2007).

Of the three action-selection systems, *Pavlovian responding* is the simplest computationally and provides fast responses. Pavlovian learning describes the association of a conditioned stimulus with a naturalistic behavior (Breland & Breland, 1961; Dayan et al., 2006; Pavlov, 1927). Experiments identify amygdala and periaqueductal gray as key players in Pavlovian responding, with nucleus accumbens shell involved in avoidance behaviors (Bravo-Rivera, Roman-Ortiz, Montesinos-Cartagena, & Quirk, 2015; Diehl et al., 2018; Diehl, Bravo-Rivera, & Quirk, 2019; Kim, Rison, & Fanselow, 1993; LeDoux, 2015; LeDoux & Daw, 2018; McNally, Johansen, & Blair, 2011). While the Pavlovian system provides the advantage of a rapid, potentially life-saving response, it does so at the expense of flexibility (Breland & Breland, 1961). Actions resulting from this system are instinctual in nature, yet, by translating simple stimuli across contexts, the Pavlovian system can act even in truly novel situations.

In situations where the structure of the environment is known, agents can use that known structure to plan novel actions within the world (O'Keefe & Nadel, 1978; Redish, 1999; Tolman, 1948). *Deliberation*, or "goal-directed action selection," involves several complex processing steps, resulting in a computationally expensive process dependent on multiple brain structures. First, it requires an understanding of the external world, including knowledge of relevant features and how to effectively transition between situations. This model of the world is known as a *cognitive map* from which actions and their consequences can be identified through a search process.

The forward sweeping activity of hippocampal place cells along prospective paths has been identified as a potential mechanism for this search process (Amemiya & Redish, 2016; Johnson & Redish, 2007; Kay et al., 2020; Papale, Zielinski, Frank, Jadhav, & Redish, 2016; Pfeiffer & Foster, 2013; Redish, 2016). After identifying candidate options, the Deliberative system must evaluate and compare their values, a process that involves the nucleus accumbens core and orbitofrontal cortex (Steiner & Redish, 2012; van der Meer & Redish, 2009). Finally, Deliberation must use this value information to converge to a single action, though how this occurs remains an open question. The modularity and wide applicability of these four steps (cognitive map, search, evaluate, and action-selection) demonstrate the flexibility and versatility of the Deliberative decision system, yet they also reveal its computational complexity highlighting it as slow and energetically costly.

Like Deliberation, *Procedural* decision-making (also known as *Habit-based* learning) has the ability to learn arbitrary responses. Like Pavlovian systems, Procedural decision-making can respond quickly. Computationally, Procedural action-selection is achieved by developing situation-action associations and chunking discrete actions into fluid motions and complex behaviors (Barnes et al., 2005; Dezfouli & Balleine, 2012; Friedman, Homma, et al., 2015; Friedman, Jomma, et al., 2015; Graybiel, 1998; Hull, 1943; Rand, Hikosaka, Miyachi, Lu, & Miyashita, 1998). Procedural decision-making is thus faster to execute than Deliberation, but more flexible than the Pavlovian system where action repertoire is limited. However, Procedural action-selection develops slowly with experience and is generally difficult to adapt to new situations. Given that it lacks a comprehensive search process, the Procedural system can respond faster with more efficient action-chains than Deliberation. But the inflexibility of habits means that they are optimized for situations where the set of relevant actions is highly reliable. Studies have found this system to depend upon dorsolateral striatum (dlStr), motor cortex, and cerebellum (Berke & Eichenbaum, 2009; Graybiel, 1998; Molinari et al., 1997; Schmitzer-Torbert & Redish, 2004; Smith & Graybiel, 2016; van der Meer, Johnson, Schmitzer-Torbert, & Redish, 2010; van der Meer, van der Meer, Johnson, Schmitzer-Torbert, & David Redish, 2010). Most notably, dlStr cells have been found to exhibit rapid bouts of activity at the initiation of an action sequence, a process termed "task bracketing" and thought to reflect the release of the complete motor sequence (Barnes et al., 2005; Regier, Amemiya, & Redish, 2015; Smith & Graybiel, 2013, 2016; Thorn, Atallah, Howe, & Graybiel, 2010).

3. Prefrontal cortex and decision systems

While relatively straightforward relationships have emerged between the three decision systems and their respective neurobiology, the situation is less clear when considering the medial prefrontal cortex (mPFC). One immediate consideration is that the medial wall of the rodent prefrontal cortex is not a homogenous structure (de Bruin, Corner, Feenstra, Van Eden, & Uylings, 1991; Groenewegen & Uylings, 2000; Kolb, 1990; Laubach, Amarante, Swanson, & White, 2018; Uylings, Groenewegen, & Kolb, 2003). Instead, anatomical findings have revealed gradations and subdivisions along the dorso-ventral axis. As such, the rodent mPFC is often segregated into the anterior cingulate cortex (ACC) dorsally, transitioning into the prelimbic cortex (PL) in the mid-region, followed by the infralimbic cortex (IL) more ventral (Heidbreder & Groenewegen, 2003; Hoover & Vertes, 2007). Collectively, these subregions have long been thought to serve a central role in higher cognitive functioning, executive control, and decision-making (Kesner & Churchwell, 2011; Laubach, 2011; Laubach et al., 2018; Miller & Cohen, 2001). However, the exact role of each subregion, how their functioning relates to different decision systems, and the degree to which they interact all remain areas of ongoing research.

Early theories of mPFC posited that a division of labor may exist between ACC, PL, and IL with each subregion dedicated to one specific role. For decision-making, PL was thought to be involved in cognitive flexibility and planning, making it a good candidate for supporting Deliberative processes (Dalley, Cardinal, & Robbins, 2004; Fuster, 1997; Killcross & Coutureau, 2003; Kolb, 1990; Ragozzino, Wilcox, Raso, & Kesner, 1999; Rich & Shapiro, 2007; Tran-Tu-Yen, Marchand, Pape, Di Scala, & Coutureau, 2009). Infralimbic cortex on the other hand seemed to be involved in cases of repeated actions characteristic of Procedural responding (Barker, Taylor, & Chandler, 2014; Coutureau & Killcross, 2003; Graybiel, 2008; Killcross & Coutureau, 2003; Ostlund & Balleine, 2009; Smith & Graybiel, 2013). Finally, ACC has been linked to processing naturalistic stimuli such as rewards or punishments, giving preliminary reason to associate it with Pavlovian action-selection (Bussey, Everitt, & Robbins, 1997; Cardinal et al., 2003; Parkinson, Willoughby, Robbins, & Everitt, 2000), but other work has associated ACC with conflict resolution (de Wit, Kosaki, Balleine, & Dickinson, 2006; Fisher, Saksida, Robbins, & Bussey, 2020) or overcoming effort (Walton, Bannerman, Alterescu, & Rushworth, 2003).

While direct association between mPFC subregions and decision systems seems attractive, subsequent evidence suggests considerably more interaction both between the anatomical areas, and between the cognitive processes themselves.

Perhaps pointing directly to its role in decision-making, a considerable portion of our understanding of mPFC subregions comes from exploring the different decision systems (Deliberative, Procedural, and Pavlovian) that drive behavior.

Deliberative decision-making has received the most attention regarding potential mPFC involvement but has primarily concentrated on the PL region, particularly the dorsal aspects of PL, along with ventral aspects of ACC (Hasz & Redish, 2020b; Hyman, Whitman, Emberly, Woodward, & Seamans, 2013; Powell & Redish, 2016; Zielinski, Shin, & Jadhav, 2019). While a wide variety of tasks have been employed, they share the common feature that optimal performance is goal-directed and necessitates future planning (Hyman, Zilli, Paley, & Hasselmo, 2010; Powell & Redish, 2016; Rich & Shapiro, 2009; Schmidt et al., 2019; Zielinski et al., 2019). Furthermore, prominent goal-directed, task-relevant activity is observed in hippocampus during these behaviors (Cei, Girardeau, Drieu, El Knabi, & Zugaro, 2014; Gupta, van der Meer, Touretzky, & Redish, 2012; Ito, Zhang, Witter, Moser, & Moser, 2015; Johnson & Redish, 2007; Kay et al., 2020; O'Keefe & Nadel, 1978; O'Keefe & Speakman, 1987; Pfeiffer & Foster, 2013; Redish, 1999; Rosenzweig, Redish, McNaughton, & Barnes, 2003; Wikenheiser & Redish, 2015) highlighting their utilization of Deliberation. In regard to mPFC, these studies have provided compelling evidence that PL is important to Deliberative processing. Manipulations of PL disrupt the development of goal-directed strategies during learning and inhibit the expression of these strategies in fully trained rats (Dalton, Wang, Phillips, & Floresco, 2016; Killcross & Coutureau, 2003; Ragozzino et al., 1999; Riaz et al., 2019; Tran-Tu-Yen et al., 2009). Physiologically, PL activity coincides with events during which Deliberation would be expected: at decision points and instances when a change in strategy becomes necessary (Balaguer-Ballester et al., 2011; Durstewitz et al., 2010; Hasz & Redish, 2020a, 2020b; Hyman et al., 2010; Powell & Redish, 2016; Rich & Shapiro, 2009; Zielinski et al., 2019). Additionally, a series of studies have directly linked activity in PL/ACC with dorsal hippocampus. Coordinated activity has been found in both unit and LFP recordings (Adhikari, Topiwala, & Gordon, 2010; Adhikari, Topwala, & Gordon, 2011; Benchenane et al., 2010; Hyman et al., 2010; Ito et al., 2015;

Ito, Moser, & Moser, 2018; Jadhav, Rothschild, Roumis, & Frank, 2016; Jones & Wilson, 2005; Wirt & Hyman, 2019; Zielinski et al., 2019), and manipulation of PL activity has led to functional changes within hippocampal representations (Guise & Shapiro, 2017; Hok, Chah, Save, & Poucet, 2013; Ito et al., 2015; Schmidt et al., 2019).

Behavioral studies have also approached mPFC functioning through Procedural decisions; here concentrating on IL. Under extended training regimens, repetitive trials lead to the development of goal-independent Procedural responding, or habits (Coutureau & Killcross, 2003; Killcross & Coutureau, 2003; Smith & Graybiel, 2013). In rats with IL lesions prior to training, Procedural responding never develops, even with extended training well past what is normally required (Killcross & Coutureau, 2003). Instead, behavior appears to remain in a goal-directed mode. In the case of fully trained rats with well-established habits, IL manipulations lead animals to abandon Procedural responding and fall back on Deliberation, or in some cases revert to older habits that had been seemingly overridden (Barker, Glen, Linsenbardt, Lapish, & Chandler, 2017; Coutureau & Killcross, 2003; Smith & Graybiel, 2013). In line with these behavioral results, recordings from IL cells in overtrained rats found activity that coincides with Procedural responding (Smith & Graybiel, 2013). Bursts of IL firing occur at movement initiation, concurrent with the task bracketing activity of dlStr (Barnes et al., 2005; Jog, Kubota, Connolly, Hillegaart, & Graybiel, 1999; Regier et al., 2015; Thorn et al., 2010). Furthermore, these bouts of IL activity develop over training, mirroring the time course of both behavioral habits and dlStr activity (Smith & Graybiel, 2013).

A third body of literature has examined the role of mPFC during Pavlovian associations, most notably using fear conditioning and extinction paradigms. At a macro level, lesion studies indicate that mPFC is central to Pavlovian fear learning (Lebron, Milad, & Quirk, 2004; LeDoux, 2000; Milad & Quirk, 2002; Morgan & LeDoux, 1995) with newer findings suggesting opposing roles of PL and IL. Prelimbic inactivation leads to impairment in freezing, and neural firing in PL is correlated with the expression of fear learning (Corcoran & Quirk, 2007; Do-Monte, Quiñones-Laracuente, & Quirk, 2015; Sierra-Mercado, Padilla-Coreano, & Quirk, 2011) highlighting its role in initial fear leaning. In contrast, infralimbic manipulations do not impact these early stage Pavlovian processes, but instead influence fear extinction or the reinstatement of freezing at remote time points (Do-Monte et al., 2015; Milad & Quirk, 2002; Quirk, Russo, Barron, & Lebron, 2000; Sierra-Mercado et al., 2011).

Despite evidence that the Pavlovian system operates beyond simple fear memory, research into mPFC in these other regimes has lagged. Newer work has begun to study PL during Pavlovian mediated approach-avoidance situations, presenting rodents with an inherent conflict between approaching food and avoiding attack by an artificial predator (Amir, Lee, Headley, Herzallah, & Pare, 2015; Choi & Kim, 2010; Kim & Jung, 2018; Mobbs & Kim, 2015; Walters, Jubran, Sheehan, Erickson, & Redish, 2019). Recordings from PL during this task found responses to the food option, the potential predator, as well as interactions between the two stimuli (Kim et al., 2018). Furthermore, correlated firing was observed between cells in PL and the lateral amygdala, highlighting strong links between PL activity and structures involved in the Pavlovian decision system.

Complementing the decision-making focused approach to mPFC, a second body of literature has sought to query the neural activity of specific mPFC subregions during behavior. While generally not testing a proposed role in decision-making directly, these findings provide valuable information from which we can infer how activity within mPFC could be utilized in decision processing.

Within ACC, a body of work has found firing rate changes associated with the resultant outcome of behavioral choices. Following correct responding, ACC cells exhibit firing increases thought to be associated with the delivery of reward and a reinforcement of the decision that produced it (Hart, Blair, O'Dell, Blair, & Izquierdo, 2020; Hillman & Bilkey, 2010; Hyman et al., 2013). In other cases, incorrect decisions lead to increases in ACC activity which is thought to reflect a lack of reward delivery or perhaps mental exploration of alternative choices (Caracheo, Grewal, & Seamans, 2018; Hyman, Holroyd, & Seamans, 2017; Mashhoori, Hashemnia, McNaughton, Euston, & Gruber, 2018). Similar results have been found in PL (Horst & Laubach, 2012, 2013; Powell & Redish, 2014), suggesting this is either a general property of mPFC or that the boundaries of these structures may not be so clearly identified (Laubach, Amarante, Swanson, & White, 2018). Collectively, this work seems to point to a role for ACC (and possibly PL) in evaluating choice outcomes to update an understanding of the world.

A second set of studies on ACC have found robust evidence for coherent representations of task states at the population level (Balaguer-Ballester et al., 2011; Hyman, Ma, Balaguer-Ballester, Durstewitz, & Seamans, 2012; Lapish, Durstewitz, Chandler, & Seamans, 2008; Ma, Hyman, Durstewitz, Phillips, & Seamans, 2016). Furthermore, these neural state representations are distinct for different contexts, with reliable neural changes coinciding with changes in task structure or behavioral responses (Caracheo et al., 2018; Durstewitz et al., 2010; Rich & Shapiro, 2009). Again, similar results have been observed within PL (Hasz & Redish, 2020a; Powell & Redish, 2016), suggesting a potential commonality between these two subregions. Thus one hypothesized role of dorsal mPFC is to establish, maintain, and utilize an internal representation of external environmental states (Fuster, 1997; Seamans & Yang, 2004; Sharpe & Killcross, 2015; Sharpe, Wikenheiser, Niv, & Schoenbaum, 2015). Such a function would support involvement in both Deliberative and Procedural decision systems as both rely on a coherent state signal and reliable transitions between states.

Unfortunately, considerably less data exist as to the potential inner workings of IL. What has been shown is that IL is involved in Procedurally driven goal-independent responding (Killcross & Coutureau, 2003), its activity develops with and is critical to the extinction of Pavlovian associations (Milad & Quirk, 2002), and during Procedural responding IL ensembles exhibit punctate bouts of firing at trial initiations (Smith & Graybiel, 2013). While the exact nature of these bouts is unclear, their developmental time course, striking similarity to striatal activity patterns, and the impact of IL manipulations on behavior all point to a role in Procedural decision processing.

There currently exists ample, though incomplete, evidence to support a prominent role for mPFC in decision processing. Furthermore, some evidence points to functionally distinct roles of ACC, PL, and IL. Yet there remain several wrinkles in this explanation that require consideration.

Firstly, while links have been drawn between PL and dorsal hippocampus (Deliberation) and between IL and dlStr (Procedural actions), neither pair share direct anatomical connections (Beckstead, 1979; Heilbronner, Rodriguez-Romaguera, Quirk, Groenewegen, & Haber, 2016; Sesack, Deutch, Roth, & Bunney, 1989; Vertes, 2004). For Deliberation, intermediary structures such as ventral hippocampus, entorhinal cortex, or nucleus reuniens could serve as relays (Adhikari et al., 2010; Hoover & Vertes, 2007; Ito et al., 2015; Jay & Witter, 1991; Swanson, 1981). For Procedural actions no obvious anatomical pathway exists. Infralimbic sends direct projections to PL which subsequently projects to dorsal striatum (Friedman, Homma, et al., 2015; Friedman, Jomma, et al., 2015; Heilbronner et al., 2016; Vertes, 2004), though given the strong association between PL and Deliberation it represents an unlikely relay point for Procedural information. Alternatively, IL could send information to the ventral striatum where internal ascending spiral loops then facilitate relay to dlStr (Haber, Fudge, & McFarland, 2000; Montaron, Deniau, Menetrey, Glowinski, & Thierry, 1996). A third option is that IL and dlStr serve complementary roles in Procedural actions.

While dlStr executes motor commands, IL may act to establish a general brain state that is permissive to the execution of habits (Smith, Virkud, Deisseroth, & Graybiel, 2012). In this way, the two brain regions could work in parallel to support Procedural decisions, but require no means of direct information transfer.

A second consideration surrounding the role of mPFC in decision processing involves how the different subregions interact. Anatomically ACC, PL, and IL are interconnected, implying functional interaction and influence across the subregions (Heidbreder & Groenewegen, 2003; van Aerde, Heistek, & Mansvelder, 2008; Vertes, 2004). Indeed results from a variety of studies that examined directed manipulations of PL and IL point to opposing roles and mutual competition between the pair (Killcross & Coutureau, 2003; Mukherjee & Caroni, 2018; Sierra-Mercado et al., 2011; Smith et al., 2012; Vidal-Gonzalez, Vidal-Gonzalez, Rauch, & Quirk, 2006). Most notably, PL lesions bias rats away from Deliberation and toward Procedural systems, while IL lesions appear to eliminate Procedural responding and instead force animals into goal-directed strategies, as would be expected in the case of direct competition between the subregions.

In actuality, competition between PL and IL, and by proxy the underlying Deliberative and Procedural decision systems, is indicative of a larger question for decision-making and the role of mPFC in it. A decision maker built from multiple independent subsystems can find itself in conflict, yet it must ultimately coalesce into a single unified action. Deconstructing this decision maker into computational and neural components has proven essential, but we must also reassemble these parts to ask how they interact in the pursuit of action-selection. Given its long established role in cognitive functioning and decision-making, along with its anatomical position as a hub for integration and processing of information, we believe that the rodent medial prefrontal cortex likely plays a central role in moderating the brain's multiple decision systems.

4. Theories for decision system integration

Current theories of decision system integration have identified three main candidate proposals of how Pavlovian, Deliberative, and Procedural decision systems converge to a single action. To this set of three candidates we add a fourth novel proposal for future consideration:

Theory 1. An independent central executive could act to delegate responsibility between the three decision systems.

Theory 2. Downstream motor control mechanisms could receive input from the three systems in parallel, with the final action reflecting the input of the most internally self-consistent signal.

Theory 3. The Deliberative system could occupy a privileged position to exert an override and/or arbitrator role over both the Pavlovian and Procedural systems.

Theory 4. The three decision systems could operate in a hierarchical structure reflecting their maturation over the course of an agent's experience with the world. In this fourth theory, actions are initially driven by ingrained Pavlovian processes, which are superseded by Deliberation as an understanding of the world develops. Deliberation is then superseded by Procedural responding as extensive training gives rise to well developed action chains.

Critical to each of these four theories is the computational structure describing how three distinct, competing decision systems interact to produce a single behavioral action. Furthermore, direct neurobiological predictions can be made from each computational theory into the nature and interactions of the medial prefrontal cortex.

4.1 Theory 1: Independent central executive

Many theories postulate that decision systems function in isolation, only coming together at a final stage to compete for behavioral control. Building from this conceptualization has led to the idea that an additional stand-alone executive system is necessary to effectively manage multiple decision systems (Freud, 1923; Fuster, 2008; Hofmann, Schmeichel, & Baddeley, 2012; Plato, 2008). Importantly, this central executive would be computationally and physiologically distinct from the other decisions systems, delegating which system directs behavior and guiding information flow, but not actively engaging in any individual system's computational processing. Specifically in the context of conflicts arising between the three decision systems (Pavlovian, Deliberative, and Procedural), the central executive model posits that a fourth entity empowers a single neural decision circuitry (amygdala, hippocampus, or dlStr) to influence action-selection (see Fig. 1).

Given its role, the requirements of this central executive suggest that it would need to contain representations of strategy while also remaining dissociated from the computational inner workings of any specific decision system. This makes neuroanatomical identification of a separate central executive challenging, though mPFC has been proposed to fulfill such a role (Fuster, 2008; Otto, Gershman, Markman, & Daw, 2013).

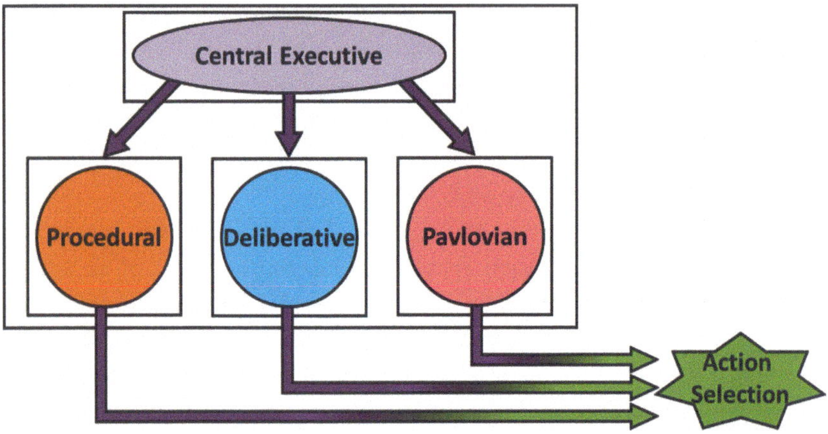

Fig. 1 Theory 1: Central executive. Procedural, Deliberative and Pavlovian systems function in parallel (black boxes) to one another with their own discrete mechanisms of action-selection (purple/green arrows). A separate central executive acts upon the three subsystems to delegate which of the three mechanisms influences action-selection (purple arrows).

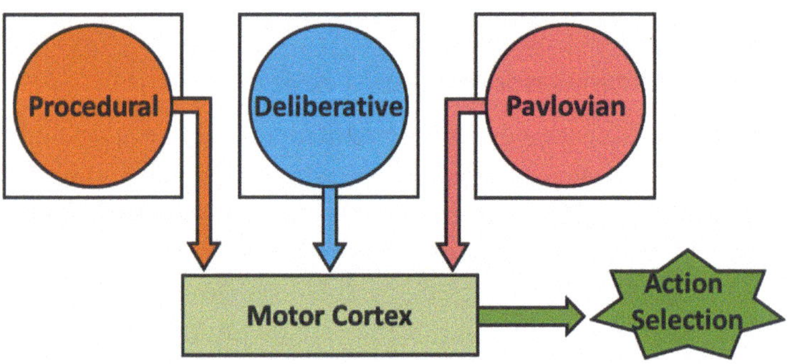

Fig. 2 Theory 2: Multiple experts. Procedural, Deliberative and Pavlovian systems function in parallel (black boxes) and use their own internal mechanisms to alter their self-consistency. Each system sends a stream of information (orange, blue, pink arrows) to downstream motor centers that produce the action (green arrow) signaled by the most self-consistent input.

4.2 Theory 2: A multiple experts model

A second theoretical approach points to a conflict resolution hypothesis that has neither a central executive nor a hierarchy. Rather, this heuristic model claims that each decision system is individually responsible for driving downstream structures according to changes in their own self-consistency (van der Meer et al., 2012) (see Fig. 2). Because neural representations are distributed (Hertz, Krogh, & Palmer, 1991; McClelland & Rumelhart, 1986; Rumelhart & McClelland, 1986), the output of each decision system can contain information about an array of actions, not just the best choice (Zemel, Dayan, & Pouget, 1998). Thus, measurement of the distribution breadth, or the representation's self-consistency (Jackson & Redish, 2003; Johnson, Jackson, & Redish, 2008), points toward the certainty of a particular action, a feature commonly proposed to drive behavior (Daw et al., 2005; Keramati, Dezfouli, & Piray, 2011).

This idea relies on the principle that each decision system acts independently, in parallel, and that within each system there exist internal dynamics that determine the self-consistency of the neural signal. These parallel decision systems then relay their neural information to a single downstream motor center where conflicts are resolved in a competitive winner-take-all process according to the input stream whose information is most self-consistent. Mechanistically, several theories have proposed that internal neural coherence can change the effectiveness of structure-to-structure connectivity, providing a means for self-consistency to regulate neural drive (Engel, Fries, & Singer, 2001; Fries, 2005; Hyman et al., 2010).

Conceptually, this second theory is related to mixture models and multiple experts models of machine learning (Bishop, 2006; Mitchell, 1997). In these models, specialized agents address subspaces of a problem. Each agent provides its best answer along with a confidence signal. Agents applying their specialized knowledge to appropriate components of the problem provide more confident answers, allowing a macro-agent to determine and utilize the most appropriate specialized sub-agent and outperform a single generalist approach.

Both the multiple experts and central executive theories are rooted in the assumption that each subsystem functions in parallel, only coming together at an upstream executive (central executive) or downstream effector region (multiple experts). In contrast, the next two theories entail direct interactions between the systems themselves.

4.3 Theory 3: Deliberative override

Deciding between decision systems when internal conflicts arise is often viewed as an executive responsibility that requires flexibility. When the brain is presented with external conflicts requiring flexible evaluation of options, Deliberation is typically utilized (Kahneman, 2011; Mischel, 2014; Redish, Schultheiss, & Carter, 2016; Regier & Redish, 2015).

Accordingly, the Deliberative decision system has long been proposed to function as an arbitrator among internal decision processing (Augustine of Hippo (Saint Augustine), 1972; Gazzaniga, 2011; Haidt, 2006). Lending credence to this proposal, Deliberative decisions are slow, computationally intensive, and thus could reflect executive overhead, a stark contrast to the impulsive and inflexible nature of Pavlovian and Procedural responses.

Classic dual-system theories oriented systems hierarchically, with simpler systems functioning autonomously and complex (deliberative, cognitive) systems monitoring ongoing activity and overriding as needed (Bechara & van der Linden, 2005; Eagleman, 2011; Evans, 2008; Gray, 2004; Haidt, 2006; McClure & Bickel, 2014; Mischel, 2014; Redish, 2013). Extending this dual-system story to our three-system story, the resultant hybrid structure retains Deliberative processing atop the hierarchy where it adopts an executive role delegating between Pavlovian and Procedural systems (see Fig. 3). This proposed interaction has been a reigning hypothesis among behavioral economists studying intrapersonal and intertemporal choice models of self-control (Bechara & van der Linden, 2005; Bernheim & Rangel, 2004; Frederick, Loewenstein, & O'Donoghue, 2002; Loewenstein, 1996, 2000; McClure & Bickel, 2014; McClure, Laibson, Loewenstein, & Cohen, 2004; Mischel, 2014).

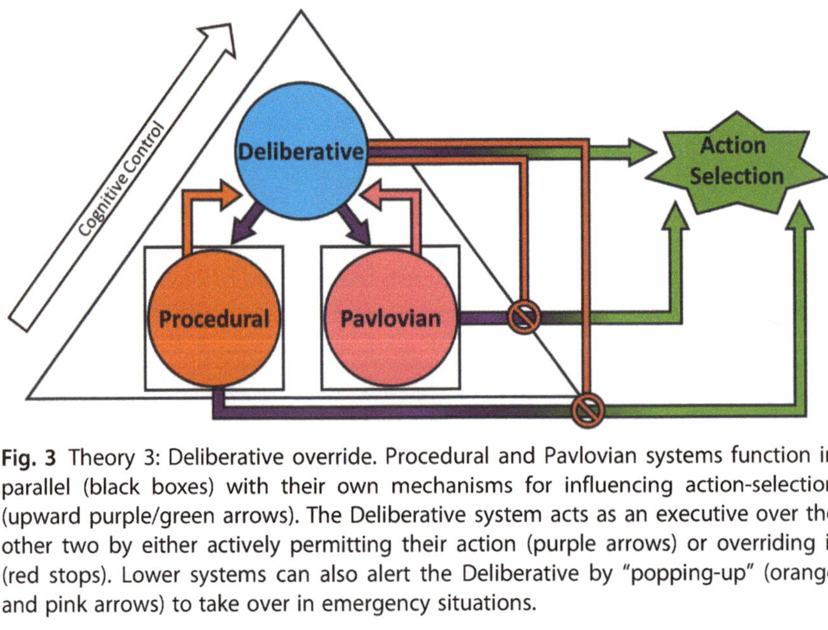

Fig. 3 Theory 3: Deliberative override. Procedural and Pavlovian systems function in parallel (black boxes) with their own mechanisms for influencing action-selection (upward purple/green arrows). The Deliberative system acts as an executive over the other two by either actively permitting their action (purple arrows) or overriding it (red stops). Lower systems can also alert the Deliberative by "popping-up" (orange and pink arrows) to take over in emergency situations.

From a neurobiology perspective, deliberative-override presents concrete mechanisms of interactions between decision systems that contrast with the two earlier theories. In particular the interactions between subsystems suggests three distinct mechanisms by which "lower-order" Pavlovian and Procedural systems could drive behavior:

1. The Deliberative executive could actively permit a lower system to function, such as in the case of indulging a craving for sweets.
2. The Deliberative executive could attempt to override a lower system but fail to do so, perhaps due to fatigue. Such a scenario may explain the feeling of "zoning-out" while driving and finding oneself inadvertently reverting to a well-worn route.
3. An emergency or highly unexpected situation could occur, such as a deer darting across a road, inducing a fast lower-order system to actively take control of decision-making and "pop-up" to overpower the Deliberative executive.

Importantly, these mechanisms provide distinct implications for the type of neural activity expected under different decision conflict scenarios, a topic which we will explore in the final section.

Despite its introspective support, the deliberative-override theory possesses notable shortcomings. In particular, the theory lacks sufficient explanation for the relationship between Pavlovian and Procedural decision-making. Are these two systems independently parallel or is there direct competitive interaction between them? Are conflicts between the two fast systems always filtered through the slow Deliberative system, even if Deliberation is otherwise not required or making a rapid decision is critical? The deliberative-override theory also rests on the premise that a decision hierarchy depends on the level of behavioral flexibility of the decision systems at the moment of action-selection.

4.4 Theory 4: Hierarchical override according to environment familiarity

Even in a completely novel situation, the ingrained nature of Pavlovian responses allow them to drive behavior. With more experience, an agent gradually learns enough about the world to develop an internal cognitive map. At this point Deliberation becomes useful for pursuing an optimal outcome based on abstract or changing goals. Under certain conditions, the same situations necessitate the same actions to reach the same outcomes. Learning these situation-action associations through repetition solidifies Procedural responding and allows for fast decision-making with low effort.

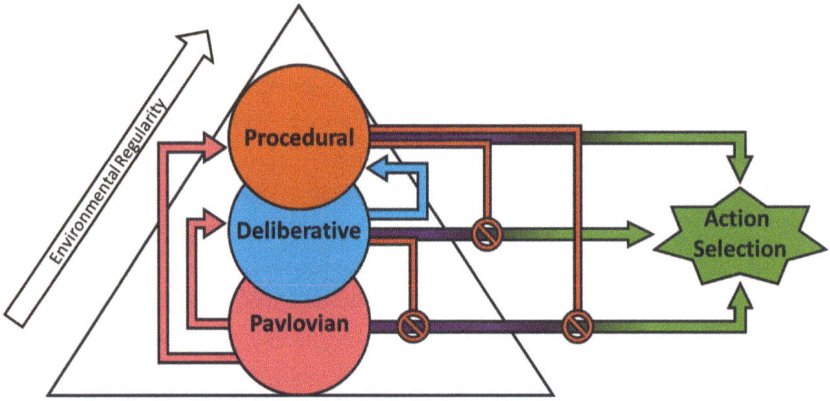

Fig. 4 Theory 4: Environmental regularity override. The three systems have their own mechanisms for influencing action-selection (purple/green arrows) that develop as environmental regularity is learned. The systems take on a hierarchical structure with Pavlovian processes driving responses under high uncertainty then Deliberative and Procedural systems override lower tiers (red stops) with increasing regularity. Instances where regularity in the environment diminishes, lower systems "pop-up" (pink and blue arrows) to take over.

This natural progression of experience with the external world and the corresponding maturation of the different decision systems suggests a novel (fourth) theory for integrating multiple decision systems: that decision systems operate in a hierarchical setting whereby Procedural action sits atop the hierarchy where it can override Deliberative processes, Deliberation sits above Pavlovian responding which it can override, and Pavlovian systems reside at the base (see Fig. 4). This theory is reminiscent of subsumption architecture, which may provide clues to its neural instantiation (Brooks, 1986). Importantly, in this theory, systems are only applicable in some scenarios, and so come online over time as experience and familiarization with the world drives the maturation of higher level decisions.

Of course even once habits are established, Pavlovian and Deliberative decision systems can drive behavior when situations change and a new world state must be learned. Additionally, unique scenarios can cause the decision systems to seemingly break their hierarchy. In a sudden moment of surprise or fear, behavior becomes driven by the Pavlovian system, ignoring any ongoing Deliberative or Procedural responses. Similarly, even during strongly ingrained habits, Deliberative precommitment can intervene to deviate from the default response. Much as in the case of the deliberative-override theory, these instances of "pop-up" whereby a lower hierarchical system dictates behavior can provide decision-making with additional flexibility to deal with rapid environment changes.

Mechanistically similar to the deliberative-override theory in its reliance on override by higher level systems and pop-up from lower ones, this new theory presents a key conceptual distinction. Deliberative-override centers on the highly flexible and energetically costly Deliberative system adopting a secondary executive role. In contrast, this new theory instead posits that interactions between decision systems are reflective of how these systems naturally adapt to and operate in new situations. Importantly, these two override based theories differ in their predictions for the underlying neurobiology of decision systems, particularly in regard to the role of mPFC. Discussion of predictions that arise from each of the four theories and how they can direct future study of mammalian decision-making are the focus of the final section.

5. Prefrontal cortex and combining decision systems

Given its long perceived role in high level cognitive functioning and executive control, it seems likely that mPFC has a central role in resolving conflicts among the brain's decision systems. However, which of the above strategies underlies how conflicts among the decisions systems are resolved has remained a difficult topic to study. By the very nature of complex decision-making, and the subtleties between the proposed conflict-resolution theories, only a small number of studies have provided good evidence to support any one idea over others, and no solid conclusions can yet be drawn. To gain significant traction in understanding how decision systems come together, and mPFC's role in this process, we believe future work must approach decision-making through the lens of multiple systems. This is a complex undertaking, but we identify several predictions of mPFC functioning that can begin to shed light on how decision systems may interact to produce a singular action. Although we believe these predictions can provide a starting point, they should be recognized as exactly that—simplified starting points that are sure to evolve and mature as data emerge.

Most notably, we believe future work into decision-making should adopt one of two approaches.

Either, behavioral tasks should aim to engage only a single decision system (Pavlovian, Deliberative, or Procedural) as relevant to performance, creating situations where the remaining two systems are explicitly irrelevant for the task. A valuable test in such circumstances would be to completely remove a central component hypothesized to be important to one or more of the systems and identify that the behavior is affected if and only if the system in question is involved. As it will be nearly impossible to construct behavioral tests that truly access only a single system, specialized behavioral analyses will likely be necessary to segregate decision systems. Classic work utilizing different training paradigms in the Morris Water Maze (Day, Weisend, Sutherland, & Schallert, 1999; Eichenbaum, Stewart, & Morris, 1990; McDonald & White, 1994; Redish, 1999) or localized lesions in the plus maze (Packard & McGaugh, 1992; Yin, Knowlton, & Balleine, 2002) may provide a good template. If such an approach can be successfully employed, it would allow for direct study of a single decision system's inner workings without the need to consider how multiple systems may influence or conflict with each other.

Alternatively, behavioral tasks should aim to engage two or even all three decision systems, ideally with discrete moments in which conflicts occur and behavior is directly reflective of the dominant decision system. Such tasks would present critical time points during which neural activity could be evaluated or manipulated and compared to a definable behavioral output (Gläscher, Daw, Dayan, & O'Doherty, 2010; Hasz & Redish, 2018; Papale, Stott, Powell, Regier, & Redish, 2012; Schmidt, Papale, Redish, & Markus, 2013; Sweis et al., 2018; van der Meer, Johnson, et al., 2010; van der Meer, van der Meer, et al., 2010).

We can identify three ways in which mPFC activity could differentially support the above theories for decision integration:

First, one could look for distinct physiologic signatures of individual decision systems localized to distinct subregions or subpopulations within mPFC.

Theory 1. The central executive theory predicts that segregated decision information should not be observed as the executive reflects an entity independent of the three decision systems.

Theory 2. The multiple experts theory predicts that independent streams should exist within mPFC, that these streams should be constituently active, but that each should only provide reliable action-selection information when driving behavior.

Theory 3. The deliberative-override theory identifies that independent streams of decision processing should exist among distinct subpopulations within the mPFC, but that there should be interactions such that the Deliberative stream can suppress the other two.

Theory 4. A hierarchical structure based on behavioral and environmental regularity predicts that independent streams should exist within mPFC and that hierarchical systems are actively engaged with higher systems becoming more active with more familiarity and behavioral regularity as lower systems fade away.

Current data may point toward distinct processing streams within mPFC (Killcross & Coutureau, 2003; Powell & Redish, 2016; Rich & Shapiro, 2009; Smith & Graybiel, 2013, 2016), though examples of a single region involved with processing of multiple decision systems can also be inferred (Do-Monte et al., 2015; Kim et al., 2018; Quirk et al., 2000; Sierra-Mercado et al., 2011).

Second, one could evaluate how manipulations of mPFC impact a decision response. We will address conflict resolution later; here, we address situations in which only a single system is engaged.

Theory 1. The central executive theory predicts that when behavior is driven by a single active system, the central executive would not be required and thus lesions or other experimental manipulations of mPFC would have no impact on behavior.

Theory 2. The multiple experts model predicts that the effect of manipulations to mPFC should be both dependent on which decision stream is impacted, and on the nature of the manipulation. If the qualitative nature of the neural signal were changed it would be expected to change the nature of the behavioral response, but if instead only the signal variability were altered it would have no behavioral consequence in the case of only a single active decision system.

Theory 3. The deliberative-override theory predicts that the impact of mPFC manipulation should be highly dependent on which subpopulation of mPFC is altered and which decision system is active. Substantive impacts on behavior should only occur when the affected subpopulation is the one involved in the decision system of interest.

Theory 4. Similar to deliberative-override, a regularity-based hierarchy predicts that impacts of mPFC manipulations should be highly dependent on the specific subpopulation affected and the respective decision situation.

Third, one could evaluate how manipulations of mPFC impact neural decision representations in other brain areas (amygdala, hippocampus, and dorsolateral striatum).

Theory 1. The central executive theory predicts that manipulation of mPFC would have no impact on downstream processing in other brain regions when only a single decision system is engaged, as the central executive is not needed.

Theory 2. The multiple experts theory also predicts mPFC manipulations should not influence processing in other brain regions. However, this is because other brain regions exist upstream from mPFC in decision processing in this theory.

Theory 3. The deliberative-override theory predicts (though not necessarily requires) that manipulation of mPFC would produce measurable impacts on decision representations elsewhere in the brain. Furthermore, this interregional impact would be localized within mPFC where specific subpopulations only impact specific brain regions.

Theory 4. A regularity-driven hierarchy would likewise predict any impacts from mPFC manipulations on other brain regions should be localized, and should depend on which subpopulation was affected.

To date, the prediction common to theories 3 and 4 has the best experimental support (Guise & Shapiro, 2017; Hok et al., 2013; Ito et al., 2015; Killcross & Coutureau, 2003; Rich & Shapiro, 2009; Schmidt et al., 2019; Smith et al., 2012).

Each theory also provides a set of predictions for multiple decision systems in conflict (see Table 1).

Theory 1. Under conditions in which multiple decision systems compete to drive behavior, the central executive theory relies on a single critical location (here proposed to be mPFC) actively relaying instructions directly to each decision system. Additionally, under this theory, the identities of the decision systems in conflict would determine which structures the central executive would be expected to communicate with. These premises elicit three prominent predictions.

First, activity in the central executive and different decision systems should be temporally linked, perhaps indicated by LFP coherence between mPFC and amygdala, hippocampus, dlStr (Adhikari et al., 2010; Benchenane et al., 2010; Hyman et al., 2010; Jones & Wilson, 2005; Kim et al., 2018).

Second, if the timing of connections between mPFC and an individual decision system were disrupted it would produce deficits in appropriately utilizing the respective system.

Table 1 Key features and predictions summary for future mPFC decision system experiments.

Theory features	Key predictions
1. *Central executive*	
Physiologically distinct executive system	No independent streams of decision system information would be observed in mPFC
Information flow is guided by empowering a single decision system to influence action	mPFC manipulations should not impact single decision behavior or downstream processing in other brain regions
All conflict resolution relies on the central executive	Any mPFC perturbation should lead to breakdowns in decision-making during internal conflict resolution
2. *Multiple experts*	
Decision systems are individually responsible for driving downstream structures	Parallel streams exist within mPFC that are consistently active and converge on a downstream target
Information flow is guided by changes in each stream's self-consistency	Impact of mPFC manipulation should depend on the altered subregion and the qualitative nature of the signal
A winner-take-all process determines action based on the stream whose information is most self-consistent	mPFC perturbation should not impact representations in other brain regions and should not result in total breakdown of conflict resolution
3. *Deliberative override*	
The deliberative decision system functions as a central executive by delegating between Pavlovian and Procedural systems	Information streams should exist within mPFC and are not consistently active
	Impact of mPFC manipulation should depend on the affected subregion and active decision system
Information flow is guided by deliberative override and lower-system pop-up mechanisms	There should be interactions between streams that show override, but not pop-up signatures of the deliberative system to suppress the lower other two
4. *Regularity-based hierarchy*	
Decision system hierarchy exists with Procedural at the top followed by Deliberative then Pavlovian at the base	Independent streams should exist within mPFC and higher systems become more active with more environmental familiarity
	Impact of mPFC manipulation depends on the affected subregion and the respective decision situation
Experience and familiarization with the world drives the maturation of decisions up the hierarchy and overrides lower system function	There should be interactions between streams that show both override and pop-up signatures of the Deliberative system to suppress lower Pavlovian and overcome higher Procedural systems
Unique scenarios can cause lower systems to pop-up and break the natural hierarchy	

Distinguishing physiological and behavioral predictions that are implied by the main computational features of each conflict-resolution theory.

Third, given that all conflict resolution relies on the central executive, any perturbation of the mPFC itself should produce a breakdown in all decision-making scenarios in which conflicts arise.

Theory 2. In the multiple experts model, distinct streams of decision processing should be present concurrently and should project to a single common source where the final action-selection can be derived based on the relative self-consistency of the streams. Accordingly, we present three predictions for experimental data.

First, distinct streams of decision information should be observable among mPFC subpopulations that share a common downstream target, and these representations should occur concurrently. Thus the requisite self-consistency information would arrive at a downstream premotor/motor center to be directly contrasted.

Second, the decision system that drives an agent's action should be directly predictable by comparing the internal self-consistency between these parallel information streams, a premise that has found experimental support (Lee, Shimojo, & O'Doherty, 2014).

Third, fine scale experimental manipulations of different aspects of decision representations within mPFC should produce different behavioral consequences. Specifically, altering the general nature of a decision signal while concurrently maintaining its overall level of self-consistency would be predicted to impact the specific action taken, but not the decision system that produced it (e.g., tweak to the Deliberative action but still a Deliberative approach). Alternatively, if an intervention reduced the self-consistency of a single system, perhaps through the addition of noise, it would decrease the likelihood of the affected decision system driving behavior.

Theories 3 and 4. Both of the proposed override-based theories resolve decision conflicts using similar mechanisms described here as *override* and *pop-up*. One prediction common to these two theories is that interactions between pairs of decision systems should be characterized by one of two fundamentally distinct physiologic signatures.

First, an "override" mechanism should be observable in which a higher status system exhibits prolonged activation, leading to suppression of activity within the corresponding lower status system.

Second, a "pop-up" mechanism should be observable in which the lower status system hijacks control. While we do not know how this pop-up would manifest physiologically, we expect it to be a relatively anomalous event that coincides with external stimuli that may indicate an emergency or signal the need for a dramatic change in decision processing.

Theories 3 and 4 differ largely in their utilization of override and pop-up mechanisms and in the implications for manipulations of individual decision systems.

Theory 3 predicts that Deliberation should override both Pavlovian and Procedural systems, while both Pavlovian and Procedural systems could pop-up over Deliberation. Expanding on this framework, behavioral actions linked with the Deliberative system would always be expected to coincide with elevated activity within Deliberative subpopulations of mPFC. Both behaviorally and physiologically, pop-up signatures should never be observed in association with Deliberative approaches. In parallel, instances of either Pavlovian or Procedural responding should coincide with a weak, and thus insufficient, override signal within the Deliberative system or in special circumstances with a pop-up signature in respective mPFC subpopulations. In regards to localized manipulations of individual decision systems, this theory predicts that disruption of Pavlovian or Procedural systems would have a restricted impact, affecting only the respective system. However, manipulations to the Deliberative executive could have profound consequences particularly during conflicts between the Pavlovian and Procedural systems. In such a case, lack of an override mediator would lead to a fundamental breakdown in general decision-making.

Theory 4 predicts that the occurrence of behavioral and physiological override and pop-up would match the hierarchical relationships among decision systems. Deliberation would override Pavlovian responding and the Procedural system would override both other decision systems. In parallel, Deliberation could be observed to pop-up over Procedural actions while Pavlovian responding could pop-up over both other systems. Importantly, override signatures would never be expected within the Pavlovian subpopulation of mPFC whereas pop-up is never expected from the Procedural subpopulation. Due to this hierarchical structure of interactions, disruption of any one decision system would be predicted to impair its ability to drive behavioral responding either in isolation or during conflict situations, but critically would have no substantive impact on the functioning or interactions of any remaining decision systems.

6. Conclusions

Decision-making is a central feature of how animals interact with their external environment. There are multiple decision systems that are computationally and neurally distinct. Though separable circuitry houses each decision system, current theories suggest that the rodent medial prefrontal cortex plays a central role in how these multiple systems come together to produce a single, unified action. Many theories have been developed to explain how integration occurs between multiple decision systems, and what structures may be involved in this process. Having clear, computational definitions for each decision subsystem has helped to derive mechanistic insights that link the psychological theories of decision-making to predictions about the underlying neurobiology particularly of the rodent medial prefrontal cortex. These predictions can frame experimental data and act as a guide for the future work necessary to fully understand how decisions are made and how mPFC activity relates to behavior. These competing theories have direct implications for what mPFC is doing for individual systems as well as multiple system integration. These theories point to different neurophysiological and behavioral predictions for which data can be collected and used to provide greater insight into mPFC's role in rodent action-selection, and thus into the internal dynamics of the decision process itself.

References

Adhikari, A., Topiwala, M. A., & Gordon, J. A. (2010). Synchronized activity between the ventral hippocampus and the medial prefrontal cortex during anxiety. *Neuron, 28*(65), 257–269.

Adhikari, A., Topwala, M., & Gordon, J. A. (2011). Single units in the medial prefrontal cortex with anxiety-related firing patterns are preferentially influenced by ventral hippocampal activity. *Neuron, 71*(5), 898–910.

Amemiya, S., & Redish, A. D. (2016). Manipulating decisiveness in decision making—Effects of clonidine on hippocampal search strategies. *Journal of Neuroscience, 36*(3), 814–827.

Amir, A., Lee, S.-C., Headley, D. B., Herzallah, M. M., & Pare, D. (2015). Amygdala signaling during foraging in a hazardous environment. *The Journal of Neuroscience: The Official Journal of the Society for Neuroscience, 35*(38), 12994–13005.

Augustine of Hippo (Saint Augustine). (1972). *The City of God* (H. Bettenson, Trans.). *Penguin Classics (Original work published 426).*

Balaguer-Ballester, E., Lapish, C. C., Seamans, J. K., & Durstewitz, D. (2011). Attracting dynamics of frontal cortex ensembles during memory-guided decision-making. *PLoS Computational Biology*, 7(5), e1002057.

Barker, J. M., Glen, W. B., Linsenbardt, D. N., Lapish, C. C., & Chandler, L. J. (2017). Habitual behavior is mediated by a shift in response-outcome encoding by infralimbic cortex. *eNeuro, 4*(6). https://doi.org/10.1523/ENEURO.0337-17.2017.

Barker, J. M., Taylor, J. R., & Chandler, L. J. (2014). A unifying model of the role of the infralimbic cortex in extinction and habits. *Learning & Memory, 21*(9), 441–448.

Barnes, T. D., Kubota, Y., Hu, D., Jin, D. Z., & Graybiel, A. M. (2005). Activity of striatal neurons reflects dynamic encoding and recoding of procedural memories. *Nature, 437*, 1158–1161.

Bechara, A., & van der Linden, M. (2005). Decision-making and impulse control after frontal lobe injuries. *Current Opinion in Neurology, 18*, 734–739.

Beckstead, R. M. (1979). An autoradiographic examination of corticocortical and subcortical projections of the mediodorsal-projection (prefrontal) cortex in the rat. *The Journal of Comparative Neurology, 184*(1), 43–62.

Benchenane, K., Peyrache, A., Khamassi, M., Tierny, P. L., Gioanni, Y., Battaglia, F. P., et al. (2010). Coherent theta oscillations and reorganization of spike timing in the hippocampal-prefrontal network upon learning. *Neuron, 66*(6), 921–936.

Berke, J. D., & Eichenbaum, H. (2009). Striatal versus hippocampal representations during win-stay maze performance. *Journal of Neurophysiology, 101*(3), 1575–1587.

Bernheim, B. D., & Rangel, A. (2004). Addiction and cue-triggered decision processes. *The American Economic Review, 94*(5), 1558–1590.

Bishop, C. M. (2006). *Pattern recognition and machine learning*. Springer.

Bouton, M. E. (2007). *Learning and behavior: A contemporary synthesis*. Sinauer Associates.

Bravo-Rivera, C., Roman-Ortiz, C., Montesinos-Cartagena, M., & Quirk, G. J. (2015). Persistent active avoidance correlates with activity in prelimbic cortex and ventral striatum. *Frontiers in Behavioral Neuroscience, 9*, 184. https://doi.org/10.3389/fnbeh.2015.00184.

Breland, K., & Breland, M. (1961). The misbehavior of organisms. *The American Psychologist, 16*(11), 682–684.

Brooks, R. (1986). A robust layered control system for a mobile robot. *IEEE Journal on Robotics and Automation, 2*(1), 14–23.

Bussey, T. J., Everitt, B. J., & Robbins, T. W. (1997). Dissociable effects of cingulate and medial frontal cortex lesions on stimulus-reward learning using a novel Pavlovian autoshaping procedure for the rat: Implications for the neurobiology of emotion. *Behavioral Neuroscience, 111*(5), 908–919.

Caracheo, B. F., Grewal, J. J. S., & Seamans, J. K. (2018). Persistent valence representations by ensembles of anterior cingulate cortex neurons. *Frontiers in Systems Neuroscience, 12*, 51.

Cardinal, R. N., Parkinson, J. A., Marbini, H. D., Toner, A. J., Bussey, T. J., Robbins, T. W., et al. (2003). Role of the anterior cingulate cortex in the control over behavior by Pavlovian conditioned stimuli in rats. *Behavioral Neuroscience, 117*(3), 566–587.

Cei, A., Girardeau, G., Drieu, C., El Knabi, K., & Zugaro, M. (2014). Reversed theta sequences of hippocampal cell assemblies during backward travel. *Nature Neuroscience, 17*, 719–724.

Choi, J.-S., & Kim, J. J. (2010). Amygdala regulates risk of predation in rats foraging in a dynamic fear environment. *Proceedings of the National Academy of Sciences of the United States of America, 107*(50), 21773–21777.

Chudasama, Y., & Robbins, T. W. (2003). Dissociable contributions of the orbitofrontal and infralimbic cortex to Pavlovian autoshaping and discrimination reversal learning: Further evidence for the functional heterogeneity of the rodent frontal cortex. *Journal of Neuroscience, 23*(25), 8771–8780.

Cochran, A. L., & Cisler, J. M. (2019). A flexible and generalizable model of online latent-state learning. *PLoS Computational Biology, 15*(9), e1007331.

Corcoran, K. A., & Quirk, G. J. (2007). Activity in prelimbic cortex is necessary for the expression of learned, but not innate, fears. *The Journal of Neuroscience: The Official Journal of the Society for Neuroscience, 27*(4), 840–844.

Coutureau, E., & Killcross, S. (2003). Inactivation of the infralimbic prefrontal cortex reinstates goal-directed responding in overtrained rats. *Behavioural Brain Research, 146*, 167–174.

Dalley, J. W., Cardinal, R. N., & Robbins, T. W. (2004). Prefrontal executive and cognitive functions in rodents: Neural and neurochemical substrates. *Neuroscience and Biobehavioral Reviews, 28*(7), 771–784.

Dalton, G. L., Wang, N. Y., Phillips, A. G., & Floresco, S. B. (2016). Multifaceted contributions by different regions of the orbitofrontal and medial prefrontal cortex to probabilistic reversal learning. *The Journal of Neuroscience: The Official Journal of the Society for Neuroscience*, *36*(6), 1996–2006.

Daw, N. D., Niv, Y., & Dayan, P. (2005). Uncertainty-based competition between prefrontal and dorsolateral striatal systems for behavioral control. *Nature Neuroscience*, *8*, 1704–1711.

Day, L. B., Weisend, M., Sutherland, R. J., & Schallert, T. (1999). The hippocampus is not necessary for a place response but may be necessary for pliancy. *Behavioral Neuroscience*, *113*(5), 914–924.

Dayan, P., Niv, Y., Seymour, B., & Daw, N. D. (2006). The misbehavior of value and the discipline of the will. *Neural Networks: The Official Journal of the International Neural Network Society*, *19*(8), 1153–1160.

de Bruin, J. P. C., Corner, M. A., Feenstra, M. G. P., Van Eden, C. G., & Uylings, H. B. M. (1991). *The prefrontal cortex: Its structure, function and pathology*. Elsevier.

de Wit, S., Kosaki, Y., Balleine, B. W., & Dickinson, A. (2006). Dorsomedial prefrontal cortex resolves response conflict in rats. *The Journal of Neuroscience: The Official Journal of the Society for Neuroscience*, *26*(19), 5224–5229.

Dezfouli, A., & Balleine, B. (2012). Habits, action sequences and reinforcement learning. *The European Journal of Neuroscience*, *35*(7), 1036–1051.

Diehl, M. M., Bravo-Rivera, C., & Quirk, G. J. (2019). The study of active avoidance: A platform for discussion. *Neuroscience & Biobehavioral Reviews*, *107*, 229–237. https://doi.org/10.1016/j.neubiorev.2019.09.010.

Diehl, M. M., Bravo-Rivera, C., Rodriguez-Romaguera, J., Pagan-Rivera, P. A., Burgos-Robles, A., Roman-Ortiz, C., et al. (2018). Active avoidance requires inhibitory signaling in the rodent prelimbic prefrontal cortex. *eLife*, 7, e34657. https://doi.org/10.7554/eLife.34657.

Do-Monte, F. H., Quiñones-Laracuente, K., & Quirk, G. J. (2015). A temporal shift in the circuits mediating retrieval of fear memory. *Nature*, *519*(7544), 460–463.

Durstewitz, D., Vittoz, N. M., Floresco, S. B., & Seamans, J. K. (2010). Abrupt transitions between prefrontal neural ensemble states accompany behavioral transitions during rule learning. *Neuron*, *66*(3), 438–448.

Eagleman, D. (2011). *Incognito: The secret lives of the brain*. Vintage.

Eichenbaum, H., Stewart, C., & Morris, R. G. (1990). Hippocampal representation in place learning. *The Journal of Neuroscience: The Official Journal of the Society for Neuroscience*, *10*(11), 3531–3542.

Engel, A. K., Fries, P., & Singer, W. (2001). Dynamic predictions: Oscillations and synchrony in top-down processing. *Nature Reviews. Neuroscience*, *2*(10), 704–716.

Euston, D. R., Gruber, A. J., & McNaughton, B. L. (2012). The role of medial prefrontal cortex in memory and decision making. *Neuron*, *76*, 1057–1070.

Evans, J. S. B. T. (2008). Dual-processing accounts of reasoning, judgment, and social cognition. *Annual Review of Psychology*, *59*, 255–278.

Fisher, B. M., Saksida, L. M., Robbins, T. W., & Bussey, T. J. (2020). Functional dissociations between subregions of the medial prefrontal cortex on the rodent touchscreen continuous performance test (rCPT) of attention. *Behavioral Neuroscience*, *134*(1), 1–14.

Frederick, S., Loewenstein, G., & O'Donoghue, T. (2002). Time discounting and time preference: A critical review. *Journal of Economic Literature*, *40*(2), 351–401.

Freud, S. (1923). *The ego and the Id*. W. W. Norton.

Friedman, A., Homma, D., Gibb, L. G., Amemori, K.-I., Rubin, S. J., Hood, A. S., et al. (2015). A corticostriatal path targeting striosomes controls decision-making under conflict. *Cell*, *161*(6), 1320–1333. https://doi.org/10.1016/j.cell.2015.04.049.

Friedman, A., Jomma, D., Gibb, L. G., Amemori, K.-I., Rubin, S. J., Hood, A. S., et al. (2015). A corticostraital path targeting striosomes controls decision-making under conflict. *Cell, 161*, 132–1333.

Fries, P. (2005). A mechanism for cognitive dynamics: Neuronal communication through neuronal coherence. *Trends in Cognitive Sciences, 9*(10), 474–480.

Fuster, J. M. (1997). *The prefrontal cortex: Anatomy, physiology, and neuropsychology of the frontal lobe* (3rd ed.). Lippincott-Raven.

Fuster, J. M. (2008). *The prefrontal cortex* (4th ed.). London, UK: Elsevier/Academic Press.

Gazzaniga, M. S. (2011). *Whos in charge. Free will and the science of the brain.* New York: Ecco.

Gershman, S. J., & Niv, Y. (2010). Learning latent structure: Carving nature at its joints. *Current Opinion in Neurobiology, 20*(2), 251–256.

Gilbert, D. T., & Wilson, T. D. (2007). Prospection: Experiencing the future. *Science, 317*(5843), 1351–1354.

Gläscher, J., Daw, N., Dayan, P., & O'Doherty, J. P. (2010). States versus rewards: Dissociable neural prediction error signals underlying model-based and model-free reinforcement learning. *Neuron, 66*(4), 585–595.

Gray, J. A. (2004). *Consciousness: Creeping up on the hard problem.* Oxford University Press.

Graybiel, A. M. (1998). The basal ganglia and chunking of action repertoires. *Neurobiology of Learning and Memory, 70*, 119–136.

Graybiel, A. M. (2008). Habits, rituals, and the evaluative brain. *Annual Review of Neuroscience, 31*, 359–387.

Groenewegen, H. J., & Uylings, H. B. (2000). The prefrontal cortex and the integration of sensory, limbic and autonomic information. *Progress in Brain Research, 126*, 3–28.

Guise, K. G., & Shapiro, M. L. (2017). Medial prefrontal cortex reduces memory interference by modifying hippocampal encoding. *Neuron, 94*(1), 183–192.e8.

Gupta, A. S., van der Meer, M. A. A., Touretzky, D. S., & Redish, A. D. (2012). Segmentation of spatial experience by hippocampal theta sequences. *Nature Neuroscience, 15*, 1032–1039.

Haber, S. N., Fudge, J. L., & McFarland, N. R. (2000). Striatonigrostriatal pathways in primates form an ascending spiral from the shell to the dorsolateral striatum. *Journal of Neuroscience, 20*(6), 2369–2382.

Haidt, J. (2006). *The happiness hypothesis.* Basic Books.

Hart, E. E., Blair, G. J., O'Dell, T. J., Blair, H. T., & Izquierdo, A. (2020). Chemogenetic modulation and single-photon calcium imaging in anterior cingulate cortex reveal a mechanism for effort-based decisions. *The Journal of Neuroscience: The Official Journal of the Society for Neuroscience, 40*(29), 5628–5643.

Hasz, B. M., & Redish, A. D. (2018). Deliberation and procedural automation on a two-step task for rats. *Frontiers in Integrative Neuroscience, 12*, 30.

Hasz, B. M., & Redish, A. D. (2020a). Dorsomedial prefrontal cortex and hippocampus represent strategic context even while simultaneously changing representation throughout a task session. *Neurobiology of Learning and Memory, 171*, 107215.

Hasz, B., & Redish, A. D. (2020b). Spatial encoding in dorsomedial prefrontal cortex and hippocampus is related during deliberation. *Hippocampus, 30*, 1194–1208.

Heidbreder, C. A., & Groenewegen, H. J. (2003). The medial prefrontal cortex in the rat: Evidence for a dorso-ventral distinction based upon functional and anatomical characteristics. *Neuroscience and Biobehavioral Reviews, 27*(6), 555–579.

Heilbronner, S. R., Rodriguez-Romaguera, J., Quirk, G. J., Groenewegen, H. J., & Haber, S. N. (2016). Circuit based cortico-striatal homologies between rat and primaterat-primate cortico-striatal homologies. *Biological Psychiatry, 80*, 509–521.

Hertz, J., Krogh, A., & Palmer, R. G. (1991). *Introduction to the theory of neural computation.* Addison-Wesley.

Hillman, K. L., & Bilkey, D. K. (2010). Neurons in the rat anterior cingulate cortex dynamically encode cost-benefit in a spatial decision-making task. *The Journal of Neuroscience: The Official Journal of the Society for Neuroscience, 30*(22), 7705–7713.

Hofmann, W., Schmeichel, B. J., & Baddeley, A. D. (2012). Executive functions and self-regulation. *TICS, 16*(3), 174–180.

Hok, V., Chah, E., Save, E., & Poucet, B. (2013). Prefrontal cortex focally modulates hippocampal place cell firing patterns. *The Journal of Neuroscience: The Official Journal of the Society for Neuroscience, 33*(8), 3443–3451.

Hoover, W. B., & Vertes, R. P. (2007). Anatomical analysis of afferent projections to the medial prefrontal cortex in the rat. *Brain Structure and Function, 212*(2), 149–179. https://doi.org/10.1007/s00429-007-0150-4.

Horst, N., & Laubach, M. (2012). Working with memory: Evidence for a role for the medial prefrontal cortex in performance monitoring during spatial delayed alternation. *Journal of Neurophysiology, 108*, 3276–3288.

Horst, N. K., & Laubach, M. (2013). Reward-related activity in the medial prefrontal cortex is driven by consumption. *Frontiers in Neuroscience*, 7, 56. https://doi.org/10.3389/fnins.2013.00056.

Hull, C. L. (1943). *Principles of behavior*. Appleton-Century-Crofts.

Hyman, J. M., Holroyd, C. B., & Seamans, J. K. (2017). A novel neural prediction error found in anterior cingulate cortex ensembles. *Neuron, 95*(2), 447–456.e3.

Hyman, J. M., Ma, L., Balaguer-Ballester, E., Durstewitz, D., & Seamans, J. K. (2012). Contextual encoding by ensembles of medial prefrontal cortex neurons. *Proceedings of the National Academy of Sciences of the United States of America, 109*(13), 5086–5091.

Hyman, J. M., Whitman, J., Emberly, E., Woodward, T. S., & Seamans, J. K. (2013). Action and outcome activity state patterns in the anterior cingulate cortex. *Cerebral Cortex, 23*(6), 1257–1268.

Hyman, J. M., Zilli, E. A., Paley, A. M., & Hasselmo, M. E. (2010). Working memory performance correlates with prefrontal-hippocampal theta interactions but not with prefrontal neuron firing rates. *Frontiers in Integrative Neuroscience, 4*, 2.

Ito, H. T., Moser, E. I., & Moser, M.-B. (2018). Supramammillary nucleus modulates spike-time coordination in the prefrontal-thalamo-hippocampal circuit during navigation. *Neuron, 99*(3), 576–587.e5.

Ito, H. T., Zhang, S.-J., Witter, M. P., Moser, E. I., & Moser, M.-B. (2015). A prefrontal-thalamo-hippocampal circuit for goal-directed spatial navigation. *Nature, 522*, 50–55.

Jackson, J. C., & Redish, A. D. (2003). Detecting dynamical changes within a simulated neural ensemble using a measure of representational quality. *Network: Computation in Neural Systems, 14*, 629–645.

Jadhav, S. P., Rothschild, G., Roumis, D. K., & Frank, L. M. (2016). Coordinated excitation and inhibition of prefrontal ensembles during awake hippocampal sharp-wave ripple events. *Neuron, 90*(1), 113–127.

Jay, T. M., & Witter, M. P. (1991). Distribution of hippocampal CA1 and subicular efferents in the prefrontal cortex of the rat studied by means of anterograde transport of Phaseolus vulgaris-leucoagglutinin. *The Journal of Comparative Neurology, 313*(4), 574–586.

Jog, M. S., Kubota, Y., Connolly, C. I., Hillegaart, V., & Graybiel, A. M. (1999). Building neural representations of habits. *Science, 286*, 1746–1749.

Johnson, A., Jackson, J., & Redish, A. D. (2008). Measuring distributed properties of neural representations beyond the decoding of local variables—Implications for cognition. In C. Hölscher, & M. H. J. Munk (Eds.), *Mechanisms of information processing in the brain: Encoding of information in neural populations and networks* (pp. 95–119). Cambridge University Press.

Johnson, A., & Redish, A. D. (2007). Neural ensembles in CA3 transiently encode paths forward of the animal at a decision point. *Journal of Neuroscience, 27*(45), 12176–12189.

Johnson, A., van der Meer, M. A. A., & Redish, A. D. (2007). Integrating hippocampus and striatum in decision-making. *Current Opinion in Neurobiology*, *17*(6), 692–697.

Jones, M. W., & Wilson, M. A. (2005). Theta rhythms coordinate hippocampal-prefrontal interactions in a spatial memory task. *PLoS Biology*, *3*(12), e402.

Kahneman, D. (2011). *Thinking, fast, and slow*. Farrar, Straus and Giroux.

Kay, K., Chung, J. E., Sosa, M., Schor, J. S., Karlsson, M. P., Larkin, M. C., et al. (2020). Constant sub-second cycling between representations of possible futures in the hippocampus. *Cell*, *180*(3), 552–567.e25.

Keramati, M., Dezfouli, A., & Piray, P. (2011). Speed/accuracy trade-off between the habitual and the goal-directed processes. *PLoS Computational Biology*, 7(5), e1002055.

Kesner, R. P., & Churchwell, J. C. (2011). An analysis of rat prefrontal cortex in mediating executive function. *Neurobiology of Learning and Memory*, *96*(3), 417–431.

Killcross, S., & Coutureau, E. (2003). Coordination of actions and habits in the medial prefrontral cortex of rats. *Cerebral Cortex*, *13*(8), 400–408.

Kim, J. J., & Jung, M. W. (2018). Fear paradigms: The times they are a-changin'. *Current Opinion in Behavioral Sciences*, *24*, 38–43.

Kim, E. J., Kong, M.-S., Park, S. G., Mizumori, S. J. Y., Cho, J., & Kim, J. J. (2018). Dynamic coding of predatory information between the prelimbic cortex and lateral amygdala in foraging rats. *Science Advances*, *4*(4), eaar7328.

Kim, J. J., Rison, R. A., & Fanselow, M. S. (1993). Effects of amygdala, hippocampus, and periaqueductal gray lesions on short- and long-term contextual fear. *Behavioral Neuroscience*, *107*(6), 1093–1098.

Kolb, B. (1990). Prefrontal cortex. In Kolb, & R. C. Tees (Eds.), *The cerebral cortex of the rat* (pp. 437–458). MIT Press.

Lapish, C. C., Durstewitz, D., Chandler, L. J., & Seamans, J. K. (2008). Successful choice behavior is associated with distinct and coherent network states in anterior cingulate cortex. *Proceedings of the National Academy of Sciences of the United States of America*, *105*(33), 11963–11968.

Laubach, M. (2011). A comparative perspective on executive and motivational control by the medial prefrontal cortex. In R. B. Mars, J. Sallet, M. F. S. Rushworth, & N. Yeung (Eds.), *Neural basis of motivational and cognitive control* (pp. 95–109). Cambridge, MA: MIT Press. https://doi.org/10.7551/mitpress/9780262016438.003.0006.

Laubach, M., Amarante, L. M., Swanson, K., & White, S. R. (2018). What, if anything, is rodent prefrontal cortex? *eNeuro*, *5*(5). ENEURO.0315-18.2018 https://doi.org/10.1523/ENEURO.0315-18.2018.

Lebron, K., Milad, M. R., & Quirk, G. J. (2004). Delayed recall of fear extinction in rats with lesions of ventral medial prefrontal cortex. *Learning and Memory*, *11*(5), 544–548.

LeDoux, J. E. (2000). Emotion circuits in the brain. *Annual Review of Neuroscience*, *23*, 155–184.

LeDoux, J. E. (2015). *Anxious: Using the brain to understand and treat fear and anxiety*. Penguin.

LeDoux, J., & Daw, N. D. (2018). Surviving threats: Neural circuit and computational implications of a new taxonomy of defensive behaviour. *Nature Reviews. Neuroscience*, *19*(5), 269–282.

Lee, S. W., Shimojo, S., & O'Doherty, J. P. (2014). Neural computations underlying arbitration between model-based and model-free learning. *Neuron*, *81*(3), 687–699.

Loewenstein, G. (1996). Out of control: Visceral influences on behavior. *Organizational Behavior and Human Decision Processes*, *65*(3), 272–292.

Loewenstein, G. (2000). Emotions in economic theory and economic behavior. *The American Economic Review*, *90*(2), 426–432.

Ma, L., Hyman, J. M., Durstewitz, D., Phillips, A. G., & Seamans, J. K. (2016). A quantitative analysis of context-dependent remapping of medial frontal cortex neurons and ensembles. *The Journal of Neuroscience: The Official Journal of the Society for Neuroscience*, *36*(31), 8258–8272.

Mashhoori, A., Hashemnia, S., McNaughton, B. L., Euston, D. R., & Gruber, A. J. (2018). Rat anterior cingulate cortex recalls features of remote reward locations after disfavoured reinforcements. *eLife*, 7, e29793. https://doi.org/10.7554/eLife.29793.

McClelland, J. L., & Rumelhart, D. E. (Eds.). (1986). PDP: Explorations in the microstructures of cognition. In *Vol. 2. Psychological and biological models* MIT Press.

McClure, S. M., & Bickel, W. K. (2014). A dual-systems perspective on addiction: Contributions from neuroimaging and cognitive training. *Annals of the New York Academy of Sciences*, *1327*(1), 62–78.

McClure, S. M., Laibson, D. I., Loewenstein, G., & Cohen, J. D. (2004). Separate neural systems value immediate and delayed monetary rewards. *Science*, *306*(5695), 503–507.

McDonald, R. J., & White, N. M. (1994). Parallel information processing in the water maze: Evidence for independent memory systems involving dorsal striatum and hippocampus. *Behavioral and Neural Biology*, *61*(3), 260–270.

McNally, G. P., Johansen, J. P., & Blair, H. T. (2011). Placing prediction into the fear circuit. *Trends in Neurosciences*, *34*, 283–292.

Milad, M. R., & Quirk, G. J. (2002). Neurons in medial prefrontal cortex signal memory for fear extinction. *Nature*, *420*, 70–74.

Miller, E. K., & Cohen, J. D. (2001). An integrative theory of prefrontal cortex function. *Annual Review of Neuroscience*, *24*, 167–202.

Mischel, W. (2014). *The marshmallow test: Mastering self-control*. Little, Brown, and Co.

Mishkin, M., & Appenzeller, T. (1987). The anatomy of memory. *Scientific American*, *256*(6), 80–89.

Mitchell, T. M. (1997). *Machine learning*. McGraw-Hill.

Mobbs, D., & Kim, J. J. (2015). Neuroethological studies of fear, anxiety, and risky decision-making in rodents and humans. *Current Opinion in Behavioral Sciences*, *5*, 8–15.

Molinari, M., Leggio, M. G., Solida, A., Ciorra, R., Misciagna, S., Silveri, M. C., et al. (1997). Cerebellum and procedural learning: Evidence from focal cerebellar lesions. *Brain: A Journal of Neurology*, *120*(Pt. 10), 1753–1762.

Montaron, M. F., Deniau, J. M., Menetrey, A., Glowinski, J., & Thierry, A. M. (1996). Prefrontal cortex inputs of the nucleus accumbens-nigro-thalamic circuit. *Neuroscience*, *71*(2), 371–382.

Morgan, M. A., & LeDoux, J. E. (1995). Differential contribution of dorsal and ventral medial prefrontal cortex to the acquisition and extinction of conditioned fear in rats. *Behavioral Neuroscience*, *109*(4), 681–688.

Mukherjee, A., & Caroni, P. (2018). Infralimbic cortex is required for learning alternatives to prelimbic promoted associations through reciprocal connectivity. *Nature Communications*, *9*(1), 2727.

Narayanan, N. S., Horst, N. K., & Laubach, M. (2006). Reversible inactivations of rat medial prefrontal cortex impair the ability to wait for a stimulus. *Neuroscience*, *139*(3), 865–876.

Niv, Y., Joel, D., & Dayan, P. (2006). A normative perspective on motivation. *Trends in Cognitive Sciences*, *10*(8), 375–381.

O'Keefe, J., & Nadel, L. (1978). *The hippocampus as a cognitive map*. Clarendon Press.

O'Keefe, J., & Speakman, A. (1987). Single unit activity in the rat hippocampus during a spatial memory task. *Experimental Brain Research. Experimentelle Hirnforschung. Experimentation Cerebrale*, *68*, 1–27.

Ostlund, S. B., & Balleine, B. W. (2009). Goal-directed behavior theories. In L. R. Squire (Ed.), *Encyclopedia of neuroscience* (pp. 943–949). Cambridge MA: Academic Press. https://doi.org/10.1016/b978-008045046-9.01570-9.

Otto, A. R., Gershman, S. J., Markman, A. B., & Daw, N. D. (2013). The curse of planning: Dissecting multiple reinforcement-learning systems by taxing the central executive. *Psychological Science*, *24*(5), 751–761.

Packard, M. G., & McGaugh, J. L. (1992). Double dissociation of fornix and caudate nucleus lesions on acquisition of two water maze tasks: Further evidence for multiple memory systems. *Behavioral Neuroscience*, *106*(3), 439–446.

Padilla-Coreano, N., Bolkan, S. S., Pierce, G. M., Blackman, D. R., Hardin, W. D., Garcia-Garcia, A. L., et al. (2016). Direct ventral hippocampal-prefrontal input is required for anxiety-related neural activity and behavior. *Neuron*, *89*(4), 857–866.

Papale, A. E., Stott, J. J., Powell, N. J., Regier, P. S., & Redish, A. D. (2012). Interactions between deliberation and delay-discounting in rats. *Cognitive, Affective, & Behavioral Neuroscience*, *12*(3), 513–526.

Papale, A. E., Zielinski, M. C., Frank, L. M., Jadhav, S. P., & Redish, A. D. (2016). Interplay between hippocampal sharp-wave-ripple events and vicarious trial and error behaviors in decision making. *Neuron*, *92*(5), 975–982.

Parkinson, J. A., Willoughby, P. J., Robbins, T. W., & Everitt, B. J. (2000). Disconnection of the anterior cingulate cortex and nucleus accumbens core impairs Pavlovian approach behavior: Further evidence for limbic cortical–ventral striatopallidal systems. *Behavioral Neuroscience*, *114*(1), 42–63.

Pavlov, I. (1927). *Conditioned reflexes*. Oxford University Press.

Pfeiffer, B. E., & Foster, D. J. (2013). Hippocampal place-cell sequences depict future paths to remembered goals. *Nature*, *497*, 74–79.

Plato. (2008). *Phaedrus* (B. Jowett, Trans.). *Project Gutenberg Plato (Original work published 370 BCE)*.

Powell, N. J., & Redish, A. D. (2014). Complex neural codes in rat prelimbic cortex are stable across days on a spatial decision task. *Frontiers in Behavioral Neuroscience*, *8*, 120.

Powell, N. J., & Redish, A. D. (2016). Representational changes of latent strategies in rat medial prefrontal cortex precede changes in behaviour. *Nature Communications*, 7, 12830.

Quirk, G. J., Russo, G. K., Barron, J. L., & Lebron, K. (2000). The role of ventromedial prefrontal cortex in the recovery of extinguished fear. *The Journal of Neuroscience: The Official Journal of the Society for Neuroscience*, *20*(16), 6225–6231.

Ragozzino, M. E., Wilcox, C., Raso, M., & Kesner, R. P. (1999). Involvement of rodent prefrontal cortex subregions in strategy switching. *Behavioral Neuroscience*, *113*, 32–41.

Rand, M. K., Hikosaka, O., Miyachi, S., Lu, X., & Miyashita, K. (1998). Characteristics of a long-term procedural skill in the monkey. *Experimental Brain Research. Experimentelle Hirnforschung. Experimentation Cerebrale*, *118*, 293–297.

Redish, A. D. (1999). *Beyond the cognitive map: From place cells to episodic memory*. MIT Press.

Redish, A. D. (2013). *The mind within the brain: How we make decisions and how those decisions go wrong*. Oxford.

Redish, A. D. (2016). Vicarious trial and error. *Nature Reviews. Neuroscience*, *17*(3), 147–159.

Redish, A. D., Jensen, S., Johnson, A., & Kurth-Nelson, Z. (2007). Reconciling reinforcement learning models with behavioral extinction and renewal: Implications for addiction, relapse, and problem gambling. *Psychological Review*, *114*(3), 784–805.

Redish, A. D., Schultheiss, N. W., & Carter, E. C. (2016). The computational complexity of valuation and motivational forces in decision-making processes. *Current Topics in Behavioral Neurosciences*, *27*, 313–333.

Regier, P. S., Amemiya, S., & Redish, A. D. (2015). Hippocampus and subregions of the dorsal striatum respond differently to a behavioral strategy change on a spatial navigation task. *Journal of Neurophysiology*, *114*, 1399–1416.

Regier, P. S., & Redish, A. D. (2015). Contingency management and deliberative decision-making processes. *Frontiers in Psychiatry/Frontiers Research Foundation*, *6*, 0076.

Riaz, S., Puveendrakumaran, P., Khan, D., Yoon, S., Hamel, L., & Ito, R. (2019). Prelimbic and infralimbic cortical inactivations attenuate contextually driven discriminative responding for reward. *Scientific Reports*, *9*(1), 3982.

Rich, E. L., & Shapiro, M. L. (2007). Prelimbic/infralimbic inactivation impairs memory for multiple task switches, but not flexible selection of familiar tasks. *The Journal of Neuroscience: The Official Journal of the Society for Neuroscience*, *27*(17), 4747–4755.

Rich, E. L., & Shapiro, M. (2009). Rat prefrontal cortical neurons selectively code strategy switches. *The Journal of Neuroscience: The Official Journal of the Society for Neuroscience*, *29*(22), 7208–7219.

Rosenzweig, E. S., Redish, A. D., McNaughton, B. L., & Barnes, C. A. (2003). Hippocampal map realignment and spatial learning. *Nature Neuroscience*, *6*(6), 609–615.

Rumelhart, D. E., & McClelland, J. L. (Eds.). (1986). PDP: Explorations in the microstructures of cognition. In *Vol. 1. Foundations* MIT Press.

Schmidt, B., Duin, A. A., & Redish, A. D. (2019). Disrupting the medial prefrontal cortex alters hippocampal sequences during deliberative decision making. *Journal of Neurophysiology*, *121*(6), 1981–2000.

Schmidt, B. J., Papale, A. E., Redish, A. D., & Markus, E. J. (2013). Conflict between place and response navigation strategies: Effects on vicarious trial and error (VTE) behaviors. *Learning and Memory*, *20*, 130–138.

Schmitzer-Torbert, N. C., & Redish, A. D. (2004). Neuronal activity in the rodent dorsal striatum in sequential navigation: Separation of spatial and reward responses on the multiple-T task. *Journal of Neurophysiology*, *91*(5), 2259–2272.

Seamans, J. K., & Yang, C. R. (2004). The principal features and mechanisms of dopamine modulation in the prefrontal cortex. *Progress in Neurobiology*, *74*, 1–57.

Sesack, S. R., Deutch, A. Y., Roth, R. H., & Bunney, B. S. (1989). Topographical organization of the efferent projections of the medial prefrontal cortex in the rat: An anterograde tract-tracing study with Phaseolus vulgaris leucoagglutinin. *The Journal of Comparative Neurology*, *290*(2), 213–242.

Sharpe, M. J., & Killcross, S. (2015). The prelimbic cortex directs attention toward predictive cues during fear learning. *Learning & Memory*, *22*(6), 289–293.

Sharpe, M. J., Wikenheiser, A. M., Niv, Y., & Schoenbaum, G. (2015). The state of the orbitofrontal cortex [review of the state of the orbitofrontal cortex]. *Neuron*, *88*(6), 1075–1077.

Sierra-Mercado, D., Padilla-Coreano, N., & Quirk, G. J. (2011). Dissociable roles of prelimbic and infralimbic cortices, ventral hippocampus, and basolateral amygdala in the expression and extinction of conditioned fear. *Neuropsychopharmacology: Official Publication of the American College of Neuropsychopharmacology*, *36*(2), 529–538.

Smith, K. S., & Graybiel, A. M. (2013). A dual operator view of habitual behavior reflecting cortical and striatal dynamics. *Neuron*, *79*(2), 361–374.

Smith, K. S., & Graybiel, A. M. (2016). Habit formation coincides with shifts in reinforcement representations in the sensorimotor striatum. *Journal of Neurophysiology*, *115*(3), 1487–1498.

Smith, K. S., Virkud, A., Deisseroth, K., & Graybiel, A. M. (2012). Reversible online control of habitual behavior by optogenetic perturbation of medial prefrontal cortex. *Proceedings of the National Academy of Sciences of the United States of America*, *109*(46), 18932–18937.

Steiner, A., & Redish, A. D. (2012). Orbitofrontal cortical ensembles during deliberation and learning on a spatial decision-making task. *Frontiers in Decision Neuroscience*, *6*, 131.

Swanson, L. W. (1981). A direct projection from Ammon's horn to prefrontal cortex in the rat. *Brain Research*, *217*(1), 150–154.

Sweis, B. M., Abram, S. V., Schmidt, B. J., Seeland, K. D., MacDonald, A. W., Thomas, M. J., et al. (2018). Sensitivity to "sunk costs" in mice, rats, and humans. *Science*, *361*(6398), 178–181.

Thorn, C. A., Atallah, H., Howe, M., & Graybiel, A. M. (2010). Differential dynamics of activity changes in dorsolateral and dorsomedial striatal loops during learning. *Neuron*, *66*(5), 781–795.

Tolman, E. C. (1948). Cognitive maps in rats and men. *Psychological Review, 55*, 189–208.
Tran-Tu-Yen, D. A. S., Marchand, A. R., Pape, J.-R., Di Scala, G., & Coutureau, E. (2009). Transient role of the rat prelimbic cortex in goal-directed behaviour. *The European Journal of Neuroscience, 30*, 464–471.
Uylings, H. B. M., Groenewegen, H. J., & Kolb, B. (2003). Do rats have a prefrontal cortex? *Behavioural Brain Research, 146*(1–2), 3–17.
van Aerde, K. I., Heistek, T. S., & Mansvelder, H. D. (2008). Prelimbic and infralimbic prefrontal cortex interact during fast network oscillations. *PLoS One, 3*(7), e2725.
van der Meer, M. A. A., Johnson, A., Schmitzer-Torbert, N. C., & Redish, A. D. (2010). Triple dissociation of information processing in dorsal striatum, ventral striatum, and hippocampus on a learned spatial decision task. *Neuron, 67*(1), 25–32.
van der Meer, M. A. A., Kurth-Nelson, Z., & Redish, A. D. (2012). Information processing in decision-making systems. *The Neuroscientist: A Review Journal Bringing Neurobiology, Neurology and Psychiatry, 18*(4), 342–359.
van der Meer, M. A. A., & Redish, A. D. (2009). Covert expectation-of-reward in rat ventral striatum at decision points. *Frontiers in Integrative Neuroscience, 3*(1), 1–15.
van der Meer, M. A. A., van der Meer, M. A. A., Johnson, A., Schmitzer-Torbert, N. C., & David Redish, A. (2010). Triple dissociation of information processing in dorsal striatum, ventral striatum, and hippocampus on a learned spatial decision task. *Neuron, 67*(1), 25–32. https://doi.org/10.1016/j.neuron.2010.06.023.
Vertes, R. P. (2004). Differential projections of the infralimbic and prelimbic cortex in the rat. *Synapse, 51*, 32–58.
Vidal-Gonzalez, I., Vidal-Gonzalez, B., Rauch, S. L., & Quirk, G. J. (2006). Microstimulation reveals opposing influences of prelimbic and infralimbic cortex on the expression of conditioned fear. *Learning & Memory, 13*(6), 728–733.
Walters, C. J., Jubran, J., Sheehan, A., Erickson, M. T., & Redish, A. D. (2019). Avoid-approach conflict behaviors differentially affected by anxiolytics: Implications for a computational model of risky decision-making. *Psychopharmacology, 236*(8), 2513–2525.
Walton, M. E., Bannerman, D. M., Alterescu, K., & Rushworth, M. F. S. (2003). Functional specialization within medial frontal cortex of the anterior cingulate for evaluating effort-related decisions. *The Journal of Neuroscience: The Official Journal of the Society for Neuroscience, 23*(16), 6475–6479.
Wikenheiser, A. M., & Redish, A. D. (2015). Hippocampal theta sequences reflect current goals. *Nature Neuroscience, 18*, 289–294.
Wirt, R. A., & Hyman, J. M. (2019). ACC theta improves hippocampal contextual processing during remote recall. *Cell Reports, 27*(8), 2313–2327.e4.
Yin, H. H., Knowlton, B. J., & Balleine, B. W. (2002). From habits to actions: Dorsolateral striatum lesions alter the content of learning. In *Proceedings of the 9th international conference on neural information processing, 2002. ICONIP'02*. https://doi.org/10.1109/iconip.2002.1202887.
Zemel, R. S., Dayan, P., & Pouget, A. (1998). Probabilistic interpretation of population codes. *Neural Computation, 10*(2), 403–430.
Zielinski, M. C., Shin, J. D., & Jadhav, S. P. (2019). Coherent coding of spatial position mediated by theta oscillations in the hippocampus and prefrontal cortex. *The Journal of Neuroscience: The Official Journal of the Society for Neuroscience, 39*(23), 4550–4565.

CHAPTER TEN

Anterior cingulate cortex and adaptive control of brain and behavior

Adam T. Brockett[a,b,*] and Matthew R. Roesch[a,b]
[a]Department of Psychology, University of Maryland, College Park, MD, United States
[b]Program in Neuroscience and Cognitive Science, University of Maryland, College Park, MD, United States
[*]Corresponding author: e-mail address: brockett@umd.edu

Contents

Abstract

Research examining the functional underpinnings of anterior cingulate cortex (ACC) and its relationship to cognitive control have been described as "perennially controversial" and a "Rorschach Test" for modern neuroscience. Although there is near universal agreement that ACC is important for the adaptation of behavior, debate, despite decades of work, stems from the exact manner in which ACC goes about doing this. This chapter provides a brief overview of the various past and present theoretical arguments and research surrounding ACC function, and highlights an emerging literature of single unit ACC recordings from several species that support these theories. We will finish the chapter by focusing on our work examining the firing of single neurons in rat dorsal medial striatum (DMS) and ACC, and examining DMS's dependency on ACC to accurately signal adaptive behavioral output. Ultimately, we will conclude that ACC carries a myriad of signals (error detection, reinforcement/feedback, value, response conflict, etc.) necessary for the modulation of attention and task-relevant/irrelevant signals so that difficult decisions can be made and action plans adapted when necessary.

International Review of Neurobiology, Volume 158
ISSN 0074-7742
https://doi.org/10.1016/bs.irn.2020.11.013

1. Introduction

Understanding the factors and strategies that govern adaptive decision-making and cognitive control has provided a near endless supply of research directions for past, present and future work in cognitive neuroscience. At its core, this research is defined by an attempt to understand which brain circuits facilitate our ability to change and adapt behavior, and how. Decision-making is one of the most pervasive aspects of human life, everything we do from when we choose to wake up to when we choose to go to sleep is in some way a decision. Decisions, much like behavior, exist on a continuum from automatic to effortful. Deciding to brush one's teeth is (hopefully) an example of a relatively automatic decision, whereas choosing a college major or where to go to college in the first place is likely more effortful. Automatic and effortful decisions blend seamlessly into our day to day, often without impeding our actual ability to carry on with daily life tasks (i.e., work, chores, socializing). Consequently, much research has been devoted to understanding how we are able to navigate the world so successfully despite a constant barrage of decisions and distractions.

As an example, consider something relatively trivial that many of us do every day, navigating a car/bike/or walking with a GPS. The logistics of operating our preferred form of transportation notwithstanding, using a GPS is fairly straightforward; you key in an address, a handful of routes populate, you usually choose the fastest one, and then "voila!" you are on your way. Already in this example we have made several decisions, first the decision to go somewhere, second to consult our GPS, and third, to decide which route to take. Often we do not even think of these steps or process them as decisions because generally speaking, they are all relatively automatic, taking just a few seconds, if that, to make.

However, as we begin moving toward our intended goal sometimes, seemingly out of the blue, our decisions must adapt to new information, as for example, when our GPS informs us that an alternative route exists. If we are aware of the fact that traffic on the current route is getting bad, or we cannot stand the idea of potentially encountering traffic, we may quickly make the adaptive low effort decision to choose the newly offered route. However, if the predicted ETA associated with choosing the alternative route is no better than the ETA associated with staying on the current route, or if the alternative routes takes us on roads with lots of potholes or requires paying a toll, then adapting behavior becomes more difficult because there is competition between multiple factors. In other words, in the absence of clear cost or benefit to either route, conflict between staying or changing course arises and will drive (pun intended) the recruitment of additional cognitive resources so that a decision can be made.

In many ways, the decision of whether to invest time or energy into overwriting a previous decision or quickly countermanding a previous plan of action when necessary parallels what is asked of subjects performing numerous cognitive control tasks that probe cognitive function. The stop-signal task for instance, asks participants to make a relatively automatic (GO) response (i.e., press a lever after hearing tone A), on a majority of trials, only to then randomly interleave a different trial type, STOP trials, on a lower percentage of trials. On STOP trials, a participant must withhold a response in the presence of a second cue (i.e., do not press a lever after hearing tone A when light is on). This task along with others, such as the Stroop Task, The Flanker Task, and Foraging Tasks, has formed the bedrock of research into cognitive control. This chapter catalogs the history of the debate over the contribution of ACC to the adaptation of behavior, and highlights recent novel work exploring the role of this brain region using single unit recordings in a variety of behavioral paradigms demonstrating that ACC signals that contribute to attention, error processing, value and effort computations, and response conflict.

2. The evolution of the conflict monitoring hypothesis

2.1 Conflict-monitoring

While the exact function of ACC continues to be debated, early network theories of cognition and control suggested ACC contributed to executive function (Mesulam, 1981; Papez, 1937; Posner & Dehaene, 1994; Posner, Petersen, Fox, & Raichle, 1988). In his theory of emotion, Papez theorized that ACC, in part due to its connections with the hypothalamus, was a "receptive region for the experiencing of emotion" (Papez, 1937). Papez speculated that the emotive process radiated from ACC to other brain regions, providing a kind of emotional context that in turn would influence or bias planned behavior (Papez, 1937). As functional theories of ACC moved away from emotion, per se, and more toward cognition, Mesulam hypothesized that ACC contained a map of motivational valence that helped direct attention and biased motor planning accordingly (Mesulam, 1981). Attentional network theorists refined this view, suggesting that ACC was critical for the selection of task or goal-relevant actions and later that

ACC was responsible for attentional awareness and implementation of control (Posner et al., 1988; Posner & Dehaene, 1994). Other evidence supporting ACC's involvement in control came from patient studies in which cingulotomies produced marked deficits in "attentional abilities" (Janer & Pardo, 1991; Turken & Swick, 1999).

The 1990s ushered in the use of neuroimaging techniques that in turn helped support and refine theories of ACC function. Using the Stroop Task, Pardo and others showed strong activation of ACC when subjects were asked to respond to the ink color in which a conflicting color name was presented (Bench et al., 1993; Carter, Mintun, & Cohen, 1995; Pardo, Pardo, Janer, & Raichle, 1990). Similarly, evidence suggested that ACC was activated more when detecting multiple color forms or motion targets was necessary in comparison to passive viewing (Corbetta, Miezin, Dobmeyer, Shulman, & Petersen, 1991), suggesting that ACC is preferentially activated by incongruence, as well as when action is required. Other work revealed that ACC was activated more strongly when the number of targets increased (Posner et al., 1988), but that this activation decreased when subjects were allowed to practice with the stimulus set (Raichle et al., 1994).

With evidence implicating ACC in functions pertaining to the ability to resolve conflicting information between relevant inputs and incongruent distractors growing, a new theoretical perspective emerged (Botvinick, Braver, Barch, Carter, & Cohen, 2001). This perspective, aptly named the "conflict-monitoring hypothesis," postulated that one role of ACC might be in the detection of instances where a subject's intended action, and the action dictated by environment were at odds (i.e., response conflict). In these instances, once detected, a "conflict-signal" would develop to help bias attention, motor planning and action selection in downstream brain regions to drive behavior toward the correct response. The theory attempted to account for a variety of different behavioral results that implicated a role for ACC in tasks requiring response override (Bench et al., 1993; Bush et al., 1998; Carter et al., 1995; Casey et al., 1997; George et al., 1994; MacLeod, 1991; Pardo et al., 1990; Paus, Petrides, Evans, & Meyer, 1993; Stroop, 1935; Taylor, Kornblum, Minoshima, Oliver, & Koeppe, 1994), tasks that studied underdetermined responding (Andreasen et al., 1995; Barch et al., 2001; Buckner et al., 1995; Deiber et al., 1991; Friston, Frith, Liddle, & Frackowiak, 1993; Frith, Friston, Liddle, & Frackowiak, 1991a, 1991b; Jueptner, Frith, Brooks, Frackowiak, & Passingham, 1997; Petersen, Fox, Posner, Mintun, & Raichle, 1988, 1989; Playford et al., 1992;

Raichle et al., 1994; Thompson-Schill, D'Esposito, Aguirre, & Farah, 1997; Yetkin et al., 1995), and tasks that involved error commission (sCarter et al., 1998; Dehaene, Posner, & Tucker, 1994; Falkenstein, Hohnsbein, & Hoormann, 1995; Falkenstein, Hohnsbein, Hoormann, & Blanke, 1991; Gehring, Goss, Coles, Meyer, & Donchin, 1993; Gratton, Coles, Sirevaag, Eriksen, & Donchin, 1988; Scheffers, Coles, Bernstein, Gehring, & Donchin, 1996), as well as studies that did not fully fit within the other three categories (Baker et al., 1996; Corbetta et al., 1991; D'Esposito et al., 1995; Grasby et al., 1993). The work created a computational framework where ACC, focused on conflict monitoring, could account for the behavioral results of almost all of these findings.

Importantly, this work did not maintain that the sole function of ACC was conflict-monitoring, rather that this was just one of several functions likely served by ACC (Botvinick, 2007; Botvinick et al., 2001; Botvinick, Cohen, & Carter, 2004). Instead, this account merely ascribed a "special status" to response conflict (Botvinick et al., 2004), suggesting that many of the behavioral phenomenon associated with ACC in the past could be accounted for using this framework. Specifically, this model held that ACC was engaged most strongly during conflicts at the level of response selectively, and that trial history or sequence effects cannot be attributed entirely to attention, perceptual, or motor priming. Indeed the computational foundation of theory shows reactive adjustments in control on trial to trial basis can be accounted for by conflict signals (Botvinick et al., 2001, 2004). This was supported by work showing that increased ACC activity on incongruent trials of the Stroop task was associated with relatively low interference on subsequent trials (Kerns et al., 2004), suggesting that ACC is responsible for shifts in cognitive control on subsequent trials.

2.2 Merging conflict-monitoring with expected outcomes

While the original crafting of the conflict-monitoring hypothesis offered a theory that accounted for many of the behavioral and neuroimaging findings related to ACC, it did not account for everything, and with an increased focus on the role of ACC, and cognitive control more generally, challenges to the conflict-monitoring hypothesis began to appear. Research emerged suggesting that ACC responds preferentially to aversive outcomes or reductions in reward (Gehring & Willoughby, 2002), suggesting a potential role for ACC in guiding action selection based on past outcomes (Hadland, Rushworth, Gaffan, & Passingham, 2003; Holroyd & Coles, 2002;

Matsumoto, Suzuki, & Tanaka, 2003; Rushworth, Walton, Kennerley, & Bannerman, 2004). Moreover, studies in rats and monkeys found that ACC lesions did not disrupt task switching (which requires conflict detection), but did impair evaluation of effort necessary to obtain reward (Dias & Aggleton, 2000; Rushworth, Hadland, Gaffan, & Passingham, 2003; Walton, Bannerman, Alterescu, & Rushworth, 2003).

At the same time, research looking at the involvement of ACC in action outcome encoding revealed that the firing of monkey ACC neurons reflected the relationship between cue and reward (Matsumoto et al., 2003), and that lesions disrupted the reward guided selection of actions (Hadland et al., 2003). Moreover, evidence from a sequence learning task suggested that populations of cells in ACC were critical for mapping action outcome relationships necessary for successful sequence learning (Procyk, Tanaka, & Joseph, 2000). Similar results were described in rats, where ACC was shown to represent errors and preparatory attention within the same behavioral sequence (Totah, Kim, Homayoun, & Moghaddam, 2009). Collectively, these findings led to the development of an alternative hypothesis that suggests the function of ACC was more directly concerned with action outcome learning, and the guiding of decision-making based on the expected value of reward, rather than conflict monitoring per se (Rushworth et al., 2004).

It is worth noting at this point that just as the conflict-monitoring hypothesis was unable to account for several new results, these alternative accounts, in turn, were also unable to fully account for some of the findings supported by conflict-monitoring. Occasionally, these alternative hypotheses even required aspects of conflict monitoring in order to appropriately model their results (Frank, Woroch, & Curran, 2005; Holroyd & Coles, 2002). As evidence for both theories continued to accrue, a reframing of the original conflict-monitoring hypothesis was proposed, one that attempted to account for both the conflict-monitoring findings and the outcome evaluation/decision-making theories (Botvinick, 2007). This hybrid model, framed as an extension of the initial conflict-monitoring hypothesis, suggests that conflict acts as a teaching signal, which in turn, drives a form of avoidance learning to bias behavioral decision making toward "cognitively efficient" strategies (Botvinick, 2007). In other words, instead of an occurrence of conflict triggering ACC, and thus shifting control to protect against future conflict, this new view suggests that the experience of conflict instead drives the avoidance of tasks or strategies that have given rise to conflict in the past (Botvinick, 2007). In this way, the experience of conflict is encoded as a negative reinforcing event that biases future behavior away from experiencing conflict again. In line with this theory, work investigating effort in rodents has suggested that lesions to ACC modulate the willingness of rats to engage in physically effortful tasks based on the value of reward (Walton et al., 2003; Walton, Bannerman, & Rushworth, 2002; Walton, Kennerley, Bannerman, Phillips, & Rushworth, 2006), and evidence from monkeys performing a foraging task suggest that ACC lesions diminish willingness to sustain actions that lead to reward (Kennerley, Walton, Behrens, Buckley, & Rushworth, 2006).

2.3 Foraging

The emphasis on effort in addition to a desire to explore the evolution of cognitive control within a more ethologically relevant perspective led to an increased interest in exploring the role of ACC in the context of foraging behavior (Kolling et al., 2016; Kolling, Behrens, Mars, & Rushworth, 2012; Shenhav, Cohen, & Botvinick, 2016). In many ways, like the GPS example at the beginning of this chapter, a foraging context emphasizes the fact that adaptive decision-making does not occur in isolation. Decision-making is a tool often employed in the service of achieving an overarching goal and/or action, such as successfully navigating to one's desired destination. Already, data has suggested that within the foraging context, the ACC of monkeys contributes to both reward and error processing, and damage to ACC diminishes the willingness to sustain rewarded actions (Kennerley et al., 2006). In an attempt to more directly incorporate these findings, the conflict-monitoring hypothesis was extensively reframed into what is now known as the expected value of control (EVC) hypothesis (Shenhav, Cohen, et al., 2016; Shenhav, Straccia, Cohen, & Botvinick, 2014).

The EVC theory subsumes previous conflict-monitoring accounts as well as offers a broader explanation for ACC functioning in general (Shenhav, Cohen, et al., 2016). Critically, EVC predicts that control signals vary along two dimensions: (1) identity, a judgment about what the appropriate action/response may be and (2) intensity, a determination of how achieving the appropriate response should be executed (Shenhav, Cohen, et al., 2016). In this view, conflict, encoded as the need to put forth effort, drives the biasing of behavior in the appropriate direction, and specifically within the context of foraging, choice difficulty (i.e., when to continue harvesting versus when to move on from a patch) is reflected as a proxy for effort/conflict. Two fMRI studies showed that when decision difficulty was highest,

ACC activity also peaked (Shenhav et al., 2014; Shenhav, Straccia, Botvinick, & Cohen, 2016). Researchers separated choice difficulty from reward value by noting that within a foraging context, when the value of staying versus leaving is equal, the decision on whether to stay or leave is more difficult, and control is required. However, when the value to stay or go, respectively, is high, then the decision is obvious, and ACC should not be involved. In line with this prediction, ACC activity was increased only when decision difficulty was also high, and not when the value of staying or going was obvious (Shenhav et al., 2014, Shenhav, Straccia, et al., 2016). While in this view, the meaning of conflict, at least as originally presented, has certainly been expanded, the advantage to EVC is that it offers a comprehensive explanation of conflict both on its own and within a foraging context, while also providing a clearer rendering of the functional role of ACC.

Unsurprisingly, the EVC theory has not been universally accepted, as the question of whether reward value or the need to put forth control rages. Consider the GPS example again. As described earlier, the difficult decision arises when you are asked to decide between staying on the same route where traffic has slowed versus taking a new route that offers a similar ETA. With the EVC theory in mind, this decision is difficult because the overall outcome (i.e., ETA) is essentially the same, suggesting a need for ACC engagement in order to exert control and drive behavior appropriately. In a foraging context, this example would be similar to the reaching the optimal point at which an animal must decide to leave one patch in order to find another. The original patch still has food to harvest, but staying too long may deplete the resources to the point that the benefits of staying and eating no longer outweigh the costs associated with going without food or resources while traveling to find a new patch. At the same time, the GPS and foraging decision could also be viewed as relatively easy. At this point in time along the decision timeline, both ETA or resource supply are equal so the law of least work might predict that an animal would stay until accruing some deficit rather than engage in the hassle of preemptively acting.

This is where an alternative account of ACC function has risen from the original effort and expected outcome work. This alternative account suggests that ACC is primarily concerned with search value (i.e., the average value of the environment) which predicts whether to continue searching or to stay put. If the average value associated with a particular action (i.e., staying in on the originally chosen GPS route or staying in the same patch) begins to decrease, an animal would then be motivated to move on (i.e., choose the alternative route or find a new patch), but if the value is relatively high or the values of going and staying are relatively equal (i.e., ETA's for both routes are the same, or there is still resource to harvest) then an animal's motivation for search would be minimal. Consistent with this theory, a recent human fMRI study showed that ACC is strongly activated by the quality, quantity and interaction of quality and quantity features of an offer, and that tracking of subjective values correlates with choice (de Berker, Kurth-Nelson, Rutledge, Bestmann, & Dolan, 2019).

3. What does ACC signal? Insights from behavioral neurophysiology

As it stands now, EVC and search value hypotheses, as well as their respective predecessors, offer explicit predictions about ACC function. One favors functions related to conflict or choice difficulty while the other suggests that ACC estimates average value of stimuli and actions in the environment. Despite the constant reframing of the conflict-monitoring hypothesis, two critical questions that have lingered over decades of cognitive and computational research has been: What information do single neurons in ACC carry? Do single ACC neurons detect conflict like BOLD signals do or does ACC influence behavior via signals more closely aligned with reinforcement/feedback and value/effort estimates? While the single neuron data showing that ACC does signal expected value is abundant (Amiez, Joseph, & Procyk, 2006; Blanchard & Hayden, 2014; Cai & Padoa-Schioppa, 2012; Hayden, Heilbronner, Pearson, & Platt, 2011; Hayden, Pearson, & Platt, 2011; Ito, Stuphorn, Brown, & Schall, 2003; Kawai, Yamada, Sato, Takada, & Matsumoto, 2015; Kennerley, Behrens, & Wallis, 2011; Kennerley, Dahmubed, Lara, & Wallis, 2009; Luk & Wallis, 2013; Nakamura, Roesch, & Olson, 2005), until recently, there has been surprisingly few single unit studies that support that ACC contributes to conflict-like mechanisms and adaptive adjustments in behavioral control outside the realm of value, reward, or effort encoding.

Early single unit work in this field focused on neural correlates of error detection (Amiez et al., 2006; (Ito et al., 2003; Quilodran, Rothe, & Procyk, 2008; Rushworth & Behren, 2008); Totah et al., 2009) and aspects of behavioral feedback related to reward and attention (Kennerley et al., 2006; Oliveira, McDonald, & Goodman, 2007; Rothé, Quilodran, Sallet, & Procyk, 2011). For example, Hayden and colleagues showed that activity in monkey ACC was high when rewards were delivered and omitted unexpectedly in a task in which rewards were delivered at predetermined probabilities (Hayden, Heilbronner, et al., 2011). Importantly, changes in ACC firing occurred regardless of valence at the single cell level suggesting that primate ACC encodes unsigned reward prediction errors consistent with theories put forth by Pearce and Hall (Hayden, Heilbronner, et al., 2011). Consistent with these findings, we have also shown that neurons in rat ACC fire similarly to appetitive reward and aversive shock suggesting that activity in ACC better reflects attention as opposed to signed value per se (Schneider, Sciarillo, Nudelman, Cheer, & Roesch, 2020). Further, we have shown that activity in ACC is elevated on behavioral trials after reward contingences changed unexpectedly for both up- and down-shifts in value, suggesting that ACC appears to not only be involved in detecting errors but in utilizing that information to drive Pearce and Hall-like attention and learning on subsequent trials (Bryden, Johnson, Tobia, Kashtelyan, & Roesch, 2011; Vázquez, Pribut, Burton, Tennyson, & Roesch, 2020).

From this work it has been clear that firing in ACC correlates well with commission and omission errors, attention, and value in the service of adaptive decision-making but for a long time it was unclear if single neuron activity or even ensemble activity over multiple neurons correlated well with conflict between competing responses as is it does in so many fMRI studies that originally fueled early theories of ACC function. Below, we describe cross species data that have shed light on this issue.

3.1 Humans

Surprisingly, much of the single-unit support for conflict-related encoding ACC has come from humans. Although human ACC recordings originally suggested attentional modulation of ACC neurons (Davis, Hutchison, Lozano, Tasker, & Dostrovsky, 2000), subsequent research using a high-conflict version of a Stroop-like task revealed elevated activity on high-conflict trials (Davis et al., 2005). These results were further supported by more recent work using a multi-source inference task suggesting that human ACC neurons are continually modulated by expected cognitive demand (Sheth et al., 2012). In this task, subjects were asked identify the position of the unique number in an array of three numbers. Researchers varied the positioning of the unique number (i.e., left, middle, or right), which relative to the distractors, modulated cognitive demand accordingly (Sheth et al., 2012). Critically, ACC neurons mirrored conflict-adaptation hypotheses, or the Gratton effect, described previously by an fMRI study (Botvinick, Nystrom, Fissell, Carter, & Cohen, 1999; Sheth et al., 2012).

Subsequent work utilized targeted dimensionality reduction characterized activity across the population of ACC neurons as reflecting the amplification of task-relevant information during conflict as opposed to a general conflict signal or an epiphenomenal signal arising from the co-activation of competing inputs (Ebitz et al., 2020).

3.2 Non-human primates

Evidence for conflict monitoring by ACC at the single neuron level has been notably harder to come by in non-human primates. In monkeys performing a saccade-based countermanding task, single units in ACC showed selectivity for reinforcement and error detection, but no conflict-like signals were observed (Ito et al., 2003). This is in contrast to a previous result showing that single neurons in supplementary eye field (SEF) did show conflict-specific responses (Stuphorn, Taylor, & Schall, 2000), leading some to speculate whether species difference in brain homology between monkeys and humans may account for these discrepancies (Botvinick et al., 2004; Cole, Yeung, Freiwald, & Botvinick, 2009; Ito et al., 2003; Rushworth et al., 2004). A second study, utilizing two behavioral tasks, a spatial incompatibility task and a reversal task, recorded from SEF and ACC in monkeys and found little to no evidence for conflict-monitoring signals in either brain region. Instead, researchers showed possible evidence for representation of both competing responses that theoretically could give rise increased BOLD signals observed in human fMRI studies, but were unable to identify a single population of cells independently encoding conflict, as predicted by the original framework of the conflict-monitoring hypothesis (Botvinick et al., 2001; Nakamura et al., 2005). Notably, recent work in rats expands on this idea suggesting that feedback-related negativity in ACC neurons is generated across neurons in a similar fashion (Hyman, Holroyd, & Seamans, 2017).

The first signs that the firing of single ACC neurons could represent conflict and associated arousal came from a study that examined cue directed saccade behavior in the presence or absence of social and nonsocial distractors (Ebitz & Platt, 2015). Social distractors significantly slowed reaction times more than non-social distractors, and when they were incongruent with the saccade target (Ebitz & Platt, 2015). Neurons in ACC were modulated both by social context and task-relevant distractors (i.e., distractors presented during the response window versus those presented during the ITI) (Ebitz & Platt, 2015). Further, firing of ACC neurons predicted conflict-related changes in pupil diameter, a measure of arousal, attentiveness and task engagement (Ebitz & Platt, 2015). Collectively, these findings suggest that firing in monkey ACC is modulated by conflicting distractors, however, since congruency affected firing in both directions a clear population level "response conflict-signal" was not evident (Ebitz & Platt, 2015).

3.3 Rats

To address this issue in rats, our lab developed a novel-variant of the stop-signal task to be used for behavioral neurophysiological studies in rats (Brockett, Tennyson, deBettencourt, Gaye, & Roesch, 2020; Bryden et al., 2019). As illustrated in Fig. 1A, rats were trained to place their nose in a central port, at which point a light cue on either the left or right side was flashed for 100 ms. On 80% of the trials (i.e., GO trials) this cue predicted which of two fluid wells the rat must respond to in order to receive a reward. On the remaining 20% of trials (i.e., STOP-change trials) within 100 ms of the first cue (i.e., GO cue) and second cue (i.e., STOP cue) was illuminated instructing the rat to cancel its initial response in the direction of the first cue, in favor of a response in the direction of the second cue in order to obtain reward. Critically, unlike traditional stop-signal tasks in which the appropriate response on STOP-trials is the refrain from any action, correct responding on this "stop-change task" requires the rat to cancel one action in favor of another.

During performance of this task, "response conflict" arises on STOP-change trials from neural signals that promote opposite actions (i.e., the response signaled by the first cue and the response signaled by the second cue). This is demonstrated behaviorally in that rats were slow and performed worse on STOP trials, and in that rats were extremely fast on STOP errors, moving quickly in the wrong direction. More interestingly, we were able to observe response conflict at the neural level by recording from motor-related neurons in dorsal medial striatum (DMS), an area whose firing is positively correlated with movement speed. This is illustrated in Fig. 1 which plots the average populations firing over 122 neurons in rats performing our STOP-change task. On GO trials (blue), activity strongly represents firing of the upcoming movement (Fig. 1B). This is a clean un-conflicted signal from start to finish, with significant differences between response directions (blue tick marks) emerging shortly after the presentation of the first cue and ending shortly after the completion of the response (arrow embedded in tick marks). On STOP-change trials (Fig. 1C), accurate response signals were slow to emerge in line with slower movement speeds and less accurate behavior

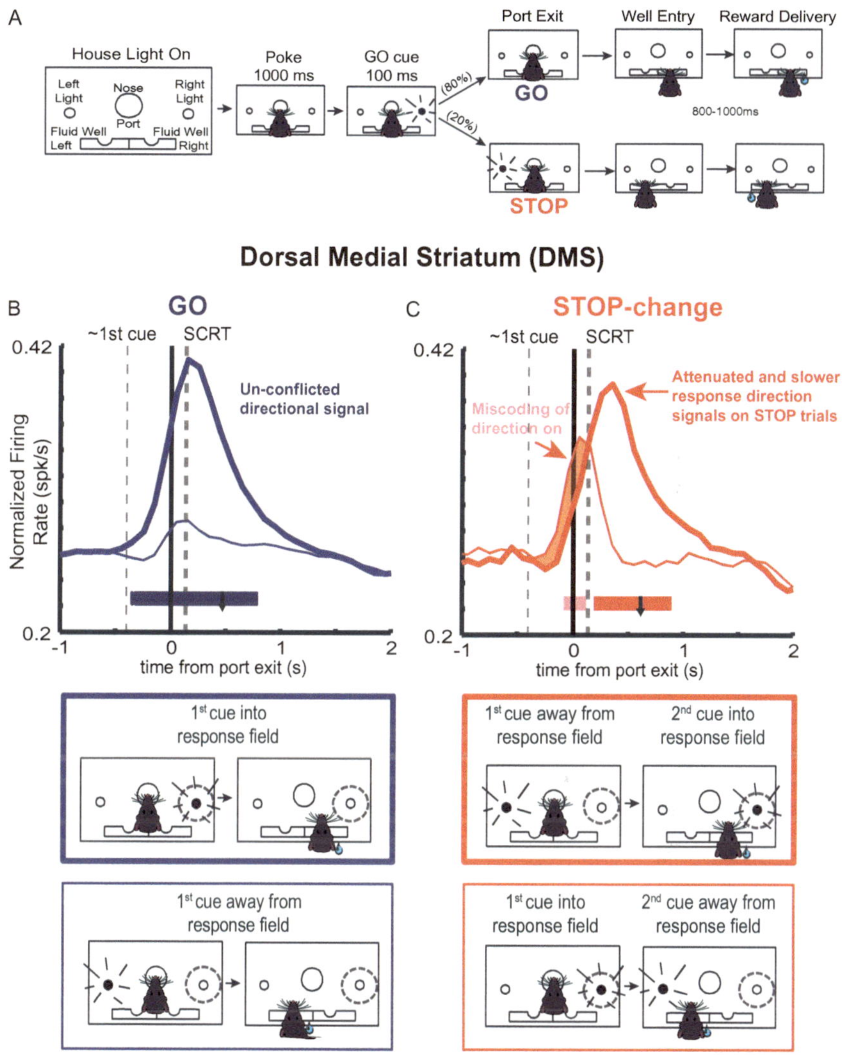

Fig. 1 Miscoding and slow development of accurate directional response signals in DMS during performance of a STOP-change task. (A) Illustration of the stop-change task (Brockett et al., 2020; Bryden, Burton, Kashtelyan, Barnett, & Roesch, 2012). Rats were instructed to hold their nose in the center port for 1 s, at which point a GO cue would be presented to either the left or right of the rat. On 80% of trials the GO cue instructed the rat as to which of the two fluid wells to move to in order to receive reward. On the remaining 20% of trials, STOP trials, after the initial GO cue, a STOP cue was presented on the opposite side, instructing the rat to inhibit its initial response in the direction of the GO cue, in favor of responding in the direction of the STOP cue. (B) Population firing of DMS neurons on GO trials aligned to center port exit. Thick and thin blue lines reflect

(Continued)

on STOP-change trials. Specifically, on trials where the first and second cues are into and then away from the response field, respectively, activity increased initially but then rapidly shuts down (thin line), whereas when the first and second cue is away from and then into the response field, respectively, activity was slow to rise (thick red line). The shaded red area in this plot reflects time when there was response conflict (i.e., competition between left and right responses). Importantly, accurate response tuning in DMS resolves early enough to enact behavioral change within a single trial as measured by the Stop Change Reaction Time (i.e., STOP minus GO reaction times; SCRT). Notably the resolution of the two conflicted signal (Fig. 1C; thick vs thin red lines; tick marks indicate significant differences) in DMS corresponded amazingly well to the SCRT, and when errors were made, activity failed to reconcile prior to the SCRT, reflecting the location of the incorrect movement (Brockett et al., 2020; Bryden et al., 2012).

When thinking about response conflict in the context of single neurons, average population level firing—as shown in Fig. 1—consists of single neurons whose firing failed, weakly, or accurately represented the correct response direction on STOP trial. Extreme examples of this continuum (i.e., fails to represent vs. accurately represents the correct response) are illustrated in Fig. 2. Both units show directional tuning on GO trials in that firing is significantly stronger for movement made in one direction versus the other. For these examples, activity was stronger for rightward GO movements, thus "right" is considered to be "into" the cell's response field (i.e., gray dashed circle in inset) or the cell's "preferred" response direction.

Now, let's consider how these same neurons fire on STOP-change trials. For Unit 1, illustrated in Fig. 1A, firing strongly reflected the action signaled

Fig. 1—Cont'd firing in the preferred and non-preferred directions, respectively. Illustrations of directionality are shown below. The time necessary to inhibit a response (stop-change reaction time; SCRT) is defined as the difference between STOP trial movement time and GO trial movement time and is depicted by the dashed line. (C) Population firing of DMS neurons on STOP-change trials aligned to center port exit. Thick and thin red lines reflect firing when responses were made into the preferred (i.e., into the response field) and non-preferred directions (i.e., away from the response field), respectively. *Figures adapted from Bryden, D.W., Burton, A.C., Kashtelyan, V., Barnett, & B.R., Roesch, M.R. (2012). Response inhibition signals and miscoding of direction in dorsomedial striatum.* Frontiers in Integrative Neuroscience, 6, *69 and Brockett, A.T., Tennyson, S.S., deBettencourt, C.A., Gaye, F., & Roesch, M.R. (2020). Anterior cingulate cortex is necessary for adaptation of action plans.* Proceedings of the National Academy of Sciences, 117, *6196–6204.*

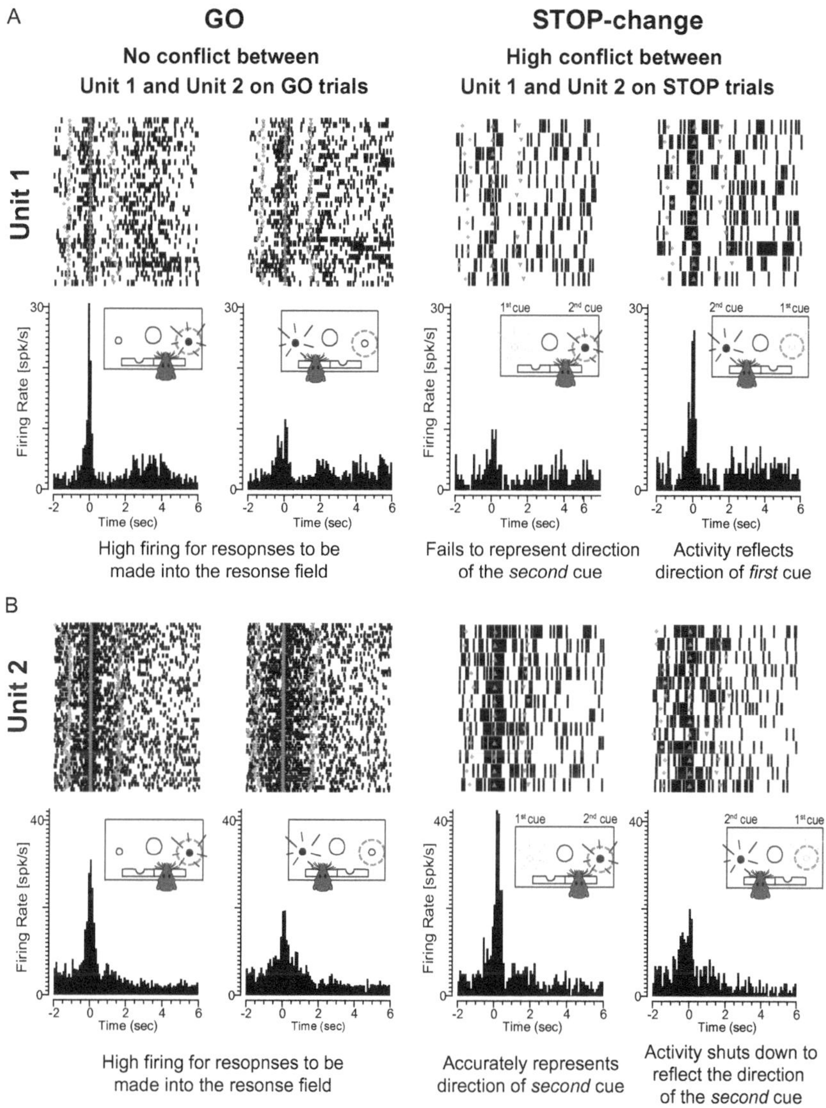

Fig. 2 Raster plots showing firing for individual neurons from DMS. Firing of example units on correct GO and STOP-change trials. High firing is observed on GO trials reflecting the direction the rat intends to move in. The direction associated with higher firing is referred to as the "preferred" direction or "into the response field". a. Unit 1 fails to represent the direction of the correct response, instead firing represents the direction of the first cue. (B) In contrast, when looking at Unit 2, firing in the direction of the 2nd cue emerges on the STOP-change trials and firing in the direction of the first cue is correctly shut down. *Figures adapted from Bryden, D.W., Burton, A.C., Kashtelyan, V., Barnett, & B.R., Roesch, M.R. (2012). Response inhibition signals and miscoding of direction in dorsomedial striatum.* Frontiers in Integrative Neuroscience, 6, 69.

by the first cue, and never accurately represents the correct response direction. One can imagine that having too many of these units active during a STOP-change trial would result in movement into the wrong direction. Although neurons like this are present in normal rats at significant levels, the majority of neuronal firing looks like Unit 2. For this neuron, the firing reflects the correct response direction on STOP-change trials (i.e., the direction of the second cue light). In examining these two units, one can see how response conflict can arise in the brain with different units simultaneously signaling opposite directions on STOP-change trials (high conflict), but not during performance of GO trials (no conflict). With these results we firmly established, both from a neural standpoint and a behavioral one that the two responses (left versus right) are in conflict with each other when rats perform our behavioral paradigm. The question that remained is whether or not ACC is able to (i.e., signal conflict) and is necessary to resolve this conflict (i.e., response conflict cannot be resolved without ACC).

To address this question we recorded from ACC in rats performing the identical task (Bryden et al., 2019). We found that neurons in ACC increased firing on STOP trials compared to GO trials in a directional fashion (see Fig. 3). Firing emerged just after the presentation of the STOP cue and well before the stop-change reaction time (Fig. 3B; SCRT), as well as before the resolution of directional signals in DMS. Further, we found that firing correlated with both movement time and accuracy, suggesting that when ACC was engaged, rats tended to slow down and perform better. Lastly, consistent with human imaging studies (Botvinick et al., 1999; Bryden et al., 2019), we found that ACC firing was strongest on high conflict trials (STOP trials that follow GO trials; i.e., incongruent that follow congruent trials) compared to lower conflict trials (STOP trials that follow STOP trials; i.e., incongruent followed by another incongruent trial; Fig. 3C).

These results demonstrate that firing in ACC can contribute to within trial adaptation of behavior and downstream motor units. This was true averaged across the entire population of recorded ACC neurons (Fig. 3B) and significant in the majority of task-related single units (Fig. 4A; Trial-type circle). Notably, this signal is unique to ACC in that counts of neurons carrying such a signal is absent from other brain regions that we have recorded from (Fig. 4B–D), including DMS (Bryden et al., 2012), lateral orbitofrontal cortex (Bryden & Roesch, 2015), and medial prefrontal cortex (Bryden et al., 2016). Also important to note is that neurons that that were modulated by conflict, also tended to be directionally selective (Fig. 2A; green).

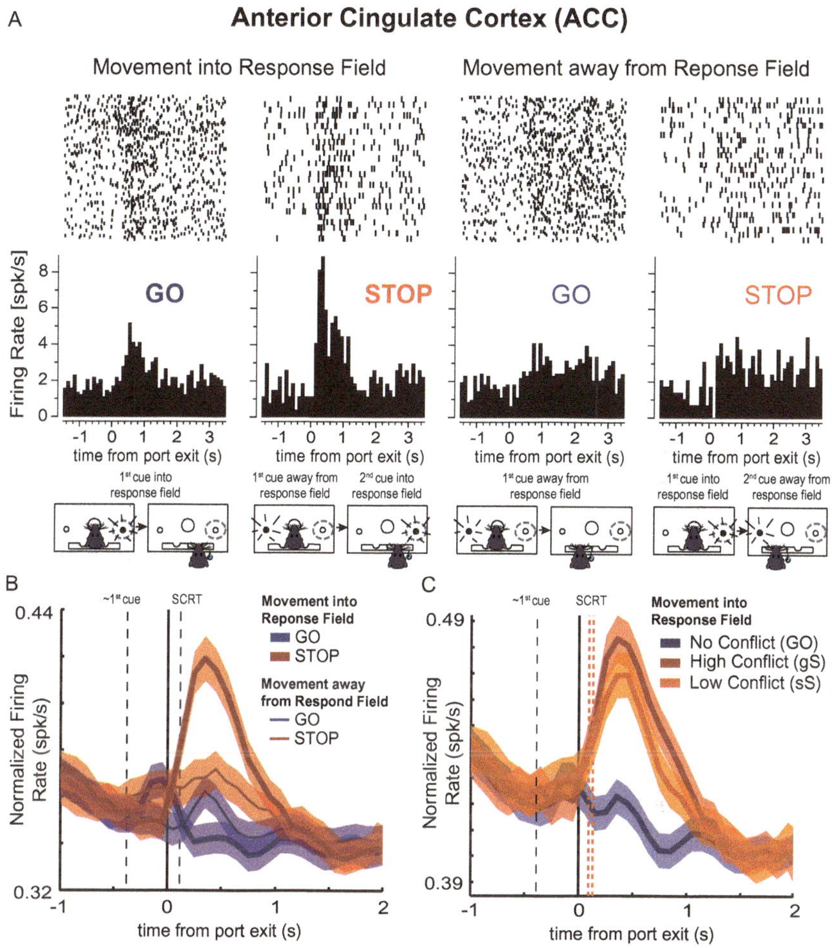

Fig. 3 ACC fires more strongly on STOP-change trials. (A) Raster plots of an example unit showing higher firing on STOP-change trials when the movement was made into the unit's response field. (B) Population histograms mirror single unit examples showing selectively higher firing on STOP-change trials when the direction of the movement was made into the response field (i.e., preferred direction). (C) ACC firing is modulated by the degree of conflict, showing higher firing on 'high conflict' trials (red) (i.e., trials where a STOP-change trial is preceded by a GO trial) relative to low conflict trials (orange) (i.e., trials where a STOP-change trials is preceded by another STOP-change trial). Note that both high and low conflict trials are higher than GO trials (blue). *Figures adapted from Bryden, D.W., Brockett, A.T., Blume, E., Heatley, K., Zhao, A., Roesch, M.R. (2019). Single neurons in anterior cingulate cortex signal the need to change action during performance of a stop-change task that induces response competition.* Cerebral Cortex *(New York, N.Y.: 1991), 29, 1020–1031.*

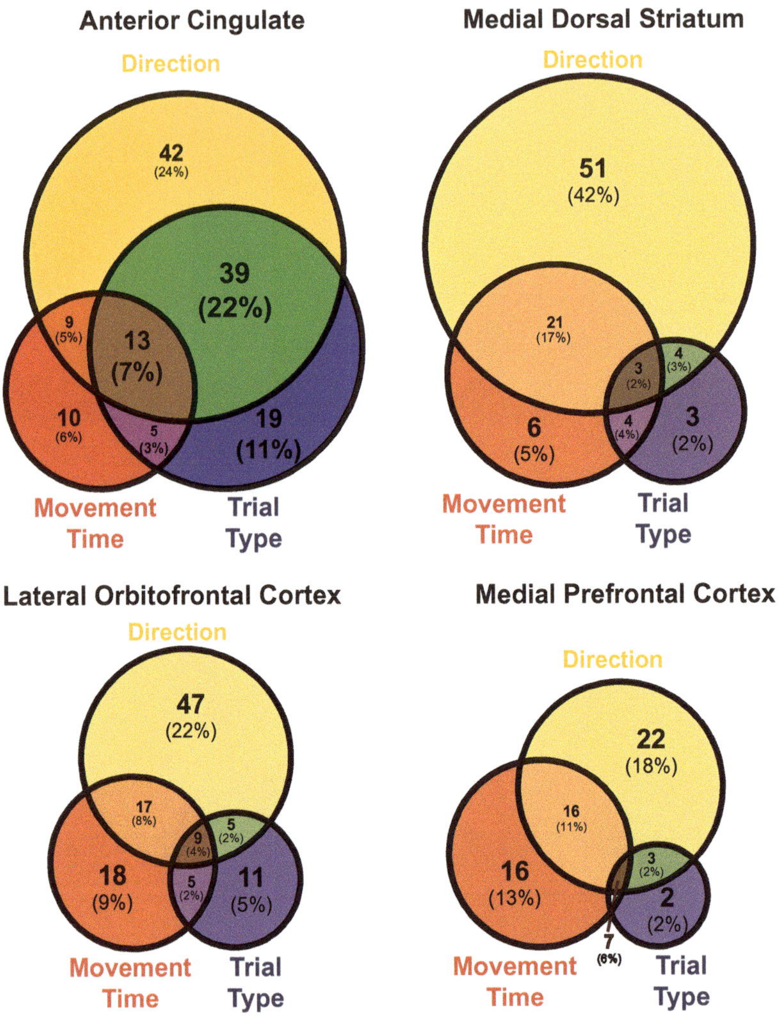

	Direction		Movement Time		Trial Type	
β-valence	+	-	+	-	+	-
mDS	***59****	20	6	***28****	9	5
lOFC	30	48	20	29	19	11
mPFC	21	24	9	***28****	8	5
ACC	46	57	22	15	***67****	9

Fig. 4 Neurons in ACC are highly selective for direction and trial-type on the stop-change task (e.g., Fig. 3A). (A–D). Multiple regression on firing during the response epoch (port exit to fluid well entry) for neurons in ACC (A), medial dorsal striatum (B), lateral orbitofrontal cortex (C), and medial prefrontal cortex (a.k.a. prelimbic region)

Thus, single ACC neurons are not globally representing the presence of conflict, but signaling the need for conflict resolution in a specific response direction. This finding harkens back to the human single neuron data demonstrating that conflict signals in ACC may serve to amplify relevant task features to adapt behavior accordingly (Ebitz et al., 2020).

To determine if these unique conflict-likes signals in rat ACC were indeed necessary for the adaptation of behavior and in downstream motor-related signals, we performed a second study where we lesioned ACC unilaterally in rats performing the identical STOP-change task while simultaneously recording from DMS in the same hemisphere. In line with the hypothesis that ACC is necessary to resolve response conflict, ACC lesions increased STOP errors and lengthened the time necessary for rats to inhibit and redirect behavior on successful STOP-change trials (Brockett et al., 2020). These deficits in behavior suggest that the brain and therefore the rat is less able to reconcile the conflict between the two competing response directions. Indeed, average response selectivity across the population of DMS neurons did not significantly discriminate between the two directions and the frequency of single neurons that accurately (i.e., like the single neuron in Fig. 2B) signaled the appropriate action were fewer after ACC lesions. Thus, without ACC, DMS failed to signal the correct action on STOP-change trials. Further, when rats made errors, activity strongly represented the location of the first cue as in example unit 1 (Fig. 2A). This over representation on the incorrect response on

(D). Circle size represents the relative proportions of neurons showing significant partial r^2 values for the labeled task parameters (direction (yellow), movement time (red), trial-type (blue)). Positive β-values indicate greater firing for the contralateral direction (direction), greater firing for slower movement time (movement time), and greater firing for STOP over GO trials (Trial-Type). Asterisks indicated significantly more β-values for one valence within a parameter (binomial sign test; $P < 0.05$). Only in ACC, were proportions of neurons that increased firing on STOP in the significant majority (67 vs 9 neurons). *Figures adapted from Bryden, D.W., Burton, A.C., Kashtelyan, V., Barnett, B.R., Roesch, M.R. (2012). Response inhibition signals and miscoding of direction in dorsomedial striatum.* Frontiers in Integrative Neuroscience, 6, *69; Bryden, D.W., Burton, A.C., Barnett, B.R., Cohen, V.J., Hearn, T.N., Jones, E.A. et al. (2016) Prenatal nicotine exposure impairs executive control signals in medial prefrontal cortex.* Neuropsychopharmacology: Official Publication of the American College of Neuropsychopharmacology, 41, *716–725; Bryden, D.W., Brockett, A.T., Blume, E., Heatley, K., Zhao, A., Roesch, M.R. (2019). Single neurons in anterior cingulate cortex signal the need to change action during performance of a stop-change task that induces response competition.* Cerebral Cortex *(New York, N.Y.: 1991), 29, 1020–1031; and Bryden, D.W., Roesch, M.R. (2015). Executive control signals in orbitofrontal cortex during response inhibition.* Journal of Neuroscience: The Official Journal of the Society for Neuroscience, 35, *3903–3914.*

STOP-change trials is illustrated across the entire population of recorded DMS neurons in Fig. 5C. Average firing quickly rose to the first cue light that was presented in the preferred direction (i.e., into the response field) and did not shut down until after the SCRT (i.e., after it was too late). In addition to

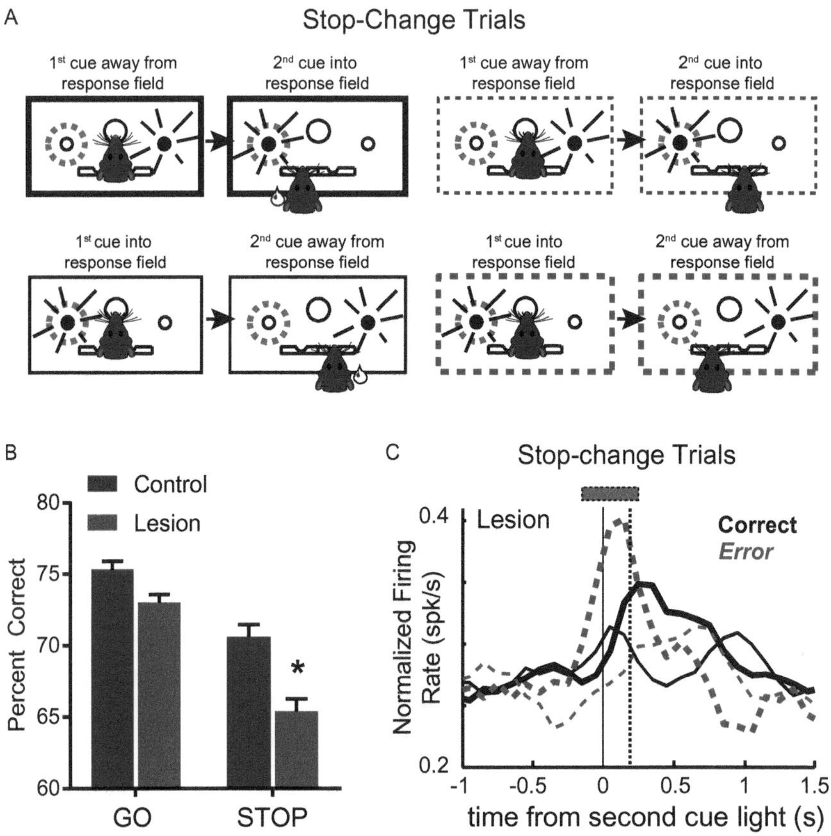

Fig. 5 Unilateral ACC lesions disrupt STOP trial performance and firing in DMS. (A) Illustrations of responding into (thick) and away from (thin) a neuron's response field on correct (solid) and errant (dashed) STOP-change trials. (B) Rats with unilateral ACC lesions perform significantly worse on STOP-change trials compared to control rats. (C) Population histogram for lesioned rats ($n = 53$) aligned to the onset of the STOP cue. Solid lines represent firing on STOP-Change trials where the rat made a correct response and dashed line represent firing on STOP-Change trials when the rat committed an error. Thick and thin line represent firing into (thick) and away from *thin) a neuron's preferred direction. Tick marks (gray bar) represent significance when comparing thick to thin lines within trial type. *Figures adapted from Brockett, A.T., Tennyson, S.S., deBettencourt, C.A., Gaye, F., & Roesch, M.R. (2020). Anterior cingulate cortex is necessary for adaptation of action plans.* Proceedings of the National Academy of Sciences, *117, 6196–6204.*

ACC being necessary to be able for directional signals in DMS to adapt mid-trial, ACC was also necessary for modulation of firing on subsequent trials during conflict adaptation that deemphasized processing of the first cue, while strengthening directional selectivity associated with the second cue (Brockett et al., 2020).

Overall these results demonstrate that ACC—which predominately fired more strongly under high-conflict trials—is necessary for the reduction and amplification of the inappropriate and appropriate response signals in DMS, respectively, consistent with the impaired ability of rats to perform STOP-change trials after ACC lesions.

4. Conclusion

The debate over ACC function has been tenuous, but at the same time, has motivated decades of elegant work, elevating the status of research involving cognitive control to arguably the pinnacle of cognitive neuroscience thus far. The debate and subsequent attempts to support or overthrow the current schools of thought also highlights the importance of theory driven research. This work has incorporated the full spectrum of neuroscience and computational techniques spanning all major methods and models of neuroscience research to show that ACC is fundamentally important for guiding and adapting behavior. While the debate is far from over, single unit recording work is beginning to uncover the neural signals that contribute to these functions. It is becoming clear across the field of behavioral neurophysiology that firing of single neurons and ensembles of neurons represent multiplexed information across multiple domains. ACC seems to be no exception, carrying signals related to error detection, reinforcement/feedback, attention, value, and conflict to name a few. Ultimately, ACC seems to use this myriad of signals to modulate attention and task-relevant/irrelevant signals in downstream structures so that decisions can be made and action adapted when plans change.

Conflict of interest

The authors declare no biomedical financial interests or potential conflicts of interest.

Acknowledgments

This work was supported by the following grants: NIMH: MH1117836 to A.T.B. and NIDA: DA031695 to M.R.R.

References

Amiez, C., Joseph, J. P., & Procyk, E. (2006). Reward encoding in the monkey anterior cingulate cortex. *Cerebral Cortex (New York, N.Y.: 1991)*, *16*, 1040–1055.

Andreasen, N. C., O'Leary, D. S., Cizadlo, T., Arndt, S., Rezai, K., Watkins, G. L., et al. (1995). Remembering the past: Two facets of episodic memory explored with positron emission tomography. *The American Journal of Psychiatry*, *152*, 1576–1585.

Baker, S. C., Rogers, R. D., Owen, A. M., Frith, C. D., Dolan, R. J., Frackowiak, R. S., et al. (1996). Neural systems engaged by planning: A PET study of the tower of London task. *Neuropsychologia*, *34*, 515–526.

Barch, D. M., Braver, T. S., Akbudak, E., Conturo, T., Ollinger, J., & Snyder, A. (2001). Anterior cingulate cortex and response conflict: Effects of response modality and processing domain. *Cerebral Cortex*, *11*, 837–848.

Bench, C. J., Frith, C. D., Grasby, P. M., Friston, K. J., Paulesu, E., Frackowiak, R. S., et al. (1993). Investigations of the functional anatomy of attention using the Stroop test. *Neuropsychologia*, *31*, 907–922.

Blanchard, T. C., & Hayden, B. Y. (2014). Neurons in dorsal anterior cingulate cortex signal postdecisional variables in a foraging task. *Journal of Neuroscience: The Official Journal of the Society for Neuroscience*, *34*, 646–655.

Botvinick, M. M. (2007). Conflict monitoring and decision making: Reconciling two perspectives on anterior cingulate function. *Cognitive, Affective, & Behavioral Neuroscience*, *7*, 356–366.

Botvinick, M. M., Braver, T. S., Barch, D. M., Carter, C. S., & Cohen, J. D. (2001). Conflict monitoring and cognitive control. *Psychological Review*, *108*, 624–652.

Botvinick, M. M., Cohen, J. D., & Carter, C. S. (2004). Conflict monitoring and anterior cingulate cortex: An update. *Trends in Cognitive Sciences*, *8*, 539–546.

Botvinick, M., Nystrom, L. E., Fissell, K., Carter, C. S., & Cohen, J. D. (1999). Conflict monitoring versus selection-for-action in anterior cingulate cortex. *Nature*, *402*, 179–181.

Brockett, A. T., Tennyson, S. S., deBettencourt, C. A., Gaye, F., & Roesch, M. R. (2020). Anterior cingulate cortex is necessary for adaptation of action plans. *Proceedings of the National Academy of Sciences*, *117*, 6196–6204.

Bryden, D. W., Brockett, A. T., Blume, E., Heatley, K., Zhao, A., & Roesch, M. R. (2019). Single neurons in anterior cingulate cortex signal the need to change action during performance of a stop-change task that induces response competition. *Cerebral Cortex (New York, N.Y.: 1991)*, *29*, 1020–1031.

Bryden, D. W., Burton, A. C., Barnett, B. R., Cohen, V. J., Hearn, T. N., Jones, E. A., et al. (2016). Prenatal nicotine exposure impairs executive control signals in medial prefrontal cortex. *Neuropsychopharmacology: Official Publication of the American College of Neuropsychopharmacology*, *41*, 716–725.

Bryden, D. W., Burton, A. C., Kashtelyan, V., Barnett, B. R., & Roesch, M. R. (2012). Response inhibition signals and miscoding of direction in dorsomedial striatum. *Frontiers in Integrative Neuroscience*, *6*, 69.

Bryden, D. W., Johnson, E. E., Tobia, S. C., Kashtelyan, V., & Roesch, M. R. (2011). Attention for learning signals in anterior cingulate cortex. *Journal of Neuroscience: The Official Journal of the Society for Neuroscience*, *31*, 18266–18274.

Bryden, D. W., & Roesch, M. R. (2015). Executive control signals in orbitofrontal cortex during response inhibition. *Journal of Neuroscience: The Official Journal of the Society for Neuroscience*, *35*, 3903–3914.

Buckner, R. L., Petersen, S. E., Ojemann, J. G., Miezin, F. M., Squire, L. R., & Raichle, M. E. (1995). Functional anatomical studies of explicit and implicit memory retrieval tasks. *Journal of Neuroscience: The Official Journal of the Society for Neuroscience*, *15*, 12–29.

Bush, G., Whalen, P. J., Rosen, B. R., Jenike, M. A., McInerney, S. C., & Rauch, S. L. (1998). The counting stroop: An interference task specialized for functional neuroimaging: Validation study with functional MRI. *Human Brain Mapping*, *6*, 270–282.
Cai, X., & Padoa-Schioppa, C. (2012). Neuronal encoding of subjective value in dorsal and ventral anterior cingulate cortex. *Journal of Neuroscience: The Official Journal of the Society for Neuroscience*, *32*, 3791–3808.
Carter, C. S., Braver, T. S., Barch, D. M., Botvinick, M. M., Noll, D., & Cohen, J. D. (1998). Anterior cingulate cortex, error detection, and the online monitoring of performance. *Science*, *280*, 747–749.
Carter, C. S., Mintun, M., & Cohen, J. D. (1995). Interference and facilitation effects during selective attention: An H215O PET study of Stroop task performance. *NeuroImage*, *2*, 264–272.
Casey, B. J., Trainor, R. J., Orendi, J. L., Schubert, A. B., Nystrom, L. E., Giedd, J. N., et al. (1997). A developmental functional MRI study of prefrontal activation during performance of a go-no-go task. *Journal of Cognitive Neuroscience*, *9*, 835–847.
Cole, M. W., Yeung, N., Freiwald, W. A., & Botvinick, M. (2009). Cingulate cortex: Diverging data from humans and monkeys. *Trends in Neurosciences*, *32*, 566–574.
Corbetta, M., Miezin, F. M., Dobmeyer, S., Shulman, G. L., & Petersen, S. E. (1991). Selective and divided attention during visual discriminations of shape, color, and speed: Functional anatomy by positron emission tomography. *Journal of Neuroscience: The Official Journal of the Society for Neuroscience*, *11*, 2383–2402.
Davis, K. D., Hutchison, W. D., Lozano, A. M., Tasker, R. R., & Dostrovsky, J. O. (2000). Human anterior cingulate cortex neurons modulated by attention-demanding tasks. *Journal of Neurophysiology*, *83*, 3575–3577.
Davis, K. D., Taylor, K. S., Hutchison, W. D., Dostrovsky, J. O., McAndrews, M. P., Richter, E. O., et al. (2005). Human anterior cingulate cortex neurons encode cognitive and emotional demands. *Journal of Neuroscience: The Official Journal of the Society for Neuroscience*, *25*, 8402–8406.
de Berker, A. O., Kurth-Nelson, Z., Rutledge, R. B., Bestmann, S., & Dolan, R. J. (2019). Computing value from quality and quantity in human decision-making. *The Journal of Neuroscience*, *39*, 163–176.
Dehaene, S., Posner, M. I., & Tucker, D. M. (1994). Localization of a neural system for error detection and compensation. *Psychological Science*, *5*, 303–305.
Deiber, M. P., Passingham, R. E., Colebatch, J. G., Friston, K. J., Nixon, P. D., & Frackowiak, R. S. (1991). Cortical areas and the selection of movement: A study with positron emission tomography. *Experimental Brain Research*, *84*, 393–402.
D'Esposito, M., Detre, J. A., Alsop, D. C., Shin, R. K., Atlas, S., & Grossman, M. (1995). The neural basis of the central executive system of working memory. *Nature*, *378*, 279–281.
Dias, R., & Aggleton, J. P. (2000). Effects of selective excitotoxic prefrontal lesions on acquisition of nonmatching- and matching-to-place in the T-maze in the rat: Differential involvement of the prelimbic-infralimbic and anterior cingulate cortices in providing behavioural flexibility. *The European Journal of Neuroscience*, *12*, 4457–4466.
Ebitz, R. B., & Platt, M. L. (2015). Neuronal activity in primate dorsal anterior cingulate cortex signals task conflict and predicts adjustments in pupil-linked arousal. *Neuron*, *85*, 628–640.
Ebitz, R. B., Smith, E. H., Horga, G., Schevon, C. A., Yates, M. J., McKhann, G. M., et al. (2020). Human dorsal anterior cingulate neurons signal conflict by amplifying task-relevant information. *BioRxiv*. 2020.03.14.991745.
Falkenstein, M., Hohnsbein, J., & Hoormann, J. (1995). Event-related potential correlates of errors in reaction tasks. *Electroencephalography and Clinical Neurophysiology. Supplement*, *44*, 287–296.

Falkenstein, M., Hohnsbein, J., Hoormann, J., & Blanke, L. (1991). Effects of crossmodal divided attention on late ERP components. II. Error processing in choice reaction tasks. *Electroencephalography and Clinical Neurophysiology*, *78*, 447–455.

Frank, M. J., Woroch, B. S., & Curran, T. (2005). Error-related negativity predicts reinforcement learning and conflict biases. *Neuron*, *47*, 495–501.

Friston, K. J., Frith, C. D., Liddle, P. F., & Frackowiak, R. S. (1993). Functional connectivity: The principal-component analysis of large (PET) data sets. *Journal of Cerebral Blood Flow and Metabolism: Official Journal of the International Society of Cerebral Blood Flow and Metabolism*, *13*, 5–14.

Frith, C. D., Friston, K., Liddle, P. F., & Frackowiak, R. S. (1991a). Willed action and the prefrontal cortex in man: A study with PET. *Proceedings of the Biological Sciences*, *244*, 241–246.

Frith, C. D., Friston, K. J., Liddle, P. F., & Frackowiak, R. S. J. (1991b). A PET study of word finding. *Neuropsychologia*, *29*, 1137–1148.

Gehring, W. J., Goss, B., Coles, M. G., Meyer, D. E., & Donchin, E. (1993). A neural system for error detection and compensation. *Psychological Science*, *4*, 385–390.

Gehring, W. J., & Willoughby, A. R. (2002). The medial frontal cortex and the rapid processing of monetary gains and losses. *Science*, *295*, 2279–2282.

George, M. S., Ketter, T. A., Parekh, P. I., Rosinsky, N., Ring, H., Casey, B. J., et al. (1994). Regional brain activity when selecting a response despite interference: An H2 (15) O PET study of the stroop and an emotional stroop. *Human Brain Mapping*, *1*, 194–209.

Grasby, P. M., Frith, C. D., Friston, K. J., Bench, C., Frackowiak, R. S., & Dolan, R. J. (1993). Functional mapping of brain areas implicated in auditory- -verbal memory function. *Brain: A Journal of Neurology*, *116*(Pt 1), 1–20.

Gratton, G., Coles, M. G., Sirevaag, E. J., Eriksen, C. W., & Donchin, E. (1988). Pre- and poststimulus activation of response channels: A psychophysiological analysis. *Journal of Experimental Psychology. Human Perception and Performance*, *14*, 331–344.

Hadland, K. A., Rushworth, M. F. S., Gaffan, D., & Passingham, R. E. (2003). The anterior cingulate and reward-guided selection of actions. *Journal of Neurophysiology*, *89*, 1161–1164.

Hayden, B. Y., Heilbronner, S. R., Pearson, J. M., & Platt, M. L. (2011). Surprise signals in anterior cingulate cortex: Neuronal encoding of unsigned reward prediction errors driving adjustment in behavior. *Journal of Neuroscience: The Official Journal of the Society for Neuroscience*, *31*, 4178–4187.

Hayden, B. Y., Pearson, J. M., & Platt, M. L. (2011). Neuronal basis of sequential foraging decisions in a patchy environment. *Nature Neuroscience*, *14*, 933–939.

Holroyd, C. B., & Coles, M. G. H. (2002). The neural basis of human error processing: Reinforcement learning, dopamine, and the error-related negativity. *Psychological Review*, *109*, 679–709.

Hyman, J. M., Holroyd, C. B., & Seamans, J. K. (2017). A novel neural prediction error found in anterior cingulate cortex ensembles. *Neuron*, *95*, 447–456. e3.

Ito, S., Stuphorn, V., Brown, J. W., & Schall, J. D. (2003). Performance monitoring by the anterior cingulate cortex during saccade countermanding. *Science*, *302*, 120–122.

Janer, K. W., & Pardo, J. V. (1991). Deficits in selective attention following bilateral anterior cingulotomy. *Journal of Cognitive Neuroscience*, *3*, 231–241.

Jueptner, M., Frith, C. D., Brooks, D. J., Frackowiak, R. S., & Passingham, R. E. (1997). Anatomy of motor learning II. Subcortical structures and learning by trial and error. *Journal of Neurophysiology*, 77, 1325–1337.

Kawai, T., Yamada, H., Sato, N., Takada, M., & Matsumoto, M. (2015). Roles of the lateral Habenula and anterior cingulate cortex in negative outcome monitoring and Behavioral adjustment in nonhuman primates. *Neuron*, *88*, 792–804.

Kennerley, S. W., Behrens, T. E. J., & Wallis, J. D. (2011). Double dissociation of value computations in orbitofrontal and anterior cingulate neurons. *Nature Neuroscience*, *14*, 1581–1589.

Kennerley, S. W., Dahmubed, A. F., Lara, A. H., & Wallis, J. D. (2009). Neurons in the frontal lobe encode the value of multiple decision variables. *Journal of Cognitive Neuroscience*, *21*, 1162–1178.

Kennerley, S. W., Walton, M. E., Behrens, T. E. J., Buckley, M. J., & Rushworth, M. F. S. (2006). Optimal decision making and the anterior cingulate cortex. *Nature Neuroscience*, *9*, 940–947.

Kerns, J. G., Cohen, J. D., MacDonald, A. W., III, Cho, R. Y., Stenger, V. A., & Carter, C. S. (2004). Anterior cingulate conflict monitoring and adjustments in control. *Science*, *303*, 1023–1026.

Kolling, N., Behrens, T. E. J., Mars, R. B., & Rushworth, M. F. S. (2012). Neural mechanisms of foraging. *Science*, *336*, 95–98.

Kolling, N., Wittmann, M. K., Behrens, T. E. J., Boorman, E. D., Mars, R. B., & Rushworth, M. F. S. (2016). Value, search, persistence and model updating in anterior cingulate cortex. *Nature Neuroscience*, *19*, 1280–1285.

Luk, C.-H., & Wallis, J. D. (2013). Choice coding in frontal cortex during stimulus-guided or action-guided decision-making. *Journal of Neuroscience: The Official Journal of the Society for Neuroscience*, *33*, 1864–1871.

MacLeod, C. M. (1991). Half a century of research on the Stroop effect: An integrative review. *Psychological Bulletin*, *109*, 163–203.

Matsumoto, K., Suzuki, W., & Tanaka, K. (2003). Neuronal correlates of goal-based motor selection in the prefrontal cortex. *Science*, *301*, 229–232.

Mesulam, M. M. (1981). A cortical network for directed attention and unilateral neglect. *Annals of Neurology*, *10*, 309–325.

Nakamura, K., Roesch, M. R., & Olson, C. R. (2005). Neuronal activity in macaque SEF and ACC during performance of tasks involving conflict. *Journal of Neurophysiology*, *93*, 884–908.

Oliveira FTP, McDonald JJ, Goodman D (2007) Performance monitoring in the anterior cingulate is not all error related: expectancy deviation and the representation of action-outcome associations. Journal of Cognitive Neuroscience 19(12): 1994–2004. https://doi.org/10.1162/jocn.2007.19.12.1994. PMID: 17892382.

Papez, J. W. (1937). A proposed mechanism of emotion. *Archives of Neurology and Psychiatry*, *38*, 725–743.

Pardo, J. V., Pardo, P. J., Janer, K. W., & Raichle, M. E. (1990). The anterior cingulate cortex mediates processing selection in the Stroop attentional conflict paradigm. *Proceedings of the National Academy of Sciences of the United States of America*, *87*, 256–259.

Paus, T., Petrides, M., Evans, A. C., & Meyer, E. (1993). Role of the human anterior cingulate cortex in the control of oculomotor, manual, and speech responses: A positron emission tomography study. *Journal of Neurophysiology*, *70*, 453–469.

Petersen, S. E., Fox, P. T., Posner, M. I., Mintun, M., & Raichle, M. E. (1988). Positron emission tomographic studies of the cortical anatomy of single-word processing. *Nature*, *331*, 585–589.

Petersen, S. E., Fox, P. T., Posner, M. I., Mintun, M., & Raichle, M. E. (1989). Positron emission tomographic studies of the processing of singe words. *Journal of Cognitive Neuroscience*, *1*, 153–170.

Playford, E. D., Jenkins, I. H., Passingham, R. E., Nutt, J., Frackowiak, R. S., & Brooks, D. J. (1992). Impaired mesial frontal and putamen activation in Parkinson's disease: A positron emission tomography study. *Annals of Neurology*, *32*, 151–161.

Posner, M. I., & Dehaene, S. (1994). Attentional networks. *Trends in Neurosciences*, *17*, 75–79.

Posner, M. I., Petersen, S. E., Fox, P. T., & Raichle, M. E. (1988). Localization of cognitive operations in the human brain. *Science, 240*, 1627–1631.

Procyk, E., Tanaka, Y. L., & Joseph, J. P. (2000). Anterior cingulate activity during routine and non-routine sequential behaviors in macaques. *Nature Neuroscience, 3*, 502–508.

Quilodran R, Rothe M, Procyk E (2008) Behavioral shifts and activation valuation in the anterior cingulate cortex. Neuron 57(2): 314–325. https://doi.org/10.1016/j.neuron.2007.11.031. PMID: 18215627.

Raichle, M. E., Fiez, J. A., Videen, T. O., MacLeod, A. M., Pardo, J. V., Fox, P. T., et al. (1994). Practice-related changes in human brain functional anatomy during nonmotor learning. *Cerebral Cortex (New York, N.Y.: 1991), 4*, 8–26.

Rothé M, Quilodran R, Sallet J, Procyk E (2011) Coordination of high gamma activity in anterior cingulate and lateral prefrontal cortical areas during adaptation. Journal of Neuroscience: The Official Journal of the Society for Neuroscience 31(31): 11110–11117. https://doi.org/10.1523/JNEUROSCI.1016-11.2011. PMID: 21813672.

Rushworth MFS & Behren TEJ (2008) Choice, uncertainty and value in prefrontal and cingulate cortex. Nature Neuroscience 11(4): 389–397. https://doi.org/10.1038/nn2066. PMID: 18368045.

Rushworth, M. F. S., Hadland, K. A., Gaffan, D., & Passingham, R. E. (2003). The effect of cingulate cortex lesions on task switching and working memory. *Journal of Cognitive Neuroscience, 15*, 338–353.

Rushworth, M. F. S., Walton, M. E., Kennerley, S. W., & Bannerman, D. M. (2004). Action sets and decisions in the medial frontal cortex. *Trends in Cognitive Sciences, 8*, 410–417.

Scheffers, M. K., Coles, M. G., Bernstein, P., Gehring, W. J., & Donchin, E. (1996). Event-related brain potentials and error-related processing: An analysis of incorrect responses to go and no-go stimuli. *Psychophysiology, 33*, 42–53.

Schneider KN, Sciarillo XA, Nudelman JL, Cheer JR, Roesch MR (2020) Anterior cingulate cortex signals attention in a social paradigm that manipulates reward and shock. Current Biology 30(19): 3724–3735. https://doi.org/10.1016/j.cub.2020.07.039. PMID: 32763169.

Shenhav, A., Cohen, J. D., & Botvinick, M. M. (2016). Dorsal anterior cingulate cortex and the value of control. *Nature Neuroscience, 19*, 1286–1291.

Shenhav, A., Straccia, M. A., Botvinick, M. M., & Cohen, J. D. (2016). Dorsal anterior cingulate and ventromedial prefrontal cortex have inverse roles in both foraging and economic choice. *Cognitive, Affective, & Behavioral Neuroscience, 16*, 1127–1139.

Shenhav, A., Straccia, M. A., Cohen, J. D., & Botvinick, M. M. (2014). Anterior cingulate engagement in a foraging context reflects choice difficulty, not foraging value. *Nature Neuroscience, 17*, 1249–1254.

Sheth, S. A., Mian, M. K., Patel, S. R., Asaad, W. F., Williams, Z. M., Dougherty, D. D., et al. (2012). Human dorsal anterior cingulate cortex neurons mediate ongoing behavioural adaptation. *Nature, 488*, 218–221.

Stroop, J. R. (1935). Studies of interference in serial verbal reactions. *Journal of Experimental Psychology, 18*, 643–662.

Stuphorn, V., Taylor, T. L., & Schall, J. D. (2000). Performance monitoring by the supplementary eye field. *Nature, 408*, 857–860.

Taylor, S. F., Kornblum, S., Minoshima, S., Oliver, L. M., & Koeppe, R. A. (1994). Changes in medial cortical blood flow with a stimulus-response compatibility task. *Neuropsychologia, 32*, 249–255.

Thompson-Schill, S. L., D'Esposito, M., Aguirre, G. K., & Farah, M. J. (1997). Role of left inferior prefrontal cortex in retrieval of semantic knowledge: A reevaluation. *Proceedings of the National Academy of Sciences of the United States of America, 94*, 14792–14797.

Totah, N. K. B., Kim, Y. B., Homayoun, H., & Moghaddam, B. (2009). Anterior cingulate neurons represent errors and preparatory attention within the same behavioral sequence. *Journal of Neuroscience: The Official Journal of the Society for Neuroscience, 29*, 6418–6426.

Turken, A. U., & Swick, D. (1999). Response selection in the human anterior cingulate cortex. *Nature Neuroscience*, *2*, 920–924.

Vázquez, D., Pribut, H. J., Burton, A. C., Tennyson, S. S., & Roesch, M. R. (2020). Prior cocaine self-administration impairs attention signals in anterior cingulate cortex. *Neuropsychopharmacology*, *45*, 833–841.

Walton, M. E., Bannerman, D. M., Alterescu, K., & Rushworth, M. F. S. (2003). Functional specialization within medial frontal cortex of the anterior cingulate for evaluating effort-related decisions. *Journal of Neuroscience: The Official Journal of the Society for Neuroscience*, *23*, 6475–6479.

Walton, M. E., Bannerman, D. M., & Rushworth, M. F. S. (2002). The role of rat medial frontal cortex in effort-based decision making. *Journal of Neuroscience: The Official Journal of the Society for Neuroscience*, *22*, 10996–11003.

Walton, M. E., Kennerley, S. W., Bannerman, D. M., Phillips, P. E. M., & Rushworth, M. F. S. (2006). Weighing up the benefits of work: Behavioral and neural analyses of effort-related decision making. *Neural Networks: The official Journal of the International Neural Network Society*, *19*, 1302–1314.

Yetkin, F. Z., Hammeke, T. A., Swanson, S. J., Morris, G. L., Mueller, W. M., McAuliffe, T. L., et al. (1995). A comparison of functional MR activation patterns during silent and audible language tasks. *AJNR. American Journal of Neuroradiology*, *16*, 1087–1092.

CHAPTER ELEVEN

Oscillations as a window into neuronal mechanisms underlying dorsal anterior cingulate cortex function

Benjamin Voloh*, Rachel Knoebl, Benjamin Y. Hayden, and Jan Zimmermann
Department of Neuroscience, University of Minnesota, Minneapolis, MN, United States
*Corresponding author: e-mail address: voloh001@umn.edu

Contents

Abstract

The function of dorsal Anterior Cingulate Cortex (dACC) remains poorly understood. While many methods, spanning bottom-up and top-down approaches, have been deployed, the view they offer is often conflicting. Integrating bottom-up and top-down approaches requires an intermediary with sufficient explanatory power, theoretical development, and empirical support. Oscillations in the local field potential (LFP) provide such a link. LFP oscillations arise from empirically well-characterized neuronal circuit motifs. Synchronizing the firing of individual units has appealing properties to bind disparate brain regions and propagate information, including gating, routing, and

International Review of Neurobiology, Volume 158
ISSN 0074-7742
https://doi.org/10.1016/bs.irn.2020.11.003

coding. Moreover, the LFP, rather than single unit activity, more closely relates to macroscale recordings, such as the electroencephalogram and functional magnetic resonance imaging. Thus, LFP oscillations are a critical link that allow for the inference of neuronal micro-circuitry underlying macroscopic brain recordings.

1. Introduction

The dorsal anterior cingulate cortex (dACC) is an enigmatic structure located on the medial wall of the cerebral cortex. It is associated with executive control, learning, adjustment, economic choice, and self-control (Heilbronner & Hayden, 2016; Morecraft et al., 2012; Paus, 2001; Shenhav et al., 2017). It has prominent anatomical connections to other structures in the prefrontal cortex, including the orbitofrontal cortex and pre/motor structures, as well as to the amygdala (Barbas & Pandya, 1989; Morecraft et al., 2012; Vogt & Pandya, 1987). It even includes direct connections to the spinal cord, as well as the hypothalamus, suggesting a role in autonomic control (Passingham & Wise, 2012). Dysregulation of the dACC is associated with several psychiatric diseases, including depression, obsessive compulsive disorder, and drug addiction, and it has been argued to play a central role in integrating and communicating information across the cortex (Goodkind et al., 2015).

Like other brain structures, dACC is amenable to study via multiple experimental methods. Single unit electrophysiology and anatomy, which we have called bottom-up methods, are two sources (Heilbronner & Hayden, 2016). Another source of information comes from top-down methods, which include functional neuroimaging (fMRI), electroencephalography (EEG), magnetoencephalography, and lesion studies. Reconciling a portrait of the dACC derived from the different methods has proven to be a challenge. Indeed, there is perhaps no other brain region wherein the disagreement over methods—epitomized by the debate over conflict detection - has proven to be so central to discussions of its function (Cole, Yeung, Freiwald, & Botvinick, 2009; Ebitz & Hayden, 2016; Holroyd, Nieuwenhuis, Mars, & Coles, 2004; Schall & Emeric, 2010).

In theory, we should be able to leverage the large number of approaches into the dACC, triangulate across methods, and obtain a better understanding of its function. Our central argument is that the consideration of multiple methods can indeed provide a better and more accurate understanding of the dACC. But for that to happen, the methods need to be compared in a sophisticated way that respects the strengths and weaknesses of each.

In particular, while the bottom-up approaches continue to provide the gold standard ground truth, they provide only a part of the story, one that is impoverished by failures to be integrated with other methods.

An important lemma of our argument will be that the local field potential (LFP) serves as a critical intellectual intermediary between top-down and bottom-up methods. The LFP is recorded intracranially, and thus, typically in animal models (although there is growing interest in human recordings (Widge, Heilbronner, & Hayden, 2019)). The LFP represents the summed transmembrane currents in a ~200 μm radius (Einevoll, Kayser, Logothetis, & Panzeri, 2013). The relative prominence of different frequency components f comprising the signal can be determined with a spectrogram. An LFP spectrogram typically has a $1/f$ structure. Frequency-specific "bumps" above this $1/f$ structure can shed insights on engaged neuronal circuits. Typically, these bumps are centered around canonical frequencies, such as theta (4–10 Hz), alpha/beta (10 – 30), or gamma (30–100 Hz). The frequency of interest can change with age (e.g., Tran, Rolle, Gazzaley, & Voytek, 2020), and does exhibit some inter-species variability (Javitt et al., 2020), though it remains convenient to discuss results in terms of these canonical bands (Javitt et al., 2020).

The present work is organized as follows. First, we briefly review insights derived from bottom-up approaches of dACC function (covered in detail in Heilbronner & Hayden, 2016). Second, starting with the observation that task-dependent changes in dACC single unit activity is modulated as a function of LFP oscillation phase, we describe circuit and network insights from LFP oscillations. Third, we elucidate key computational properties of phase modulated spiking. Finally, we describe efforts to reconcile micro- and macro-scale insights that leverage LFP recordings.

2. Insights from single unit studies: dACC as an economic structure

Neurons in the dACC clearly participate in economic computations (Heilbronner & Hayden, 2016). Key findings include the fact that firing rates of neurons in dACC encode the values of offered options and of chosen options, as well as values of obtained options (Amiez, Joseph, & Procyk, 2006; Azab & Hayden, 2018; Cai & Padoa-Schioppa, 2012; Hayden, Heilbronner, Pearson, & Platt, 2011; Rudebeck, Mitz, Chacko, & Murray, 2013; Seo & Lee, 2007). Those responses are modulated by the prior likelihood of the outcome and by the surprisingness of the outcome (Hayden et al., 2011). It is well established that dACC has a strong motor repertoire; consequently, it is not surprising that dACC neurons track the positions of offers and of choices (Hayden & Platt, 2010; Strait et al., 2016).

Recent work from our lab has further clarified the economic computations of the dACC (Azab & Hayden, 2017, 2020). That work puts dACC precisely in the middle of two key value-related computations, namely, value formation and value comparison. Specifically, we examined responses of dACC neurons in an economic choice task in which subjects chose between two staggered gambles defined by varying probabilities. Behavior depended systematically on the values of both outcomes, as did firing rates of single neurons (Azab & Hayden, 2020). This type of pattern has been observed before, and has been used to argue that neurons represent an *abstract* value signal, that is, one that is independent of the factors that influence value.

However, we used some novel statistical techniques to ask whether dACC neurons carry a *pure* value signal—that is, whether dACC neurons had fully shed information about the components of value. Surprisingly, we found that value-related encoding in dACC is only partial. That is, dACC responses partially integrate the values of the two offers, much as if they reflect the middle of a hierarchical process that transforms discrete information channels into integrated abstract value signals (Balasubramani, Moreno-Bote, & Hayden, 2018; Yoo & Hayden, 2018).

A complementary study used the same analysis technique to ask about value comparison signals. We had initially identified an anti-correlation between regression weights for the values of two offers, and argued that these weights were signatures of value comparison processes (Azab & Hayden, 2018; Strait, Blanchard, & Hayden, 2014; Strait, Sleezer, & Hayden, 2015). These anti-correlations reflect frank encoding of the key decision variable predicted from value comparison processes, namely, value difference. We had previously argued that such signals reflect the completion of a value comparison process. However, we showed that this comparison process is only partially complete, even relatively late in the trial (Azab & Hayden, 2017). That is, the outputs of dACC reflect only partial comparison of values, suggesting, as before, occupation of a middle stage in a hierarchical value comparison process. These works, together with other studies showing similar responses in other value-sensitive regions, suggests that dACC does not have a specific and unique role in economic choice, but is better understood as part of a larger distributed hierarchy for economic decision-making, and thus reflects the interaction of multiple distributed processes (Chen & Stuphorn, 2015; Cisek & Kalaska, 2010; Hunt & Hayden, 2017; Kolling, Behrens, Wittmann, & Rushworth, 2016; Maisson, Cash-padgett, & Hayden, 2020).

2.1 Linking economic function with other roles of dACC

The economic view of dACC function emphasizes its role in value formation and comparison. Critically, the values of options may change, and must be updated. This dovetails with a second viewpoint that sees dACC as critical to *learning* option values. In support of this, neurons in dACC encode feature-specific reward prediction errors, and do so earlier than the lateral prefrontal cortex (LPFC) or striatum (Oemisch et al., 2019; Voloh, Oemisch, & Womelsdorf, 2020). Furthermore, dACC neurons show individual selectivity for errors with different sources (Shen et al., 2015). These studies thus suggest a neural substrate for attributing general outcomes to their specific attendant sources.

A second viewpoint links valuation to concepts from foraging theory, where dACC is thought to reflect switching from a default option to a new one (e.g., exploring a new patch, (Hayden et al., 2011; Kolling, Behrens, Mars, & Rushworth, 2012)). Indeed, neurons in ACC have a higher degree of task selectivity than LPFC when switching to a new task set (Johnston et al., 2007). When multiple options are available (for example, multiple patches that can be (re)visited), the dACC must be able to maintain a memory trace for different options, even if they do not affect choice on the current trial. Indeed, neurons in dACC can track correct targets even when the trial was ultimately unsuccessful (Westendorff, Kaping, Everling, & Womelsdorf, 2016). This is in line with evidence that the dACC tracks counter-factual outcomes (what could have been), as has been demonstrated in macaques in both single unit (Hayden, Pearson, & Platt, 2009), and macroscopic BOLD (Fouragnan et al., 2019) studies. Presumably, the dACC evaluates if a different option could have had a better outcome, and generates a signal to switch to it.

Taken together, single unit studies have identified economic computations that are likely expressions of a broader fronto-striatal circuitry designed for the selection and instantiation of action that maximize future rewards.

3. Oscillations as a window into circuit interactions

To optimize behavior, the computations performed by dACC neurons must be coordinated and disseminated with other brain regions (Heilbronner & Hayden, 2016; Kolling, Wittmann, et al., 2016; Shenhav, Cohen, & Botvinick, 2016; Womelsdorf & Everling, 2015). A key feature of coordination between ACC and distal structures is that spikes in ACC are concentrated to certain phases of distal oscillations. Many studies of primates have shown such task-specific spike-phase synchrony, including between the ACC and frontal eye fields (Babapoor-Farrokhran, Vinck, Womelsdorf, & Everling, 2017), ACC and LPFC (Smith et al., 2019; Voloh, Valiante, Everling, & Womelsdorf, 2015; Voloh & Womelsdorf, 2017; Womelsdorf, Ardid, Everling, & Valiante, 2014), ACC and ventro-medial PFC (Womelsdorf, Ardid, et al., 2014), ACC and striatum (Voloh et al., 2020), and ACC and amygdala (Taub, Perets, Kahana, & Paz, 2018). Further evidence from rodents suggests that the ACC also synchronizes to hippocampus (Benchenane et al., 2010; Fujisawa & Buzsáki, 2011; Hyman, Ma, Balaguer-Ballester, Durstewitz, & Seamans, 2012; Sirota et al., 2008), ventral tegmental area (Fujisawa & Buzsáki, 2011), and infra-limbic cortex (Totah, Jackson, & Moghaddam, 2013), although care must be taken in translating results from rodents to primates (Heilbronner & Hayden, 2016). Such synchrony selectively emerges when attentional (Voloh & Womelsdorf, 2017; Womelsdorf, Ardid, et al., 2014), or control (Smith et al., 2019) demands are high. This suggests that communication between ACC and distal structures may be modulated and controlled by band-limited, population-level oscillatory activity, arising from distinct neuronal circuits.

3.1 Oscillatory timescales arise from specific dynamic circuit motifs

Local circuits of cells can generate oscillatory activity. The period of this activity is determined by the specific configuration of inhibitory and excitatory cells, the kinetics of transmembrane channels, axonal delays, and the nature of arriving inputs (Womelsdorf, Valiante, Sahin, Miller, & Tiesinga, 2014). For example, a well-characterized circuit motif that generates activity in the gamma (30–100 Hz) band is the pyramidal-interneuron gamma (PING) motif. Depolarizing pyramidal cells excite inhibitory parvalbumin-positive (PV +) interneurons, which subsequently inhibit those same pyramidal cells. The frequency of the generated oscillation is thus determined by the timescale of $GABA_A$-ergeric inhibition, and how quickly pyramidal cells can recover from inhibition. It is important to note that this sort of circuit can thus generate oscillatory activity even when the input is non-oscillatory. Different circuit motifs can generate oscillations with different periods (reviewed in (Womelsdorf, Valiante, et al., 2014) as well as (Wang, 2010)).

Oscillatory circuit motifs are likely computational units that are conserved across brain areas (Fries, 2015; Womelsdorf, Valiante, et al., 2014), yet still support processing of region-dependent information. For example, in dACC and LPFC, electrophysiological classification of neuronal subtypes revealed a similar interneuron type in both dACC and LPFC, likely corresponding to PV+ interneurons (Boroujeni, Tiesinga, & Womelsdorf, 2020). This neuron type encoded different information in the two areas, with choice probabilities being encoded in LPFC, and reward prediction errors in dACC. These same cells also showed gamma band synchrony that mirrored the cells' encoding preferences. Thus, the same neuronal circuits, conserved across dACC and LPFC, may nevertheless be leveraged to process area-specific information.

Importantly, activity at multiple frequency bands can be generated by the same circuit, with the faster oscillation being nested within the slower one. This can be evident as "phase-amplitude coupling," where the amplitude of a fast oscillation is modulated by the phase of a slow one (*see* Hyafil, Giraud, Fontolan, & Gutkin, 2015 for a review of this and other forms of cross frequency interactions). At the circuit level, different instantiations are possible. One option, most extensively studied in the hippocampus, suggests that a separate population of cells, whose activity is dominated by a slow Ca^{2+} I_h hyper-polarizing current, provides theta-pulsed inhibition (Malerba & Kopell, 2013; Neymotin et al., 2013; Stark et al., 2013; Tort, Rotstein, Dugladze, Gloveli, & Kopell, 2007). Alternatively, phasic and/or tonic excitation of inhibitory and excitatory populations in the PING motif (described above) can generate different phase-amplitude coupling profiles (Onslow, Jones, & Bogacz, 2014). This implies that information contained in high-frequency oscillations can be organized according to a low-frequency scaffold (Lisman & Jensen, 2013; Panzeri, Brunel, Logothetis, & Kayser, 2010; Voloh & Womelsdorf, 2016).

The maintenance and generation of oscillatory activity does not only depend on neuronal cell types. Even within the same neurons, different spike profiles have a differential impact on local oscillations. This is evident when comparing the relation of regular spiking vs burst spiking (spike occurring <5 ms apart) to local LFP activity in dACC and LPFC (Voloh &

Womelsdorf, 2017). Bursts of putative interneuron cells are associated with increased local theta power that peaks after the burst has occurred, as well as increased beta phase synchronization, compared to non-bursts. On the other hand, bursts, rather than non-bursts, of putative pyramidal cells were associated with a stronger phase-dependent variation in beta power. These results are suggestive of the neuronal circuits at play, such as a role for Martinotti cells in theta maintenance, and/or the coactivation of proximal and dendritic inputs of a network of excitatory and inhibitory cells for beta generation (*see* Voloh & Womelsdorf, 2017 for detailed discussion).

In summary, the frequency of a particular rhythm being engaged in a task-dependent manner is suggestive of the neuronal circuitry at play.

3.2 Oscillations index inter- and intra-laminar coordination

Intriguingly, frequency specific activity can often be localized to specific cortical layers (Voloh & Womelsdorf, 2016; Wang, 2010; Womelsdorf, Valiante, et al., 2014). Slower oscillations are prevalent in infra-granular layers, whereas faster ones are reported in granular and supragranular layers (Bastos, Loonis, Kornblith, Lundqvist, & Miller, 2018; Bollimunta, Chen, Schroeder, & Ding, 2008; Buffalo, Fries, Landman, Buschman, & Desimone, 2011; Ninomiya, Dougherty, Godlove, Schall, & Maier, 2015; Van Kerkoerle et al., 2014; Wang, 2010). Varying task demands changes the strength of activation but not the pattern of functional connectivity, suggesting such activity propagation proceeds along conserved anatomical paths (Van Kerkoerle et al., 2014). Intriguingly, this segregation of frequency by layer also indexes feedforward and feedback connectivity between (Van Kerkoerle et al., 2014; Voloh & Womelsdorf, 2016; Bastos et al., 2018), and within (Van Kerkoerle et al., 2014; Voloh & Womelsdorf, 2016) columns. Indeed, it is possible to recapitulate a large swath of the visual hierarchy based on oscillatory coordination alone, both in macaques (Bastos et al., 2014) and humans (Michalareas et al., 2016).

Evidence suggests that even within a column, oscillations between layers can interact (Bollimunta et al., 2008; Florez et al., 2013; Lee, Whittington, & Kopell, 2013; McGinn & Valiante, 2014; Ninomiya et al., 2015; Spaak, Bonnefond, Maier, Leopold, & Jensen, 2012). For example, in human neocortical slices, theta activity is prevalent in supra and infra-granular layers (Florez et al., 2013). The phase of the former lags behind that of the latter, suggesting a directional influence. Importantly, gamma amplitude in supragranular layers is modulated by the phase of theta. Taken together, both inter- and intra- columnar coordination may be realized via oscillatory interactions.

Visual areas are a major source of evidence for a role of oscillations in widespread coordination, but have a unique architectonic fingerprint, raising the question of whether these results can be extended to agranular cortex such as the dACC (Voloh & Womelsdorf, 2016). An important piece of evidence is offered by a study of macaque SEF (Ninomiya et al., 2015), an agranular structure much like dACC (Barbas, 2015). In SEF, unlike in V1, deep layer slow oscillations did not modulate superficial layer fast oscillations (Ninomiya et al., 2015). Instead, the interaction was localized to layer 3. As the authors note, this may be due to the propensity of inputs and output to be localized to the (same) layer 3 (Ninomiya et al., 2015). Thus, a more general principle of oscillatory organization may be that oscillations index the input and output of local circuits, rather than simply the depth of the layer.

The inter-laminar patterns of oscillatory coordination may support a role for the dACC in the modulation of local computations in downstream regions (Helfrich & Knight, 2016; Womelsdorf & Everling, 2015). This possibility was probed in a study of selective attention in macaques (Voloh et al., 2015). Monkeys were cued to attend to a target, and had to report the direction of rotation, while ignoring a distractor. After the attention cue came on, but not before, high-frequency gamma amplitudes was modulated as a function of low-frequency theta phase (Fig. 1A and B). Critically, such theta-gamma phase-amplitude coupling selectively emerged when monkeys made a subsequent correct choice, but not an erroneous one (Fig. 1B). This directly suggests that coordination of oscillatory activity across timescales can support the successful deployment of attention. Importantly, theta and gamma activity was anatomically distinct; theta phases were more likely to arise in dACC, whereas gamma amplitudes were more likely to be found in LPFC (Fig. 1C). This is in line with the putative functional distinction between dACC as a controlling region that modulates task-relevant representations in LPFC (Kolling, Wittmann, et al., 2016; Shenhav et al., 2016). Further evidence directly implicates theta phases in the modulation of gamma amplitudes. First, a theta-phase reset was evident on correct, but not incorrect, trials (Fig. 1D). Second, gamma amplitudes were more tightly coupled to theta phases than the attention cue itself. Taken together, this suggests that dACC modulation of downstream LPFC is instantiated via theta phasic influence of gamma amplitudes, a suggestion borne out by computational modeling (Verguts, 2017).

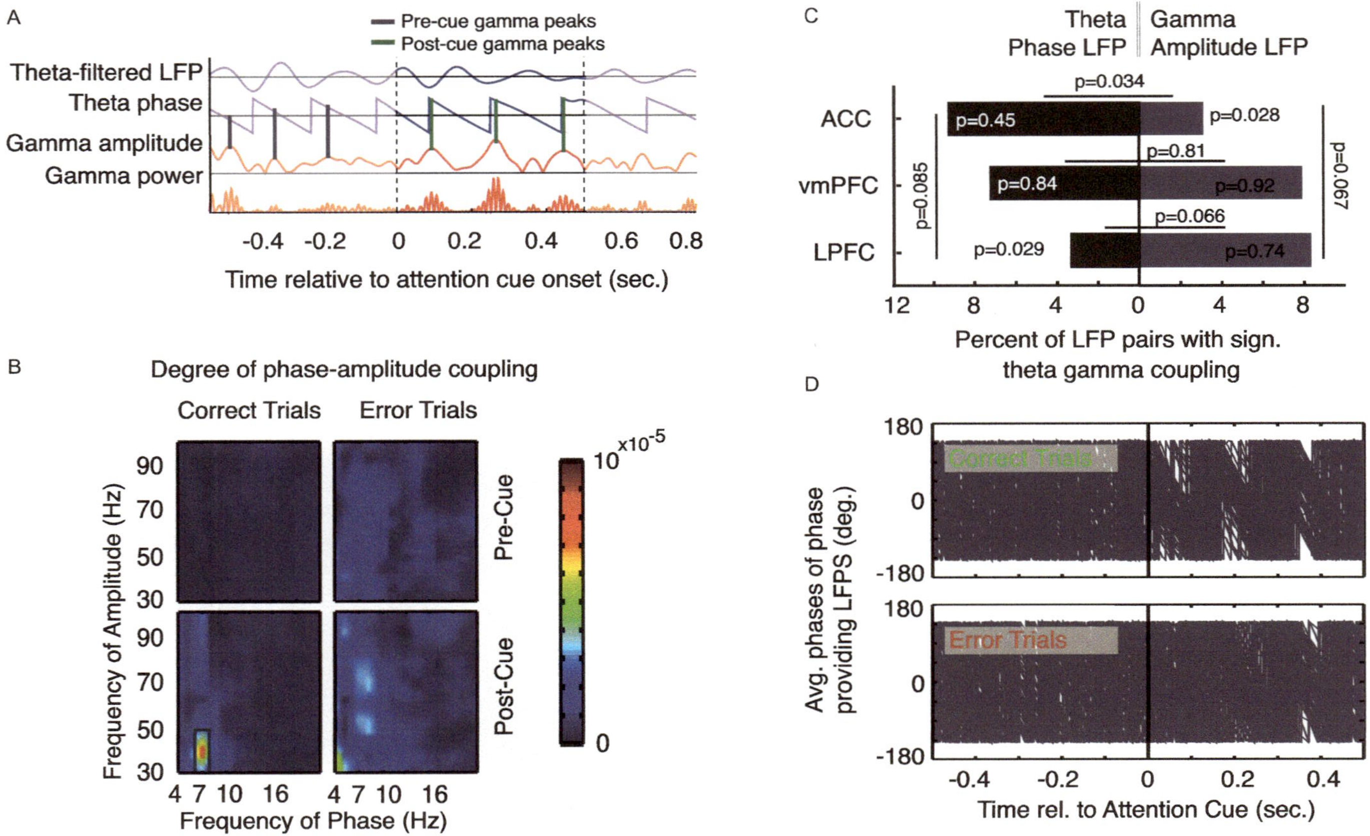

Fig. 1 See figure legend on opposite page.

4. Computational principles of oscillatory coordination

The excitatory-inhibitory push-pull of local circuits creates periods where spiking activity is more or less likely to occur (Fries, 2015; Voloh & Womelsdorf, 2016; Wang, 2010; Womelsdorf, Valiante, et al., 2014). This has a number of important implications.

4.1 Gating input

Information transmission can be gated to the excitable periods of the oscillation cycle. In this scenario, incoming spikes only have an effect on the downstream neurons if these neurons are not prohibitively inhibited. Optogenetic experiments provide some of the strongest, causal evidence for this suggestion. Gamma oscillations can be optogenetically induced by specifically targeting fast-spiking interneurons (Cardin et al., 2009). Importantly, pyramidal cell activity, induced by a single whisker deflection, strongly depended on the gamma phase at which it was delivered. Thus, local oscillations gate input in such a way that may facilitate (or inhibit) incoming information.

Perhaps more surprisingly, it seems that neurons have inherent mechanisms that can sensitize them to synchronized inputs. Evidence for this comes from in vivo recordings in cat striate cortex (Azouz & Gray, 2003). Slow depolarization is associated with increasing spike threshold, indicating that the probability of firing decreases when cells are depolarized slowly. On the other hand, the same mechanism *decreased* the spike threshold in response to fast depolarizations, as occur during synchronized spike input. This adaptive spike threshold mechanism can help tune task-relevant representations,

Fig. 1 Theta-gamma coupling predicts successful attentional deployment. (A) Example theta-gamma phase-amplitude coupling on one trial. Blue and red traces correspond to LFPs from two different sites. After attention cue onset, gamma amplitudes (red) were modulated as a function of theta phases (blue) (gamma peaks highlighted with vertical gray/green lines). (B) Phase-amplitude coupling, dissociated by frequency, task epoch, and outcome. Theta-gamma coupling increased in the post-cue epoch relative to the pre-cue epoch, but only for correct trials (black box). (C) Likelihood of a phase-providing channel (values left from zero) and an amplitude-providing channel (right from zero) in the vmPFC, dACC, and LPFC during cross-area theta-gamma coupling. (D) Progression of the average phase (y-axis) for all phase- providing LFP channels engaging in significant theta-gamma coupling around the time of the attention cue onset (x-axis). Each gray line represents the average phase across trials of one such LFP. Top and bottom panels show the progression of mean phases on correct trials and on error trials, respectively. More channels exhibited a cue-induced reset on correct, rather than error, trials.

and is tightly correlated with gamma band activity (Azouz & Gray, 2003). Although such a correspondence has not, to our knowledge, been shown in the frontal cortex, a similar mechanism may be in play. An interesting future direction would be to determine if synchronous firing in dACC leverages such intrinsic cellular adaptation to tune downstream, task-relevant representations.

4.2 Gating output

If neuronal output is synchronized, the impact on downstream targets can be maximized. This follows from the well-known phenomenon of temporal summation; if spikes from different neurons arrive around the same time, the probability of depolarization of their target is increased. For example, periods of higher coordination in earlier visual areas increases the probability of spiking in later visual areas (Zandvakili & Kohn, 2015). This effect is specific to retino-topically aligned, rather than non-aligned, cells in V1 and V2 (Zandvakili & Kohn, 2015).

Likewise, rodent studies of the medial prefrontal cortex (likely an analogue to primate dACC) also suggest synchronizing output can have a powerful gating effect. A clear example of this is from an optogenetic study of conditioned fear responses in rodents (Courtin et al., 2013). Here, optogenetic inhibition (or activation) of inhibitory interneurons resulted in the synchronization of local pyramidal cells, also evident as an increase in phase-locking to local theta activity. The synchronization of pyramidal output could be traced to a theta-phase reset, similarly induced either via optogenetic manipulation, or by the onset of the conditioning cue. Furthermore, antidromic stimulation revealed that the synchronized output of pyramidal cells had preferential downstream targets involved in fear expression. The results of (Courtin et al., 2013) are shadowed by those of (Voloh et al., 2015) (described earlier), who also found that cue-related theta-phase resetting is predictive of behavioral outcome. Taken together, these results suggest that synchronizing dACC output mediates a powerful effect on downstream effector regions.

4.3 Routing

By aligning the phase of distal oscillations, information can be selectively routed. Intuitively, this follows from the points above; spikes from a sending region that arrive at an excitable phase can be processed, while those arriving at an inhibited phase cannot. Thus, manipulating the relative phase between sending and receiving regions may mediate the flow of information

(Fries, 2015; Hahn, Ponce-Alvarez, Deco, Aertsen, & Kumar, 2019; Palmigiano, Geisel, Wolf, & Battaglia, 2017). Testing this proposition requires simultaneous recording from two separate upstream regions that project to the same target. This has been done in macaque visual cortex during a selective attention task (Bosman et al., 2012). The investigators found that while gamma activity in V1 was prevalent at sites with different receptive fields, downstream sites in V4 whose receptive field encompassed both upstream sites were only coherent with the attended receptive field sites (Bosman et al., 2012). Thus, relative phase differences between sites can be leveraged to selectively route information. This is true even if oscillations are short-lived and occur stochastically (Palmigiano et al., 2017), as seems to be the case in in vivo recordings in frontal cortex (Bastos et al., 2018; Lundqvist et al., 2016). One question that arises is how phase differences may be controlled in order to route information. Modeling suggests that increasing tonic firing rate activity of the target site is sufficient to modify relative phases and route information transmission (Palmigiano et al., 2017). While this has not yet been demonstrated in the dACC, we hypothesize that such routing may underlie the ability of the dACC to selectively modulate downstream, task-relevant targets such as in the LPFC. This would require careful dissection of network level variables that determine information propagation via synchronous (or asynchronous) firing (Hahn et al., 2019; Kumar, Rotter, & Aertsen, 2010).

4.4 Coding

One consequence of phase-specific sensitivity to input is that information can be encoded as a function of phase. Put another way, a downstream reader can deduce the informational content of spikes based on which phase they arrive at. Phase-of-firing coding has been shown to increase the informational content of spikes (Kayser, Montemurro, Logothetis, & Panzeri, 2009; Montemurro, Rasch, Murayama, Logothetis, & Panzeri, 2008), and to carry information about stimulus orientation (Vinck et al., 2010; Womelsdorf et al., 2012), object in memory (Siegel, Warden, & Miller, 2009), spatial navigation parameters (Huxter, Senior, Allen, & Csicsvari, 2008; Turesson, Logothetis, & Hoffman, 2012), and learning variables (Hawellek, Wong, & Pesaran, 2016; Voloh et al., 2020) (*see* below for further details). This may be achieved through a rate-to-phase conversion (Buzsáki & Draguhn, 2004; Fries, Nikolić, & Singer, 2007; Mehta, Lee, & Wilson, 2002); neurons that are activated most strongly will be able to escape oscillatory inhibition earliest, thus firing on earlier phases. Importantly, this also provides a mechanism to multiplex different streams of information (Akam & Kullmann, 2014). A downstream reader can decode a packet of information by determining the phase (relative to a reference oscillation) at which it arrived (Lisman & Jensen, 2013; Akam & Kullmann, 2014; *see* Akam & Kullmann, 2014 for other oscillation-based multiplexing schemes). Such a reference oscillation may be instantiated by a phase reset of ongoing oscillations (Voloh & Womelsdorf, 2016), thus bringing sending and receiving regions into phase alignment.

Recent evidence suggests that, like primary and associative areas reviewed above, spike-phase interactions in dACC (as well as LPFC and striatum) also boost information content (Voloh et al., 2020). Monkeys performed a reversal learning task, whereby they had to learn the color associated with reward in a block of trials. All considered brain areas encoded key learning variables, including outcome, reward prediction error (RPE), and the history of outcomes. The critical question was whether the phase at which spikes occurred conveyed information about these three critical learning variables; if they did, differences in firing rate would be modulated as a function of phase, whereas if a pure firing rate code was in effect, differences would be flat across phases (Fig. 2A). Indeed, we found evidence of the former for all three learning variables (Fig. 2B and C). This effect was specific to a 10–25 Hz beta band (Fig. 2D), and only for sites that synchronized to distal LFPs. Importantly, the encoding phase gain was greatest in the dACC as compared to the LPFC or the striatum (Fig. 2E). Similar phase-of-firing coding has been observed in humans when control demands are high (Smith et al., 2019). Thus, phase-of-firing encoding of key learning variables is prevalent across the dACC, LPFC, and striatum, and is strongest in the dACC.

One might expect that spikes concentrated on a cell's preferred firing phase are the richest carriers of information. However, this was not the case in the aforementioned study. Specifically, for neurons encoding a reward prediction error, maximal information gain occurred on approximately the anti-preferred firing phase (Fig. 2F). It is unclear how this may be mechanistically instantiated. One possibility, suggested by previous rodent work (Benchenane et al., 2010), is that a dopaminergic influence may shift encoding phases during learning. Alternatively, keeping in mind that a prediction error is a comparison of two signals (expected and observed outcomes), such a phase offset may be the result of a latency difference between these incoming signals (*see* Voloh et al., 2020 for further discussion). These and other studies suggest

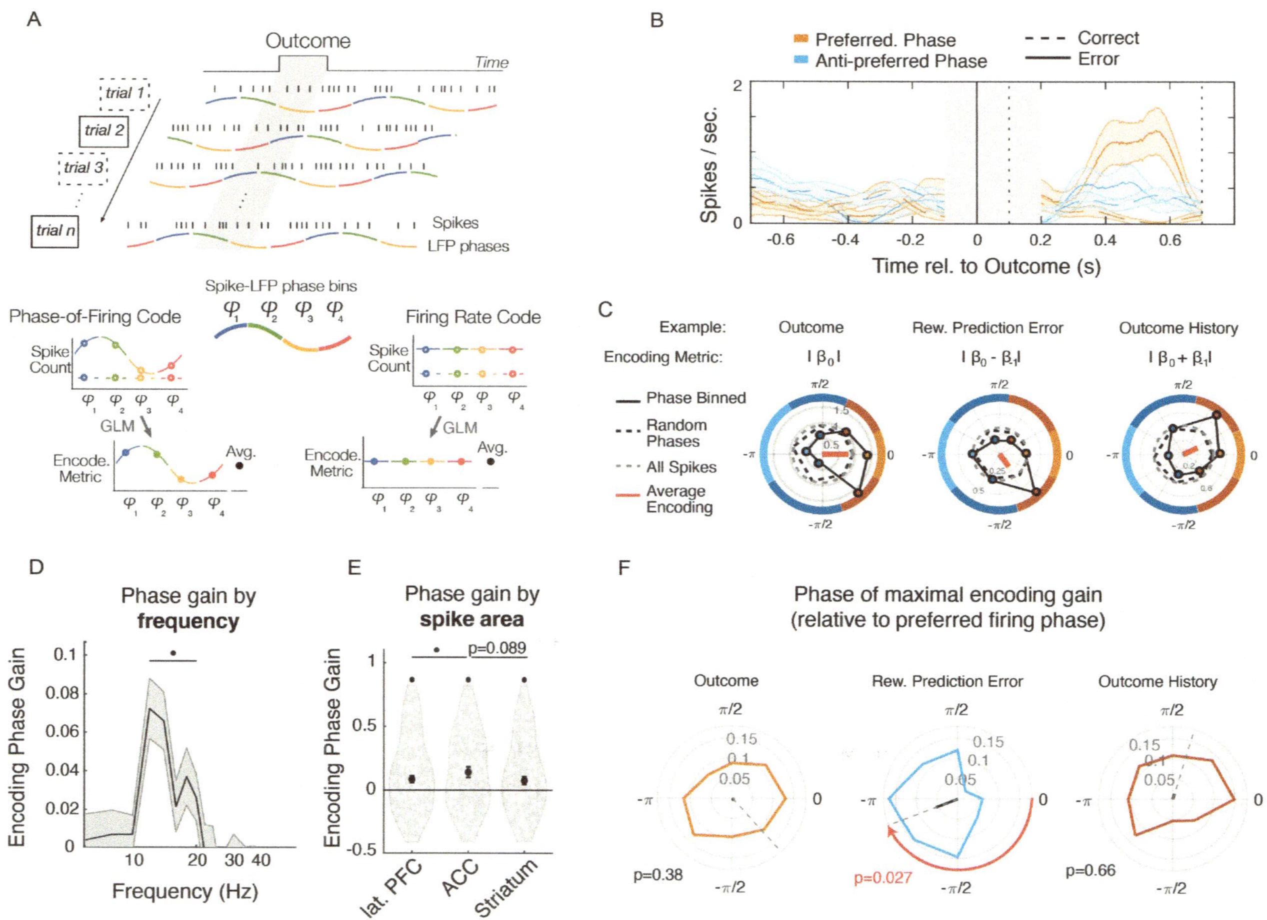

Fig. 2 See figure legend on next page.

that spiking activity in dACC synchronized to LFP phases have an important role in disseminating task-relevant information.

Do these results indicate a bona-fide phase-code, or rather a phase gain of rate encoding? Indeed, both may be true. Outcome cells showed a significant difference in phase synchronization between conditions, suggesting phasic modulation of rate encoding. On the other hand, RPE and Outcome History cells tended to synchronize on different phases between conditions, consistent with a bona-fide phase code (Voloh et al., 2020). This suggests that different phase coding schemes may co-exist. This is in line with evidence that, in monkey visual cortex (in the same dataset), phase may both modulate rate encoding (Womelsdorf et al., 2012), as well as act as a scaffold for bona-fide phase-coding (Vinck et al., 2010). Future studies should critically assess the mechanisms and conditions under which different phase-coding schemes are dominant.

Fig. 2 Spike-phase encoding in dACC. (A) Phase-dependent encoding analysis. Spikes were segregated according to the phase of the LFP. In a pure firing rate code (right, middle), firing rates on different conditions (solid vs dotted lines) would be consistent across phase bins. This would emerge as a flat firing rate difference (i.e., encoding strength, derived from a Generalized Linear Model (GLM) analysis) across phase bins (right, bottom). Alternatively, a phase-of-fire code would show phasic encoding strength (left, middle, bottom). (B) Example cell showing phase-dependent outcome encoding, comparing correct (dashed lines) and error (solid) trials. Color represents firing on the preferred (orange) or anti-preferred (blue) encoding phase. The gray box between [−0.1 0.2] was not analyzed. Firing rate differences emerged after reward onset, but only on preferred firing phases. (C) Example of phase-of-firing encoding for *Outcome*, *Reward Prediction Error*, and *Outcome History* cells. Colored dots reflect the encoding metric for a corresponding phase bin (derived from GLM model coefficients). Zero radians correspond to the preferred (mean) firing phase. Numbers on concentric circles are the value of the encoding metric. The gray dotted line represents the encoding metric estimated using all spikes, whereas the black dotted line represents the average across many permutations of spike phases. The red line is the average direction. Colored borders represent phase bin size. These cells show stronger encoding near the 0 rad phase, and weaker encoding on opposite phases. (D) Median encoding phase gain per frequency for encoding and locking cells. Phase gain was above chance at [1020] Hz. (E)Violin plots of the encoding phase gain, split by the spike area. Black circle and vertical bars depict the median and standard error. All areas showed some evidence of significant phase gain, with stronger encoding in ACC rather than LPFC or striatum. (F)Distribution of the relative phase of maximal encoding of cells (0 corresponds to the preferred phase), split by functional cell types. *Outcome* and *Outcome History* cells did not show an encoding phase preference, whereas *RPE* cells did.

5. Oscillations in dACC as a link between the micro and macro

Up to this point, we have reviewed evidence that band-limited oscillations emerge from specific neuronal circuit motifs, index anatomical and functional connectivity, provide a key substrate for neural coding, and are functionally relevant. These properties allow building links between insights using micro- and macro-scale recordings.

Two major macro-scale recording techniques, particularly prevalent in human studies due to their non-invasive nature, are the EEG and fMRI. Both have a strong relationship with the LFP that can be leveraged to build links between the micro- and macro-scale.

5.1 Linking EEG theta to neuronal circuits

A rich source of evidence implicating the dACC in cognitive control is based on EEG in humans, where frontal midline theta power increases when cognitive control demands are high (Cohen, 2014; Cavanaugh and Cohen, 2014). Theta as an index of control is stable throughout life (Cooper et al., 2019). Overall theta power is associated with negative prediction errors (thus disambiguating the valence of errors), but can also signal the degree of surprise (Cavanagh, Figueroa, Cohen, & Frank, 2012). Such a surprise signal may serve to instantiate top-down control, or focus attentional resources (Cavanagh & Frank, 2014; Womelsdorf & Everling, 2015). In so doing, if theta is indeed a medium by which dACC exerts control over downstream circuits, we would expect to see dACC synchrony with such regions. This has indeed been shown to be true, both in humans (*see* Cavanagh & Frank, 2014) and non-human primates (Phillips, Vinck, Everling, & Womelsdorf, 2014).

The control-indexing theta evident in human EEG has also been observed in macaques (Womelsdorf, Johnston, Vinck, & Everling, 2010). A promising line of research uses inverse modeling techniques (critically reviewed in (Cohen, 2017)) to determine how theta rhythmogenesis may arise from a columnar circuit of layer 5 pyramidal cells and inhibitory Martinotti cells (Herrera, Sajad, Woodman, Schall, & Riera, 2020). In this work, conflict detection is conceptualized as coincidence detection between an efferent motor copy and a task rule. Activated L5 pyramidal cells then activate Martinotti cells, which then (indirectly) inhibit the apical dendrites of the L5 cells. The dynamics of this circuit instantiate deep layer theta, which may be detected at the level of scalp recordings. This is in line with optogenetic work that has shown that optogenetic activation of even a single Martinotti cell can generate synchronized activity in nearby pyramidal cells (Hilscher, Leão, Edwards, Leão, & Kullander, 2017). Leveraging such work, in conjunction with detailed laminar encoding preferences (Sajad, Godlove, & Schall, 2019) and translational studies of EEG in macaques (e.g., Godlove et al., 2011), is critical for understanding the micro-circuitry of dACC (Cohen, 2014; Sajad et al., 2019).

5.2 BOLD and LFP

A second major source of evidence of dACC function comes from human fMRI experiments, which measure the blood-oxygenation level dependent (BOLD) response. The bi-directional translation of knowledge from invasive experiments in animals on the one hand, and non-invasive BOLD measurements on the other, is of great importance when studying complex cognitive processing in associative cortex such as the dACC. BOLD responses are widely employed not only to map brain activation, but also as a means for studying the dynamics of neural networks across spatial and temporal scales (Biswal, Zerrin Yetkin, Haughton, & Hyde, 1995; Fox & Raichle, 2007; He, Snyder, Zempel, Smyth, & Raichle, 2008; Logothetis, Pauls, Augath, Trinath, & Oeltermann, 2001). The BOLD response seems to correlate most closely with low-frequency activity (Shi et al., 2019; Wang, Saalmann, Pinsk, Arcaro, & Kastner, 2012), although other investigators find a closer correspondence to high-frequency activity (Logothetis et al., 2001). This may depend on whether recordings are task-dependent or obtained from resting state (Wilson, Yang, Gore, & Chen, 2016). Critically, the BOLD response is more closely correlated with LFP activity than with single- or multi-unit activity (Logothetis et al., 2001). These findings have led the field to assume that the BOLD contrast mechanism reflects the input and intracortical processing of a given area rather than its spiking output. Linking BOLD activity to band-limited LFP activity further allows us to infer the neuronal circuits at play.

6. Conclusion

The breadth of research into the function of the dACC is impressive and accelerating. It spans multiple recording hierarchies, and is supported by multiple computational theories. Such richness of evidence has a cost; namely, that paradoxes and conflict between competing theories/methodologies naturally arise. This is fertile ground to build a more nuanced understanding of dACC function. LFP oscillations are a critical link between micro- and macroscopic recording techniques; they are both grounded in specific inter-neuronal interactions, and correlate more closely with EEG and BOLD measurements than do individual neurons. Importantly, oscillatory coordination between regions has key and attractive properties for gating, propagation, and encoding of information. Thus, they offer a unique perspective on understanding the function of the dACC within the broader fronto-striatal circuits.

References

Akam, T., & Kullmann, D. M. (2014). Oscillatory multiplexing of population codes for selective communication in the mammalian brain. *Nature Reviews. Neuroscience*, *15*, 111–122.

Amiez, C., Joseph, J. P., & Procyk, E. (2006). Reward encoding in the monkey anterior cingulate cortex. *Cerebral Cortex*, *16*, 1040–1055.

Azab, H., & Hayden, B. Y. (2017). Correlates of decisional dynamics in the dorsal anterior cingulate cortex. *PLoS Biology*, *15*, 1–25.

Azab, H., & Hayden, B. Y. (2018). Correlates of economic decisions in the dorsal and subgenual anterior cingulate cortices. *The European Journal of Neuroscience*, *47*, 979–993.

Azab, H., & Hayden, B. (2020). Dorsal anterior cingulate cortex neurons only partially integrate determinants of value. *bioRxiv*. 2020.01.01.892380.

Azouz, R., & Gray, C. M. (2003). Adaptive coincidence detection and dynamic gain control in visual cortical neurons in vivo. *Neuron*, *37*, 513–523.

Babapoor-Farrokhran, S., Vinck, M., Womelsdorf, T., & Everling, S. (2017). Theta and beta synchrony coordinate frontal eye fields and anterior cingulate cortex during sensorimotor mapping. *Nature Communications*, *8*, 13967.

Balasubramani, P. P., Moreno-Bote, R., & Hayden, B. Y. (2018). Using a simple neural network to delineate some principles of distributed economic choice. *Frontiers in Computational Neuroscience*, *12*, 1–13.

Barbas, H. (2015). General cortical and special prefrontal connections: Principles from structure to function. *Annual Review of Neuroscience*, *38*, 269–289.

Barbas, H., & Pandya, D. N. (1989). Architecture and intrinsic connections of the prefrontal cortex in the rhesus monkey. *The Journal of Comparative Neurology*, *286*, 353–375.

Bastos, A. M., Loonis, R., Kornblith, S., Lundqvist, M., & Miller, E. K. (2018). Laminar recordings in frontal cortex suggest distinct layers for maintenance and control of working memory. *Proceedings of the National Academy of Sciences*, *115*, 201710323.

Bastos, A. M., Vezoli, J., Bosman, C. A., Schoffelen, J.-M. J.-M., Oostenveld, R., Dowdall, J. R., et al. (2014). Visual areas exert feedforward and feedback influences through distinct frequency channels. *Neuron*, *85*, 390–401.

Benchenane, K., Peyrache, A., Khamassi, M., Tierney, P. L., Gioanni, Y., Battaglia, F. P., et al. (2010). Coherent Theta oscillations and reorganization of spike timing in the hippocampal-prefrontal network upon learning. *Neuron*, *66*, 921–936.

Biswal, B., Zerrin Yetkin, F., Haughton, V. M., & Hyde, J. S. (1995). Functional connectivity in the motor cortex of resting human brain using echo-planar MRI. *Magnetic Resonance in Medicine*, *34*, 537–541.

Bollimunta, A., Chen, Y., Schroeder, C. E., & Ding, M. (2008). Neuronal mechanisms of cortical alpha oscillations in awake-behaving macaques. *The Journal of Neuroscience*, *28*, 9976–9988.

Boroujeni, K. B., Tiesinga, P., & Womelsdorf, T. (2020). Interneuron specific gamma synchronization encodes uncertain cues and prediction errors in lateral prefrontal and anterior cingulate cortex. *bioRxiv*. doi:2020.07.24.220319.

Bosman, C. A., Schoffelen, J.-M., Brunet, N., Oostenveld, R., Bastos, A. M., Womelsdorf, T., et al. (2012). Attentional stimulus selection through selective synchronization between monkey visual areas. *Neuron, 75*, 875–888.

Buffalo, E. A., Fries, P., Landman, R., Buschman, T. J., & Desimone, R. (2011). Laminar differences in gamma and alpha coherence in the ventral stream. *Proceedings of the National Academy of Sciences, 108*, 11262–11267.

Buzsáki, G., & Draguhn, A. (2004). Neuronal oscillations in cortical networks. *Science, 304*, 1926–1929.

Cai, X., & Padoa-Schioppa, C. (2012). Neuronal encoding of subjective value in dorsal and ventral anterior cingulate cortex. *The Journal of Neuroscience, 32*, 3791–3808.

Cardin, J. A., Carlén, M., Meletis, K., Knoblich, U., Zhang, F., Deisseroth, K., et al. (2009). Driving fast-spiking cells induces gamma rhythm and controls sensory responses. *Nature, 459*, 663–667.

Cavanagh, J. F., Figueroa, C. M., Cohen, M. X., & Frank, M. J. (2012). Frontal theta reflects uncertainty and unexpectedness during exploration and exploitation. *Cerebral Cortex, 22*, 2575–2586.

Cavanagh, J. F., & Frank, M. J. (2014). Frontal theta as a mechanism for cognitive control. *Trends in Cognitive Sciences, 18*, 414–421.

Chen, X., & Stuphorn, V. (2015). Sequential selection of economic good and action in medial frontal cortex of macaques during value-based decisions. *eLife, 4*, 1–24.

Cisek, P., & Kalaska, J. F. (2010). Neural mechanisms for interacting with a world full of action choices. *Annual Review of Neuroscience, 33*, 269–298.

Cohen, M. X. (2014). A neural microcircuit for cognitive conflict detection and signaling. *Trends in Neurosciences, 37*, 480–490.

Cohen, M. X. (2017). Where does EEG come from and what does it mean? *Trends in Neurosciences, 40*, 208–218.

Cole, M. W., Yeung, N., Freiwald, W. A., & Botvinick, M. (2009). Cingulate cortex: Diverging data from humans and monkeys. *Trends in Neurosciences, 32*, 566–574.

Cooper, P. S., Karayanidis, F., McKewen, M., McLellan-Hall, S., Wong, A. S. W., Skippen, P., et al. (2019). Frontal theta predicts specific cognitive control-induced behavioural changes beyond general reaction time slowing. *NeuroImage, 189*, 130–140.

Courtin, J., Chaudun, F., Rozeske, R. R., Karalis, N., Gonzalez-Campo, C., Wurtz, H., et al. (2013). Prefrontal parvalbumin interneurons shape neuronal activity to drive fear expression. *Nature, 505*, 92–96.

Ebitz, R. B., & Hayden, B. Y. (2016). Dorsal anterior cingulate: A Rorschach test for cognitive neuroscience. *Nature Neuroscience, 19*, 1278–1279.

Einevoll, G. T., Kayser, C., Logothetis, N. K., & Panzeri, S. (2013). Modelling and analysis of local field potentials for studying the function of cortical circuits. *Nature Reviews. Neuroscience, 14*, 770–785.

Florez, C. M., McGinn, R. J., Lukankin, V., Marwa, I., Sugumar, S., Dian, J., et al. (2013). In vitro recordings of human neocortical oscillations. *Cerebral Cortex*, 578–597.

Fouragnan, E. F., Chau, B. K. H., Folloni, D., Kolling, N., Verhagen, L., Klein-Flügge, M., et al. (2019). The macaque anterior cingulate cortex translates counterfactual choice value into actual behavioral change. *Nature Neuroscience, 22*, 797–808.

Fox, M. D., & Raichle, M. E. (2007). Spontaneous fluctuations in brain activity observed with functional magnetic resonance imaging. *Nature Reviews. Neuroscience, 8*, 700–711.

Fries, P. (2015). Rhythms for cognition: Communication through coherence. *Neuron, 88*, 220–235.

Fries, P., Nikolić, D., & Singer, W. (2007). The gamma cycle. *Trends in Neurosciences, 30*, 309–316.

Fujisawa, S., & Buzsáki, G. (2011). A 4 Hz oscillation adaptively synchronizes prefrontal, VTA, and hippocampal activities. *Neuron, 72*, 153–165.
Godlove, D. C., Emeric, E. E., Segovis, C. M., Young, M. S., Schall, J. D., & Woodman, G. F. (2011). Event-related potentials elicited by errors during the stop-signal task. I. macaque monkeys. *The Journal of Neuroscience, 31*, 15640–15649.
Goodkind, M., et al. (2015). Identification of a common neurobiological substrate for mental illness. *JAMA Psychiatry, 72*, 305–315.
Hahn, G., Ponce-Alvarez, A., Deco, G., Aertsen, A., & Kumar, A. (2019). Portraits of communication in neuronal networks. *Nature Reviews. Neuroscience, 20*, 117–127.
Hawellek, D. J., Wong, Y. T., & Pesaran, B. (2016). Temporal coding of reward-guided choice in the posterior parietal cortex. *Proceedings of the National Academy of Sciences, 113*, 13492–13497.
Hayden, B. Y., Heilbronner, S. R., Pearson, J. M., & Platt, M. L. (2011). Surprise signals in anterior cingulate cortex: Neuronal encoding of unsigned reward prediction errors driving adjustment in behavior. *The Journal of Neuroscience, 31*, 4178–4187.
Hayden, B. Y., Pearson, J. M., & Platt, M. L. (2009). Fictive reward signals in the anterior cingulate cortex. *Science, 324*, 948–950.
Hayden, B. Y., & Platt, M. L. (2010). Neurons in anterior cingulate cortex multiplex information about reward and action. *The Journal of Neuroscience, 30*, 3339–3346.
He, B. J., Snyder, A. Z., Zempel, J. M., Smyth, M. D., & Raichle, M. E. (2008). Electrophysiological correlates of the brain's intrinsic large-scale functional architecture. *Proceedings of the National Academy of Sciences, 105*, 16039–16044.
Heilbronner, S. R., & Hayden, B. Y. (2016). Dorsal anterior cingulate cortex: A bottom-up view. *Annual Review of Neuroscience, 39*, 149–170.
Helfrich, R. F., & Knight, R. T. (2016). Oscillatory dynamics of prefrontal cognitive control. *Trends in Cognitive Sciences, 20*(12), 916–930. https://doi.org/10.1016/j.tics.2016.09.007.
Herrera, B., Sajad, A., Woodman, G. F., Schall, J. D., & Riera, J. J. (2020). A minimal biophysical model of neocortical pyramidal cells: Implications for frontal cortex microcircuitry and field potential generation. *bioRxiv*. 019882:2020.01.29.925180.
Hilscher, M. M., Leão, R. N., Edwards, S. J., Leão, K. E., & Kullander, K. (2017). Chrna2-Martinotti cells synchronize layer 5 type a pyramidal cells via rebound excitation. Bacci A, ed *PLoS Biology, 15*, e2001392.
Holroyd, C. B., Nieuwenhuis, S., Mars, R. B., & Coles, M. G. H. (2004). Anterior cingulate cortex, selection for action, and error processing. In M. I. Posner (Ed.), *Cognitive neuroscience of attention* (1st ed., pp. 219–231). New York, NY: The Guilford Press.
Hunt, L. T., & Hayden, B. Y. (2017). A distributed, hierarchical and recurrent framework for reward-based choice. *Nature Reviews. Neuroscience, 18*, 172–182.
Huxter, J. R., Senior, T. J., Allen, K., & Csicsvari, J. (2008). Theta phase-specific codes for two-dimensional position, trajectory and heading in the hippocampus. *Nature Neuroscience, 11*, 587–594.
Hyafil, A., Giraud, A., Fontolan, L., & Gutkin, B. (2015). Neural cross-frequency coupling: Connecting architectures, mechanisms, and functions. *Trends in Neurosciences, 38*, 725–741.
Hyman, J. M., Ma, L., Balaguer-Ballester, E., Durstewitz, D., & Seamans, J. K. (2012). Contextual encoding by ensembles of medial prefrontal cortex neurons. *Proceedings of the National Academy of Sciences, 109*, 5086–5091.
Javitt, D. C., Siegel, S. J., Spencer, K. M., Mathalon, D. H., Hong, L. E., Martinez, A., et al. (2020). A roadmap for development of neuro-oscillations as translational biomarkers for treatment development in neuropsychopharmacology. *Neuropsychopharmacology, 45*, 1411–1422.
Johnston, K., et al. (2007). Top-down control-signal dynamics in anterior cingulate and prefrontal cortex neurons following task switching. *Neuron, 53*, 453–462.

Kayser, C., Montemurro, M. A., Logothetis, N. K., & Panzeri, S. (2009). Spike-phase coding boosts and stabilizes the information carried by spatial and temporal spike patterns. 1. supplemental material and methods *Neuron*, *61*, 597–608.

Kolling, N., Behrens, T. E. J., Mars, R. B., & Rushworth, M. F. S. (2012). Neural mechanisms of foraging. *Science*, *336*, 95–98.

Kolling, N., Behrens, T. E. J., Wittmann, M. K., & Rushworth, M. F. S. (2016). Multiple signals in anterior cingulate cortex. *Current Opinion in Neurobiology*, *37*, 36–43.

Kolling, N., Wittmann, M. K., Behrens, T. E. J., Boorman, E. D., Mars, R. B., & Rushworth, M. F. S. (2016). Value, search, persistence and model updating in anterior cingulate cortex. *Nature Neuroscience*, *19*, 1280–1285.

Kumar, A., Rotter, S., & Aertsen, A. (2010). Spiking activity propagation in neuronal networks: Reconciling different perspectives on neural coding. *Nature Reviews. Neuroscience*, *11*, 615–627.

Lee, J. H., Whittington, M. A., & Kopell, N. J. (2013). Top-down beta rhythms support selective attention via interlaminar interaction: A model. Coombes S, ed *PLoS Computational Biology*, *9*, e1003164.

Lisman, J., & Jensen, O. (2013). The theta-gamma neural code. *Neuron*, 77, 1002–1016.

Logothetis, N. K., Pauls, J., Augath, M., Trinath, T., & Oeltermann, A. (2001). Neurophysiological investigation of the basis of the fMRI signal. *Nature*, *412*, 150–157.

Lundqvist, M., Rose, J., Herman, P., Brincat, S. L., Buschman, T. J., & Miller, E. K. (2016). Gamma and beta bursts underlie working memory. *Neuron*, *90*, 152–164.

Maisson, D. J., Cash-padgett, T. V., & Hayden, B. Y. (2020). A functional hierarchy for choice in medial prefrontal cortex. *boiRxiv*. 2020.07.28.226043.

Malerba, P., & Kopell, N. (2013). Phase resetting reduces theta-gamma rhythmic interaction to a one-dimensional map. *Journal of Mathematical Biology*, *66*, 1361–1386.

McGinn, R. J., & Valiante, T. A. (2014). Phase-amplitude coupling and interlaminar synchrony are correlated in human neocortex. *The Journal of Neuroscience*, *34*, 15923–15930.

Mehta, M. R., Lee, A. K., & Wilson, M. A. (2002). Role of experience and oscillations in transforming a rate code into a temporal code. *Nature*, *417*, 741–746.

Michalareas, G., Vezoli, J., van Pelt, S., Schoffelen, J. M., Kennedy, H., & Fries, P. (2016). Alpha-Beta and Gamma rhythms subserve feedback and feedforward influences among human visual cortical areas. *Neuron*, *89*, 384–397.

Montemurro, M. A., Rasch, M. J., Murayama, Y., Logothetis, N. K., & Panzeri, S. (2008). Phase-of-firing coding of natural visual stimuli in primary visual cortex. *Current Biology*, *18*, 375–380.

Morecraft, R. J., Stilwell-Morecraft, K. S., Cipolloni, P. B., Ge, J., McNeal, D. W., & Pandya, D. N. (2012). Cytoarchitecture and cortical connections of the anterior cingulate and adjacent somatomotor fields in the rhesus monkey. *Brain Research Bulletin*, *87*, 457–497.

Neymotin, S. A., Hilscher, M. M., Moulin, T. C., Skolnick, Y., Lazarewicz, M. T., & Lytton, W. W. (2013). Ih tunes theta/gamma oscillations and cross-frequency coupling in an in silico CA3 model. Cymbalyuk G, ed *PLoS One*, *8*, e76285.

Ninomiya, T., Dougherty, K., Godlove, D. C., Schall, J. D., & Maier, A. (2015). Microcircuitry of agranular frontal cortex: Contrasting laminar connectivity between occipital and frontal areas. *Journal of Neurophysiology*, *113*, 3242–3255.

Oemisch, M., Westendorff, S., Azimi, M., Hassani, S. A., Ardid, S., Tiesinga, P., et al. (2019). Feature-specific prediction errors and surprise across macaque fronto-striatal circuits. *Nature Communications*, *10*, 1–15.

Onslow, A. C. E., Jones, M. W., & Bogacz, R. (2014). A canonical circuit for generating phase-amplitude coupling. *PLoS One*, *9*, e102591.

Palmigiano, A., Geisel, T., Wolf, F., & Battaglia, D. (2017). Flexible information routing by transient synchrony. *Nature Neuroscience*, *20*, 1014–1022.

Panzeri, S., Brunel, N., Logothetis, N. K., & Kayser, C. (2010). Sensory neural codes using multiplexed temporal scales. *Trends in Neurosciences*, *33*, 111–120.

Passingham, R. E., & Wise, S. P. (2012). *The neurobiology of the prefrontal cortex: Anatomy, evolution, and the origin of insight*. Oxford University Press.

Paus, T. (2001). Primate anterior cingulate cortex: Where motor control, drive and cognition interface. *Nature Reviews. Neuroscience*, *2*, 417–424.

Phillips, J. M., Vinck, M., Everling, S., & Womelsdorf, T. (2014). A long-range Fronto-parietal 5- to 10-Hz network predicts "top-down" controlled guidance in a task-switch paradigm. *Cerebral Cortex*, *24*, 1996–2008.

Rudebeck, P. H., Mitz, A. R., Chacko, R. V., & Murray, E. A. (2013). Effects of amygdala lesions on reward-value coding in orbital and medial prefrontal cortex. *Neuron*, *80*, 1519–1531.

Sajad, A., Godlove, D. C., & Schall, J. D. (2019). Cortical microcircuitry of performance monitoring. *Nature Neuroscience*, *22*, 265–274.

Schall, J. D., & Emeric, E. E. (2010). Conflict in cingulate cortex function between humans and macaque monkeys: More apparent than real. *Brain, Behavior and Evolution*, *75*, 237–238.

Seo, H., & Lee, D. (2007). Temporal filtering of reward signals in the dorsal anterior cingulate cortex during a mixed-strategy game. *The Journal of Neuroscience*, *27*, 8366–8377.

Shen, C., Ardid, S., Kaping, D., Westendorff, S., Everling, S., & Womelsdorf, T. (2015). Anterior cingulate cortex cells identify process-specific errors of attentional control prior to transient prefrontal-cingulate inhibition. *Cerebral Cortex*, *25*, 2213–2228.

Shenhav, A., Cohen, J. D., & Botvinick, M. M. (2016). Dorsal anterior cingulate cortex and the value of control. *Nature Neuroscience*, *19*, 4–6.

Shenhav, A., Musslick, S., Lieder, F., Kool, W., Griffiths, T. L., Cohen, J. D., et al. (2017). Toward a rational and mechanistic account of mental effort. *Annual Review of Neuroscience*, *40*, 99–124.

Shi, Z., Wilkes, D. M., Yang, P. F., Wang, F., Wu, R., Wu, T. L., et al. (2019). On the relationship between MRI and local field potential measurements of spatial and temporal variations in functional connectivity. *Scientific Reports*, *9*, 1–11.

Siegel, M., Warden, M. R., & Miller, E. K. (2009). Phase-dependent neuronal coding of objects in short-term memory. *Proceedings of the National Academy of Sciences*, *106*, 21341–21346.

Sirota, A., Montgomery, S., Fujisawa, S., Isomura, Y., Zugaro, M., & Buzsáki, G. (2008). Entrainment of neocortical neurons and gamma oscillations by the hippocampal Theta rhythm. *Neuron*, *60*, 683–697.

Smith, E. H., Horga, G., Yates, M. J., Mikell, C. B., Banks, G. P., Pathak, Y. J., et al. (2019). Widespread temporal coding of cognitive control in the human prefrontal cortex. *Nature Neuroscience*, *22*, 1883–1891.

Spaak, E., Bonnefond, M., Maier, A., Leopold, D. A., & Jensen, O. (2012). Layer-specific entrainment of gamma-band neural activity by the alpha rhythm in monkey visual cortex. *Current Biology*, *22*, 2313–2318.

Stark, E., Eichler, R., Roux, L., Fujisawa, S., Rotstein, H. G., & Buzsáki, G. (2013). Inhibition-induced theta resonance in cortical circuits. *Neuron*, *80*, 1263–1276.

Strait, C. E., Blanchard, T. C., & Hayden, B. Y. (2014). Reward value comparison via mutual inhibition in ventromedial prefrontal cortex. *Neuron*, *82*, 1357–1366.

Strait, C. E., Sleezer, B. J., Blanchard, T. C., Azab, H., Castagno, M. D., & Hayden, B. Y. (2016). Neuronal selectivity for spatial positions of offers and choices in five reward regions. *Journal of Neurophysiology*, *115*, 1098–1111.

Strait, C. E., Sleezer, B. J., & Hayden, B. Y. (2015). Signatures of value comparison in ventral striatum neurons. *PLoS Biology*, *13*, 1–22.

Taub, A. H., Perets, R., Kahana, E., & Paz, R. (2018). Oscillations synchronize amygdala-to-prefrontal primate circuits during aversive learning. *Neuron*, *97*, 291–298. e3.

Tort, A. B. L., Rotstein, H. G., Dugladze, T., Gloveli, T., & Kopell, N. J. (2007). On the formation of gamma-coherent cell assemblies by oriens lacunosum-moleculare interneurons in the hippocampus. *Proceedings of the National Academy of Sciences*, *104*, 13490–13495.

Totah, N. K. B., Jackson, M. E., & Moghaddam, B. (2013). Preparatory attention relies on dynamic interactions between prelimbic cortex and anterior cingulate cortex. *Cerebral Cortex*, *23*, 729–738.

Tran, T. T., Rolle, C. E., Gazzaley, A., & Voytek, B. (2020). Linked sources of neural noise contribute to age-related cognitive decline. *Journal of Cognitive Neuroscience*, *32*, 1–110.

Turesson, H. K., Logothetis, N. K., & Hoffman, K. L. (2012). Category-selective phase coding in the superior temporal sulcus. *Proceedings of the National Academy of Sciences*, *109*, 19438–19443.

Van Kerkoerle, T., Self, M. W., & Dagnino, B. (2014). Alpha and gamma oscillations characterize feedback and feed- forward processing in monkey visual cortex. *Proceedings of the National Academy of Sciences*, *111*, 14332–14341.

Verguts, T. (2017). Binding by random bursts: A computational model of cognitive control. *Journal of Cognitive Neuroscience*, *29*, 1103–1118.

Vinck, M., Lima, B., Womelsdorf, T., Oostenveld, R., Singer, W., Neuenschwander, S., et al. (2010). Gamma-phase shifting in awake monkey visual cortex. *The Journal of Neuroscience*, *30*, 1250–1257.

Vogt, B. A., & Pandya, D. N. (1987). Cingulate cortex of the rhesus monkey: II. Cortical afferents. *The Journal of Comparative Neurology*, *262*, 271–289.

Voloh, B., Oemisch, M., & Womelsdorf, T. (2020). Phase of firing coding of learning variables across the fronto-striatal network during feature-based learning. *Nature Communications*, *11*, 4669.

Voloh, B., Valiante, T. A., Everling, S., & Womelsdorf, T. (2015). Theta-gamma coordination between anterior cingulate and prefrontal cortex indexes correct attention shifts. *Proceedings of the National Academy of Sciences*, *112*, 8457–8462.

Voloh, B., & Womelsdorf, T. (2016). A role of phase-resetting in coordinating large scale neural networks during attention and goal-directed behavior. *Frontiers in Systems Neuroscience*, *10*, 18.

Voloh, B., & Womelsdorf, T. (2017). Cell-type specific burst firing interacts with Theta and Beta activity in prefrontal cortex during attention states. *Cerebral Cortex*, *28*, 4348–4364.

Wang, X. (2010). Neuophysiological and computational principles of cortical rhythms in cognition. *Physiological Reviews*, *90*, 1195–1268.

Wang, L., Saalmann, Y. B., Pinsk, M. A., Arcaro, M. J., & Kastner, S. (2012). Electrophysiological low-frequency coherence and cross-frequency coupling contribute to BOLD connectivity. *Neuron*, *76*, 1010–1020.

Westendorff, S., Kaping, D., Everling, S., & Womelsdorf, T. (2016). Prefrontal and anterior cingulate cortex neurons encode attentional targets even when they do not apparently bias behavior. *Journal of Neurophysiology*, *116*, 796–811.

Widge, A. S., Heilbronner, S. R., & Hayden, B. Y. (2019). Prefrontal cortex and cognitive control: New insights from human electrophysiology [version 1; peer review: 3 approved]. *F1000Research*, *8*, 1–10.

Wilson, G. H., Yang, P. F., Gore, J. C., & Chen, L. M. (2016). Correlated inter-regional variations in low frequency local field potentials and resting state BOLD signals within S1 cortex of monkeys. *Human Brain Mapping*, *37*, 2755–2766.

Womelsdorf, T., Ardid, S., Everling, S., & Valiante, T. A. (2014). Burst firing synchronizes prefrontal and anterior cingulate cortex during attentional control. *Current Biology*, *24*, 2613–2621.

Womelsdorf, T., & Everling, S. (2015). Long-range attention networks: Circuit motifs underlying endogenously controlled stimulus selection. *Trends in Neurosciences*, *38*, 682–700.

Womelsdorf, T., Johnston, K., Vinck, M., & Everling, S. (2010). Theta-activity in anterior cingulate cortex predicts task rules and their adjustments following errors. *Proceedings of the National Academy of Sciences*, *107*, 5248–5253.

Womelsdorf, T., Lima, B., Vinck, M., Oostenveld, R., Singer, W., Neuenschwander, S., et al. (2012). Orientation selectivity and noise correlation in awake monkey area V1 are modulated by the gamma cycle. *Proceedings of the National Academy of Sciences*, *109*, 4302–4307.

Womelsdorf, T., Valiante, T., Sahin, N. T., Miller, K. J., & Tiesinga, P. (2014). Dynamic circuit motifs underlying rhythmic gain control, gating and integration. *Nature Neuroscience*, *17*, 1031–1039.

Yoo, S. B. M., & Hayden, B. Y. (2018). Economic choice as an untangling of options into actions. *Neuron*, *99*, 434–447.

Zandvakili, A., & Kohn, A. (2015). Coordinated neuronal activity enhances corticocortical communication. *Neuron*, *87*, 827–839.

CHAPTER TWELVE

Neuronal oscillations and the mouse prefrontal cortex

Felix Jung[a] and Marie Carlén[a,b,*]
[a]Department of Biosciences and Nutrition, Karolinska Institutet, Stockholm, Sweden
[b]Department of Neuroscience, Karolinska Institutet, Stockholm, Sweden
*Corresponding author: e-mail address: marie.carlen@ki.se

Contents

Abstract

The mouse prefrontal cortex (PFC) encompasses a collection of agranual brain regions in the rostral neocortex and is considered to be critically involved in the neuronal computations underlying intentional behaviors. Flexible behavioral responses demand coordinated integration of sensory inputs with state, goal and memory information in brain-wide neuronal networks. Neuronal oscillations are proposed to provide a temporal scaffold for coordination of neuronal network activity and routing of information. In the present book chapter, we review findings on the role neuronal oscillations in prefrontal functioning, with a specific focus on research in mice. We discuss discoveries pertaining to local prefrontal processing, as well to interactions with other brain regions. We also discuss how the recent discovery of brain-wide respiration-entrained rhythms (RR) warrant re-evaluation of certain findings on slow oscillations (<10 Hz) in prefrontal functioning.

1. Introduction

The prefrontal cortex (PFC) is an umbrella term for rostrally located neocortical brain regions (see Box 1). In mice, the PFC plays an essential role in processes linked to intentional behaviors, including attention

International Review of Neurobiology, Volume 158
ISSN 0074-7742
https://doi.org/10.1016/bs.irn.2020.11.005

BOX 1 Anatomy and connectivity of the prefrontal cortex

The mouse PFC is an umbrella term for rostrally located cortical brain regions and while the delineation of the mouse PFC vary across studies and atlases, the following regions have been defined as prefrontal (nomenclature according to the Allen Brain Atlas): the secondary motor cortex (MOs), the dorsal and ventral anterior cingulate cortex (d/vACA), the pre- and infralimbic cortices (PL/IL), the orbital areas (ORB) and the agranular insular areas (AI). Cytoarchitectonically, all these regions are defined by the absence of a granular layer IV in rodents. This is of particular interest as the granular layer IV was historically used to define the PFC in primates (Carlén, 2017). The lack of layer IV in frontal regions in rodents has lead to discussions whether a mouse (or rat) PFC exists (see also: Carlén, 2017; Laubach, Amarante, Swanson, & White, 2018; Passingham & Wise, 2012; Preuss, 1995; Uylings, Groenewegen, & Kolb, 2003; Wise, 2008). The PFC has also been defined across species based on projections from the mediodorsal nucleus of the thalamus (MD) as proposed by Rose and Woosley (1948). Guldin, Pritzel, and Markowitsch (1981) could show that these projections also exist in mice. The PFC is a densely interconnected brain region receiving strong input from a large number of cortical and subcortical brain regions A recent study analyzing the input-output connectivity of cortical regions (and the thalamus) suggests that the neocortex is hierarchically organized, with the PFC at the top of the hierarchy (Harris et al., 2019). Two recent retrograde tracing studies using a transsynaptic pseudotyped rabies virus strategy in combination with transgenic Cre-driver lines, could show that the afferent input to the mouse mPFC is at large homogeneously distributed among CamKII, VIP, PV and SOM-positive neurons (Ährlund-Richter et al., 2019; Sun et al., 2019).

(Bissonette & Powell, 2012; Kim, Ährlund-Richter, Wang, Deisseroth, & Carlén, 2016; Koike et al., 2016), working memory (Abbas et al., 2018; Kim et al., 2016; Spellman et al., 2015), decision-making (Allen et al., 2017; Lak et al., 2020; Le Merre et al., 2018; Pinto & Dan, 2015), spatial navigation (Negrón-Oyarzo et al., 2018), preparation of motor activity (Chen, Li, Daie, & Svoboda, 2017; Guo et al., 2017, 2014; Inagaki, Fontolan, Romani, & Svoboda, 2019; Inagaki, Inagaki, Romani, & Svoboda, 2018), processing of value and reward signals (Kim et al., 2017; Lak et al., 2020; Otis et al., 2017) and defensive behaviors (Courtin et al., 2014; Dejean et al., 2016; Rozeske & Herry, 2018; Senn et al., 2014; Vander Weele et al., 2018; Vogel, Krabbe, Gründemann, Wamsteeker Cusulin, & Lüthi, 2016) and possesses rich efferent and afferent brain-wide interconnectivity (Ährlund-Richter et al. (2019); Sun et al. (2019); Zingg et al. (2014); see Box 1). Building on the concept of a hierarchical organization of the mouse neocortex where the PFC resides at the very top (Harris et al., 2019), it is assumed that sensory information flows from the periphery via primary to associative cortical regions, and further to the PFC, in processes increasing the level of complexity of stimulus representation. In the PFC, sensory information is integrated with experience, state, and goal information—a process essential for the generation of intentional actions (Fuster, 2015). Intentional behaviors engage activity in spatially distributed neuronal networks, and so-called neuronal oscillations—rhythmic fluctuations in the extracellular field potential—are proposed to provide the means for brain-wide, spatio-temporal coordination of neuronal activity (Buzsáki, 2006; Canolty & Knight, 2010; Hyafil, Giraud, Fontolan, & Gutkin, 2015; Jensen & Colgin, 2007; Varela, Lachaux, Rodriguez, & Martinerie, 2001). We here review findings on the role of neuronal oscillations in the functions of the PFC, both in local prefrontal processing and in interactions with other brain regions, with a focus on discoveries in mice.

2. The origins of neuronal oscillations

Ultimately, neuronal oscillations are a result of transmembrane currents, such as the in- and efflux of Na^+ and K^+ ions, occurring during neuronal activity (reviewed in Buzsáki, Anastassiou, & Koch, 2012). The resulting extracellular potential is conventionally termed *local field potential (LFP)* and its main contributor are excitatory postsynaptic currents (EPSCs; Buzsáki et al., 2012). In addition to EPSCs, fast sodium action potentials (often termed "spikes" when extracellularly recorded; Koch, 1999) from groups of synchronously active neurons can have a substantial impact on high frequency (>100 Hz) components of LFP signals (Belluscio, Mizuseki, Schmidt, Kempter, & Buzsáki, 2012; Buzsáki, Anastassiou, & Koch, 2012). Likewise, though less well understood is the contribution of calcium spikes (Helmchen, Svoboda, Denk, & Tank, 1999), intrinsic currents (Llinas, 1988), resonance properties of distinct classes of excitatory (Leung & Yim, 1991; Silva, Amitai, & Connors, 1991) or inhibitory neurons (Cardin et al., 2009, 2010) and phenomena such as spike after hyperpolarization (Buzsáki et al., 1988) to the LFP (Buzsáki et al., 2012; see also Herreras, 2016).

The LFP signal can be recorded intracranially—using e.g. microelectrodes—as well as extracranially—using e.g. electroencephalography. A central discovery on the properties of the LFP signal was already made by Hans Berger, the developer of the electroencephalogram (EEG) in the 1920s and 1930s (for a historical overview on the development of the EEG see: Coenen, Fine, and Zayachkivska (2014)). Berger discovered that the recorded LFP signal displayed oscillatory fluctuations, which can be categorically classified. Berger was able to demonstrate that a distinct class of oscillations with a frequency of 8–12 Hz appeared when the participants—some of his patients, his son Klaus, and Berger himself—closed their eyes, while faster oscillations of 15–30 Hz occurred when the participants had opened their eyes and were in an attentive state (Berger, 1929). Berger termed the first, slower rhythm *alpha* and the second, faster rhythm *beta* (Berger, 1931). However, it was not until 1974 that a first classification system for neuronal oscillations was proposed by the International Federation of Societies for Electroencephalography and Clinical Neurophysiology (Chatrian et al., 1974). This classification system followed Berger's tradition in terming individual rhythms with Greek letters but relied mostly on classifying them based on their respective frequency bands. Even today, as our understanding of the mechanisms underlying the rhythmogenesis of most classes of neuronal oscillations is still very rudimentary, the frequency of rhythms remains the most popular definition criterion (Buzsáki, 2006; Buzsáki, Logothetis, & Singer, 2013). However, the recent discovery of the co-occurrence of neuronal oscillations with overlapping frequency bands (Tort, Brankačk, & Draguhn, 2018) illustrate the limitations of a solely frequency-based taxonomy. Regardless, oscillatory patterns and their behavioral correlates are highly conserved across non-volant, mammalian species (Buzsáki et al., 2013; Eliav et al., 2018). One essential feature of all neuronal oscillations is that amplitude and frequency are inversely correlated (Penttonen & Buzsáki, 2003), indicating that slower extracellular events result from the activity of larger groups of neurons while faster extracellular events result from smaller populations (Steriade, 2001).

Despite recent advances in our understanding of neuronal oscillations, one question has not been answered: *why* do neuronal oscillations occur? To date, it is in a sense more a *brute fact* that collective neuronal activity self-organizes into neuronal oscillations (Draguhn & Buzsáki, 2004). Historically, oscillatory patterns were mainly treated as an epiphenomenon, and used for read-out for neuronal activity of a given brain region or as a tool in clinical diagnosis (Schomer & Lopes da Silva, 2017). However, accumulative evidence indicates that neuronal oscillations influence individual neurons via *ephaptic coupling* (Anastassiou, Montgomery, Barahona, Buzsáki, & Koch, 2010; Anastassiou, Perin, Markram, & Koch, 2011;

Francis, Gluckman, & Schiff, 2003; Fröhlich & McCormick, 2010; Jefferys, 1995). Hence, neuronal oscillations are not only a result of the internal dynamics of the brain, but also constrain the activity of the individual constituents they emerge from. It can, thus, be assumed that individual neuron activity and neuronal oscillations stand in a *synergetic* relationship to each other where upward and downward causation form an inseparable entity (Buzsáki, 2006; Haken, 1992).

3. Gamma oscillations in the prefrontal cortex

Alongside so-called sharp-wave ripple (SPW-R; *see below*) oscillations (Buzsáki, 2015), the arguably best understanding of its rhythmogenesis exists for gamma oscillations (30–80 Hz) which have been extensively studied in rodents, non-human and human primates in the past 30 years (Buzsáki & Wang, 2012; Gray, 1994; Tiesinga & Sejnowski, 2009; Wang, 2010). Historically, the term "gamma waves" was first used by Jasper and Andrews (1938) for frequencies between 35 and 40 Hz in humans. However, it was not until the 1980s until the term "gamma rhythm" (Bressler & Freeman, 1980)—coined to describe the frequency range between 30 and 80 Hz—gained popularity (Buzsáki & Wang, 2012). Importantly, throughout this book chapter the term gamma oscillations will exclusively refer to oscillations in the frequency range of 30–80 Hz. We refer to 80–300 oscillations as epsilon oscillations and not as "high" or "fast" gamma oscillations (e.g. Csicsvari, Hirase, Czurko, Mamiya, & Buzsaki, 1999; Zhong et al., 2017) as it can be assumed that the generation of gamma versus epsilon oscillations are mechanistically distinct (Freeman, 2007).

In their 1989 hallmark study, Gray and colleagues demonstrated a putative functional role of gamma oscillations as a proposed solution to the *feature binding problem* in the cat primary visual cortex (Gray, Konig, Engel, & Singer, 1989; Singer & Gray, 1995). In subsequent years, evidence accumulated demonstrating that synchronization of *cell assemblies* via gamma oscillations is a fundamental mechanism of cognitive processing in the brain (reviewed in e.g. Buzsáki & Wang, 2012; Engel, Fries, & Singer, 2001; Fries, 2009; Wang, 2010). In brief, a cell assembly emerges by temporally compressing the spiking of excitatory neurons into individual gamma cycles (10–30 ms) which matches the critical time window for induction of spike-timing dependent plasticity (Magee & Johnston, 1997; Markram, Lübke, Frotscher, & Sakmann, 1997). Further, only this temporal compression of assembly spiking allows for a summation of AMPA-receptor mediated EPSPs necessary for depolarizing the downstream integrator neurons (Buzsáki, 2010; Harris, Csicsvari, Hirase, Dragoi, & Buzsáki, 2003; Johnston & Wu, 1995). However, this fast gamma synchronization of excitatory neuron spiking activity relies on effective inhibition that is provided by perisomatic inhibitory postsynaptic currents (Lytton & Sejnowski, 1991) from basket interneurons in cortical networks (Freund, 2003; Freund & Katona, 2007), later identified to be parvalbumin-positive (PV+; Tamás, Buhl, Lörincz, & Somogyi, 2000). It was found that the spiking activity of PV+ interneurons phase locks to gamma oscillations (Buzsáki, Lai-Wo, & Vanderwolf, 1983; Soltesz & Deschenes, 1993; Whittington, Traubtt, & Jefferys, 1995) and temporally lags the spiking activity of the pyramidal cells in the hippocampus of awake rats (Bragin et al., 1995) as well as in the PFC of anesthetized ferrets (Hasenstaub et al., 2005; but see e.g. Perrenoud, Pennartz, & Gentet, 2016; Vinck et al., 2015). Based on this collective work, perisomatic inhibition has become a central theme in the majority of models of gamma rhythmogenesis which can be divided into two classes: models postulating that gamma oscillation emerge from reciprocally connected excitatory and inhibitory neurons ("*E-I* models"; e.g. Brunel & Wang, 2003; Freeman, 1975; Geisler, Brunel, & Wang, 2005; Tiesinga & Sejnowski, 2009; Wilson & Cowan, 1972) or from mutually connected interneurons driven by non-specific excitatory inputs ("*I-I* models"; e.g. Traub, Whittington, Stanford, & Jefferys, 1996; Van Vreeswijk, Abbott, & Bard Ermentrout, 1994; Wang & Buzsáki, 1996). Until a decade ago, or rather, until the development of optogenetics (Boyden, Zhang, Bamberg, Nagel, & Deisseroth, 2005), it was not possible to causally test the role of PV+ interneurons in gamma rhythmogenesis *in vivo*. In 2009, it was found that gamma oscillations can be robustly induced by driving channelrhodopsin (ChR2)-expressing PV+ interneurons at gamma frequencies in the mouse barrel cortex (Cardin et al., 2009) and PFC (Sohal, Zhang, Yizhar, & Deisseroth, 2009). The observation that the increase in gamma power was highest for driving PV+ interneurons at 40 Hz indicates that cortical circuits possess the property to resonate at gamma frequencies (Cardin et al., 2009). Supportingly, it was found that broadband (random 5–200 Hz) stimulation of ChR2-expressing PV+ interneurons induces gamma oscillations (Cardin et al., 2009). Subsequent research could show that optogenetic activation of pyramidal neurons in the mouse barrel cortex is able to evoke gamma oscillations (Adesnik & Scanziani, 2010), supporting the hypothesis that "*E-I*" mechanisms are the main source of gamma rhythmogenesis in the neocortex (Cardin, 2016; Soltesz & Deschenes, 1993).

Studies in human and non-human primates have identified prefrontal gamma oscillations to be a pivotal constituent of prefrontal computations underlying cognitive processes such as attention (Baluch & Itti, 2011; Miller & Buschman, 2013; Noudoost, Chang, Steinmetz, & Moore, 2010). In the past decade, studies in mice utilizing the diversity of tools for cell-type specific dissection of neuronal circuits (Luo, Callaway, & Svoboda, 2018) added important insight to the underlying mechanisms. Using a combination of electrophysiology and optogenetics in mice performing an attention task, the circuit underpinnings of PFC's control of attentional processing have been elucidated, revealing prefrontal computations that could serve selective enhancement of relevant representations in downstream regions, and a causal link between gamma oscillations and cognitive behavior (Kim, Ährlund-Richter, et al., 2016). It was found that successful allocation of attention is characterized by elevated firing of medial PFC (mPFC) PV+ interneurons, and increased gamma power. The enhanced gamma was signified by increased phase-locking of both PV+ firing and firing of local excitatory neurons, with a strong phase concentration of the PV+ spiking in the trough of the gamma cycle, consistent with the dynamics of gamma driven by fast-spiking PV+ interneurons (Buzsáki & Wang, 2012; Pritchett, Siegle, Deister, & Moore, 2015). The gamma-synchronous PV+ firing resulted in increased temporal precision and synchronization of the activities of local excitatory neurons, supporting the view that gamma rhythm supports the formation of neuronal assemblies with coordinated firing (Hasenstaub et al., 2005), a mechanism suggested to foster relay of relevant information and drive of activity in the proper targets with higher probability (Buzsáki & Watson, 2012; Salinas & Sejnowski, 2000; Salinas et al., 2001). Optogenetic silencing, or disruptive synchronization, of PV+ interneurons had negative effects on attentional processing, while optogenetic synchronization of PV+ interneurons in a typical gamma frequency (30–40 Hz) enhanced attentional processing and goal-directed behavior (Kim, Ährlund-Richter, et al., 2016). Consistent with a constitutive role of PV+ interneurons in gamma oscillations, reduced prefrontal gamma (and epsilon) power have been observed in $Dlx5/6^{+/-}$ mice, the Dlx5/6 gene being a regulator of interneuron development (Cho et al., 2015). The abnormal gamma oscillations were linked to impaired cognitive flexibility, as optogenetic restoration of gamma oscillations could rescue the phenotype (Cho et al., 2015).

In addition to the intra-regional synchronization of neuronal activity by gamma oscillations, inter-regional gamma synchronization is assumed to play essential roles in cortical processing and transfer of information (Buzsáki & Schomburg, 2015; Harris & Gordon, 2015; Salinas et al., 2001; Singer, 1999; Von der Malsburg, 1981, but see: Ray & Maunsell, 2010, 2015). In this context, it has been proposed that gamma synchronization via communication through coherence (CTC; Fries, 2005, 2015) could serve a role in prefrontal computations underlying e.g. attention, by optimizing information transfer involving prefrontal networks (Benchenane, Tiesinga, & Battaglia, 2011). The CTC hypothesis postulates that information transfer is most favorable when the gamma cycles are highly synchronized between brain regions (taking into account the transmission delays between the regions; Fries, 2005, 2015). This would allow the arrival of presynaptic spikes at the postsynaptic structures in a temporal window of maximum excitability. Importantly, shifting of the relative LFP gamma phases between the pre- and postsynaptic circuits affects the input gain, i.e., the effectiveness of the transmission. Bosman et al. (2012) performed triplet recordings, two recording sites in V1 (with distinct receptive fields) and one in V4, in non-human primates performing a visual stimulus selection task. The stimuli were presented at defined locations corresponding to the respective receptive fields of the two V1 recording sites. It was found the gamma oscillations synchronized specifically between V4 and the V1 site with the receptive field in which the stimulus was presented (V1 to V4 directionality). Further, stimulus presentation (two stimuli) in the receptive fields of both V1 sites was shown to be associated with gamma oscillatory activity in both V1 sites. However, only the gamma oscillations in the V1 recording site whose receptive field was attended synchronized with the V4 gamma oscillations, indicating that selective interregional gamma synchronization provides the means for effective information transfer between brain regions (here V1 to V4; Bosman et al., 2012). In a similar non-human primate study involving the PFC (Gregoriou, Gotts, Zhou, & Desimone, 2009), it was shown that attending a stimulus presented in a joint receptive field of frontal (frontal eye field (FEF; part of the PFC) and visual recording sites induces interregional gamma synchronization between the PFC and visual cortex, putatively enhancing the information transfer between the regions (Gregoriou et al., 2009). In mice, however, the functional role of inter-regional gamma synchronization has been explored to a lesser extent. Very recently cross-hemispheric gamma synchrony between the right and left mPFC was shown to be linked to cognitive flexibility in mice (Cho et al. (2020). The study by Cho et al. did not investigate the role of gamma oscillations in information transfer between cortical regions of differential hierarchy, but shows that gamma synchrony, as measured by novel voltage indicators, between PV+ interneurons in the two mPFCs increases during learning of rule shifts. Optogenetic perturbations setting the gamma synchrony between the PV+ interneurons in the two mPFCs out of phase disrupted rule shifting, demonstrating that naturally occurring neuronal gamma synchrony is necessary for normal behavior. Studies in mice have in addition found that prefrontal-hippocampal gamma synchrony increases across days during the acquisition of a spatial reference memory necessary for solving a Barnes maze (Barnes, 1979). It was found that the prefrontal-hippocampal gamma synchronization was most pronounced during the navigation in the maze (when compared to gamma synchronization the minute before, and after navigation in the maze, respectively; Negrón-Oyarzo et al., 2018). Negrón-Oyarzo and colleagues found that mice adopted two different strategies for solving the maze, either a spatial strategy where the distant visual cues were used, or a non-spatial trial-and-error strategy not relying on visual cues. It was found that phase-locking of a subset of PFC units to local gamma and theta oscillations was higher when mice adopted the spatial strategy. Moreover, the most pronounced modulation of PFC units by local gamma oscillations was found in the navigation phase of the task. However, no modulation of PFC units by hippocampal gamma oscillations was found (Negrón-Oyarzo et al., 2018). Taken together these results indicate that prefrontal-hippocampal gamma synchronization evolves during the learning of a task relying on the integration of spatial and goal information. Of relevance, Spellman et al. (2015) observed in mice that PFC units are entrained by hippocampal gamma oscillations during spatial working memory (sWM; Spellman et al., 2015; see also Tamura, Mukai, Gordon, & Gogos, 2016). Nevertheless, a clear demonstration of CTC involving mouse prefrontal networks is yet awaiting.

To summarize, an increasing amount of data highlights the involvement of gamma oscillations in prefrontal functions in mice (Sohal, 2016). However, more research is needed in order to elucidate potential roles of gamma oscillations in PFC computations. A substantial body of studies in human and non-human primates have linked prefrontal gamma oscillations to non-spatial working memory (Bastos, Loonis, Kornblith, Lundqvist, & Miller, 2018; Honkanen, Rouhinen, Wang, Palva, & Palva, 2015; Howard et al., 2003; Kornblith, Buschman, & Miller, 2016; Roux, Wibral, Mohr, Singer, & Uhlhaas, 2012), with more recent studies showing that the spiking thought to underlie working memory is organized in bursts of gamma oscillations (Lundqvist et al., 2016; Miller, Lundqvist, & Bastos, 2018). The relationship between prefrontal gamma and non-spatial working in the mouse is at large unknown. In the following section, the interaction of

PFC with other networks and the constitutive role of neuronal oscillations will be examined in more detail.

4. Neuronal oscillations in the prefrontal-hippocampal system

The brain processes underlying memory consolidation and retrieval require temporal coordination of activity in spatially distributed neuronal networks. The PFC and the hippocampus (HC) are critically involved in memory processes, and accumulating evidence highlights the role of neuronal oscillations in prefrontal-hippocampal communication (Eichenbaum, 2017).

During active behavior, the hippocampal LFP is dominated by theta oscillations (5–10 Hz; Colgin, 2013). Although originating from the HC, theta oscillations are a global rhythm present in a multitude of brain regions, including the PFC. Theta oscillations are suggested to create a temporal scaffold for coordinated transfer and integration of information by local networks (Lisman & Jensen, 2013). In relation to prefrontal function, it has been found that local theta oscillations critically shape prefrontal dynamics in behavior demanding sWM. In sWM tasks, performance depends on the retention and utilization of spatial information. Studies in mice have demonstrated that theta power in the PFC is increased during correct solving of a sWM task (O'Neill, Gordon, & Sigurdsson, 2013) and that theta synchrony between the PFC and HC increases when retained spatial information is correctly utilized (Benchenane et al., 2010). Moreover, correct performance in sWM tasks has been demonstrated to be associated with increased phase-locking of PFC units to hippocampal theta (Myroshnychenko, Seamans, Phillips, & Lapish, 2017; O'Neill et al., 2013) as well as to hippocampal gamma oscillations (Spellman et al., 2015; see also Negrón-Oyarzo et al., 2018). The hippocampal theta rhythm also modulates PFC activity via a phenomenon termed cross-frequency coupling (CFC, Canolty & Knight, 2010; Hyafil et al., 2015). In one form of CFC, so-called phase-amplitude coupling (Tort, Komorowski, Eichenbaum, & Kopell, 2010; Tort, Rotstein, Dugladze, Gloveli, & Kopell, 2007), the phase of a slower oscillation (e.g. theta) modulates the amplitude of faster oscillations such as gamma or epsilon oscillations (e.g. Sirota et al., 2008). Thus, the activity of local cell assemblies, reflected in the faster rhythms, can be coordinated across brain regions, aiding the information transfer between distant brain regions, including between the PFC and HC. It was found in rats that theta-gamma coupling increases in the PFC during correct sWM task performance (Li, Bai, Liu, Yi, & Tian, 2012), and studies in mice have shown that theta-gamma coupling strength positively correlates with sWM task difficulty in mice (Tamura, Spellman, Rosen, Gogos, & Gordon, 2017). Taken together, these studies illustrate that during sWM processes that involve the integration of retained information with state- and goal-information, prefrontal and hippocampal networks are tightly coupled via neuronal oscillations.

Further evidence for a causal role of neuronal oscillations in facilitating prefrontal-hippocampal function is provided by mouse models possessing disease-related phenotypes. Sigurdsson, Stark, Karayiorgou, Gogos, and Gordon (2010) found that the LFP coherence between the HC and PFC is significantly decreased in delta (1–4 Hz) and low theta (4–6 Hz) and descriptively decreased in the high theta (7–10 Hz) and gamma (30–80 Hz) bands in a mouse model of schizophrenia ($Df(16)A^{+/-}$; Mukai et al., 2008). Concordantly, it was observed that dopamine—whose receptors are targeted by a variety of antipsychotic drugs used for the treatment of schizophrenia (Carlsson, Waters, & Carlsson, 1999)—influences the synchronization of the prefrontal-hippocampal system (Benchenane et al., 2010, 2011) and that altered prefrontal dopamine signaling affects sWM performance (Duvarci et al., 2018). It was also observed that the CFC between delta (1–4 Hz) and epsilon oscillations increases in the rat PFC as a result of increased dopamine availability (Andino-Pavlovsky et al., 2017). With respect to cell-types, studies in mice have shown that the coherence between the PFC and HC critically depends on the activity of prefrontal somatostatin-positive (SOM+) interneurons (Abbas et al., 2018)—a cell type known to be affected in schizophrenia and mood disorders (Lin & Sibille, 2013; Morris, Hashimoto, & Lewis, 2008). The specific inhibition of SOM+ interneurons in the PFC decreases the theta coherence between the PFC and the HC and reduces the number of neurons coupled to the theta rhythm in the PFC. Of relevance, inhibition of prefrontal PV+ interneurons did not affect theta's influence on prefrontal neuronal activity (Abbas et al., 2018).

While theta oscillations are the dominating oscillatory pattern in the HC during active behavior, so-called sharp-wave ripples (SPW-R) prominently occur during passive behaviors, i.e. sleep or consummatory behavior (Buzsáki, 2015; see Box 2 *for details*). SPW-R are thought to underlie transfer of information from hippocampal into neocortical networks, thereby regulating the formation of stable memory engrams (Buzsáki, 2015). The exact effects of hippocampal SPW-R activity on neocortical networks in general, and prefrontal networks in particular, are a topic of ongoing research. Only recently, Khodagholy, Gelinas, and Buzsáki (2017) were able to detect ripple

BOX 2 Navigation in space and mind—the role of neuronal oscillations in the hippocampal formation

The hippocampal formation is composed of the hippocampus (HC), dentate gyrus, subicular complex, and the entorhinal cortex and is closely linked to brain functions involved in spatial navigation and memory formation (Andersen, Morris, Amaral, Bliss, & O'Keefe, 2007). It is assumed that both spatial navigation and memory formation emerge from the same neuronal mechanism: the navigation in "mental" and "physical" space, respectively (reviewed in Buzsáki & Moser, 2013). The two-stage model of memory formation (Buzsáki, 1989) postulates that neuronal oscillations play an integral part in the computations underlying hippocampal-cortical memory transformation. According to this model, the process of memory formation is divided into an active stage which is dominated by theta oscillations (5–10 Hz; the generation of theta oscillations is reviewed in: Buzsáki, 2002; Pignatelli, Beyeler, & Leinekugel, 2012). In this stage an animal samples and encodes information, i.e. forms a labile memory trace, a prominent example being the structured firing of hippocampal place cells in the cornu ammonis region 1 (CA1) representing the physical space an animal has traversed (Dragoi & Buzsáki, 2006; O'Keefe & Recce, 1993; Skaggs, McNaughton, Wilson, & Barnes, 1996). This labile memory trace needs to be consolidated in the subsequent passive phase, i.e. sleep or consummatory behavior (Buzsáki, 2015), a process aided by a highly synchronized activity pattern—a so-called sharp-wave (SPW; 1–50 Hz)—emerging from heavily recurrently connected excitatory neurons of the cornu ammonis region 3 (CA3) of the HC. The SPW-induced strong excitatory drive originating in CA3 triggers an ultrafast, superimposed oscillatory pattern (ripples (R), 150–200 Hz) in the CA1 region, forming so-called sharp-wave ripple (SPW-R) complexes. During SPW-R, place cell sequences representing the traversed space, i.e. the labile memory trace, are replayed in a temporally compressed form and transferred to neocortical regions (Girardeau & Zugaro, 2011). The repeating SPW-R events potentiate the synapses of target neurons in the neocortex, thereby generating a stable memory engram. It has been shown that SPW-R de-potentiate intrahippocampal synapses (Norimoto et al., 2018), thereby "making room for new memories" (Draguhn, 2018). Concordant with the two-stage model of memory formation is the observation that the disruption of SPW-R impairs memory formation (Ego-Stengel & Wilson, 2010; Girardeau, Benchenane, Wiener, Buzsáki, & Zugaro, 2009; Jadhav, Kemere, German, & Frank, 2012) leading to the proposal that SPW-R are a biomarker of memory consolidation (Buzsáki, 2015).

events in the PFC of the rat using novel high-density microelectrode arrays enabling electrophysiological recordings across the dorsal cortical surface (NeuroGrid; Khodagholy et al., 2015), while no reports on ripples in the PFC of mice exist to-date. Nevertheless, an informative, collective body of research exists on the impact of hippocampal SPW-R events on PFC activity and their functional implications in mice and rats. It has been observed in rats that hippocampal SPW-R events consistently elicit strong, short latency (~100 ms) responses in PFC neurons during slow-wave sleep (SWS) (Wierzynski, Lubenov, Gu, & Siapas, 2009), preferentially modulating the activity of excitatory units in deep PFC layers (Peyrache, Battaglia, & Destexhe, 2011). In rats, SPW-Rs coordinate the SWS replay of prefrontal activity patterns observed during awake rule learning behavior, indicating a role of hippocampal SPW-R in coordinating PFC activity during memory consolidation (Peyrache, Khamassi, Benchenane, Wiener, & Battaglia, 2009). Concurrent recordings in the PFC and the HC of rats have shown that individual SPW-R events coordinate the excitation of PFC units linked to a specific hippocampal representation while suppressing the activity of PFC units linked to different hippocampal representations (Jadhav, Rothschild, Roumis, & Frank, 2016). Thus, SPW-R not only mediates excitation of PFC cells, but also inhibition.

As hippocampal SPW-Rs occur during SWS (and other passive behaviors), identifying a putative interaction between SPW-R and neocortical oscillatory patterns occurring during SWS could help to elucidate the neuronal mechanisms of hippocampal-cortical memory transformation. One prominent prefrontal oscillatory pattern occurring during SWS are sleep spindles (7–14 Hz) which have been shown to temporally correlate with hippocampal SPW-R events in rats (Siapas & Wilson, 1998) and mice (Wang & Ikemoto, 2016). Peyrache et al. (2011) found that in the rat PFC, both sleep spindles and SPW-R modulate prefrontal interneuron activity, indicating a "functional 'deafferentiation'" (Peyrache et al., 2011) of the PFC from hippocampal inputs during sleep spindle activity and therefore, a possible competition between sleep spindles and SPW-R for entraining PFC activity (Buzsáki, 2015). Interestingly, Xia et al. (2017) observed that the correlation of hippocampal SPW-R and prefrontal sleep spindles itself is modulated by PV+ interneurons as it was found that inhibition of PV+ interneurons in either the PFC or the HC disrupted the coupling between hippocampal SPW-R and prefrontal spindles without affecting their respective incidence (Xia et al., 2017). Taken together, SPW-R prominently influence PFC activity, thereby gating hippocampal inputs to the PFC (Peyrache et al., 2011). Of interest in this context are observations made by Spellman et al. (2015) who found that disrupting hippocampal efferents to the PFC (irrespective of SPW-R) disrupts the encoding, but not the maintenance, nor the retrieval, of information, indicating that the prefrontal-hippocampal system coordinates the storage of information but does not store the information itself.

In addition to the above discussed evidence for a temporally structured flow of information from the HC to the PFC, evidence for a bidirectional information transfer between the PFC and the HC also exists. In this context, the nucleus reuniens of the thalamus has been identified as a critical relay structure for information transfer from the PFC to the HC (Ito, Zhang, Witter, Moser, & Moser, 2015). Ito et al. (2015) found that both the PFC and the HC encode trajectory specific information in rats performing a T-maze alternation task. After lesioning of the nucleus reuniens, trajectory specific information in the PFC was preserved while being severely reduced in the HC. Further, the supramamillary nucleus of the thalamus has been attributed to play a modulating role in this form of information transfer, with optogenetic silencing of the supramamillary nucleus reducing PFC-nucleus reuniens-HC coherence (Ito, Moser, & Moser, 2018). With respect to neuronal oscillations, Roy and colleagues identified two distinct forms of information transfer within the prefrontal-hippocampal system that depend on slow oscillations and that are routed via the nucleus reuniens in rats (Roy, Svensson, Mazeh, & Kocsis, 2017). Information from the HC to the PFC is routed via the theta rhythms while information in the opposite direction is routed via a 2–5 Hz rhythm that is distinct from theta. As remarked by Lockmann and Tort (2018), the later slow rhythm is likely to be the respiration-entrained rhythm (RR) (see Bagur & Benchenane, 2018; Kocsis, Pittman-Polletta, & Roy, 2018; Lockmann & Tort, 2018 for discussion), which is the topic of the following section in the current chapter. Noteworthy, as noted by Bagur and Benchenane (2018), identifying the 2–5 Hz rhythms as the RR poses some challenges on the assumption that information is routed from the PFC to HC via the nucleus reuniens as the nucleus reuniens does not project to the dentate gyrus of the HC where the most prominent RR activity is found (Chi et al., 2016) In summary, a more detailed understanding of prefrontal-hippocampal interactions exists for rats than mice, especially during passive behaviors. However, given the increasing amount of information on the connectivity of the mouse PFC (Ährlund-Richter et al., 2019; Harris et al., 2019; Sun et al., 2019; Zingg et al., 2014) in combination with novel tools for cell-type specific dissection of neural circuits (Luo, Callaway, & Svoboda, 2018), an investigation of the prefrontal-hippocampal system in mice could help to better understand the precise mechanisms underlying the systems' computations as well as the role of the involved neuronal oscillations.

5. Respiration-entrained rhythms and the prefrontal cortex

Since pioneering work performed by Edgar Adrian in the 1940s in hedgehogs, it has been known that the LFP signal in the olfactory bulb (OB) can follow nasal respiration. This entrainment of neuronal activity by respiration is causally driven by the stimulation of afferent mechanosensitive cells (olfactory sensitive neurons (OSNs), Grosmaitre, Santarelli, Tan, Luo, & Ma, 2007) in the olfactory epithelium by the in- and exhaled air (Adrian, 1942). Later research identified the presence of RR in olfactory brain regions such as the piriform cortex, where it is suggested to aid the integration of orofacial sensory information, i.e. the perceptual binding of different sensory modalities such as smell and touch (Fontanini & Bower, 2006; Kleinfeld, Deschênes, Wang, & Moore, 2014). However, only very recently it was shown that nasal respiration entrains neuronal activity throughout the brain, including in the PFC (reviewed in Heck et al., 2017; Tort, Brankačk, & Draguhn, 2018), thereby adding another aspect to the complex picture of neuronal oscillations in the PFC.

The main reason why the omnipresent RR activity was overlooked for a comparably long time is that mice breathe at frequencies <5 Hz during sleep or under anesthesia, and between 6 and 14 Hz during active behavior. Hence, the respiration frequency spectra overlap with delta (1–4 Hz) and theta rhythms (5–10 Hz) impeding its detection without proper monitoring of the animals' respiratory activity. The presence of the RR is operationally defined by (i) a peak in the LFP *power spectrum* and (ii) a peak in the respiration-LFP *coherence* spectrum at the respiration frequency (Tort, Brankačk, & Draguhn, 2018).

The hypothesis that the RR is originally generated by mechanic stimulation of cells in the olfactory epithelium is further supported by experiments showing that the RR dissolves after bulbectomy (Biskamp, Bartos, & Sauer, 2017), after tracheotomy when nasal respiration is bypassed (Yanovsky, Ciatipis, Draguhn, Tort, & Brankačk, 2014) and chemogenetic inhibition of OB activity using DREADDs (Liu, McAfee, & Heck, 2017).

However, a recent preprint by Karalis and Sirota (2018) highlights a potential role of an efference copy signal from the preBötzinger complex—the brainstem nucleus generating the breathing rhythm—to the HC in the local generation of the RR in the dentate gyrus of the HC (Karalis & Sirota, 2018).

The neuroanatomically defined origin of the RR raises the question of its transmission into neocortical brain areas such as the PFC as no direct synaptic connection between the PFC and the OB exists (Igarashi et al., 2012). Two opposing ideas are currently discussed, and it has yet to be experimentally determined whether the RR is transmitted via *volume conduction* or synaptic conduction into the PFC. Arguing in favor of volume conduction is the observation that the RR power follows a rostrocaudal gradient that decreases with distance from the OB (Tort et al., 2018). In contrast, some evidence exists, doubting this notion and suggesting a synaptic transmission (and local generation in the PFC) of the RR from the OB to the PFC via e.g. secondary olfactory structures such as the piriform cortex. Classic indicators of the local generation of neuronal oscillations are the modulation of unit activity, *current source density (CSD) analyses* and recordings with bipolar electrodes (Tort, Ponsel, et al., 2018). Biskamp et al. (2017) recently demonstrated that the activity of both (putative) pyramidal neurons and interneurons in the mPFC, respectively, is modulated by the RR (see also Kőszeghy, Lasztóczi, Forro, & Klausberger, 2018; Zhong et al., 2017). The same observation was made by Kőszeghy et al. (2018) for putative pyramidal and interneurons in the orbital PFC (oPFC). Additionally, CSD analyses in the HC (Chi et al., 2016; Yanovsky et al., 2014) as well as in the PFC (Karalis & Sirota, 2018) could show clear *sinks and sources* for the RR. Lastly, the RR could be detected using bipolar electrodes in the posterior parietal cortex of mice (Tort, Ponsel, et al., 2018). Taken together, these observations suggest that the RR is at least partially locally generated and not purely transmitted via volume conduction.

As mentioned above, phase-locking of individual neurons to the RR—irrespective of putative neuronal subtype—has been reported in the PFC. Interestingly, it was observed that both the proportion of coupled neurons and the coupling strength is increased in the orbital compared to the medial subdivisions of the PFC (Kőszeghy et al., 2018). Within the oPFC itself, it was observed that more neurons are stronger coupled to RR in superficial than deep layers (Kőszeghy et al., 2018). In addition, it was found that a slow epsilon band (80–120 Hz) is selectively entrained by the RR (Zhong et al. (2017); Biskamp et al. (2017); similar to observations in the OB

(Kay, 2003)), while the prominent coupling to gamma (30–80 Hz) and fast epsilon oscillations (120–180 Hz) was absent (Zhong et al., 2017). The underlying mechanisms of this selective coupling remain elusive at present. Tort, Brankačk, and Draguhn (2018) speculated that the RR-low epsilon coupling might be a result of GABAergic transmission by local interneurons that are phase-locked to the respiration circle.

Given the recency of intensified research on the RR in the PFC, very little is known about a putative functional role of this slow rhythm. It is speculated that the brain-wide prominence of the RR aids the temporal integration of information in the PFC by synchronizing its inputs from distant brain areas (Biskamp et al., 2017; Tort, Brankačk, & Draguhn, 2018). In line with this, Fujisawa and Buzsáki (2011) found that a 4 Hz rhythm is coordinating PFC, HC and ventral tegmental area activity during a working memory task in rats. Although no direct evidence exits, it can be speculated that the observed 4 Hz rhythm is identical with the RR. However, the presence of RR has been demonstrated in the rat PFC (Kocsis et al., 2018; Lockmann & Tort, 2018; Mofleh & Kocsis, 2020).

Importantly, the RR has also been found to be present in humans, and has been detected in the piriform cortex as well as in the amygdala and HC using electrocorticography (Herrero, Khuvis, Yeagle, Cerf, & Mehta, 2018; Zelano et al., 2016). Interestingly, Zelano et al. (2016) observed that the respiration phase modulates response times in an emotion discrimination task, indicating a putative role of the RR in coordinating the dynamics of the limbic system. In addition, it has been shown that long-term meditators—a practice highly relying on the control of respiration—self-induce high-amplitude gamma oscillations in frontoparietal networks during meditation (Lutz, Greischar, Rawlings, Ricard, & Davidson, 2004; see also: Banquet, 1973; Das & Gastaut, 1955). In turn, meditation is proposed to be linked to attention and emotion regulatory processes (Lutz, Slagter, Dunne, & Davidson, 2008) and meditation practices have been suggested as a treatment strategy for anxiety disorders and depression (Brown & Gerbarg, 2005). Nevertheless, further research is needed in order to evaluate a putative causal role of the RR in these observations.

Collectively, these points illustrate that comparably little is known about the functional role of the newest member of the "oscillatory zoo" (Bagur & Benchenane, 2018) of the PFC. Liu et al. (2017) observed by recording hippocampal LFP signals in mice that the RR entrains SPW-R, implicating a possible role of the RR in modulating memory processes (see Heck, Kozma, & Kay, 2019). In addition to such specific hypotheses on influences of the RR on cognitive processes, Corcoran, Pezzulo, and Hohwy (2018) suggested that the RR could be interpreted within a predictive coding framework. In this view, RR is speculated to reduce uncertainty of the state of the perceived environment by temporally pre-structuring the sensation of environmental information. Hence, the RR might be beneficial by reducing the cognitive load as well as for selecting actions matching the affordances of the environment (Corcoran et al., 2018; Friston et al., 2015). Given the close link of sensory perception (i.e. olfactory information) and motor signals (i.e. the breathing), Varga and Heck (2017) discussed the possible interpretation of the RR in the context of *embodied cognition* (e.g. Shapiro, 2011; Varela, Rosch, & Thompson, 1991), which could aid understanding of brain-body interactions in cognition. Of interest—but beyond the scope of this book chapter—are current discussions on the impact of other endogenous rhythms such as the cardiac rhythm on cognitive processes (Critchley & Garfinkel, 2018). Nevertheless, further research is needed to formulate and test specific hypotheses on the role of the RR in (mouse) PFC function.

From a technical perspective, the overlap of the RR with delta and theta frequencies advises for a parallel tracking of respiration in order to avoid confounds in experiments evaluating the role of slow oscillations in the PFC (Tort, Brankačk, & Draguhn, 2018). In the subsequent passage, we discuss experimental findings in mice pertaining to RR, explicitly or by retrospect inference.

6. Cross-region synchronization by neuronal oscillations during fear behavior

The generation of fear behaviors, e.g. freezing, and fight or flight behavior, requires the coordination of activity in a variety of brain regions, including the PFC, and it is assumed that slow neuronal oscillations temporally structure the activity of the spatially distributed brain regions. However, it is only known since very recently that the RR is a dominating rhythm in prefrontal networks—especially during freezing behaviors. In the following section, research results will be discussed where respiration was mostly not tracked, thus it is not possible to identify RR in retrospect. However, in light of more recent data it became conclusive that the observed results relate to RR, and not to a distinct class of neuronal oscillations. In their hallmark paper, Karalis et al. (2016) were able to identify a slow rhythm occurring at 4 Hz in prefrontal networks during freezing behaviors in an auditory fear conditioning task. This rhythm was demonstrated to be distinct from the hippocampal theta rhythm, and the on- and offset of this slow rhythm was found to be time-locked to the on- and offset of freezing behaviors. Optogenetic induction of the "4-Hz rhythm" in the PFC synchronized prefrontal LFP oscillations with single neuron activity in the basolateral amygdala (BLA) (Karalis et al., 2016)—a central brain structure for valence-related processing (Janak & Tye, 2015). Directionality analysis showed the causal influence of the PFC on BLA activity (Karalis et al., 2016). Most interestingly, the induction of a prefrontal "4-Hz rhythm" induced freezing behavior in mice, suggesting a causal link between the induced neuronal dynamics and the observed behavior. The "4-Hz rhythm" temporally structures prefrontal neuronal ensembles emerging during freezing, with the ensembles being activated in the ascending phase of the "4-Hz rhythm" (Dejean et al., 2016). The functional implication of these ensembles and the temporal organization by the "4-Hz rhythm" have been highlighted by closed loop optogenetic experiments. Here, inhibition of the PFC during the ascending phase of the "4-Hz rhythm" reduced freezing, while PFC inhibition in the descending phase of the "4-Hz rhythm" prolonged freezing episodes (Dejean et al., 2016). These findings exemplify the central role of slow neuronal oscillations in the temporal coordination of PFC cell assembly activity and their effects on downstream limbic structures such as the BLA.

More recently three independent studies could show that the "4-Hz rhythm" is identical with the RR (Bagur et al., 2018; Karalis & Sirota, 2018; Moberly et al., 2018). It was shown that bulbectomy (Bagur et al., 2018), methimamizole-induced depletion of the olfactory epithelium (Karalis & Sirota, 2018; Moberly et al., 2018), nasal occlusion and tetrodotoxin infusion in the OB (both Moberly et al., 2018) disrupt the occurrence of the "4-Hz rhythm." Nevertheless, further research is needed to understand the exact contributions of respiration on freezing behavior as no coherent pattern on freezing behavior upon disruption of olfactory outputs emerged from the studies. Upon disruption of the "4-Hz rhythm" Moberly and colleagues observed significantly prolonged freezing, Bagur and colleagues significantly shortened freezing, and Karalis & Sirota did not find statistically significant changes, but descriptively reduced occurrence of, and shorter, freezing periods. One intriguing observation was made by Bagur and colleagues who—based on optogenetic stimulation of the OB—suggested that the PFC network possesses a 4 Hz resonance property enabling a strong synchronization of the PFC network in response to inputs at that particular frequency. The reason why the breathing of mice synchronizes at 4 Hz during freezing remains elusive.

Not related to RR, it has been shown that freezing behavior is controlled by a disinhibitory circuit motif in the PFC: upon presentation of a fear-conditioned stimulus, a subclass of PV+ interneurons is inhibited which results in disinhibition of prefrontal BLA-projecting pyramidal neurons (Courtin, Chaudun, et al., 2014). In addition, it was observed that the phasic inhibition of PV+ interneurons leads to a resetting of prefrontal theta oscillations (8–12 Hz; Courtin, Chaudun, et al., 2014). Later research identified these interneurons as PV+ chandelier neurons (Lu et al., 2017; Rozeske & Herry, 2018). With respect to the neuronal activity in the BLA, it was observed that the power of gamma (Courtin, Karalis, Gonzalez-Campo, Wurtz, & Herry, 2014; Stujenske, Likhtik, Topiwala, & Gordon, 2014) and epsilon oscillations (Stujenske et al., 2014) increase during presentation of a fear-conditioned stimulus. Further, both classes of fast oscillations couple to a slower BLA rhythm in the theta band (broadly defined by 4–12 Hz in Stujenske et al. (2014)). However, only the epsilon-theta coupling was found to be increased during the presentation of a fear-conditioned stimulus (Stujenske et al., 2014). It was also found that the frequency of the theta band with an increased coupling to epsilon oscillations was between 5 and 8 Hz. These results possibly indicate that the broadly defined theta oscillations (4–12 Hz) are contaminated with the RR as RR has been shown to couple to epsilon oscillations in the OB, olfactory cortices and the neocortex (Kay, 2003; Zhong et al., 2017). However, Karalis and Sirota did not observe any coupling of epsilon—but gamma—oscillations to the RR in the BLA (Karalis & Sirota, 2018). Together the observations on the interactions of slow and fast oscillations in the BLA during fear-related behaviors argue for further investigations to carefully dissect the distinct contributions of theta oscillations and the RR, respectively. Further, Likhtik, Stujenske, Topiwala, Harris, and Gordon (2014) made the intriguing discovery that in the presence of safety predicting cues, a prefrontal slow oscillation in the broad theta range (4–12 Hz) is causally driving spiking activity in the BLA. The theta entrained BLA spiking was found to be negatively correlated with freezing behavior, indicating that the PFC signals safety (Likhtik et al., 2014). The synchronization of distributed networks in the broad theta range (4–12 Hz) is not restricted to PFC-BLA circuits, but extents to other limbic structures such as the hippocampus (Adhikari, Topiwala, & Gordon, 2010; Lesting et al., 2011; Seidenbecher, Laxmi, Stork, & Pape, 2003).

Taken together the reviewed observations argue for a prominent role of slow neuronal oscillations in coordinating the activity of prefrontal and limbic networks during fear-related behaviors. However, the role of distinct classes of slow oscillations needs to be carefully re-evaluated in light of recent discoveries on the role of RR in these behaviors.

7. Conclusion

The PFC serves a crucial role in brain function by integrating sensory inputs with state, goal and memory information in order to create adequate output signals to downstream brain regions, ultimately leading to the generation of flexible behavioral responses. The dense interconnectivity of the PFC with virtually all parts of the brain provides the structural backbone of these processes. However, the connectivity only serves as infrastructure for the information transfer. In analogy to a complex traffic system, brain-connectivity operates as the net of rails, streets and highways on which the trains, cars and busses commute. These individual actors of traffic all need to be temporally coordinated by e.g. traffic lights and rules of transport in order to avoid utter chaos and the collapse of the system itself. In the brain, neuronal oscillations serve this function by spatio-temporally scaffolding neuronal activity. It is important to note that neuronal oscillations themselves are the result of the collective activity of individual neurons, but it is assumed that they serve as *order parameters* that influence lower-level entities (e.g. individual neurons) via downward causation (Haken, 1984, 1992). Therefore, neuronal oscillations and individual neuron activity build one interconnected system—or in other words: neuronal oscillations are "unavoidable and useful" (Singer, 2018). This usefulness of neuronal oscillations in PFC functioning is observed on the local level where neuronal oscillations structure ensemble activity enhancing cognitive operations such as attention deployment (Kim, Ährlund-Richter, et al., 2016) and cognitive flexibility (Cho et al., 2020, 2015). More globally, slow neuronal oscillations such as the hippocampal theta rhythm and the RR help to coordinate the integration of information in the PFC, processes involving the utilization of spatial information or, as in case of the RR, the coordination of a multitude of output structures in order to generate adequate behaviors in the presence of a threat. Taken together, neuronal oscillations in the PFC are an essential feature of PFC function or simply said: without neuronal oscillations the PFC would not be what it is—an integral part of higher brain function.

Glossary

Cell assemblies Cell assemblies (also known as neuronal ensembles) are groups of cooperatively active, recurrently connected excitatory neurons. It is assumed that cell assemblies play a crucial role in neuronal computation.

Current source density analysis Current source density (CSD) analysis requires multi-side LFP recordings in order to estimate the origin of transmembrane currents.

Coherence Estimator of the stability of the phase-lag between two signals of the same frequency band. Two signals are highly coherent if the phase-lag between the two signals is stable. The phase-lag can be non-zero.

Embodied cognition An umbrella term for a plurality of theories highlighting the constitutive role of the physical body of an agent (and the interaction with the environment) for cognitive processes.

Ephaptic coupling Describes the non-synaptic coupling of individual neurons to fluctuations in the LFP. Ephaptic effects were shown to occur under pathological conditions such as during epilepsy or as a result of external perturbations (transcranial electrical stimulation). The role of ephaptic coupling under physiological conditions is less well understood.

Feature binding problem The feature binding problem describes the question on how different features of a perceived object or scene are "bound" together in order to evoke a single experience.

Local field potential (LFP) Electrical potential V_e that can be measured by intracranial electrodes in the extracellular space of the brain, but also by non-invasive methods such as EEG and MEG. The origins of the LFP are reviewed in Buzsáki et al. (2012).

Power spectrum The power spectrum or power spectral density is the frequency by incidence illustration of an LFP signal (or any other time varying signal) after signal decomposition using e.g. Fourier analyses. An increased power of a defined frequency band indicates a relative dominance of that particular frequency in the signal.

Sinks and sources Sinks and sources describe locations along neuronal membranes at which positive charges flow into (sink) or out of (source) neurons.

Synergetics Synergetics (according to Hermann Haken) is a theory describing the processes of self-organization in far-from equilibrium systems. One central theme is the assumption of circular causation between higher order parameters affecting the lower level constituents they emerge from via downward causation leading to reduced degrees of freedom of the whole system (Haken, 1984).

Volume conduction Also called "volume transmission" or "far field effects," describing the transmission of an electric field through biological tissue such as the brain. Volume conduction results from return currents from an electrical dipole. Therefore, some LFP signals can be recorded in brain areas spatially distant from the source of the LFP signal without being locally generated at the recording site.

References

Abbas, A. I., Sundiang, M. J. M., Henoch, B., Morton, M. P., Bolkan, S. S., Park, A. J., et al. (2018). Somatostatin interneurons facilitate hippocampal-prefrontal synchrony and prefrontal spatial encoding. *Neuron*, *100*(4), 926–939.e3. https://doi.org/10.1016/j.neuron.2018.09.029.

Adesnik, H., & Scanziani, M. (2010). Lateral competition for cortical space by layer-specific horizontal circuits. *Nature, 464*(7292), 1155–1160. https://doi.org/10.1038/nature08935.
Adhikari, A., Topiwala, M. A., & Gordon, J. A. (2010). Synchronized activity between the ventral hippocampus and the medial prefrontal cortex during anxiety. *Neuron, 65*(2), 257–269. https://doi.org/10.1016/j.neuron.2009.12.002.
Adrian, E. D. (1942). Olfactory reactions in the brain of the hedgehog. *The Journal of Physiology, 100*(4), 459–473. https://doi.org/10.1113/jphysiol.1942.sp003955.
Ährlund-Richter, S., Xuan, Y., van Lunteren, J. A., Kim, H., Ortiz, C., Pollak Dorocic, I., et al. (2019). A whole-brain atlas of monosynaptic input targeting four different cell types in the medial prefrontal cortex of the mouse. *Nature Neuroscience, 22*(4), 657–668. https://doi.org/10.1038/s41593-019-0354-y.
Allen, W. E., Kauvar, I. V., Chen, M. Z., Richman, E. B., Yang, S. J., Chan, K., et al. (2017). Global representations of goal-directed behavior in distinct cell types of mouse neocortex. *Neuron, 94*(4), 891–907.e6. https://doi.org/10.1016/j.neuron.2017.04.017.
Anastassiou, C. A., Montgomery, S. M., Barahona, M., Buzsáki, G., & Koch, C. (2010). The effect of spatially inhomogeneous extracellular electric fields on neurons. *Journal of Neuroscience, 30*(5), 1925–1936. https://doi.org/10.1523/JNEUROSCI.3635-09.2010.
Anastassiou, C. A., Perin, R., Markram, H., & Koch, C. (2011). Ephaptic coupling of cortical neurons. *Nature Neuroscience, 14*(2), 217–224. https://doi.org/10.1038/nn.2727.
Andersen, P., Morris, R., Amaral, D., Bliss, T., & O'Keefe, J. (2007). *The hippocampus book* (1st ed., p. 832). Oxford University Press.
Andino-Pavlovsky, V., Souza, A. C., Scheffer-Teixeira, R., Tort, A. B. L., Etchenique, R., & Ribeiro, S. (2017). Dopamine modulates delta-gamma phase-amplitude coupling in the prefrontal cortex of behaving rats. *Frontiers in Neural Circuits, 11*(May), 1–29. https://doi.org/10.3389/fncir.2017.00029.
Bagur, S., & Benchenane, K. (2018). Taming the oscillatory zoo in the hippocampus and neo-cortex: a review of the commentary of Lockmann and Tort on Roy et al. *Brain Structure and Function, 223*(1), 5–9. https://doi.org/10.1007/s00429-017-1569-x.
Bagur, S., Lefort, J. M., Lacroix, M. M., de Lavilléon, G., Herry, C., Billand, C., … Benchenane, K. (2018). Dissociation of fear initiation and maintenance by breathing-driven prefrontal oscillations. *bioRxiv*, 1–32. https://doi.org/10.1101/468264.
Baluch, F., & Itti, L. (2011). Mechanisms of top-down attention. *Trends in Neurosciences, 34*(4), 210–224. https://doi.org/10.1016/j.tins.2011.02.003.
Banquet, J. P. (1973). Spectral analysis of the EEG in meditation. *Electroencephalography and Clinical Neurophysiology, 35*(2), 143–151. https://doi.org/10.1016/0013-4694(73)90170-3.
Barnes, C. A. (1979). Memory deficits associated with senescence: A neurophysiological and behavioral study in the rat. *Journal of Comparative Psychology, 93*(1), 74–104. https://doi.org/10.1037/h0077579.
Bastos, A. M., Loonis, R., Kornblith, S., Lundqvist, M., & Miller, E. K. (2018). Laminar recordings in frontal cortex suggest distinct layers for maintenance and control of working memory. *Proceedings of the National Academy of Sciences of the United States of America, 115*(5), 1117–1122. https://doi.org/10.1073/pnas.1710323115.
Belluscio, M. A., Mizuseki, K., Schmidt, R., Kempter, R., & Buzsáki, G. (2012). Cross-frequency phase-phase coupling between theta and gamma oscillations in the hippocampus. *Journal of Neuroscience, 32*(2), 423–435. https://doi.org/10.1523/JNEUROSCI.4122-11.2012.
Benchenane, K., Peyrache, A., Khamassi, M., Tierney, P. L., Gioanni, Y., Battaglia, F. P., et al. (2010). Coherent theta oscillations and reorganization of spike timing in the hippocampal-prefrontal network upon learning. *Neuron, 66*(6), 921–936. https://doi.org/10.1016/j.neuron.2010.05.013.

Benchenane, K., Tiesinga, P. H., & Battaglia, F. P. (2011). Oscillations in the prefrontal cortex: A gateway to memory and attention. *Current Opinion in Neurobiology*, *21*(3), 475–485. https://doi.org/10.1016/j.conb.2011.01.004.

Berger, H. (1929). Über das Elektrenkephalogramm des Menschen. *Archiv für Psychiatrie und Nervenkrankheiten*, *87*, 527–570.

Berger, H. (1931). Über das Elektrenkephalogramm des Menschen. Dritte Mitteilung. *Archiv für Psychiatrie und Nervenkrankheiten*, *94*, 16–60. https://doi.org/10.1007/BF01835097.

Biskamp, J., Bartos, M., & Sauer, J. F. (2017). Organization of prefrontal network activity by respiration-related oscillations. *Scientific Reports*, 7, 1–11. https://doi.org/10.1038/srep45508.

Bissonette, G. B., & Powell, E. M. (2012). Reversal learning and attentional set-shifting in mice. *Neuropharmacology*, *62*(3), 1168–1174. https://doi.org/10.1016/j.neuropharm.2011.03.011.

Bosman, C. A., Schoffelen, J. M., Brunet, N., Oostenveld, R., Bastos, A. M., Womelsdorf, T., et al. (2012). Attentional stimulus selection through selective synchronization between monkey visual areas. *Neuron*, *75*(5), 875–888. https://doi.org/10.1016/j.neuron.2012.06.037.

Boyden, E. S., Zhang, F., Bamberg, E., Nagel, G., & Deisseroth, K. (2005). Millisecond-timescale, genetically targeted optical control of neural activity. *Nature Neuroscience*, *8*(9), 1263–1268. https://doi.org/10.1038/nn1525.

Bragin, A., Jandó, G., Nádasdy, Z., Hetke, J., Wise, K., & Buzsáki, G. (1995). Gamma (40–100 Hz) oscillation in the hippocampus of the behaving rat. *Journal of Neuroscience*, *15*(1 I), 47–60. https://doi.org/10.1523/jneurosci.15-01-00047.1995.

Bressler, S. L., & Freeman, W. J. (1980). Frequency analysis of olfactory system EEG in cat, rabbit, and rat. *Electroencephalography and Clinical Neurophysiology*, *50*(1-2), 19–24. https://doi.org/10.1016/0013-4694(80)90319-3.

Brown, R. P., & Gerbarg, P. L. (2005). Shardarshan yogic kriya breathing. Anxiety, and depression: Part I—Neurophysiologic model. *The Journal of Alternative and Complementary Medicine*, *11*(1), 189–201.

Brunel, N., & Wang, X. J. (2003). What determines the frequency of fast network oscillations with irregular neural discharges? I. Synaptic dynamics and excitation-inhibition balance. *Journal of Neurophysiology*, *90*(1), 415–430. https://doi.org/10.1152/jn.01095.2002.

Buzsáki, G. (1989). Two-stage model of memory trace formation: A role for "noisy" brain states. *Neuroscience*, *31*(3), 551–570. https://doi.org/10.1016/0306-4522(89)90423-5.

Buzsáki, G. (2002). Theta oscillations in the hippocampus. *Neuron*, *33*(3), 325–340. https://doi.org/10.1016/S0896-6273(02)00586-X.

Buzsáki, G. (2006). *Rhythms of the Brain, vol. 1*, 1–464. https://doi.org/10.1093/acprof:oso/9780195301069.001.0001.

Buzsáki, G. (2010). Neural syntax: Cell assemblies, synapsembles, and readers. *Neuron*, *68*(3), 362–385. https://doi.org/10.1016/j.neuron.2010.09.023.

Buzsáki, G. (2015). Hippocampal sharp wave-ripple: A cognitive biomarker for episodic memory and planning. *Hippocampus*, *25*(10), 1073–1188. https://doi.org/10.1002/hipo.22488.

Buzsáki, G., Anastassiou, C. A., & Koch, C. (2012). The origin of extracellular fields and currents-EEG, ECoG, LFP and spikes. *Nature Reviews Neuroscience*, *13*(6), 407–420. https://doi.org/10.1038/nrn3241.

Buzsáki, G., Bickford, R. G., Ponomareff, G., Thal, L. J., Mandel, R., & Gage, F. H. (1988). Nucleus basalis and thalamic control of neocortical activity in the freely moving rat. *Journal of Neuroscience*, *8*(11), 4007–4026. https://doi.org/10.1523/jneurosci.08-11-04007.1988.

Buzsáki, G., Lai-Wo, S. L., & Vanderwolf, C. H. (1983). Cellular bases of hippocampal EEG in the behaving rat. *Brain Research Reviews*, *6*(2), 139–171. https://doi.org/10.1016/0165-0173(83)90037-1.

Buzsáki, G., Logothetis, N., & Singer, W. (2013). Scaling brain size, keeping timing: Evolutionary preservation of brain rhythms. *Neuron*, *80*(3), 751–764. https://doi.org/10.1016/j.neuron.2013.10.002.

Buzsáki, G., & Moser, E. I. (2013). Memory, navigation and theta rhythm in the hippocampal-entorhinal system. *Nature Neuroscience*, *16*(2), 130–138. https://doi.org/10.1038/nn.3304.

Buzsáki, G., & Schomburg, E. W. (2015). What does gamma coherence tell us about inter-regional neural communication? *Nature Neuroscience*, *18*(4), 484–489. https://doi.org/10.1038/nn.3952.

Buzsáki, G., & Wang, X.-J. (2012). Mechanisms of gamma oscillations. *Annual Review of Neuroscience*, *35*(1), 203–225. https://doi.org/10.1146/annurev-neuro-062111-150444.

Buzsáki, G., & Watson, B. O. (2012). Brain rhythms and neural syntax: Implications for efficient coding of cognitive content and neuropsychiatric disease. *Dialogues in Clinical Neuroscience*, *14*(4), 345–367. https://doi.org/10.31887/dcns.2012.14.4/gbuzsaki.

Canolty, R. T., & Knight, R. T. (2010). The functional role of cross-frequency coupling. *Trends in Cognitive Sciences*, *14*(11), 506–515. https://doi.org/10.1016/j.tics.2010.09.001.

Cardin, J. A. (2016). Snapshots of the brain in action: Local circuit operations through the lens of γ oscillations. *Journal of Neuroscience*, *36*(41), 10496–10504. https://doi.org/10.1523/JNEUROSCI.1021-16.2016.

Cardin, J. A., Carlén, M., Meletis, K., Knoblich, U., Zhang, F., Deisseroth, K., et al. (2009). Driving fast-spiking cells induces gamma rhythm and controls sensory responses. *Nature*, *459*(7247), 663–667. https://doi.org/10.1038/nature08002.

Cardin, J. A., Carlén, M., Meletis, K., Knoblich, U., Zhang, F., Deisseroth, K., et al. (2010). Targeted optogenetic stimulation and recording of neurons in vivo using cell-type-specific expression of Channelrhodopsin-2. *Nature Protocols*, *5*(2), 247–254. https://doi.org/10.1038/nprot.2009.228.

Carlén, M. (2017). What constitutes the prefrontal cortex? *Science*, *358*(6362), 478–482. https://doi.org/10.1126/science.aan8868.

Carlsson, A., Waters, N., & Carlsson, M. L. (1999). Neurotransmitter interactions in schizophrenia—Therapeutic implications. *Biological Psychiatry*, *46*(10), 1388–1395. https://doi.org/10.1016/S0006-3223(99)00117-1.

Chatrian, G. E., Bergamini, L., Dondey, M., Klass, D. W., Lennox-Buchthal, M., & Petersén, I. (1974). A glossary of terms most commonly used by clinical electroencephalographers. *Electroencephalography and Clinical Neurophysiology*, *37*(5), 538–548. https://doi.org/10.1016/0013-4694(74)90099-6.

Chen, T. W., Li, N., Daie, K., & Svoboda, K. (2017). A map of anticipatory activity in mouse motor cortex. *Neuron*, *94*(4), 866–879.e4. https://doi.org/10.1016/j.neuron.2017.05.005.

Chi, V. N., Müller, C., Wolfenstetter, T., Yanovsky, Y., Draguhn, A., Tort, A. B. L., et al. (2016). Hippocampal respiration-driven rhythm distinct from theta oscillations in awake mice. *Journal of Neuroscience*, *36*(1), 162–177. https://doi.org/10.1523/JNEUROSCI.2848-15.2016.

Cho, K. K. A., Davidson, T. J., Bouvier, G., Marshall, J. D., Schnitzer, M. J., & Sohal, V. S. (2020). Cross-hemispheric gamma synchrony between prefrontal parvalbumin interneurons supports behavioral adaptation during rule shift learning. *Nature Neuroscience*, *23*, 892–902. https://doi.org/10.1038/s41593-020-0647-1.

Cho, K. K. A., Hoch, R., Lee, A. T., Patel, T., Rubenstein, J. L. R., & Sohal, V. S. (2015). Gamma rhythms link prefrontal interneuron dysfunction with cognitive inflexibility in Dlx5/6+/- mice. *Neuron*, *85*(6), 1332–1343. https://doi.org/10.1016/j.neuron.2015.02.019.

Coenen, A., Fine, E., & Zayachkivska, O. (2014). Adolf Beck: A forgotten pioneer in electroencephalography. *Journal of the History of the Neurosciences*, *23*(3), 276–286. https://doi.org/10.1080/0964704X.2013.867600.

Colgin, L. L. (2013). Mechanisms and functions of theta rhythms. *Annual Review of Neuroscience*, *36*(1), 295–312. https://doi.org/10.1146/annurev-neuro-062012-170330.

Corcoran, A. W., Pezzulo, G., & Hohwy, J. (2018). Commentary: Respiration-entrained brain rhythms are global but often overlooked. *Frontiers in Systems Neuroscience*, *12*(June), 10–13. https://doi.org/10.3389/fnsys.2018.00025.

Courtin, J., Chaudun, F., Rozeske, R. R., Karalis, N., Gonzalez-Campo, C., Wurtz, H., et al. (2014). Prefrontal parvalbumin interneurons shape neuronal activity to drive fear expression. *Nature*, *505*(7481), 92–96. https://doi.org/10.1038/nature12755.

Courtin, J., Karalis, N., Gonzalez-Campo, C., Wurtz, H., & Herry, C. (2014). Persistence of amygdala gamma oscillations during extinction learning predicts spontaneous fear recovery. *Neurobiology of Learning and Memory*, *113*, 82–89. https://doi.org/10.1016/j.nlm.2013.09.015.

Critchley, H. D., & Garfinkel, S. N. (2018). The influence of physiological signals on cognition. *Current Opinion in Behavioral Sciences*, *19*, 13–18. https://doi.org/10.1016/j.cobeha.2017.08.014.

Csicsvari, J., Hirase, H., Czurko, A., Mamiya, A., & Buzsaki, G. (1999). Fast network oscillations in the hippocampal CA1 region of the behaving rat. *Journal of Neuroscience*, *19*, RC20.

Das, N. N., & Gastaut, H. (1955). Variations de l''activite electrique du cerveau, du coeur et des muscles squellettiques au cours de la meditation et de l''extase yogique. *Electroencephalography and Clinical Neurophysiology*, *6*, 211–219.

Dejean, C., Courtin, J., Karalis, N., Chaudun, F., Wurtz, H., Bienvenu, T. C. M., et al. (2016). Prefrontal neuronal assemblies temporally control fear behaviour. *Nature*, *535*(7612), 420–424. https://doi.org/10.1038/nature18630.

Dragoi, G., & Buzsáki, G. (2006). Temporal encoding of place sequences by hippocampal cell assemblies. *Neuron*, *50*(1), 145–157. https://doi.org/10.1016/j.neuron.2006.02.023.

Draguhn, A. (2018). Making room for new memories. *Science*, *359*(6383), 1461–1462. https://doi.org/10.1126/science.aat1493.

Draguhn, A., & Buzsáki, G. (2004). Neuronal oscillations in cortical networks. *Science*, *304*(June), 1926–1930.

Duvarci, S., Simpson, E. H., Schneider, G., Kandel, E. R., Roeper, J., & Sigurdsson, T. (2018). Impaired recruitment of dopamine neurons during working memory in mice with striatal D2 receptor overexpression. *Nature Communications*, *9*(1). https://doi.org/10.1038/s41467-018-05214-4.

Ego-Stengel, V., & Wilson, M. A. (2010). Disruption of ripple-associated hippocampal activity during rest impairs spatial learning in the rat. *Hippocampus*, *20*(1), 1–10. https://doi.org/10.1002/hipo.20707.

Eichenbaum, H. (2017). Prefrontal-hippocampal interactions in episodic memory. *Nature Reviews Neuroscience*, *18*(9), 547–558. https://doi.org/10.1038/nrn.2017.74.

Eliav, T., Geva-Sagiv, M., Yartsev, M. M., Finkelstein, A., Rubin, A., Las, L., & Ulanovsky, N. (2018). Nonoscillatory phase coding and synchronization in the bat hippocampal formation. *Cell*, *175*, 1119–1130. https://doi.org/10.1016/j.cell.2018.09.017.

Engel, A. K., Fries, P., & Singer, W. (2001). Dynamic predictions: Oscillations and synchrony in top–down processing. *Nature Reviews Neuroscience*, *2*(10), 704–716. https://doi.org/10.1038/35094565.

Fontanini, A., & Bower, J. M. (2006). Slow-waves in the olfactory system: An olfactory perspective on cortical rhythms. *Trends in Neurosciences*, *29*(8), 429–437. https://doi.org/10.1016/j.tins.2006.06.013.

Francis, J. T., Gluckman, B. J., & Schiff, S. J. (2003). Sensitivity of neurons to weak electric fields. *Journal of Neuroscience*, *23*(19), 7255–7261. https://doi.org/10.1523/jneurosci.23-19-07255.2003.

Freeman, W. J. (1975). *Mass action in the nervous system*. New York: Academic.

Freeman, W. J. (2007). Definitions of state variables and state space for brain-computer interface: Part 1. Multiple hierarchical levels of brain function. *Cognitive Neurodynamics*, *1*(1), 3–14. https://doi.org/10.1007/s11571-006-9001-x.
Freund, T. F. (2003). Interneuron diversity series: Rhythm and mood in perisomatic inhibition. *Trends in Neurosciences*, *26*(9), 489–495. https://doi.org/10.1016/S0166-2236(03)00227-3.
Freund, T. F., & Katona, I. (2007). Perisomatic inhibition. *Neuron*, *56*(1), 33–42. https://doi.org/10.1016/j.neuron.2007.09.012.
Fries, P. (2005). A mechanism for cognitive dynamics: Neuronal communication through neuronal coherence. *Trends in Cognitive Sciences*, *9*(10), 474–480. https://doi.org/10.1016/j.tics.2005.08.011.
Fries, P. (2009). Neuronal gamma-band synchronization as a fundamental process in cortical computation. *Annual Review of Neuroscience*, *32*(1), 209–224. https://doi.org/10.1146/annurev.neuro.051508.135603.
Fries, P. (2015). Rhythms for cognition: Communication through coherence. *Neuron*, *88*(1), 220–235. https://doi.org/10.1016/j.neuron.2015.09.034.
Friston, K., Rigoli, F., Ognibene, D., Mathys, C., Fitzgerald, T., & Pezzulo, G. (2015). Active inference and epistemic value. *Cognitive Neuroscience*, *6*(4), 187–214. https://doi.org/10.1080/17588928.2015.1020053.
Fröhlich, F., & McCormick, D. A. (2010). Endogenous electric fields may guide neocortical network activity. *Neuron*, *67*(1), 129–143. https://doi.org/10.1016/j.neuron.2010.06.005.
Fujisawa, S., & Buzsáki, G. (2011). A 4 Hz oscillation adaptively synchronizes prefrontal, VTA, and hippocampal activities. *Neuron*, *72*(1), 153–165. https://doi.org/10.1016/j.neuron.2011.08.018.
Fuster, J. M. (2015). *The prefrontal cortex* (5th ed., p. 460). Academic Press.
Geisler, C., Brunel, N., & Wang, X. J. (2005). Contributions of intrinsic membrane dynamics to fast network oscillations with irregular neuronal discharges. *Journal of Neurophysiology*, *94*(6), 4344–4361. https://doi.org/10.1152/jn.00510.2004.
Girardeau, G., Benchenane, K., Wiener, S. I., Buzsáki, G., & Zugaro, M. B. (2009). Selective suppression of hippocampal ripples impairs spatial memory. *Nature Neuroscience*, *12*(10), 1222–1223. https://doi.org/10.1038/nn.2384.
Girardeau, G., & Zugaro, M. (2011). Hippocampal ripples and memory consolidation. *Current Opinion in Neurobiology*, *21*(3), 452–459. https://doi.org/10.1016/j.conb.2011.02.005.
Gray, C. M. (1994). Synchronous oscillations in neuronal systems: Mechanisms and functions. *Journal of Computational Neuroscience*, *1*(1–2), 11–38. https://doi.org/10.1007/BF00962716.
Gray, C. M., Konig, P., Engel, A. K., & Singer, W. (1989). Oscillatory responses in cat visual cortex exhibit inter-columnar global stimulus properties. *Nature*, *338*(March), 334–337.
Gregoriou, G. G., Gotts, S. J., Zhou, H., & Desimone, R. (2009). High-frequency, long-range coupling between prefrontal and visual cortex during attention. *Science*, *324*(5931), 1207–1210. https://doi.org/10.1126/science.1171402.
Grosmaitre, X., Santarelli, L. C., Tan, J., Luo, M., & Ma, M. (2007). Dual functions of mammalian olfactory sensory neurons as odor detectors and mechanical sensors. *Nature Neuroscience*, *10*(3), 348–354. https://doi.org/10.1038/nn1856.
Guldin, W. O., Pritzel, M., & Markowitsch, H. J. (1981). Prefrontal cortex of the mouse defined as cortical projection area of the thalamic mediodorsal nucleus. *Brain, Behavior and Evolution*, *19*(3–4), 93–107. https://doi.org/10.1159/000121636.
Guo, Z. V., Inagaki, H. K., Daie, K., Druckmann, S., Gerfen, C. R., & Svoboda, K. (2017). Maintenance of persistent activity in a frontal thalamocortical loop. *Nature*, *545*(7653), 181–186. https://doi.org/10.1038/nature22324.

Guo, Z. V., Li, N., Huber, D., Ophir, E., Gutnisky, D., Ting, J. T., et al. (2014). Flow of cortical activity underlying a tactile decision in mice. *Neuron, 81*(1), 179–194. https://doi.org/10.1016/j.neuron.2013.10.020.

Haken, H. (1984). *The science of structure: Synergetics* (p. 255). Van Nostrand Reinhold.

Haken, H. (1992). Synergetics of the brain: An outline of some basic ideas. *Induced Rhythms in the Brain*, 417–421. https://doi.org/10.1007/978-1-4757-1281-0_23.

Harris, K. D., Csicsvari, J., Hirase, H., Dragoi, G., & Buzsáki, G. (2003). Organization of cell assemblies in the hippocampus. *Nature, 424*(6948), 552–556. https://doi.org/10.1038/nature01834.

Harris, A. Z., & Gordon, J. A. (2015). Long-range neural synchrony in behavior. *Annual Review of Neuroscience, 38*(1), 171–194. https://doi.org/10.1146/annurev-neuro-071714-034111.

Harris, J. A., Mihalas, S., Hirokawa, K. E., Whitesell, J. D., Choi, H., Bernard, A., et al. (2019). Hierarchical organization of cortical and thalamic connectivity. *Nature, 575*(7781), 195–202. https://doi.org/10.1038/s41586-019-1716-z.

Hasenstaub, A., Shu, Y., Haider, B., Kraushaar, U., Duque, A., & McCormick, D. A. (2005). Inhibitory postsynaptic potentials carry synchronized frequency information in active cortical networks. *Neuron, 47*(3), 423–435. https://doi.org/10.1016/j.neuron.2005.06.016.

Heck, D. H., Kozma, R., & Kay, L. M. (2019). The rhythm of memory: How breathing shapes memory function. *Journal of Neurophysiology, 122*(2), 563–571. https://doi.org/10.1152/jn.00200.2019.

Heck, D. H., McAfee, S. S., Liu, Y., Babajani-Feremi, A., Rezaie, R., Freeman, W. J., & Kozma, R. (2017). Breathing as a fundamental rhythm of brain function. *Frontiers in Neural Circuits, 10*, 1–8. https://doi.org/10.3389/fncir.2016.00115.

Helmchen, F., Svoboda, K., Denk, W., & Tank, D. W. (1999). In vivo dendritic calcium dynamics in deep-layer cortical pyramidal neurons. *Nature Neuroscience, 2*(11), 989–996. https://doi.org/10.1038/14788.

Herreras, O. (2016). Local field potentials: Myths and misunderstandings. *Frontiers in Neural Circuits, 10*(Dec), 1–16. https://doi.org/10.3389/fncir.2016.00101.

Herrero, J. L., Khuvis, S., Yeagle, E., Cerf, M., & Mehta, A. D. (2018). Breathing above the brain stem: Volitional control and attentional modulation in humans. *Journal of Neurophysiology, 119*(1), 145–159. https://doi.org/10.1152/jn.00551.2017.

Honkanen, R., Rouhinen, S., Wang, S. H., Palva, J. M., & Palva, S. (2015). Gamma oscillations underlie the maintenance of feature-specific information and the contents of visual working memory. *Cerebral Cortex, 25*(10), 3788–3801. https://doi.org/10.1093/cercor/bhu263.

Howard, M. W., Rizzuto, D. S., Caplan, J. B., Madsen, J. R., Lisman, J., Aschenbrenner-Scheibe, R., et al. (2003). Gamma oscillations correlate with working memory load in humans. *Cerebral Cortex, 13*(12), 1369–1374. https://doi.org/10.1093/cercor/bhg084.

Hyafil, A., Giraud, A. L., Fontolan, L., & Gutkin, B. (2015). Neural cross-frequency coupling: Connecting architectures, mechanisms, and functions. *Trends in Neurosciences, 38*(11), 725–740. https://doi.org/10.1016/j.tins.2015.09.001.

Igarashi, K. M., Ieki, N., An, M., Yamaguchi, Y., Nagayama, S., Kobayakawa, K., et al. (2012). Parallel mitral and tufted cell pathways route distinct odor information to different targets in the olfactory cortex. *Journal of Neuroscience, 32*(23), 7970–7985. https://doi.org/10.1523/JNEUROSCI.0154-12.2012.

Inagaki, H. K., Fontolan, L., Romani, S., & Svoboda, K. (2019). Discrete attractor dynamics underlies persistent activity in the frontal cortex. *Nature, 566*(7743), 212–217. https://doi.org/10.1038/s41586-019-0919-7.

Inagaki, H. K., Inagaki, M., Romani, S., & Svoboda, K. (2018). Low-dimensional and monotonic preparatory activity in mouse anterior lateral motor cortex. *Journal of Neuroscience*, *38*(17), 4163–4185. https://doi.org/10.1523/JNEUROSCI.3152-17.2018.

Ito, H. T., Moser, E. I., & Moser, M. B. (2018). Supramammillary nucleus modulates spike-time coordination in the prefrontal-thalamo-hippocampal circuit during navigation. *Neuron*, *99*(3), 576–587.e5. https://doi.org/10.1016/j.neuron.2018.07.021.

Ito, H. T., Zhang, S. J., Witter, M. P., Moser, E. I., & Moser, M. B. (2015). A prefrontal-thalamo-hippocampal circuit for goal-directed spatial navigation. *Nature*, *522*(7554), 50–55. https://doi.org/10.1038/nature14396.

Jadhav, S. P., Kemere, C., German, P. W., & Frank, L. M. (2012). Awake hippocampal sharp-wave ripples support spatial memory. *Science*, *336*(6087), 1454–1458. https://doi.org/10.1126/science.1217230.

Jadhav, S. P. P., Rothschild, G., Roumis, D. K. K., & Frank, L. M. M. (2016). Coordinated excitation and inhibition of prefrontal ensembles during awake hippocampal sharp-wave ripple events. *Neuron*, *90*(1), 113–127. https://doi.org/10.1016/j.neuron.2016.02.010.

Janak, P. H., & Tye, K. M. (2015). From circuits to behaviour in the amygdala. *Nature*, *517*(7534), 284–292. https://doi.org/10.1038/nature14188.

Jasper, H. H., & Andrews, H. L. (1938). Brain potentials and voluntary muscle activity in man. *Journal of Neurophysiology*, *1*, 87–100.

Jefferys, J. G. R. (1995). Nonsynaptic modulation of neuronal activity in the brain: Electric currents and extracellular ions. *Physiological Reviews*, *75*(4), 689–723. https://doi.org/10.1152/physrev.1995.75.4.689.

Jensen, O., & Colgin, L. L. (2007). Cross-frequency coupling between neuronal oscillations. *Trends in Cognitive Sciences*, *11*(7), 267–269. https://doi.org/10.1016/j.tics.2007.05.003.

Johnston, D., & Wu, S. M.-S. (1995). *Foundations of cellular neurophysiology* (p. 676). https://doi.org/10.1097/00004691-199703000-00009.

Karalis, N., Dejean, C., Chaudun, F., Khoder, S., Rozeske, R. R., Wurtz, H., et al. (2016). 4-Hz oscillations synchronize prefrontal-amygdala circuits during fear behavior. *Nature Neuroscience*, *19*(4), 605–612. https://doi.org/10.1038/nn.4251.

Karalis, N., & Sirota, A. (2018). Breathing coordinates limbic network dynamics underlying memory consolidation. *bioRxiv*, 1–29. https://doi.org/10.2139/ssrn.3283711.

Kay, L. M. (2003). Two species of gamma oscillations in the olfactory bulb: Dependence on behavioral state and synaptic interactions. *Journal of Integrative Neuroscience*, *2*, 31–44.

Khodagholy, D., Gelinas, J. N., & Buzsáki, G. (2017). Learning-enhanced coupling between ripple oscillations in association cortices and hippocampus. *Science*, *358*(October), 369–372.

Khodagholy, D., Gelinas, J. N., Thesen, T., Doyle, W., Devinsky, O., Malliaras, G. G., et al. (2015). NeuroGrid: Recording action potentials from the surface of the brain. *Nature Neuroscience*, *18*(2), 310–315. https://doi.org/10.1038/nn.3905.

Kim, H., Ährlund-Richter, S., Wang, X., Deisseroth, K., & Carlén, M. (2016). Prefrontal parvalbumin neurons in control of attention. *Cell*, *164*(1–2), 208–218. https://doi.org/10.1016/j.cell.2015.11.038.

Kim, D., Jeong, H., Lee, J., Ghim, J. W., Her, E. S., Lee, S. H., et al. (2016). Distinct roles of parvalbumin- and somatostatin-expressing interneurons in working memory. *Neuron*, *92*(4), 902–915. https://doi.org/10.1016/j.neuron.2016.09.023.

Kim, C. K., Ye, L., Jennings, J. H., Pichamoorthy, N., Tang, D. D., Yoo, A. C. W., et al. (2017). Molecular and circuit-dynamical identification of top-down neural mechanisms for restraint of reward seeking. *Cell*, *170*(5), 1013–1027.e14. https://doi.org/10.1016/j.cell.2017.07.020.

Kleinfeld, D., Deschênes, M., Wang, F., & Moore, J. D. (2014). More than a rhythm of life: Breathing as a binder of orofacial sensation. *Nature Neuroscience*, *17*(5), 647–651. https://doi.org/10.1038/nn.3693.

Koch, C. (1999). *Biophysics of computations. Information processing in single neurons* (p. 562). Oxford University Press.

Kocsis, B., Pittman-Polletta, B. R., & Roy, A. (2018). Respiration-coupled rhythms in prefrontal cortex: Beyond if, to when, how, and why. *Brain Structure and Function*, *223*(1), 11–16. https://doi.org/10.1007/s00429-017-1587-8.

Koike, H., Demars, M. P., Short, J. A., Nabel, E. M., Akbarian, S., Baxter, M. G., et al. (2016). Chemogenetic inactivation of dorsal anterior cingulate cortex neurons disrupts attentional behavior in mouse. *Neuropsychopharmacology*, *41*(4), 1014–1023. https://doi.org/10.1038/npp.2015.229.

Kornblith, S., Buschman, T. J., & Miller, E. K. (2016). Stimulus load and oscillatory activity in higher cortex. *Cerebral Cortex*, *26*(9), 3772–3784. https://doi.org/10.1093/cercor/bhv182.

Kőszeghy, Á., Lasztóczi, B., Forro, T., & Klausberger, T. (2018). Spike-timing of orbitofrontal neurons is synchronized with breathing. *Frontiers in Cellular Neuroscience*, *12*(April), 1–12. https://doi.org/10.3389/fncel.2018.00105.

Lak, A., Okun, M., Moss, M. M., Gurnani, H., Farrell, K., Wells, M. J., et al. (2020). Dopaminergic and prefrontal basis of learning from sensory confidence and reward value. *Neuron*, *105*(4), 700–711.e6. https://doi.org/10.1016/j.neuron.2019.11.018.

Laubach, M., Amarante, L. M., Swanson, K., & White, S. R. (2018). What, if anything, is rodent prefrontal cortex? *eNeuro*, *5*(5), 1–14. https://doi.org/10.1523/ENEURO.0315-18.2018.

Le Merre, P., Esmaeili, V., Charrière, E., Galan, K., Salin, P. A., Petersen, C. C. H., et al. (2018). Reward-based learning drives rapid sensory signals in medial prefrontal cortex and dorsal hippocampus necessary for goal-directed behavior. *Neuron*, *97*(1), 83–91.e5. https://doi.org/10.1016/j.neuron.2017.11.031.

Lesting, J., Narayanan, R. T., Kluge, C., Sangha, S., Seidenbecher, T., & Pape, H. C. (2011). Patterns of coupled theta activity in amygdala-hippocampal-prefrontal cortical circuits during fear extinction. *PLoS One*, *6*(6), e21714. https://doi.org/10.1371/journal.pone.0021714.

Leung, L. W. S., & Yim, C. Y. C. (1991). Intrinsic membrane potential oscillations in hippocampal neurons in vitro. *Brain Research*, *553*(2), 261–274. https://doi.org/10.1016/0006-8993(91)90834-I.

Li, S., Bai, W., Liu, T., Yi, H., & Tian, X. (2012). Increases of theta-low gamma coupling in rat medial prefrontal cortex during working memory task. *Brain Research Bulletin*, *89*(3–4), 115–123. https://doi.org/10.1016/j.brainresbull.2012.07.012.

Likhtik, E., Stujenske, J. M., Topiwala, M. A., Harris, A. Z., & Gordon, J. A. (2014). Prefrontal entrainment of amygdala activity signals safety in learned fear and innate anxiety. *Nature Neuroscience*, *17*(1), 106–113. https://doi.org/10.1038/nn.3582.

Lin, L. C., & Sibille, E. (2013). Reduced brain somatostatin in mood disorders: A common pathophysiological substrate and drug target? *Frontiers in Pharmacology*, *4*(September), 1–12. https://doi.org/10.3389/fphar.2013.00110.

Lisman, J. E., & Jensen, O. (2013). The theta-gamma neural code. *Neuron*, 77(6), 1002–1016. https://doi.org/10.1016/j.neuron.2013.03.007.

Liu, Y., McAfee, S. S., & Heck, D. H. (2017). Hippocampal sharp-wave ripples in awake mice are entrained by respiration. *Scientific Reports*, 7(1), 1–9. https://doi.org/10.1038/s41598-017-09511-8.

Llinas, R. R. (1988). Intrinsic electrophysiological properties central nervous system function. *Science*, *242*(3), 1654–1664.

Lockmann, A. L. V., & Tort, A. B. L. (2018). Nasal respiration entrains delta-frequency oscillations in the prefrontal cortex and hippocampus of rodents. *Brain Structure and Function*, *223*(1), 1–3. https://doi.org/10.1007/s00429-017-1573-1.

Lu, J., Tucciarone, J., Padilla-Coreano, N., He, M., Gordon, J. A., & Josh Huang, Z. (2017). Selective inhibitory control of pyramidal neuron ensembles and cortical subnetworks by chandelier cells. *Nature Neuroscience*, *20*(10), 1377–1383. https://doi.org/10.1038/nn.4624.

Lundqvist, M., Rose, J., Herman, P., Brincat, S. L. L., Buschman, T. J. J., & Miller, E. K. K. (2016). Gamma and beta bursts underlie working memory. *Neuron*, *90*(1), 152–164. https://doi.org/10.1016/j.neuron.2016.02.028.

Luo, L., Callaway, E. M., & Svoboda, K. (2018). Genetic dissection of neural circuits: A decade of progress. *Neuron*, *98*, 256–281. https://doi.org/10.1016/j.neuron.2018.03.040.

Lutz, A., Greischar, L. L., Rawlings, N. B., Ricard, M., & Davidson, R. J. (2004). Long-term meditators self-induce high-amplitude gamma synchrony during mental practice. *Proceedings of the National Academy of Sciences of the United States of America*, *101*(46), 16369–16373. https://doi.org/10.1073/pnas.0407401101.

Lutz, A., Slagter, H. A., Dunne, J. D., & Davidson, R. J. (2008). Attention regulation and monitoring in meditation. *Trends in Cognitive Sciences*, *12*(4), 163–169. https://doi.org/10.1016/j.tics.2008.01.005.

Lytton, W. W., & Sejnowski, T. J. (1991). Simulations of cortical pyramidal neurons synchronized by inhibitory interneurons. *Journal of Neurophysiology*, *66*(3), 1059–1079. https://doi.org/10.1152/jn.1991.66.3.1059.

Magee, J. C., & Johnston, D. (1997). A synaptically controlled, associative signal for Hebbian plasticity in hippocampal neurons. *Science*, *275*(5297), 209–213. https://doi.org/10.1126/science.275.5297.209.

Markram, H., Lübke, J., Frotscher, M., & Sakmann, B. (1997). Regulation of synaptic efficacy by coincidence of postsynaptic APs and EPSPs. *Science*, *275*(5297), 213–215. https://doi.org/10.1126/science.275.5297.213.

Miller, E. K., & Buschman, T. J. (2013). Cortical circuits for the control of attention. *Current Opinion in Neurobiology*, *23*(2), 216–222. https://doi.org/10.1016/j.conb.2012.11.011.

Miller, E. K., Lundqvist, M., & Bastos, A. M. (2018). Working memory 2.0. *Neuron*, *100*(2), 463–475. https://doi.org/10.1016/j.neuron.2018.09.023.

Moberly, A. H., Schreck, M., Bhattarai, J. P., Zweifel, L. S., Luo, W., & Ma, M. (2018). Olfactory inputs modulate respiration-related rhythmic activity in the prefrontal cortex and freezing behavior. *Nature Communications*, *9*(1), 1528. https://doi.org/10.1038/s41467-018-03988-1.

Mofleh, R., & Kocsis, B. (2020). Delta-range coupling between prefrontal cortex and hippocampus supported by respiratory rhythmic input from the olfactory bulb. *bioRxiv*, 1–11.

Morris, H. M., Hashimoto, T., & Lewis, D. A. (2008). Alterations in somatostatin mRNA expression in the dorsolateral prefrontal cortex of subjects with schizophrenia or schizoaffective disorder. *Cerebral Cortex*, *18*(7), 1575–1587. https://doi.org/10.1093/cercor/bhm186.

Mukai, J., Dhilla, A., Drew, L. J., Stark, K. L., Cao, L., MacDermott, A. B., et al. (2008). Palmitoylation-dependent neurodevelopmental deficits in a mouse model of 22q11 microdeletion. *Nature Neuroscience*, *11*(11), 1302–1310. https://doi.org/10.1038/nn.2204.

Myroshnychenko, M., Seamans, J. K., Phillips, A. G., & Lapish, C. C. (2017). Temporal dynamics of hippocampal and medial prefrontal cortex interactions during the delay period of a working memory-guided foraging task. *Cerebral Cortex*, *27*(11), 5331–5342. https://doi.org/10.1093/cercor/bhx184.

Negrón-Oyarzo, I., Espinosa, N., Aguilar, M., Fuenzalida, M., Aboitiz, F., & Fuentealba, P. (2018). Coordinated prefrontal–hippocampal activity and navigation strategy-related prefrontal firing during spatial memory formation. *Proceedings of the National Academy of Sciences of the United States of America*, *115*(27), 7123–7128. https://doi.org/10.1073/pnas.1720117115.

Norimoto, H., Makino, K., Gao, M., Shikano, Y., Okamoto, K., Ishikawa, T., et al. (2018). Hippocampal ripples down-regulate synapses. *Science*, *359*(6383), 1524–1527. https://doi.org/10.1126/science.aao0702.

Noudoost, B., Chang, M. H., Steinmetz, N. A., & Moore, T. (2010). Top-down control of visual attention. *Current Opinion in Neurobiology*, *20*(2), 183–190. https://doi.org/10.1016/j.conb.2010.02.003.

O'Keefe, J., & Recce, M. L. (1993). Phase relationship between hippocampal place units and the EEG theta rhythm. *Hippocampus*, *3*(3), 317–330. https://doi.org/10.1002/hipo.450030307.

O'Neill, P. K., Gordon, J. A., & Sigurdsson, T. (2013). Theta oscillations in the medial prefrontal cortex are modulated by spatial working memory and synchronize with the hippocampus through its ventral subregion. *Journal of Neuroscience*, *33*(35), 14211–14224. https://doi.org/10.1523/JNEUROSCI.2378-13.2013.

Otis, J. M., Namboodiri, V. M. K., Matan, A. M., Voets, E. S., Mohorn, E. P., Kosyk, O., et al. (2017). Prefrontal cortex output circuits guide reward seeking through divergent cue encoding. *Nature*, *543*(7643), 103–107. https://doi.org/10.1038/nature21376.

Passingham, R. E., & Wise, S. P. (2012). *The neurobiology of the prefrontal cortex. Anatomy, evolution, and the origin of insight* (p. 424). Oxford University Press.

Penttonen, M., & Buzsáki, G. (2003). Natural logarithmic relationship between brain oscillators. *Thalamus & Related Systems*, *2*(02), 145–152. https://doi.org/10.1017/s1472928803000074.

Perrenoud, Q., Pennartz, C. M. A., & Gentet, L. J. (2016). Membrane potential dynamics of spontaneous and visually evoked gamma activity in V1 of awake mice. *PLoS Biology*, *14*(2), 1–21. https://doi.org/10.1371/journal.pbio.1002383.

Peyrache, A., Battaglia, F. P., & Destexhe, A. (2011). Inhibition recruitment in prefrontal cortex during sleep spindles and gating of hippocampal inputs. *Proceedings of the National Academy of Sciences of the United States of America*, *108*(41), 17207–17212. https://doi.org/10.1073/pnas.1103612108.

Peyrache, A., Khamassi, M., Benchenane, K., Wiener, S. I., & Battaglia, F. P. (2009). Replay of rule-learning related neural patterns in the prefrontal cortex during sleep. *Nature Neuroscience*, *12*(7), 919–926. https://doi.org/10.1038/nn.2337.

Pignatelli, M., Beyeler, A., & Leinekugel, X. (2012). Neural circuits underlying the generation of theta oscillations. *Journal of Physiology, Paris*, *106*(3–4), 81–92. https://doi.org/10.1016/j.jphysparis.2011.09.007.

Pinto, L., & Dan, Y. (2015). Cell-type-specific activity in prefrontal cortex during goal-directed behavior. *Neuron*, *87*(2), 437–450. https://doi.org/10.1016/j.neuron.2015.06.021.

Preuss, T. M. (1995). Do rats have prefrontal cortex? The Rose-Woolsey-Akert program reconsidered. *Journal of Cognitive Neuroscience*, 7(1), 1–24. https://doi.org/10.1162/jocn.1995.7.1.1.

Pritchett, D. L., Siegle, J. H., Deister, C. A., & Moore, C. I. (2015). For things needing your attention: The role of neocortical gamma in sensory perception. *Current Opinion in Neurobiology*, *31*, 254–263. https://doi.org/10.1016/j.conb.2015.02.004.

Ray, S., & Maunsell, J. H. R. (2010). Differences in gamma frequencies across visual cortex restrict their possible use in computation. *Neuron*, *67*(5), 885–896. https://doi.org/10.1016/j.neuron.2010.08.004.

Ray, S., & Maunsell, J. H. R. (2015). Do gamma oscillations play a role in cerebral cortex? *Trends in Cognitive Sciences*, *19*(2), 78–85. https://doi.org/10.1016/j.tics.2014.12.002.
Rose, C. N., & Woosley, J. E. (1948). Structure and relations of limbic cortex and anterior thalamic nuclei in rabbit and cat. *Journal of Comparative Neurology*, *89*(3), 279–347.
Roux, F., Wibral, M., Mohr, H. M., Singer, W., & Uhlhaas, P. J. (2012). Gamma-band activity in human prefrontal cortex codes for the number of relevant items maintained in working memory. *Journal of Neuroscience*, *32*(36), 12411–12420. https://doi.org/10.1523/JNEUROSCI.0421-12.2012.
Roy, A., Svensson, F. P., Mazeh, A., & Kocsis, B. (2017). Prefrontal-hippocampal coupling by theta rhythm and by 2–5 Hz oscillation in the delta band: The role of the nucleus reuniens of the thalamus. *Brain Structure and Function*, *222*(6), 2819–2830. https://doi.org/10.1007/s00429-017-1374-6.
Rozeske, R., & Herry, C. (2018). Neuronal coding mechanisms mediating fear behavior. *Current Opinion in Neurobiology*, *52*, 60–64.
Salinas, E., & Sejnowski, T. J. (2000). Impact of correlated synaptic input on output firing rate and variability in simple neuronal models. *Journal of Neuroscience*, *20*(16), 6193–6209. https://doi.org/10.1523/JNEUROSCI.20-16-06193.2000.
Salinas, E., Sejnowski, T. J., Medical, H. H., Torrey, N., Road, P., & Carolina, N. (2001). Correlated neuronal activity and the flow of neural information. *Nature Reviews Neuroscience*, *2*(August), 539–550.
Schomer, D. L., & Lopes da Silva, F. H. (Eds.). (2017). *Niedermeyer's electroencephalography: Basic principles, clinical applications, and related fields* (7th ed.). Oxford University Press. https://doi.org/10.1093/med/9780190228484.001.0001.
Seidenbecher, T., Laxmi, T. R., Stork, O., & Pape, H. C. (2003). Amygdalar and hippocampal theta rhythm synchronization during fear memory retrieval. *Science*, *301*(5634), 846–850. https://doi.org/10.1126/science.1085818.
Senn, V., Wolff, S. B. E., Herry, C., Grenier, F., Ehrlich, I., Gründemann, J., et al. (2014). Long-range connectivity defines behavioral specificity of amygdala neurons. *Neuron*, *81*(2), 428–437. https://doi.org/10.1016/j.neuron.2013.11.006.
Shapiro, L. (2011). *Embodied cognition. New problems in philosophy* (pp. 1–237). Routledge.
Siapas, A. G., & Wilson, M. A. (1998). Coordinated interactions between hippocampal ripples and cortical spindles during slow-wave sleep. *Neuron*, *21*, 1123–1128.
Sigurdsson, T., Stark, K. L., Karayiorgou, M., Gogos, J. A., & Gordon, J. A. (2010). Impaired hippocampal-prefrontal synchrony in a genetic mouse model of schizophrenia. *Nature*, *464*(7289), 763–767. https://doi.org/10.1038/nature08855.
Silva, L. R., Amitai, Y., & Connors, B. W. (1991). Intrinsic oscillations of neocortex generated by layer 5 pyramidal neurons. *Science*, *251*(4992), 432–435. https://doi.org/10.1126/science.1824881.
Singer, W. (1999). Neuronal synchrony: A versatile code for the definition of relations? *Neuron*, *24*, 49–65.
Singer, W. (2018). Neuronal oscillations: Unavoidable and useful? *European Journal of Neuroscience*, *48*(7), 2389–2398. https://doi.org/10.1111/ejn.13796.
Singer, W., & Gray, C. M. (1995). Visual feature integration and the temporal correlation hypothesis. *Annual Review of Neuroscience*, *18*(1), 555–586. https://doi.org/10.1146/annurev.neuro.18.1.555.
Sirota, A., Montgomery, S., Fujisawa, S., Isomura, Y., Zugaro, M., & Buzsáki, G. (2008). Entrainment of neocortical neurons and gamma oscillations by the hippocampal theta rhythm. *Neuron*, *60*(4), 683–697. https://doi.org/10.1016/j.neuron.2008.09.014.
Skaggs, W., McNaughton, B. L., Wilson, M. A., & Barnes, C. A. (1996). Theta phase precession in hippocampal neuronal populations and the compression of temporal sequences.

Hippocampus, *6*(2), 149–172. https://doi.org/10.1002/(SICI)1098-1063(1996)6:2<149::AID-HIPO6>3.0.CO;2-K.

Sohal, V. S. (2016). How close are we to understanding what (if anything) oscillations do in cortical circuits? *Journal of Neuroscience*, *36*(41), 10489–10495. https://doi.org/10.1523/JNEUROSCI.0990-16.2016.

Sohal, V. S., Zhang, F., Yizhar, O., & Deisseroth, K. (2009). Parvalbumin neurons and gamma rhythms enhance cortical circuit performance. *Nature*, *459*(7247), 698–702. https://doi.org/10.1038/nature07991.

Soltesz, I., & Deschenes, M. (1993). Low- and high-frequency membrane potential oscillations during theta activity in CA1 and CA3 pyramidal neurons of the rat hippocampus under ketamine-xylazine anesthesia. *Journal of Neurophysiology*, *70*(1), 97–116.

Spellman, T., Rigotti, M., Ahmari, S. E., Fusi, S., Gogos, J. A., & Gordon, J. A. (2015). Hippocampal-prefrontal input supports spatial encoding in working memory. *Nature*, *522*(7556), 309–314. https://doi.org/10.1038/nature14445.

Steriade, M. (2001). Impact of network activities on neuronal properties in corticothalamic systems. *Journal of Neurophysiology*, *86*(1), 1–39. https://doi.org/10.1152/jn.2001.86.1.1.

Stujenske, J. M., Likhtik, E., Topiwala, M. A., & Gordon, J. A. (2014). Fear and safety engage competing patterns of theta-gamma coupling in the basolateral amygdala. *Neuron*, *83*(4), 919–933. https://doi.org/10.1016/j.neuron.2014.07.026.

Sun, Q., Li, X., Ren, M., Zhao, M., Zhong, Q., Ren, Y., et al. (2019). A whole-brain map of long-range inputs to GABAergic interneurons in the mouse medial prefrontal cortex. *Nature Neuroscience*, *22*(8), 1357–1370. https://doi.org/10.1038/s41593-019-0429-9.

Tamás, G., Buhl, E. H., Lörincz, A., & Somogyi, P. (2000). Proximally targeted GABAergic synapses and gap junctions synchronize cortical interneurons. *Nature Neuroscience*, *3*(4), 366–371. https://doi.org/10.1038/73936.

Tamura, M., Mukai, J., Gordon, J. A., & Gogos, J. A. (2016). Developmental inhibition of Gsk3 rescues behavioral and neurophysiological deficits in a mouse model of schizophrenia predisposition. *Neuron*, *89*(5), 1100–1109. https://doi.org/10.1016/j.neuron.2016.01.025.

Tamura, M., Spellman, T. J., Rosen, A. M., Gogos, J. A., & Gordon, J. A. (2017). Hippocampal-prefrontal theta-gamma coupling during performance of a spatial working memory task. *Nature Communications*, *8*(1), 2182. https://doi.org/10.1038/s41467-017-02108-9.

Tiesinga, P., & Sejnowski, T. J. (2009). Cortical enlightenment: Are attentional gamma oscillations driven by ING or PING? *Neuron*, *63*(6), 727–732. https://doi.org/10.1016/j.neuron.2009.09.009.

Tort, A. B. L., Brankačk, J., & Draguhn, A. (2018). Respiration-entrained brain rhythms are global but often overlooked. *Trends in Neurosciences*, *41*(4), 186–197. https://doi.org/10.1016/j.tins.2018.01.007.

Tort, A. B. L., Komorowski, R., Eichenbaum, H., & Kopell, N. (2010). Measuring phase-amplitude coupling between neuronal oscillations of different frequencies. *Journal of Neurophysiology*, *104*(2), 1195–1210. https://doi.org/10.1152/jn.00106.2010.

Tort, A. B. L., Ponsel, S., Jessberger, J., Yanovsky, Y., Brankačk, J., & Draguhn, A. (2018). Parallel detection of theta and respiration-coupled oscillations throughout the mouse brain. *Scientific Reports*, *8*(1), 1–14. https://doi.org/10.1038/s41598-018-24629-z.

Tort, A. B. L., Rotstein, H. G., Dugladze, T., Gloveli, T., & Kopell, N. J. (2007). On the formation of gamma-coherent cell assemblies by oriens lacunosum-moleculare interneurons in the hippocampus. *Proceedings of the National Academy of Sciences of the United States of America*, *104*(33), 13490–13495. https://doi.org/10.1073/pnas.0705708104.

Traub, R. D., Whittington, M. A., Stanford, I. M., & Jefferys, J. G. R. (1996). A mechanism for generation of long-range synchronous fast oscillations in the cortex. *Nature*, *383*(6601), 621–624. https://doi.org/10.1038/383621a0.

Uylings, H. B. M., Groenewegen, H. J., & Kolb, B. (2003). Do rats have a prefrontal cortex? *Behavioural Brain Research*, *146*(1–2), 3–17. https://doi.org/10.1016/j.bbr.2003.09.028.
Van Vreeswijk, C., Abbott, L. F., & Bard Ermentrout, G. (1994). When inhibition not excitation synchronizes neural firing. *Journal of Computational Neuroscience*, *1*(4), 313–321. https://doi.org/10.1007/BF00961879.
Vander Weele, C. M., Siciliano, C. A., Matthews, G. A., Namburi, P., Izadmehr, E. M., Espinel, I. C., et al. (2018). Dopamine enhances signal-to-noise ratio in cortical-brainstem encoding of aversive stimuli. *Nature*, *563*(7731), 397–401. https://doi.org/10.1038/s41586-018-0682-1.
Varela, F., Lachaux, J. P., Rodriguez, E., & Martinerie, J. (2001). The brainweb: Phase synchronization and large-scale integration. *Nature Reviews Neuroscience*, *2*(4), 229–239. https://doi.org/10.1038/35067550.
Varela, F. J., Rosch, E., & Thompson, E. (1991). *The embodied mind. Cognitive science and human experience* (pp. 1–328). MIT Press.
Varga, S., & Heck, D. H. (2017). Rhythms of the body, rhythms of the brain: Respiration, neural oscillations, and embodied cognition. *Consciousness and Cognition*, *56*(September), 77–90. https://doi.org/10.1016/j.concog.2017.09.008.
Vinck, M., Bos, J. J., Mourik-Donga, L. A., van Oplaat, K. T., Klein, G. A., Jackson, J. C., et al. (2015). Cell-type and state-dependent synchronization among rodent somatosensory, visual, perirhinal cortex, and hippocampus CA1. *Frontiers in Systems Neuroscience*, *9*, 1–25. https://doi.org/10.3389/fnsys.2015.00187.
Vogel, E., Krabbe, S., Gründemann, J., Wamsteeker Cusulin, J. I., & Lüthi, A. (2016). Projection-specific dynamic regulation of inhibition in amygdala micro-circuits. *Neuron*, *91*(3), 644–651. https://doi.org/10.1016/j.neuron.2016.06.036.
Von der Malsburg, C. (1981). The correlation theory of brain function. In *MPI biophysical chemistry internal report 81–2, July 1981* (pp. 1–26).
Wang, X. J. (2010). Neurophysiological and computational principles of cortical rhythms in cognition. *Physiological Reviews*, *90*(3), 1195–1268. https://doi.org/10.1152/physrev.00035.2008.
Wang, X. J., & Buzsáki, G. (1996). Gamma oscillation by synaptic inhibition in a hippocampal interneuronal network model. *Journal of Neuroscience*, *16*(20), 6402–6413. https://doi.org/10.1523/jneurosci.16-20-06402.1996.
Wang, D. V., & Ikemoto, S. (2016). Coordinated interaction between hippocampal sharp-wave ripples and anterior cingulate unit activity. *Journal of Neuroscience*, *36*(41), 10663–10672. https://doi.org/10.1523/JNEUROSCI.1042-16.2016.
Whittington, M. A., Traubtt, R. D., & Jefferys, B. J. G. R. (1995). Synchronized oscillations in interneuron networks glutamate receptor activation. *Nature*, *373*(February), 612–615.
Wierzynski, C. M., Lubenov, E. V., Gu, M., & Siapas, A. G. (2009). State-dependent spike-timing relationships between hippocampal and prefrontal circuits during sleep. *Neuron*, *61*(4), 587–596. https://doi.org/10.1016/j.neuron.2009.01.011.
Wilson, H. R., & Cowan, J. D. (1972). Excitatory and inhibitory interactions in localized populations of model neurons. *Biophysical Journal*, *12*(1), 1–24. https://doi.org/10.1016/S0006-3495(72)86068-5.
Wise, S. P. (2008). Forward frontal fields: phylogeny and fundamental function. *Trends in Neurosciences*, *31*(12), 599–608. https://doi.org/10.1016/j.tins.2008.08.008.
Xia, F., Richards, B. A., Tran, M. M., Josselyn, S. A., Takehara-Nishiuchi, K., & Frankland, P. W. (2017). Parvalbumin-positive interneurons mediate neocortical-hippocampal interactions that are necessary for memory consolidation. *eLife*, *6*, 1–25. https://doi.org/10.7554/eLife27868.
Yanovsky, Y., Ciatipis, M., Draguhn, A., Tort, A. B. L., & Brankačk, J. (2014). Slow oscillations in the mouse hippocampus entrained by nasal respiration. *Journal of Neuroscience*, *34*(17), 5949–5964. https://doi.org/10.1523/JNEUROSCI.5287-13.2014.

Zelano, C., Jiang, H., Zhou, G., Arora, N., Schuele, S., Rosenow, J., et al. (2016). Nasal respiration entrains human limbic oscillations and modulates cognitive function. *Journal of Neuroscience, 36*(49), 12448–12467. https://doi.org/10.1523/JNEUROSCI.2586-16.2016.

Zhong, W., Ciatipis, M., Wolfenstetter, T., Jessberger, J., Müller, C., Ponsel, S., et al. (2017). Selective entrainment of gamma subbands by different slow network oscillations. *Proceedings of the National Academy of Sciences of the United States of America, 114*(17), 4519–4524. https://doi.org/10.1073/pnas.1617249114.

Zingg, B., Hintiryan, H., Gou, L., Song, M. Y., Bay, M., Bienkowski, M. S., et al. (2014). Neural networks of the mouse neocortex. *Cell, 156*(5), 1096–1111. https://doi.org/10.1016/j.cell.2014.02.023.

CHAPTER THIRTEEN

Prefrontal contributions to action control in rodents

Stefanie Hardung[a], Zoe Jäckel[a], and Ilka Diester[b,*]
[a]Optophysiology, University of Freiburg, Faculty of Biology, Freiburg, Germany
[b]Optophysiology, University of Freiburg, Faculty of Biology, BrainLinks-BrainTools, Bernstein Center, Intelligent Machine-Brain Interfacing Technology (IMBIT), Freiburg, Germany
*Corresponding author: e-mail address: ilka.diester@biologie.uni-freiburg.de

Contents

Abstract

The rodent medial prefrontal cortex (mPFC) is typically considered to be involved in cognitive aspects of action control, e.g., decision making, rule learning and application, working memory and generally guiding adaptive behavior (Euston, Gruber, & McNaughton, 2012). These cognitive aspects often occur on relatively slow time scales, i.e., in the order of several trials within a block structure (Murakami, Shteingart, Loewenstein, & Mainen, 2017). In this way, the mPFC is able to set up a representational memory (Goldman-Rakic, 1987). On the other hand, the mPFC can also impact action control more directly (i.e., more on the motoric and less cognitive side). This impact on motor control manifests on faster time scales, i.e., on a single trial level (Hardung et al., 2017). While the more cognitive aspects have been reviewed previously as well as in other subchapters of this book, we explicitly focus on the latter aspect in this chapter, particularly on movement inhibition. We discuss models of prefrontal motor interactions, the impact of the behavioral paradigm, evidences for mPFC involvement in action control, and the anatomical connections between mPFC and motor cortex.

International Review of Neurobiology, Volume 158
ISSN 0074-7742
https://doi.org/10.1016/bs.irn.2020.11.010

1. Introduction

1.1 Nature of action control functions of the mPFC

The rat medial PFC (mPFC) is classically divided into four main divisions, which from dorsal to ventral are (1) the medial agranular (AGm) (or medial precentral, and secondary motor cortex), (2) anterior cingulate (ACC), (3) prelimbic (PL), and (4) infralimbic (IL) cortex (Vertes, 2006). The various subdivisions appear to serve separate and distinct functions, with the more dorsal and caudal regions (AGm and ACC) linked to specific motor behaviors, while ventral and rostral regions (PL and IL), also described as the rostral medial wall regions of the cortex, are functionally linked to the limbic system (Laubach, Amarante, Swanson, & White, 2018; Vertes, 2006). Tetrode recordings from all four mPFC subdivisions show a remarkable diversity of neural responses. In a single task, up to 78 separable unit correlates have been described (Jung, Qin, McNaughton, & Barnes, 1998). Often, firing rates of mPFC units change as a function of task context and reward magnitude even for the same movements. In addition, the same neurons can code different movements in different behavioral paradigms. In line with the adaptive coding hypothesis, this clearly argues for a rather abstract involvement of mPFC in movement control (Duncan, 2001; Euston, Gruber, & McNaughton, 2012). Recent cell type specific findings from a Ca2+ imaging approach, provided evidence that the response heterogeneity is based primarily on pyramidal neurons and parvalbumin-positive neurons. The inhibitory interneuron subtype of somatostatin-positive neurons signaled motor action, while vasoactive intestinal peptide-positive neurons responded strongly to action outcomes (Pinto & Dan, 2015). Thus, neuronal subpopulations within the mPFC seem to encode motor responses directly.

When thinking about action control, motor cortical areas are a natural candidate to consider. In motor and premotor cortex (in rodents also labeled as primary (M1) and secondary (M2) motor cortex, caudal (CFA) and rostral forelimb area (RFA), or lateral (AGl) and medial agranular cortex (AGm), respectively), neuronal responses are primarily correlated to motor responses. However, they also encompass a broad spectrum of other response properties (Narayanan, Horst, & Laubach, 2006; Narayanan & Laubach, 2009a, 2009b; Sul, Jo, Lee, & Jung, 2011). There is some controversy whether the secondary motor cortex is a prefrontal or a motor region (Barthas & Kwan, 2017; Laubach et al., 2018). It is not the aim of this chapter to clarify the confusions about definitions of rodent mPFC or motor cortex. Therefore, to avoid vagueness we will simply use the following nomenclature in the further course of this chapter: (1) M2 as a premotoric structure of the motor cortical areas, and (2) mPFC subdivided into the caudally located ACC, and the more rostral areas PL and IL.

M2 as a premotoric structure is also involved in planning of movements (Sul et al., 2011). A significant fraction of M2 neurons encode the waiting time in single trials. In contrast, mPFC, a major input structure of M2 (Vertes, 2004), contains very few neurons encoding the trial-to-trial waiting times (Murakami, Shteingart, Loewenstein, & Mainen, 2017). Instead, mPFC neurons only encode a slowly fluctuating component of waiting-time variability, a decision bias for the waiting duration, which fluctuates over >10 trials, and is modulated by previous experience. This finding is in line with the idea that higher-order areas, such as the mPFC, can integrate information over time as well as space (Hasson, Yang, Vallines, Heeger, & Rubin, 2008; Murray, O'Doherty, & Schoenbaum, 2007), and extends this idea to a motor domain: The mPFC contains information about decision biases or strategies that are adjusted slowly over time according to previously integrated experiences and internal states, but it cares less about planning imminent actions. Instead, more dedicated motor areas, such as M2, are critical in planning what to do next and when to execute, based on top-down influences from prefrontal areas.

1.2 Models and timescales of interactions between mPFC and motor cortex

The top-down influence of mPFC on motor cortex is described by one of the two extreme models about the interactions between these two brain areas in motor control. The *dualist model* (alternatively called *top-down model*) states that mPFC is the "prime mover," sending the decision to motor cortex (Goldman-Rakic, 1987). Motor cortex receives that decision and executes those movements including kinematic details. Narayanan and colleagues found data supporting this top-down model by comparing the delay related activity of PL and M1 neurons (Narayanan & Laubach, 2009a, 2009b). They observed a decrease in delay related firing of motor cortex neurons upon PL inactivation, indicating a top-down control of the PL on motor neurons (Narayanan & Laubach, 2006). The opposing view is based on the *connectionist model* which states that PFC and motor cortex are simultaneously active (Cisek & Kalaska, 2010). Evidence from recordings in premotor cortex support that view as premotor cortex also reflects decision making components (Sul et al., 2011; Wang et al., 2019).

While the two models sound extremely different, they might have more common ground than expected: Smith and colleagues introduced a compelling hypothesis of two competing processes which might occur simultaneously during the delay period of a motor preparation task: (1) proactive inhibition, mediated by the mPFC, and (2) prepotent excitation (increase of the motor readiness), mediated by the RFA (Smith, Horst, Liu, Caetano, & Laubach, 2010). More precisely, Narayanan et al. proposed that the processes of prepotent excitation and inhibition are regulated by two distinct pools of neurons (Narayanan & Laubach, 2006). According to the "Deadline Model" (see Box 1), these two pools of neurons would exert recurrent excitation and simultaneously inhibit the other pool during the delay period until a certain deadline has been reached. While a complete separation of these neuronal pools according to cortical areas seems unlikely, there might be a bias for cells of the dorsal mPFC to be in the pool for proactive inhibition while RFA neurons predominately represent the pool for prepotent excitation (Mars, 2012).

The time scales on which the mPFC acts, further softens the line between the dualist and connectionist model. If we take an animal which is involved in a block like trial structure, the mPFC sets the behavioral bias for a couple of trials (Murakami et al., 2017). It is not necessary to interact on a ms level with motor cortex. Thus, mPFC and M2 can be active simultaneously as predicted by the connectionist model. In this scenario, the influence of mPFC on M2 is slow (Fig. 1A). If we take an animal in a task without block structure which relies on fast responses to a sensory cue, the dualist model could be closer to reality. Only if mPFC sets the Go signal, will motor cortex execute the movement.

Just as relevant as the times scales on the level of trial blocks, might be the length of the delay period in single trials. In long delay periods, internal processes become more prominent leading, e.g., to proactive responses (Fig. 1B). The prominent cue and anticipatory correlates have led to the idea that the delay activity reflects the retrieval of a motor plan based on reward expectancy (Batuev, Kursina, & Shutov, 1990; Seamans, Lapish, & Durstewitz, 2008). Similarly, if learning is involved or the cognitive demands are high, mPFC seems to play a role on a single trial level (Gilmartin, Miyawaki, Helmstetter, & Diba, 2013; Rossi et al., 2012). However, if an established rule has to be applied consistently within a block, mPFC seems to only set the context. In these cases only chronic lesions might affect the performance within a behavioral block. Single trial manipulations might not lead to an effect unless mPFC is executing some motor control.

BOX 1 Behavioral Models of Response Inhibition.

Horse-Race Model. According to this model, there is a "horse race" between a go process, i.e., crossing a street, and a stop process, i.e., interrupting the walking action as a car comes around the corner. This model of response inhibition particularly refers to the stop-signal reaction time (SSRT) in which subjects can only successfully inhibit their action if the stop-signal delay (SSD), the time between the go and stop stimulus, is short. However, if the stop stimulus occurs close to the moment of response execution the stopping process would start too late to cancel the intended action (Lappin & Eriksen, 1966; Verbruggen & Logan, 2009; Vince, 1948).

Independent Horse-Race Model. The Independent Horse-Race Model formalizes the horse race concept in an independent race model that describes the probability of responding to a stop-signal (Logan & Cowan, 1984; Matzke, Verbruggen, & Logan, 2018; Verbruggen & Logan, 2009). The precise mathematical description of the race allows distribution-free measures of the efficiency and latency of the stop-process (Matzke et al., 2018).

Deadline Model. According to the deadline model, at the beginning of each trial, subjects start a time estimation process by setting a response deadline until they will inhibit their motor response (Ollman & Billington, 1972; Verbruggen & Logan, 2009). If the signal would occur before the deadline, i.e., in short delay trials, it will give rise to detection responses, just as if there would be no deadline. On long delay trials, the signal will lead to an anticipation response with reduced mean reaction time, as result of instances where responses are produced prior to the detection latent period. Interestingly, after the maximum delay duration, i.e., the "deadline," elapsed, the subject should respond even if the trigger stimulus has not occurred.

Inhibitory Default State Model. Based on response speed, this model states that proactive inhibition is the default state of the brain's executive control (Criaud et al., 2012). If the target indicating that there is no more uncertainty in the task appears too soon, i.e., while inhibitory control is still on, longer reaction times are expected. Conversely, if the target is presented after a sufficiently long delay, i.e., after inhibitory control has been released, short RTs are expected.

1.3 Choosing the appropriate task to reveal action control in mPFC

In the past, motor control has been investigated often with rather complex tasks addressing several behavioral concepts (Box 2). An extreme case might be the 5-choice serial reaction time task (5CSRTT) which tests for perseverance, attention, impulsivity, and reaction times (Robbins, 2002).

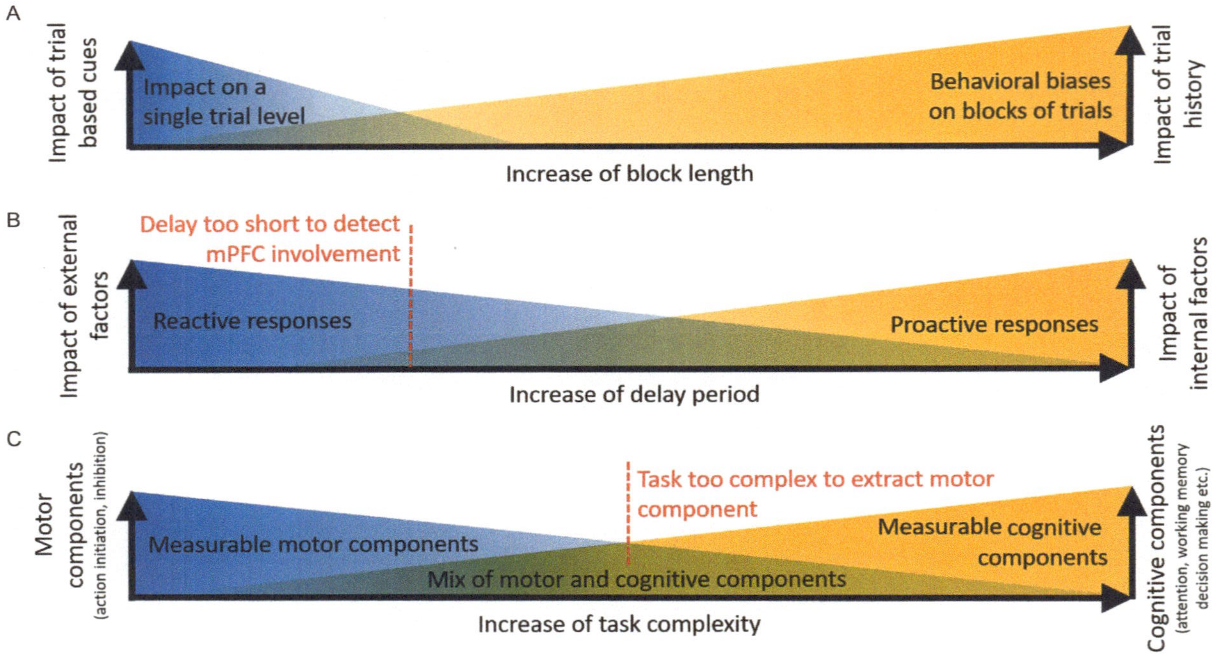

Fig. 1 Diagram of the major task related factors influencing which motor and cognitive components within mPFC can be measured during action control. (A) In simple response preparation/inhibition paradigms, the block length influences to which degree mPFC is involved on a motor and cognitive level. The longer the blocks become, the more the trial history plays a role and the more cognitive the response type becomes. (B) Staying in the realm of simple response preparation/inhibition paradigms, the length of the delay period influences the degree to which mPFC is involved in the task. For very short delay periods with purely reactive response components, mPFC is not involved in a measureable manner. For longer delay periods, action control components are observable in mPFC as response components become more and more proactive. (C) The more the complexity of a behavior increases from simple response preparation/inhibition tasks to complex tasks covering multiple cognitive components (e.g., a 2-Alternative Choice Task (2ACT) or a 5CSRTT), the more difficult it is to extract a motor related response in mPFC as motor components are masked by cognitive components.

BOX 2 Tasks for Investigating Motor Control in the mPFC.

Reversal Learning Task (*RLT*). The RLT was originally used to study motor inhibition, as one previously learned response had to be withheld after a rule change (Jones & Mishkin, 1972). This view has been revised as this task involves many cognitive processes enabling behavioral flexibility, in particular decision-making and rule learning, but also allows studying impulsivity and compulsivity (Izquierdo, 2017; Izquierdo, Brigman, Radke, Rudebeck, & Holmes, 2017; Schoenbaum, Roesch, Stalnaker, & Takahashi, 2009). RLT experiments assess behavioral flexibility by evaluating adaptive responding to changing stimulus-outcome or response-outcome contingencies (Izquierdo et al., 2017). The task typically involves a repeated pairing of an action with an outcome, which has to be learned by the subject. Subjects are trained to discriminate between two stimuli or spatial locations, one of which is rewarded every time it is chosen (Schoenbaum, Chiba, & Gallagher, 2000). Over the course of the experiment, the reversal learning is implemented by reversing the two stimuli associated with the outcomes (Izquierdo et al., 2017). This task allows studying a wide variety of cognitive and motor processes simultaneously (Izquierdo & Jentsch, 2012).

The 5-Choice Serial Reaction Time Task (*5CSRTT*). Another powerful paradigm for studying ongoing cognitive and motor processes is the 5CSRTT procedure which was originally developed as a rodent homologue to the continuous performance task for humans (Robbins, 2002). The basic form of the 5CSRTT requires the subject to scan a horizontal array of five apertures for the presentation of brief flashes of light presented in a pseudorandom order in one of the five locations. Once the subject detects the cue it must poke its nose inside the illuminated port to receive a reward. Although the task was originally designed to assess attention, it is nowadays widely used to analyze a variety of cognitive and motor processes, such as decision-making, impulsive as well as compulsive behavior (Asinof & Paine, 2014).

The 2-Alternative Choice Task (*2ACT*). The 2ACT for rodents, also often referred as gambling task, is inspired by the bandit task for humans and characterized by changing reward contingencies of two options. Those choice options can be given either temporal (Murakami et al., 2017) or spatial (Passecker et al., 2019). In either case, the subject's choice classically relies on its experience-based inner evaluation of different options with probabilistic outcome distributions (Passecker et al., 2019). In both, the temporal and spatial version of the task, the subject is free to choose between (1) the safe option, which delivers an immediate small reward or (2) a larger reward following longer delay period (Murakami et al., 2017) or in the risky area with smaller reward probability (Passecker et al., 2019). The 2ACT is not only used in models for value-based decision making and risk-taking, but also an essential instrument for understanding the complex mechanisms underlying impulsive behavior.

The Go/NoGo Task (*GNGT*). The GNGT is a simplified task for studying impulse control (Donders, 1969a). The GNGT might be seen as a less sensitive tool for research because of the limited number of available task conditions compared to two-choice or even five-choice paradigms. However, regarding specificity this

Continued

BOX 2 Tasks for Investigating Motor Control in the mPFC.—cont'd

might be a particular positive aspect (Gomez, Ratcliff, & Perea, 2007). The simple format of the GNGT allows for examination of response inhibition under conditions in which other cognitive/behavioral processes are minimized (Donders, 1969b; Gordon, 1983; Gordon & Caramazza, 1982). In a typical GNGT experiment, subjects are required to respond to one of the stimuli (go cue), usually by pressing a response key, but must withhold a response to the other alternative (no-go cue). In contrast to tasks like the 5CSRTT, where the subject has to select the correct port following the go cue, the GNGT requires only a response when the go cue is presented.

The Stop-Signal Task (SST). Another powerful paradigm widely used in basic and clinical research to study inhibitory control is the SST (Matzke et al., 2018). Similar to the GNGT, the SST also features a "go" signal. In contrast to the GNGT, however, the "go" signal is occasionally followed by a stop-signal, which instructs subjects to cancel their already planned or ongoing response (Verbruggen & Logan, 2008). The index of inhibitory control in the SST is the duration of the stopping process, the stop-signal reaction time (SSRT) which must be estimated from a stochastic model, such as the *Independent Horse Race Model* (Box 1). The SSRT has proven to be an important measure of cognitive control mechanisms involved in inhibitory processes (Mayse, Nelson, Park, Gallagher, & Lin, 2014; Verbruggen & Logan, 2008). On the down side, the SST does not allow to distinguish between inhibitory aspects (e.g., reactive stopping, evoked by external stimuli versus proactive mechanisms, developed according to specific goals) in one behavioral setup. While proactive response control is introduced as soon as uncertainty about the next event prevails (Criaud et al., 2012), it is difficult to extract the proactive component as it is overlaid by reactive components in classic stimulus-response paradigms.

The Simple Reaction Time Task (SRTT). Investigating proactive mechanisms requires a task which is divided into conditions that differ in the degree of preparatory control (i.e., one type of trial where proactive motor control is developed and another type of trial which purely relies on reactive mechanisms). The SRTT, is directed toward this goal. It is characterized by a randomized presentation of variable response delays (Hardung et al., 2017; Näätänen, 1970; Narayanan & Laubach, 2009a, 2009b; Risterucci et al., 2003). Subjects must continue a response during a waiting or delay period until a signal occurs. The time period from the trigger stimulus to the lever release is measured as reaction times. Two distinct errors can occur: (1) Subjects which are not able to respond to the signal within a given time window perform late responses. (2) In premature release trials, subjects react prematurely by releasing the lever before a signal was given. Uncertainty is introduced by varying the delay periods. By comparing both trial types, (i.e., short tone delays, where no proactive preparation is needed and long tone delays with a preparatory phase) effects of proactive inhibitory control on the entire motor response can be analyzed (Criaud et al., 2012; Jaffard et al., 2008; Jaffard, Benraiss, Longcamp, Velay, & Boulinguez, 2007).

However, it can prove difficult to target an underlying motor function if there are multiple ongoing unidentified cognitive processes (e.g., selective attention, memory retrieval, emotions) active during the task which could interfere with the overall behavioral effect (Fig. 1C). Therefore, to investigate motoric aspects in mPFC, the behavioral task has to be designed as simply as possible with minimal interference of cognitive processes.

Studies based on behavioral paradigms for flexible and goal-directed behavior revealed similar neural pathways involved in cognitive processes such as decision-making and action control (Aron, 2011; Mars, 2012). While both of these functions are necessary to adapt behavior according to environmental changes, they are two different mental activities (Coutlee & Huettel, 2012; Kitajima & Toyota, 2013; Montgomery, 1998). Decision-making is often referred to as a process of choosing a preferred option, inhibiting the temptation of certain outcomes, choosing an action from a set of alternatives, and projecting choice consequences onto certain goals. In contrast, cognitive control of actions is described as selecting "what to do next" in a changing environment, execution of actions necessary to reach desired decisions or inhibiting unimportant stimuli (Coutlee & Huettel, 2012; Kitajima & Toyota, 2013). Yet, despite a striking mechanistic and neural overlap of both cognitive processes, decision preferences are not necessarily linked to actions (Montgomery, 1998). Tasks, like the reversal learning task (RLT), 5CSRTT, as well as the 2-Choice task (Box 2) and their modifications, require subjects to make repeated and meaningful decisions, i.e., in which port and when to perform the nose poke. In such tasks, the high level of ongoing attentional and decisive processes might hamper the targeting of the functionality behind decision versus action.

1.4 Behavioral mechanisms of response inhibition

Forms of action control can be subdivided by goal, and the respective neural pathways guiding the response (reviewed in Aron, 2011). Traditionally, adaptive motor behavior can occur either proactively to anticipate task requirements, or reactively in response to external stimuli (Aron, 2011; Stuphorn & Emeric, 2012). To date, most research has focused on evaluating neural circuits involved in reactive stopping while little is known about proactive inhibitory control. However, several studies suggest that reactive motor control, evoked by external stimuli, is limited as a model for impulse control disorders since most inhibitory control used in everyday life is specified according to one's motives, not external triggers (Aron, 2011). Compared to reactive response inhibition, proactive motor control develops according to specific goals instead of external triggers. Based on the foreknowledge of upcoming demands, the proactive mechanism is activated while subjects prepare to inhibit an upcoming response tendency until the right time point (Braver, 2012; Lavallee, Meemken, Herrmann, & Huster, 2014). Since proactive inhibition works as the motor break until a certain signal is given, it is often referred as the "Hold-your-Horses" mechanism (Frank, 2006) (Box 1). Apart from the possibility to adapt actions flexibly in response to environmental changes, preparing in advance to inhibit motor actions also enables subjects to respond faster and with greater precision (Aron, 2011; Hardung et al., 2017; Näätänen, 1970).

2. Contributions of PFC subsections to response inhibition

Previous anatomical studies revealed a wide variety of cortical and subcortical connections of the individual PFC subsections (Hoover & Vertes, 2011; Vertes, 2004). Distinct cognitive functions have been assigned to subparts of the PFC (Euston et al., 2012; Miller, 2000; Seamans et al., 2008; Wallis, 2012). However, less attention has been paid to the attribution of non-cognitive aspects to prefrontal subsections. In a recent study, we employed a response-preparation task, combined with neuronal recordings and optogenetics in rats (Hardung et al., 2017). By mapping a functional gradient in the rodent PFC supporting inhibitory control we aimed to contribute to a deeper understanding of the functional heterogeneity of PFC subsections in guiding action control.

2.1 The response-preparation paradigm targets proactive and reactive motor inhibition

We employed a response-preparation paradigm similar to the simple reaction time task (SRTT) that covered both reactive and proactive behavioral elements (Fig. 2A and B) (Hardung et al., 2017; Mayse, Nelson, Park, Gallagher, & Lin, 2014). Consistent with previous studies, reaction times of long-delay trials were significantly faster than those in short-delay trials (Fig. 2C) (Eagle et al., 2008; Hardung et al., 2017; Näätänen, 1970; Narayanan et al., 2006; Narayanan & Laubach, 2009a, 2009b; Risterucci, Terramorsi, Nieoullon, & Amalric, 2003). This behavioral consistency with previous studies supports the notion that longer delay periods enable subjects to prepare their motor system to upcoming actions (Criaud, Wardak, Hamed, Ballanger, & Boulinguez, 2012; Jaffard et al., 2008; Näätänen, 1970).

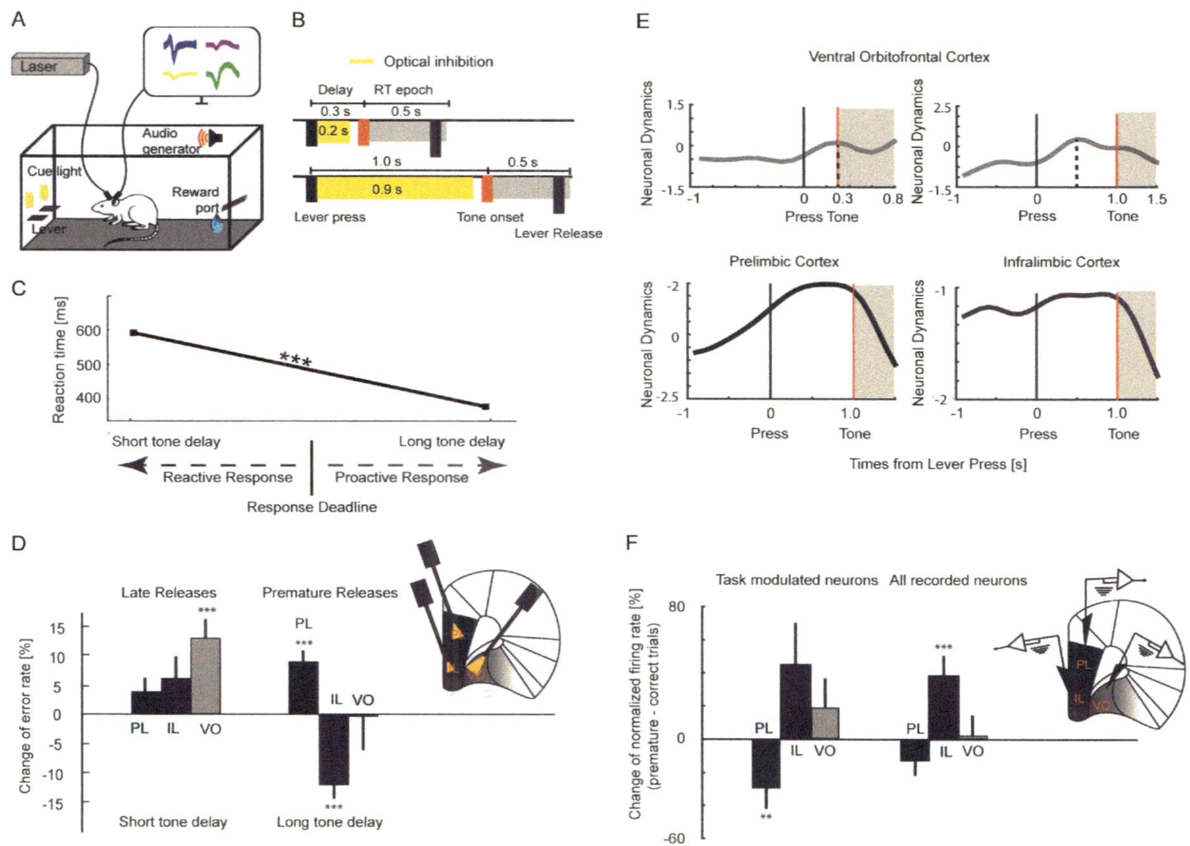

Fig. 2 See figure legend on next page.

The reasons for delay-dependent reaction time speeding following randomized and increased waiting periods preceding the response signal might be related to the subject's use of conditional probability: the more time elapses, the more likely it is that the longer delay period is being applied to that trial, and the more information the subject has to prepare its impending motor activity (Näätänen, 1970; Risterucci et al., 2003). During the delay period rats are required to maintain their readiness to respond until the signal is presented. Since the entire system is already prepared, a much faster response is possible (Fig. 2C). On the other hand, very short delay periods prevent the system from preparing, resulting in slower reaction times (Smith et al., 2010). This notion supports the *Deadline Model* (Ollman & Billington, 1972) (Box 1). Tasks with lower temporal uncertainty, i.e., increasing but non randomized delays, do not cause reaction time speeding (Näätänen, 1970). Therefore, it might be possible that subjects performing tasks with lower temporal uncertainty simply use the temporal information about the signal to activate the motor system just in time. This leads to faster reaction times and energy efficient responses. The *Inhibitory Default State Model* (Criaud et al., 2012) (Box 1) offers a different explanation of the observed reaction time increase from long to short delay periods (Fig. 2C). According to this model proactive inhibition of a response is the default mode of executive control. If the target indicating that there is no more uncertainty in the task appears too soon, i.e., while inhibitory control is still on in trials with short delay periods, then longer reaction times are to be expected. Conversely, if the target is presented after a sufficiently long delay, i.e., after inhibitory control has been released and the motor system is already prepared, short reaction times are to be expected.

Fig. 2 Contributions of PFC subsections in a response preparation task. (A) Behavioral setup for a response preparation task combined with bilateral optogenetics, or unilateral electrophysiological recordings. (B) Laser stimulation protocol for bilateral inhibition of one of the PFC subareas during the delay period in 10% of the trials. (C) The mean reaction time was significantly longer after short tone delays than after long tone delays. (D) Changes in error rate following optogenetic inhibition revealed opposing effects of PFC subsections on task performance. Inset: Schematic of implanted optical fibers in ventral orbitofrontal cortex (VO), prelimbic (PL) as well infralimbic (IL) cortex. (E) Analysis of firing pattern using principal-component analysis in VO, PL as well as IL. Shaded areas, reaction time window. Bin size is 100 ms. (F) Activity changes during the preparatory phase before early releases in long-delay trials of either all recorded (middle figure) or task modulated units (middle figure). $^{***}P < 0.001$, $^{**}P < 0.01$; error bars, SEM.

2.2 Inhibition of mPFC subareas orthogonally influences the ability to wait for a stimulus

To investigate the impact of delay related activity of mPFC (specifically PL and IL) neurons on the ability to inhibit inappropriate responses, we combined neuronal ensemble recordings and optogenetics (Fig. 2A and B). We observed two main effects following reversible optogenetic inactivation of three key subareas of the rat mPFC and orbitofrontal cortex (OFC) in our response-preparation task: (1) inactivation of mPFC subregions induced drastic changes during proactive behavior, and (2) OFC inhibition impaired the rats' reactive motor control indicated by significantly increased reaction times in short-delay trials upon optical inhibition of eNpHR3.0 expressing ventral orbitofrontal cortex (VO) neurons (Fig. 2D) (Hardung et al., 2017). Specifically, in agreement with previous studies (Narayanan & Laubach, 2009a, 2009b; Risterucci et al., 2003), PL inhibition induced an increase of premature responses. Surprisingly, IL inactivation caused an opposing effect, namely, a decreased number of early lever releases, and thus and increased ability to withhold a response until the appropriate time point. Former studies which combined a stop-signal task (SST, see Box 2) with either pharmacological mPFC inactivation (Eagle & Robbins, 2003; Narayanan et al., 2006) or lesions (Risterucci et al., 2003) in rats provided evidence that IL inactivation led to increased impulsive behavior (similar to PL). However, recent fear-related paradigms in rodents with lesioned or reversible inactivated IL and PL provided evidence of both functioning in opposite direction (Giustino & Maren, 2015; Gourley & Taylor, 2016). Further, our study (Hardung et al., 2017) together with electrophysiological recordings and pharmacological inactivated mPFC in a reward seeking and extinction task (Gourley & Taylor, 2016; Moorman & Aston-Jones, 2015) point toward antagonistic roles of both areas (Cassaday, 2014; Euston et al., 2012). Differential roles of IL and PL in controlling motor output is also supported by several anatomical studies (reviewed by Gourley & Taylor, 2016). For example, within the ventral striatum, the PL largely innervates the nucleus accumbens (NAc) core, while the IL preferentially innervates the NAc shell compartment, with both, core and shell, shown to exert opposing influences on motivated behavior (Mailly, Aliane, Groenewegen, Haber, & Deniau, 2013; Vertes, 2004; Yang et al., 2018).

The fact that we only obtained a significant effect of PL and IL inhibition on trials with delay periods of 1 s, but not with 300 ms delay periods indicates that the impact of PL and IL is not directly related to reactive motor behavior. Instead, the mPFC subsections are involved in the evaluation of internal factors which play a role during the inhibition of a movement.

2.3 Neuronal activity in PFC is task modulated

We conducted neuronal ensemble recordings in PL, IL, or VO. In all of these three areas, we found event related units (either linked to press, delay, tone, release, or reward), without any systematic preference per area for one particular trial phase (Hardung et al., 2017). However, functional grouping via principal-component analysis (Narayanan & Laubach, 2009a, 2009b) revealed neuronal differences between the three prefrontal areas with respect to their modulation during delay and tone representation (Hardung et al., 2017). The dynamics of the principal components indicated that VO was preparing for a task cue signal after a short delay whereas PL and IL populations were more involved in long delay trials with PL having a more consistent and IL a more complex activity pattern (Fig. 2E). Specifically, in short trials, all VO components showed strong modulation during the time of stop-signal presentation as well as the reaction time window (Fig. 2E, top). Interestingly, in long trials a "bump" or "dip" was visible during the delay around the time when the tone was expected in short delay trials. Several studies showed connectivity of the ventral orbital, but not the medial areas, with sensory and motor areas, including the posterior parietal cortex and the medial agranular cortex which both support reactive processing (Giustino & Maren, 2015; Hoover & Vertes, 2007, 2011; Laubach, Caetano, & Narayanan, 2015; Narayanan & Laubach, 2017; Reep, Goodwin, & Corwin, 1990; Vertes, 2004). These findings, together with our behavioral results as well as the existence of connections between VO and premotor areas (Smith et al., 2010) further support the notion of an involvement of VO in task cue representation during the reactive phase of our response-preparation task.

2.4 Neuronal activity specifically encodes trial outcomes

Presuming that neuronal activity during the delay period enables subjects to await a stimulus, we compared neural activity in error and correct trials of rats performing our response-preparation task. Here, we focus on one error type: early releases in long-delay trials (Fig. 2F) (Hardung et al., 2017). In accordance with our optogenetic results which revealed an increase of premature errors upon optical PL inhibition we found a specialized sub-population of task modulated PL neurons which significantly decreased their activity in premature trials as compared to correct trials (Fig. 2F, left). Additionally, optogenetic inhibition of eNpHR3.0 expressing IL neurons led to a significant decrease of early release trials. Corresponding to this, electrophysiology recordings of IL ensembles showed a significantly increased activity in early releases as compared to correct trials (Fig. 2F, middle).

A prerequisite for a proactive component is the ability to predict the trial outcome of the task based on characteristic neuronal firing patterns. To test this, we applied a statistical classification using the regularized discriminant analysis. In accordance with previous studies (Narayanan & Laubach, 2006) we found a strong predictive capability for the trial outcome of delay-modulated PFC neurons, with PL and VO both predicting early releases significantly better than correct trials, and IL showing opposite results (in Hardung et al., 2017).

3. Anatomical connections of the mPFC to motor output structures

While evidence accumulates that mPFC is involved in motor control, it still remains unclear how exactly the motor relevant information reaches motor cortex from mPFC. While a direct connection between mPFC and M1 has been described anatomically (Hoover & Vertes, 2007), a direct functional connection seems unlikely, given the long latencies between mPFC and M1 neuronal activity (Narayanan & Laubach, 2009a, 2009b). Similarly, even though a direct connection to M2 exists (Vertes, 2004), when inhibiting PL the delays until a significant change in neuronal activity can be observed in M2 are too long (in the order of 250 ms, Deubner, 2020) to be sufficient for an effective communication. The cortico-cortical connections are often paralleled by a transthalamic circuits (Sherman, 2016). While originally a strict separation of cortico-thalamic loops between associative, motoric, and limbic systems has been postulated, more recent evidence points to overlapping neuronal population in the thalamus. mPFC sends projections to both mediodorsal (MD) and ventromedial (VM) thalamus (Collins, Anastasiades, Marlin, & Carter, 2018). While MD has been thought to be involved in cognitive aspects (Bolkan et al., 2017; Cross, Brown, Aggleton, & Warburton, 2013; Mitchell & Chakraborty, 2013; Mitchell et al., 2014; Ouhaz, Ba-M'hamed, & Bennis, 2017; Ouhaz, Ba-M'hamed, Mitchell, Elidrissi, & Bennis, 2015; Ouhaz, Fleming, & Mitchell, 2018; Parnaudeau et al., 2013, 2015; Schmitt et al., 2017), VM has traditionally been considered to be a motor thalamic nucleus (Gaidica, Hurst, Cyr, & Leventhal, 2018). The overlap of the projections in these thalamic nuclei as gate of the communication of mPFC with M2 has so far only been sparsely investigated (Collins et al., 2018). These thalamic nuclei might play a crucial role in transferring action control signals from mPFC to motor cortex. Further a multisynaptic, unidirectional cortico-subcortical route from mPFC to M1/M2 via basal ganglia has recently been described (Aoki et al., 2019) as well as direct connections to subthalamic nucleus (STN) from mPFC (Li, Nguyen, Ma, & Dan, 2020), and also to the mid brain (Franklin et al., 2017), both motoric structures. Thus, mPFC has multiple pathways via which motor related information can be transferred. Which information is sent via which pathway remains to be determined.

4. Conclusion

Here, we argued that mPFC, in particular PL and IL, can be involved in motor tasks if the settings are appropriately chosen. If the task is too complex, one convolutes the impact of mPFC on cognitive and motor components. If trials are presented in blocks, the trial history relevance and with that the cognitive components increase, which might also mask motor components. If the task is simple and the delay period is too short, introducing purely reactive responses, no involvement of mPFC manifests. Only if the delay period is sufficiently long and the block length is appropriate, can one measure how mPFC contributes to the active withholding and propagation of an action (Fig. 1). As putative routes for this information transfer to motor output structures both cortico-cortical and cortico-subcortical connections might play a role.

References

Aoki, S., Smith, J. B., Li, H., Yan, X., Igarashi, M., Coulon, P., et al. (2019). An open cortico-basal ganglia loop allows limbic control over motor output via the nigrothalamic pathway. *eLife*, *8*, e49995. https://doi.org/10.7554/eLife.49995.

Aron, A. R. (2011). From reactive to proactive and selective control: Developing a richer model for stopping inappropriate responses. *Biological Psychiatry*, *69*, e55–e68. Elsevier https://doi.org/10.1016/j.biopsych.2010.07.024.

Asinof, S. K., & Paine, T. A. (2014). The 5-choice serial reaction time task: A task of attention and impulse control for rodents. *Journal of Visualized Experiments*, (90). https://doi.org/10.3791/51574.

Barthas, F., & Kwan, A. C. (2017). Secondary motor cortex: Where 'sensory' meets 'motor' in the rodent frontal cortex. *Trends in Neurosciences*, *40*(3), 181–193. https://doi.org/10.1016/j.tins.2016.11.006.

Batuev, A. S., Kursina, N. P., & Shutov, A. P. (1990). Unit activity of the medial wall of the frontal cortex during delayed performance in rats. *Behavioural Brain Research*, *41*(2), 95–102. https://doi.org/10.1016/0166-4328(90)90145-5.

Bolkan, S. S., Stujenske, J. M., Parnaudeau, S., Spellman, T. J., Rauffenbart, C., Abbas, A. I., et al. (2017). Thalamic projections sustain prefrontal activity during working memory maintenance. *Nature Neuroscience, 20*(7), 987–996. https://doi.org/10.1038/nn.4568.
Braver, T. S. (2012). The variable nature of cognitive control: A dual mechanisms framework. *Trends in Cognitive Sciences, 16*(2), 106–113. https://doi.org/10.1016/j.tics.2011.12.010.
Cassaday, H. J. (2014). From attention to memory along the dorsal-ventral axis of the medial prefrontal cortex: Some methodological considerations. *Frontiers in Systems Neuroscience, 8*, 160. https://doi.org/10.3389/fnsys.2014.00160.
Cisek, P., & Kalaska, J. F. (2010). Neural mechanisms for interacting with a world full of action choices. *Annual Review of Neuroscience, 33*, 269–298. https://doi.org/10.1146/annurev.neuro.051508.135409.
Collins, D. P., Anastasiades, P. G., Marlin, J. J., & Carter, A. G. (2018). Reciprocal circuits linking the prefrontal cortex with dorsal and ventral thalamic nuclei. *Neuron, 98*(2), 366–379.e4. https://doi.org/10.1016/j.neuron.2018.03.024.
Coutlee, C. G, & Huettel, S. A. (2012). The functional neuroanatomy of decision making: Prefrontal control of thought and action. *Brain Research, 1428*(C), 3–12. https://doi.org/10.1016/j.brainres.2011.05.053.
Criaud, M., Wardak, C., Hamed, S. B., Ballanger, B., & Boulinguez, P. (2012). Proactive inhibitory control of response as the default state of executive control. *Frontiers in Psychology, 3*(MAR), 59. https://doi.org/10.3389/fpsyg.2012.00059.
Cross, L., Brown, M. W., Aggleton, J. P., & Warburton, E. C. (2013). The medial dorsal thalamic nucleus and the medial prefrontal cortex of the rat function together to support associative recognition and recency but not item recognition. *Learning and Memory, 20*(1), 41–50. https://doi.org/10.1101/lm.028266.112.
Deubner, J. (2020). *Role of prelimbic excitatory neurons and secondary motor cortex in response inhibition*. Albert Ludwigs University of Freiburg.
Donders, F. C. (1969a). On the speed of mental processes. *Acta Psychologica, 30*(C), 412–431. https://doi.org/10.1016/0001-6918(69)90065-1.
Donders, F. C. (1969b). Two instruments for determining the time required for mental processes. *Acta Psychologica, 30*(C), 432–435. https://doi.org/10.1016/0001-6918(69)90066-3.
Duncan, J. (2001). An adaptive coding model of neural function in prefrontal cortex. *Nature Reviews Neuroscience, 2*(11), 820–829. https://doi.org/10.1038/35097575.
Eagle, D. M., Baunez, C., Hutcheson, D. M., Lehmann, O., Shah, A. P., & Robbins, T. W. (2008). Stop-signal reaction-time task performance: Role of prefrontal cortex and subthalamic nucleus. *Cerebral Cortex, 18*(1), 178–188. https://doi.org/10.1093/cercor/bhm044.
Eagle, D. M., & Robbins, T. W. (2003). Inhibitory control in rats performing a stop-signal reaction-time task: Effects of lesions of the medial striatum and d-amphetamine. *Behavioral Neuroscience, 117*(6), 1302–1317. https://doi.org/10.1037/0735-7044.117.6.1302.
Euston, D. R., Gruber, A. J., & McNaughton, B. L. (2012). The role of medial prefrontal cortex in memory and decision making. *Neuron, 76*(6), 1057–1070. https://doi.org/10.1016/j.neuron.2012.12.002.
Frank, M. J. (2006). Hold your horses: A dynamic computational role for the subthalamic nucleus in decision making. *Neural Networks, 19*(8), 1120–1136. https://doi.org/10.1016/j.neunet.2006.03.006.
Franklin, T. B., Silva, B. A., Perova, Z., Marrone, L., Masferrer, M. E., Zhan, Y., et al. (2017). Prefrontal cortical control of a brainstem social behavior circuit. *Nature Neuroscience, 20*(2), 260–270. https://doi.org/10.1038/nn.4470.
Gaidica, M., Hurst, A., Cyr, C., & Leventhal, D. K. (2018). Distinct populations of motor thalamic neurons encode action initiation, action selection, and movement vigor. *Journal of Neuroscience, 38*(29), 6563–6573. https://doi.org/10.1523/JNEUROSCI.0463-18.2018.

Gilmartin, M. R., Miyawaki, H., Helmstetter, F. J., & Diba, K. (2013). Prefrontal activity links nonoverlapping events in memory. *Journal of Neuroscience*, *33*(26), 10910–10914. https://doi.org/10.1523/JNEUROSCI.0144-13.2013.

Giustino, T. F., & Maren, S. (2015). The role of the medial prefrontal cortex in the conditioning and extinction of fear. *Frontiers in Behavioral Neuroscience*, *9*(November), 1–20. https://doi.org/10.3389/fnbeh.2015.00298.

Goldman-Rakic, P. S. (1987). Motor control function of the prefrontal cortex. *Ciba Foundation Symposium*, *132*, 187–200.

Gomez, P., Ratcliff, R., & Perea, M. (2007). A model of the go/no-go task. *Journal of Experimental Psychology*, *136*(3), 389–413. https://doi.org/10.1037/0096-3445.136.3.389.

Gordon, B. (1983). Lexical access and lexical decision: Mechanisms of frequency sensitivity. *Journal of Verbal Learning and Verbal Behavior*, *22*(1), 24–44. https://doi.org/10.1016/S0022-5371(83)80004-8.

Gordon, B., & Caramazza, A. (1982). Lexical decision for open- and closed-class words: Failure to replicate differential frequency sensitivity. *Brain and Language*, *15*(1), 143–160. https://doi.org/10.1016/0093-934X(82)90053-0.

Gourley, S. L., & Taylor, J. R. (2016). Going and stopping: Dichotomies in behavioral control by the prefrontal cortex. *Nature Neuroscience*, *19*(6), 656–664. https://doi.org/10.1038/nn.4275.

Hardung, S., Epple, R., Jäckel, Z., Eriksson, D., Uran, C., Senn, V., et al. (2017). A functional gradient in the rodent prefrontal cortex supports behavioral inhibition. *Current Biology*, *27*(4), 549–555. https://doi.org/10.1016/j.cub.2016.12.052.

Hasson, U., Yang, E., Vallines, I., Heeger, D. J., & Rubin, N. (2008). A hierarchy of temporal receptive windows in human cortex. *Journal of Neuroscience*, *28*(10), 2539–2550. https://doi.org/10.1523/JNEUROSCI.5487-07.2008.

Hoover, W. B., & Vertes, R. P. (2007). Anatomical analysis of afferent projections to the medial prefrontal cortex in the rat. *Brain Structure and Function*, *212*(2), 149–179. https://doi.org/10.1007/s00429-007-0150-4.

Hoover, W. B., & Vertes, R. P. (2011). Projections of the medial orbital and ventral orbital cortex in the rat. *Journal of Comparative Neurology*, *519*(18), 3766–3801. https://doi.org/10.1002/cne.22733.

Izquierdo, A, Brigman, J L, Radke, A K, Rudebeck, P H, & Holmes, A. (2017). The neural basis of reversal learning: An updated perspective. *Neuroscience*, *345*, 12–26. https://doi.org/10.1016/j.neuroscience.2016.03.021.

Izquierdo, A. (2017). Functional heterogeneity within rat orbitofrontal cortex in reward learning and decision making. *Journal of Neuroscience*, *37*(44), 10529–10540. https://doi.org/10.1523/JNEUROSCI.1678-17.2017.

Izquierdo, A., & Jentsch, J. D. (2012). Reversal learning as a measure of impulsive and compulsive behavior in addictions. *Psychopharmacology*, *219*(2), 607–620. https://doi.org/10.1007/s00213-011-2579-7.

Jaffard, M., Longcamp, M., Velay, J. L., Anton, J. L., Roth, M., Nazarian, B., et al. (2008). Proactive inhibitory control of movement assessed by event-related fMRI. *NeuroImage*, *42*(3), 1196–1206. https://doi.org/10.1016/j.neuroimage.2008.05.041.

Jaffard, M., Benraiss, A., Longcamp, M., Velay, J. L., & Boulinguez, P. (2007). Cueing method biases in visual detection studies. *Brain Research*, *1179*(1), 106–118. https://doi.org/10.1016/j.brainres.2007.08.032.

Jones, B., & Mishkin, M. (1972). Limbic lesions and the problem of stimulus-reinforcement associations. *Experimental Neurology*, *36*(2), 362–377. https://doi.org/10.1016/0014-4886(72)90030-1.

Jung, M. W., Qin, Y., McNaughton, B. L., & Barnes, C. A. (1998). Firing characteristics of deep layer neurons in prefrontal cortex in rats performing spatial working memory tasks. *Cerebral Cortex*, *8*(5), 437–450. https://doi.org/10.1093/cercor/8.5.437.

Kitajima, M., & Toyota, M. (2013). Decision-making and action selection in two minds: An analysis based on model human processor with realtime constraints (MHP/RT). *Biologically Inspired Cognitive Architectures*, *5*, 82–93. https://doi.org/10.1016/j.bica.2013.05.003.

Lappin, J. S., & Eriksen, C. W. (1966). Use of a delayed signal to stop a visual reaction-time response. *Journal of Experimental Psychology*, *72*(6), 805–811. https://doi.org/10.1037/h0021266.

Laubach, M., Amarante, L. M., Swanson, K., & White, S. R. (2018). What, if anything, is rodent prefrontal cortex? *eNeuro*, *5*(5), 315–333. Society for Neuroscience https://doi.org/10.1523/ENEURO.0315-18.2018.

Laubach, M., Caetano, M. S., & Narayanan, N. S. (2015). Mistakes were made: Neural mechanisms for the adaptive control of action initiation by the medial prefrontal cortex. *Journal of Physiology, Paris*, *109*(1–3), 104–117. https://doi.org/10.1016/j.jphysparis.2014.12.001.

Lavallee, C. F., Meemken, M. T., Herrmann, C. S., & Huster, R. J. (2014). When holding your horses meets the deer in the headlights: Time-frequency characteristics of global and selective stopping under conditions of proactive and reactive control. *Frontiers in Human Neuroscience*, *8*, 994. https://doi.org/10.3389/fnhum.2014.00994.

Li, B., Nguyen, T. P., Ma, C., & Dan, Y. (2020). Inhibition of impulsive action by projection-defined prefrontal pyramidal neurons. *Proceedings of the National Academy of Sciences of the United States of America*, *117*(29), 17278–17287. https://doi.org/10.1073/pnas.2000523117.

Logan, G. D., & Cowan, W. B. (1984). On the ability to inhibit thought and action: A theory of an act of control. *Psychological Review*, *91*(3), 295–327. https://doi.org/10.1037/0033-295X.91.3.295.

Mailly, P., Aliane, V., Groenewegen, H. J., Haber, S. N., & Deniau, J.-M. (2013). The rat prefrontostriatal system analyzed in 3D: Evidence for multiple interacting functional units. *Journal of Neuroscience*, *33*(13), 5718–5727. https://doi.org/10.1523/JNEUROSCI.5248-12.2013.

Mars, R. B. (2012). *Neural basis of motivational and cognitive control*. MIT Press.

Matzke, D., Verbruggen, F., & Logan, G. D. (2018). The stop-signal paradigm. *Stevens' handbook of experimental psychology and cognitive neuroscience*, *5*, 1–45. https://doi.org/10.1002/9781119170174.epcn510.

Mayse, J. D., Nelson, G. M., Park, P., Gallagher, M., & Lin, S. C. (2014). Proactive and reactive inhibitory control in rats. *Frontiers in Neuroscience*, *8*(104), 1–16. https://doi.org/10.3389/fnins.2014.00104.

Miller, E. K. (2000). The prefontral cortex and cognitive control. *Nature Reviews Neuroscience*, *1*(1), 59–65. https://doi.org/10.1038/35036228.

Mitchell, A. S., & Chakraborty, S. (2013). What does the mediodorsal thalamus do? *Frontiers in Systems Neuroscience*, 7(JUL), 1–19. https://doi.org/10.3389/fnsys.2013.00037.

Mitchell, A. S., Murray Sherman, S., Sommer, M. A., Mair, R. G., Vertes, R. P., & Chudasama, Y. (2014). Advances in understanding mechanisms of thalamic relays in cognition and behavior. *Journal of Neuroscience*, *34*(46), 15340–15346. https://doi.org/10.1523/JNEUROSCI.3289-14.2014.

Montgomery, H. (1998). Decision making and action. In M. Kofta, G. Weary, & G. Sedek (Eds.). *The Springer series in social clinical psychology. Personal control in action* (pp. 279–298). Boston, MA: Springer.

Moorman, D. E., & Aston-Jones, G. (2015). Prefrontal neurons encode context-based response execution and inhibition in reward seeking and extinction. *Proceedings of the National Academy of Sciences*, *112*(30), 9472–9477. https://doi.org/10.1073/pnas.1507611112.

Murakami, M., Shteingart, H., Loewenstein, Y., & Mainen, Z. F. (2017). Distinct sources of deterministic and stochastic components of action timing decisions in rodent frontal cortex. *Neuron*, *94*(4), 908–919.e7. https://doi.org/10.1016/j.neuron.2017.04.040.

Murray, E. A., O'Doherty, J. P., & Schoenbaum, G. (2007). What we know and do not know about the functions of the orbitofrontal cortex after 20 years of cross-species studies. *Journal of Neuroscience*, *27*(31), 8166–8169. https://doi.org/10.1523/JNEUROSCI.1556-07.2007.

Näätänen, R. (1970). The diminishing time-uncertainty with the lapse of time after the warning signal in reaction-time experiments with varying fore-periods. *Acta Psychologica*, *34*(C), 399–419. https://doi.org/10.1016/0001-6918(70)90035-1.

Narayanan, N. S., Horst, N. K., & Laubach, M. (2006). Reversible inactivations of rat medial prefrontal cortex impair the ability to wait for a stimulus. *Neuroscience*, *139*(3), 865–876. https://doi.org/10.1016/j.neuroscience.2005.11.072.

Narayanan, N. S., & Laubach, M. (2006). Top-down control of motor cortex ensembles by Dorsomedial prefrontal cortex. *Neuron*, *52*(5), 921–931. https://doi.org/10.1016/j.neuron.2006.10.021.

Narayanan, N. S., & Laubach, M. (2009a). Delay activity in rodent frontal cortex during a simple reaction time task. *Journal of Neurophysiology*, *101*(6), 2859–2871. https://doi.org/10.1152/jn.90615.2008.

Narayanan, N. S., & Laubach, M. (2009b). Methods for studying functional interactions among neuronal populations. *Methods in Molecular Biology*, *489*, 135–165. https://doi.org/10.1007/978-1-59745-543-5_7.

Narayanan, N. S., & Laubach, M. (2017). Inhibitory control: Mapping medial frontal cortex. *Current Biology*, *27*, R148–R150. https://doi.org/10.1016/j.cub.2017.01.010.

Ollman, R. T., & Billington, M. J. (1972). The deadline model for simple reaction times. *Cognitive Psychology*, *3*(2), 311–336. https://doi.org/10.1016/0010-0285(72)90010-2.

Ouhaz, Z., Ba-M'hamed, S., & Bennis, M. (2017). Morphological, structural, and functional alterations of the prefrontal cortex and the basolateral amygdala after early lesion of the rat mediodorsal thalamus. *Brain Structure and Function*, *222*(6), 2527–2545. https://doi.org/10.1007/s00429-016-1354-2.

Ouhaz, Z., Ba-M'hamed, S., Mitchell, A. S., Elidrissi, A., & Bennis, M. (2015). Behavioral and cognitive changes after early postnatal lesions of the rat mediodorsal thalamus. *Behavioural Brain Research*, *292*, 219–232. https://doi.org/10.1016/j.bbr.2015.06.017.

Ouhaz, Z., Fleming, H., & Mitchell, A. S. (2018). Cognitive functions and neurodevelopmental disorders involving the prefrontal cortex and mediodorsal thalamus. *Frontiers in Neuroscience*, *12*(33), 1–18. Frontiers Media S.A. https://doi.org/10.3389/fnins.2018.00033.

Parnaudeau, S., O'Neill, P. K., Bolkan, S. S., Ward, R. D., Abbas, A. I., Roth, B. L., et al. (2013). Inhibition of mediodorsal thalamus disrupts thalamofrontal connectivity and cognition. *Neuron*, *77*(6), 1151–1162. https://doi.org/10.1016/j.neuron.2013.01.038.

Parnaudeau, S., Taylor, K., Bolkan, S. S., Ward, R. D., Balsam, P. D., & Kellendonk, C. (2015). Mediodorsal thalamus hypofunction impairs flexible goal-directed behavior. *Biological Psychiatry*, *77*(5), 445–453. https://doi.org/10.1016/j.biopsych.2014.03.020.

Passecker, J., Mikus, N., Malagon-Vina, H., Anner, P., Dimidschstein, J., Fishell, G., et al. (2019). Activity of prefrontal neurons predict future choices during gambling. *Neuron*, *101*(1), 152–164.e7. https://doi.org/10.1016/j.neuron.2018.10.050.

Pinto, L., & Dan, Y. (2015). Cell-type-specific activity in prefrontal cortex during goal-directed behavior. *Neuron*, *87*(2), 437–450. https://doi.org/10.1016/J.NEURON.2015.06.021.

Reep, R. L., Goodwin, G. S., & Corwin, J. V. (1990). Topographic organization in the corticocortical connections of medial agranular cortex in rats. *Journal of Comparative Neurology*, *294*(2), 262–280. https://doi.org/10.1002/cne.902940210.

Risterucci, C., Terramorsi, D., Nieoullon, A., & Amalric, M. (2003). Excitotoxic lesions of the prelimbic-infralimbic areas of the rodent prefrontal cortex disrupt motor preparatory processes. *European Journal of Neuroscience*, *17*(7), 1498–1508. https://doi.org/10.1046/j.1460-9568.2003.02541.x.

Robbins, T. W. (2002). The 5-choice serial reaction time task: Behavioural pharmacology and functional neurochemistry. *Psychopharmacology*, *163*, 362–380. https://doi.org/10.1007/s00213-002-1154-7.

Rossi, M. A., Hayrapetyan, V. Y., Maimon, B., Mak, K., Shawn Je, H., & Yin, H. H. (2012). Prefrontal cortical mechanisms underlying delayed alternation in mice. *Journal of Neurophysiology*, *108*(4), 1211–1222. https://doi.org/10.1152/jn.01060.2011.

Schmitt, L. I., Wimmer, R. D., Nakajima, M., Happ, M., Mofakham, S., & Halassa, M. M. (2017). Thalamic amplification of cortical connectivity sustains attentional control. *Nature*, *545*(7653), 219–223. https://doi.org/10.1038/nature22073.

Schoenbaum, G., Chiba, A. A., & Gallagher, M. (2000). Changes in functional connectivity in orbitofrontal cortex and basolateral amygdala during learning and reversal training. *Journal of Neuroscience*, *20*(13), 5179–5189. https://doi.org/10.1523/jneurosci.20-13-05179.2000.

Schoenbaum, G., Roesch, M. R., Stalnaker, T. A., & Takahashi, Y. K. (2009). A new perspective on the role of the orbitofrontal cortex in adaptive behaviour. *Nature Reviews Neuroscience*, *10*(12), 885–892. https://doi.org/10.1038/nrn2753.

Seamans, J. K., Lapish, C. C., & Durstewitz, D. (2008). Comparing the prefrontal cortex of rats and primates: Insights from electrophysiology. *Neurotoxicity Research*, *14*(2–3), 249–262. https://doi.org/10.1007/BF03033814.

Sherman, S. M. (2016). Thalamus plays a central role in ongoing cortical functioning. *Nature Neuroscience*, *19*, 533–541. Nature Publishing Group https://doi.org/10.1038/nn.4269.

Smith, N. J., Horst, N. K., Liu, B., Caetano, M. S., & Laubach, M. (2010). Reversible inactivation of rat premotor cortex impairs temporal preparation, but not inhibitory control, during simple reaction-time performance. *Frontiers in Integrative Neuroscience*, *4*(124), 1–13. https://doi.org/10.3389/fnint.2010.00124.

Stuphorn, V., & Emeric, E. E. (2012). Proactive and reactive control by the medial frontal cortex. *Frontiers in Neuroengineering*, *5*(June), 1–11. https://doi.org/10.3389/fneng.2012.00009.

Sul, J. H., Jo, S., Lee, D., & Jung, M. W. (2011). Role of rodent secondary motor cortex in value-based action selection. *Nature Neuroscience*, *14*(9), 1202–1210. https://doi.org/10.1038/nn.2881.

Verbruggen, F., & Logan, G. D. (2008). Response inhibition in the stop-signal paradigm. *Trends in Cognitive Sciences*, *12*(11), 418–424. https://doi.org/10.1016/j.tics.2008.07.005.

Verbruggen, F., & Logan, G. D. (2009). Models of response inhibition in the stop-signal and stop-change paradigms. *Neuroscience and Biobehavioral Reviews*, *33*(5), 647–661. https://doi.org/10.1016/j.neubiorev.2008.08.014.

Vertes, R. P. (2004). Differential projections of the infralimbic and prelimbic cortex in the rat. *Synapse*, *51*(1), 32–58. https://doi.org/10.1002/syn.10279.

Vertes, R. P. (2006). Interactions among the medial prefrontal cortex, hippocampus and midline thalamus in emotional and cognitive processing in the rat. *Neuroscience*, *142*(1), 1–20. https://doi.org/10.1016/j.neuroscience.2006.06.027.

Vince, M. A. (1948). The intermittency of control movements and the psychological refractory period. *British Journal of Psychology*, *38*(3), 149–157. https://doi.org/10.1111/j.2044-8295.1948.tb01150.x.

Wallis, J. D. (2012). Cross-species studies of orbitofrontal cortex and value-based decision-making. *Nature Neuroscience*, *15*(1), 13–19. https://doi.org/10.1038/nn.2956.

Wang, M., Montanède, C., Chandrasekaran, C., Peixoto, D., Shenoy, K. V., & Kalaska, J. F. (2019). Macaque dorsal premotor cortex exhibits decision-related activity only when specific stimulus–response associations are known. *Nature Communications*, *10*(1), 1–16. https://doi.org/10.1038/s41467-019-09460-y.

Yang, H., de Jong, J. W., Tak, Y. E., Peck, J., Bateup, H. S., & Lammel, S. (2018). Nucleus accumbens subnuclei regulate motivated behavior via direct inhibition and disinhibition of VTA dopamine subpopulations. *Neuron*, *97*(2), 434–449.e4. https://doi.org/10.1016/j.neuron.2017.12.022.

CHAPTER FOURTEEN

The midcingulate cortex and temporal integration

Emmanuel Procyk[a,*], Vincent Fontanier[a], Matthieu Sarazin[b], Bruno Delord[b], Clément Goussi[a], and Charles R.E. Wilson[a,*]
[a]Univ Lyon, Université Claude Bernard Lyon 1, Inserm, Stem Cell and Brain Research Institute U1208, Bron, France
[b]Institute of Intelligent Systems and Robotics (ISIR), Sorbonne Université, Centre National de la Recherche Scientifique, UMR 7222, Paris, France
*Corresponding authors: e-mail address: emmanuel.procyk@inserm.fr; charles.wilson@inserm.fr

Contents

Abstract

The ability to integrate information across time at multiple timescales is a vital element of adaptive behavior, because it provides the capacity to link events separated in time, extract useful information from previous events and actions, and to construct plans for behavior over time. Here we make the argument that this information integration capacity is a central function of the midcingulate cortex (MCC), by reviewing the anatomical, intrinsic network, neurophysiological, and behavioral properties of MCC. The MCC is the region of the medial wall situated dorsal to the corpus callosum and sometimes referred to as dACC. It is positioned within the densely connected core network of the primate brain, with a rich diversity of cognitive, somatomotor and autonomic connections. Furthermore, the MCC shows strong local network inhibition which appears to control the metastability of the region—an established feature of many cortical networks in which the neural dynamics move through a series of quasi-stationary states. We propose that the strong local inhibition in MCC leads to particularly long dynamic state durations, and so less frequent transitions. Apparently as a result of these anatomical features and synaptic and ionic determinants, the MCC cells display the longest

International Review of Neurobiology, Volume 158
ISSN 0074-7742
https://doi.org/10.1016/bs.irn.2020.12.004

neuronal timescales among a range of recorded cortical areas. We conclude that the anatomical position, intrinsic properties, and local network interactions of MCC make it a uniquely positioned cortical area to perform the integration of diverse information over time that is necessary for behavioral adaptation.

1. On time and behavior

Integrating information across time at multiple timescales is a key feature of higher order cognition. Information integration across time permits us to link events and actions separated in time; to draw out useful evidence from a history of behavioral and environmental reactions; and to construct serial, time-extended plans that provide both retrospective and prospective contexts to behavior. This capacity is thus linked to the fact that the environment itself is structured along multiple timescales. This capacity appears phenomenologically in the structure of behaviors of animal species that have advanced adaptive abilities and problem-solving skills. Primates, for instance, can plan extended routes to seek resources; and they can decide, reorganize, and use detours in the face of intervening events and knowledge acquired from past events (Janmaat, Byrne, & Zuberbühler, 2006; Noser & Byrne, 2007). Certain primates and birds also use the history of interactions with objects to build tools (Gruber, Zuberbühler, & Neumann, 2016). In laboratory settings the ability of monkey species to extract adaptive performance rules over a long task history is well established (Harlow, 1949). These capacities can be considered together as capacities of behavioral adaptation. In its essence, the adaptation of behavior requires integration of multiple information sources over multiple time scales, to integrate action, outcomes, and episodic information. Adapting to real life situations is mostly non-Markovian because the history of behavioral interaction is crucial.

In a highly dynamical system like the brain, performing functions and calculations at multiple concurrent timescales requires specialized algorithms and machinery. This is therefore a difficult task requiring a complex neurobiological implementation, but in highly adaptive species the development of such capacities is clearly worthwhile (Pearson, Watson, & Platt, 2014). Several lines of research suggest that certain frontal areas have specific roles in the elaboration of behaviors over long timescales (Passingham & Wise, 2012). In this chapter we collect evidence from multiple domains suggesting that the midcingulate cortex (MCC) of primates contributes to the integration of information over long timescales. We propose that the physiological properties of the MCC network provide the dynamical properties necessary for this temporal integration, and that in turn these properties permit the multiple functions that have been ascribed to MCC, i.e., learning, feedback adaptation, and regulation of exploratory behavior.

The first lines of evidence for a role of frontal medial cortex regions in temporal integration come from studies focused on adaptive behavior. Indeed, efficient adaptation relies on extracting knowledge, structures as well as changes, from a series of interactions with the environment. In experimental setups this corresponds to learning from outcomes and feedback after a series of choices. In human and non-human primates a large-scale network encompassing frontal and posterior associative cortical areas contributes to adaptive behaviors (Duncan, 2010; Mitchell et al., 2016; Premereur, Janssen, & Vanduffel, 2018). Moreover, fMRI consistently reveals activations in one region in the medial wall in relation to outcome monitoring and the apparent regulation of cognitive control (Amiez, Sallet, Procyk, & Petrides, 2012; Bush et al., 2002; Jueptner et al., 1997; Kouneiher, Charron, & Koechlin, 2009; Passingham, 1996). The region, often named dACC in the primate literature, corresponds to the anatomically defined midcingulate cortex (MCC, see Fig. 1). Studies sometimes show extensions of activation into preSMA or SMA and in more or less anterior parts of the MCC, and the dissociation of these activations remains an important task for the field (Amiez et al., 2013). MCC activation is observed when subjects actively seek information and/or rewards, for example, when they are exploring or foraging. The MCC seems sensitive to the volatility of the environment, for example, to the temporal stability of outcomes, as well

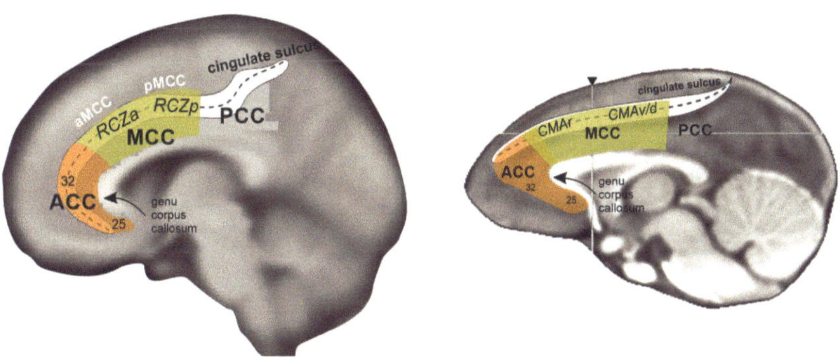

Fig. 1 Schematic representation of the cingulate regions MCC and ACC in human (left) and monkey (right). In both species the MCC contains cingulate motor areas, CMAr, CMAd and CMV in monkeys, and RCZa and RCZp in humans.

as to the propensity of subjects to make choices based on information integrated over multiple trials (Behrens, Woolrich, Walton, & Rushworth, 2007; Meder et al., 2017). The activation of this region in particular during exploratory decisions reveals its particular role in encoding values and prospective information relevant to adaptation (Kolling et al., 2016; Kolling, Behrens, Mars, & Rushworth, 2012). Studies suggest, then, that evaluation and temporal integration of information in or through the MCC might serve to guide and sustain selected behaviors in the face of current and changing features of the environment (Kolling, Wittmann, et al., 2016).

Testing in patients for clinical investigations has provided compelling data on the behavioral effect but also on the subjective feeling induced by electrical microstimulation of the MCC (Caruana et al., 2018; Parvizi, Rangarajan, Shirer, Desai, & Greicius, 2013; Talairach et al., 1973). This stimulation regularly leads to localized effects on natural goal-directed behaviors like exploratory scanning of the environment with the eyes or the head, tactile search, kneading or palpation directed toward the body or toward nearby objects (Caruana et al., 2018; Talairach et al., 1973). Patients also report general feelings of an urge to act. These results could be interpreted as reflecting MCC's contribution to generate the incentive for behaviors directed toward information search. The effects of microstimulation on search-oriented behaviors concern mostly the anterior part of the MCC, the equivalent of CMAr in monkeys.

Studies in human subjects reveal a wide array of correlative roles for the MCC, but there is among these a clear case for a role for the MCC in the temporal integration of diverse signals necessary for adaptive behavior. Our proposition here is that the best way to understand this role, and the way in which seemingly diverse responses emerge from the same region, is to consider in detail the mechanisms within MCC and its position within the anatomical hierarchy. The rest of this chapter therefore considers the evidence, largely from animal studies, for mechanisms of temporal integration in MCC.

2. The neurobiological source of temporal integration in MCC

Multiple neuroanatomical and physiological features suggest that the MCC holds a special position within the executive and more global networks, and that it is well placed and constructed to integrate diverse information in the temporal domain.

2.1 Anatomical specificities of the MCC

In most monkey recording studies referred to in this chapter, the regions of interest have been the cortex lying within the cingulate sulcus, most often the dorsal bank, and at rostro-caudal levels anterior to the genu of the arcuate sulcus (Procyk et al., 2016). We refer to this region as MCC. The entire MCC region is subject to debate in part because of the multiple versions of labeling used to describe its subdivisions. Regions' names as well as cytoarchitectonic labels are used in different ways across multiple papers, reviews, and even brain atlases. Major labeling issues are the definitions of "ACC" or "24c," which have evolved. Here we provide a definition of the term MCC we use.

The MCC is an anatomical entity defined by Vogt and colleagues, based on connectivity, cytoarchitectonics, and receptor mappings. It is evidenced in humans and monkeys, with putative equivalents in rodents, and it appears to have a separate functional identity (Vogt, 2016; Vogt, Nimchinsky, Vogt, & Hof, 1995; Vogt & Paxinos, 2012).

2.1.1 Cytoarchitecture

In humans, the MCC includes the functional subdivision frequently referred to as dACC. It is positioned dorsal to the corpus callosum and, in humans, posterior to the level of the genu of the corpus callosum. MCC cyto-architecture can be contrasted to the cytoarchitecture of the ACC (anterior cingulate cortex) which is rostral to MCC, and to the cytoarchitecture of the PCC (posterior cingulate cortex). MCC is composed of agranular cortex, which like ACC lacks a granular layer IV (Palomero-Gallagher, Vogt, Schleicher, Mayberg, & Zilles, 2009). In Vogt's nomenclature, MCC includes cytoarchitectonic areas a24c', b', and a', areas p24c', b' and a' as well as area 24d. The posterior part of MCC, containing area 24d, has the largest pyramidal cells which contribute to cingulate spinal projections and form the caudal cingulate motor areas in monkeys. Debates regarding comparative assessments in monkeys and humans relate to the dorsal and anterior parts of the MCC. Anteriorly, the limit with ACC is not clearly identified and could contain an intermediate (rACC) region. The dorsal limit of the sulcus, in terms of architecture and function is also debated (Sallet et al., 2011). Histological work from multiple groups show that indeed the dorsal limit position might vary along the rostral-caudal axis and potentially vary with species and individuals. More precise work is required to solve these issues. In this chapter we include the dorsal bank of the cingulate sulcus because the cytoarchitecture, the connectivity patterns and the extension of cingulate motor areas (see below) suggest an anatomo-functional link to the rest of MCC (Petrides & Pandya, 1994; Procyk et al., 2016; Sallet et al., 2011).

2.1.2 Receptor mapping

In humans and non-human primates, the MCC region and its subdivisions are also characterized by specific patterns of receptor distributions. We cite here two remarkable features. The anterior MCC contains higher levels of dopamine receptor D1 binding in the superficial layers, whereas area 24d has virtually none in superficial and deep layers. Also of note is the relative level of GABAergic receptor mappings: anterior MCC contains relatively lower levels of GABA-B receptors and slightly higher levels of GABA-A, compared to posterior MCC. The distribution of GABA receptors differs also clearly between anterior MCC and ACC (Bozkurt et al., 2005; Palomero-Gallagher, Mohlberg, Zilles, & Vogt, 2008; Palomero-Gallagher et al., 2009).

2.2 MCC in the anatomical hierarchy

The MCC also sits at a remarkable anatomical position, first within the large-scale network formed by cortical areas, second within the somato-motor systems, and third in relation to the autonomic system. The large-scale cortical network is, in primates, a system with heterogeneous densities of inter-area connectivity (Markov et al., 2013). It forms a so-called bow-tie organization including a core network with high density (>90%) of connections between areas, linked to side systems by low density connectivity patterns (Markov et al., 2013). The structure has some similarities with the early model of global workspace architecture that contained a central general workspace network connected to hierarchically distant modules processing specific information (Dehaene, Kerszberg, & Changeux, 1998). Within the core network lies the MCC and its interconnections with the lateral prefrontal cortex, as well as other cortical regions that were shown to form the Multiple-demand network (Duncan, 2013). As we shall see, MCC's position within this architecture, combined with more local intrinsic properties, appears crucial for its functional identity.

The MCC displays other anatomical features that, we think, must be considered to comprehend its functional specificity and contributions to cognition. First the MCC has anatomical links with the sensorimotor system, and contains, in monkeys and humans, two sensorimotor maps named CMAr and CMAc in monkeys (M3 and M4 in Morecraft, Schroeder, & Keifer, 1996) and RCZa and RCZp in humans (Amiez & Petrides, 2014; Picard & Strick, 1996). These maps are somatotopically organized and are connected to corresponding fields in supplementary premotor, lateral premotor, motor and spinal regions, with a rostro-caudal gradient displaying more primary and spinal connections posteriorly and more premotor and prefrontal connections anteriorly (Dum & Strick, 1991, 2002; He, Dum, & Strick, 1995; Loh, Hadj-Bouziane, Petrides, Procyk, & Amiez, 2018). Face, eye, upper and lower limbs seem to be represented, with one face representation positioned at the most anterior limit of the MCC. The most rostral and dorsal MCC region has been shown active in multiple studies involving behavioral adaptation in one way or another. The region mentioned in the human brain imaging literature is often referred to as dACC or dmPFC, with terms associated to significantly variable locations of activation within the human brain. In fact, the anterior and posterior subdivisions of the MCC (aMCC and pMCC) seem to be activated differently during cognitive tasks and more precise descriptions of brain activations, potentially on a subject by subject basis, are required to really grasp the functional specificity within this region (Amiez et al., 2013). We have suggested recently that the somatotopic organization of MCC reflects fields devoted to the detection and evaluation of domain-specific feedback relevant for adaptations (Loh et al., 2020; Procyk et al., 2016). Because of the specificity of connectivity patterns of MCC maps it is possible that they contribute to different functional uses of feedback, i.e., contributing to trigger different adaptive cognitive or behavioral reactions.

Finally, the MCC is, within the medial frontal cortex of primates, one main source and target of connections with the autonomic system. Recent investigations have shown that, contrary to rodents, the primate medial frontal cortex is 3–4 synapses away from an organ like the adrenal medulla (Dum, Levinthal, & Strick, 2016, 2019). Neurons trans-synaptically labeled after injection in the adrenal medulla are notably located within the somato-motor fields of the CMAr and CMAc. A review of human brain activations testing covariations of brain activity with autonomic measures showed that the MCC might have subdivisions related to the sympathetic and parasympathetic systems respectively (Amiez & Procyk, 2019).

In sum, the most rostral subdivisions of MCC that contain somatomotor fields also have connections with the lateral prefrontal cortex and with the sympathetic system. Such overlap of cognitive, somatomotor and autonomic connectivity is, we propose, a crucial property of the MCC. This connectivity and position within the cortical core bring rich and diverse inputs to the MCC, permitting it to perform functions that extend beyond any specific modality of information type, in particular favoring an integrative function. Our argument here is that a particularly important part of this role is the integration across time that can be carried out with this diversity of information. Central to this claim is the idea that the intrinsic properties of the cortical cells themselves, the construction of the local circuits, and in particular the nature of local inhibition, allow MCC to perform temporal integration over longer timescales than other cortical regions, and to do so with diverse information provided by its position in the cortical hierarchy.

2.3 Intrinsic cortical features of MCC

The cell physiological properties of MCC can provide clues to its dynamical specificities and position in the temporal hierarchy. Specific physiological data in monkeys for a region at the border between ACC and MCC show that the nature of local inhibition might be an important feature in MCC (Medalla, Gilman, Wang, & Luebke, 2017). Spontaneous inhibitory postsynaptic currents are more frequent, long and large in cingulate layer III neurons compared to other prefrontal regions (Medalla et al., 2017). Data from human MCC shows that GABA-B receptors are dense, especially in its posterior part, pMCC (Zilles & Palomero-Gallagher, 2017). Robust inhibition and long inhibitory time constants should contribute to the extension of the time window for signal summation and thus extend local temporal receptive fields. In artificial hierarchical networks, environmental uncertainty can be dynamically captured by variations of the E/I tone (Pettine, Louie, Murray, & Wang, 2020). In humans, dynamical integration of environmental uncertainty is circumscribed to the MCC (Behrens et al., 2007). A recent study also linked subjects' ability to integrate information in time and their glutamate and GABA concentrations in the MCC (Scholl, Kolling, Nelissen, Stagg, & Harmer, 2017). In general terms, inhibition is a strong determinant of network activity (Mongillo, Rumpel, & Loewenstein, 2018). The diversity of synaptic inhibition can flexibly structure network dynamics by gating pyramidal inputs depending on the current task demand (Tremblay, Lee, & Rudy, 2016; Wang, 2020; Womelsdorf, Valiante, Sahin, Miller, & Tiesinga, 2014). For example, MCC inhibitory activity has been proposed as a mechanism for transient network disengagement in response to errors (Rothé, Quilodran, Sallet, & Procyk, 2011; Shen et al., 2015). Inhibitory controlled circuits are more robust to excitatory volatility (Mongillo et al., 2018). This robustness might be particularly important for areas receiving converging multimodal inputs, and that need to integrate that information across time. Indeed, there is a macroscopic gradient of synaptic inhibition from sensory cortices to associative areas (Wang, 2020). MCC slow inhibitory decay times generate low frequency rhythmic oscillatory activity (Kopell, Kramer, Malerba, & Whittington, 2010; Medalla et al., 2017) supporting communication between distant cortical areas (Hahn, Bujan, Frégnac, Aertsen, & Kumar, 2014). MCC excitatory and inhibitory spike trains synchronize to different frequency bands (Voloh & Womelsdorf, 2018). Such cell type specific oscillations could be a means for flexible long-range tuning of coherent network activities according to the current demand. A recent study reported that MCC stimulation elicits recurrent activity in the LPFC (Nácher, Hassani, & Womelsdorf, 2019), an activity pattern that has been associated with working memory temporal maintenance (Mongillo et al., 2018; Wasmuht, Spaak, Buschman, Miller, & Stokes, 2018). In this framework, MCC activity and in particular inhibition would contribute to network synchronization and structuration of local and distant cortical areas.

We propose therefore that MCC inhibitory tone determines the activity of local units and leads to more stable or slowly evolving network dynamics favoring multimodal associations in the temporal domain. And these properties appear to be particularly exaggerated within MCC relative to other areas of association cortex, placing MCC not just at the heart of the dense core of associative cortical areas (Markov et al., 2013), but also as new analyses and modeling suggest, at the pinnacle of a hierarchy of cortical timescales.

Based on anatomical data and large-scale network modeling, Wang and colleagues have recently shown that the anatomical organization and hierarchies within the large-scale cortical network was accompanied by temporal hierarchies (Chaudhuri, Knoblauch, Gariel, Kennedy, & Wang, 2015; Murray et al., 2014). Individual neuron spiking shows some level of autocorrelation that reflects its propensity to keep a firing mode. The autocorrelation decay, referred to as spiking timescale (Murray et al., 2014), measured at the population level in different areas increases from posterior (sensory) areas to anterior frontal regions, forming a temporal hierarchy. Within the set of areas investigated in this study MCC displayed the longest timescale (about 300 ms), compared to LPFC (190 ms) or area MT (65 ms) (Cavanagh, Towers, Wallis, Hunt, & Kennerley, 2018; Murray et al., 2014). Medial frontal regions also show longer timescales in rats (Murakami, Shteingart, Loewenstein, & Mainen, 2017). Such differences in timescale might provide areas with different information processing capability, and in particular the capacity to integrate information over long temporal scales, like rewards or feedback encountered across trials of a cognitive task (Bernacchia, Seo, Lee, & Wang, 2011). Indeed very recent work proposes that cortical dynamics are modulated by a range of timescales, potentially generated by independent mechanisms, some of which directly link to relevant task parameters, while others like the general intrinsic timescale remain independent (Spitmaan, Seo, Lee, & Soltani, 2020). Large scale modeling leads to the proposition that the specific timescale of cortical areas emerges both from the large-scale pattern of inter-areal connectivity and from intrinsic properties of areas, including local recurrent excitatory connectivity which was found to increase in density in parallel to the temporal hierarchy (Chaudhuri et al., 2015). Other important features that appear to change across the cortical hierarchy and may relate to these changes in timescale include the ratio T1w/T2w in structural MRI scans, and the level of tonic inhibition (Wang, 2020). We explore the interplay between the macroscopic gradient of synaptic inhibition and timescale hierarchy (possibly via disinhibition; Wang, 2020) in the final section.

The outcome of this work is that the MCC is situated at the core of a cortical network, with the longest neuronal timescale of recorded cortical areas, relatively strong local network inhibition, and a rich diversity of cognitive, somatomotor and autonomic connections. We argue that these are the building blocks for a system whose neurophysiological role will be to integrate diverse signals over time, a role particularly relevant in the context of behavioral adaptation. In the following sections we discuss the neurophysiological correlates of behavior in MCC and relate them to the properties described above. We propose that the diversity of neural responses in MCC makes particular sense in the light of this overarching function of temporal integration for adaptation.

3. Neurophysiological and causal correlates of temporal integration in MCC

Neurophysiological studies in rodent and non-human primate models provide details of the neural processes implemented by MCC and reveal neural correlates of the history of outcomes and values in a task, as well as the switching between behavioral states. These point to a role for MCC neurons in functions dedicated to gathering information over time for subsequent regulation of behavior, a role that a limited number of intervention studies appear to confirm. Similar properties have been observed in the rat medial frontal cortex and while the anatomical correspondence in the rat brain is still in discussion, we will refer here to relevant data from medial frontal cortex (MFC) areas IL, PL and AC (Mars et al., 2018; Schaeffer et al., 2020; Vogt & Paxinos, 2012).

3.1 History of outcomes and values

MCC and MFC single unit activity encodes for reward-based action selection in simple tasks in rats, monkeys and humans (Shima & Tanji, 1998; Sul, Kim, Huh, Lee, & Jung, 2010; Williams, Bush, Rauch, Cosgrove, & Eskandar, 2004), but also for the enactment of serial actions where each action is coded relative to future outcomes (Procyk & Joseph, 2001; Procyk, Tanaka, & Joseph, 2000; Shidara & Richmond, 2002). Intervention studies in the MCC are rare, but those that exist confirm that MCC is directly contributing to the regulation of decisions based on rewards (Amiez, Joseph, & Procyk, 2006; Kennerley, Walton, Behrens, Buckley, & Rushworth, 2006; Shima & Tanji, 1998). Shima and Tanji (1998), for example, showed that neurons in the cingulate motor areas within MCC signaled a reduction of reward that indicated that a change in action should be subsequently chosen, but only when the action was indeed successfully changed. Such activity is necessarily a response to a recent reward history over several trials, rather than a single outcome. The alteration of this MCC activity by local muscimol injection leads to a deterioration in the adaptation of action selection based on the diminishing reward size (Shima & Tanji, 1998). These responses in MCC are therefore driven by more information than the immediate action to be committed—they include information about the temporal context of that action. This temporal context is of course mostly of use to ensure that the action is well adapted, and importantly MCC activity also encodes feedback of actions or choices that is relevant to adaptation, i.e., in situations where action and feedback are the focus of attention and where it can reduce uncertainty.

It is long established that cells in MCC respond to both rewards and errors (Amiez, Joseph, & Procyk, 2005), and that responses to these events are modulated by their predictability and potentially their behavioral relevance (Quilodran, Rothé, & Procyk, 2008). In addition, MCC neurons represent the expected magnitude or the probability of an upcoming reward on trial-and-error learning tasks (Kennerley, Behrens, & Wallis, 2011). While initial reports made a straight link between immediate reward and cell activity, it is now clear that the recent history of reward alters the firing of these cells (Kawai, Yamada, Sato, Takada, & Matsumoto, 2015; Seo & Lee, 2007), and that this effect will occur with varying timescales across neurons (Bernacchia et al., 2011). For example, Kawai et al. (2015) showed that MCC cells, and in particular putative inhibitory neurons (Kawai, Yamada, Sato, Takada, & Matsumoto, 2019), not only maintained the information about the last outcome but also stored outcome experiences from several past trials. MCC cells in this protocol also signaled a subsequent shift in strategy, activity that is discussed in more detail below.

The ability of MCC activity to reflect outcome history has an impact on behavior. Monkeys with bilateral MCC sulcus lesions have difficulties in integrating and building reward-action history in an uncertain context and in adapting choice behavior (Kennerley et al., 2006). Effectively the lesion appears to reduce the span over which recent outcomes are used to adjust decisions. MCC appears to have a central role in updating action values based on the outcome history, helping to improve behavioral performance toward a specific goal (Kennerley et al., 2006; Seo, Barraclough, & Lee, 2007). In general therefore these deficits seem to involve a loss of exploratory behaviors—the observed effects appear to combine altered valuation of outcomes and a reduced capacity to integrate the implications of those outcomes over time, the consequence on adaptive behavior being quite consistent.

Rodent pharmacological inactivation and chemogenetic perturbation also show that medial frontal regions in rats have a causal role in integrating outcome history and in regulating explore/exploit behaviors (Tervo et al., 2020). Indeed, the distributed ensemble dynamics might be tuned by monoaminergic afferents. Noradrenergic (NA) inputs are proposed to contribute to abrupt network reconfiguration and to promote transitions between exploitation and exploration (Bouret & Sara, 2005), and we discuss these state transitions in more detail below. NA inputs to the MCC would allow the dynamical modulation of learning rates (Silvetti, Vassena, Abrahamse, & Verguts, 2018). Consistent with those views, selective NA input enhancement in the rat MCC artificially increases behavioral stochasticity and leads to an increase in response variability. In turn, NA input suppression restores the ability to weight previously received feedback for choice regulation (Tervo et al., 2014).

In addition rare studies can provide information on the effect of cingulotomy in human patients, supporting a role for ongoing behavioral adaptation based on context from prior trials, both in the context of decisions to act (Williams et al., 2004) and in more cognitive contexts where previous trials might provide interfering information (Sheth et al., 2012). Wider lesions to the cingulate region have been reported to lead to global reductions of action and speech, often characterized as akinetic mutism (Devinsky, Morrell, & Vogt, 1995; Németh, Hegedüs, & Molnár, 1988). Such deficits have been proposed consistent with much of the above evidence if we consider the role of the MCC as acting to motivate time-extended behaviors, rather than driving trial-by-trial adaptations (Holroyd & Yeung, 2012).

Recent work proposes how the MCC goes about using outcome history to drive upcoming behavior: it is proposed that MCC neurons encode expected outcomes as a latent state representation of actual outcomes (Hyman, Holroyd, & Seamans, 2017). The interaction between expected and actual outcome representations on trials where they do not match may drive a dynamical shift that we often record as an error signal. MCC cells certainly do show error signals—they respond to errors in the form of absent rewards (Niki & Watanabe, 1979). Again this error activity is modulated in a manner that reflects the temporal context, for example, how much or how close a reward was before the error (Amiez et al., 2005). Cells in MCC also encode both the overall task value in terms of reward, as well as the actual reward obtained (Amiez et al., 2006; Sallet et al., 2007). This activity has a causal impact on behavior as alteration of local MCC activity by local muscimol injection leads to a loss of efficiency in exploration for more rewarded stimuli in a choice task (Amiez et al., 2006).

These comparative feedback responses translate to the population level, where error- and feedback-related potentials (such as the error-related negativity [ERN] and feedback potentials [FRPs]) have been widely shown to differentiate outcome valences when recorded over the medial part of the frontal lobe in electroencephalography (EEG) (Falkenstein et al., 2001; Gehring, Goss, Coles, & Meyer, 1993; Miltner, Braun, & Coles, 2001), electrocorticography (EcoG; Wilson et al., 2016), and the local field potential (LFP) (Gemba, Sasaki, & Brooks, 1986). Of course these are not standalone signals, rather they appear to provide information about the value of the feedback in terms of behavioral adaptation, be it for directly driving adaptation on subsequent trials (Khamassi, Quilodran, Enel, Dominey, & Procyk, 2015; Monosov, 2017; Quilodran et al., 2008; Shima & Tanji, 1998), or for motivating more extended behaviors beyond simple trial-to-trial adaptation (Holroyd & Yeung, 2012; Walsh & Anderson, 2011). It is this concept of time-extended behaviors that is of particular relevance here, therefore, and it is this time-extended information that is necessary to signal the necessity to change state or strategy in a task.

3.2 Neural and behavioral state switches

Monkey MCC and rat MFC show phasic neural changes at behaviorally relevant events together with global neural state changes between different behavioral states or strategies separated by switches, like exploratory behavior and repetitive behavior (Enel, Procyk, Quilodran, & Dominey, 2016; Karlsson, Tervo, & Karpova, 2012). For example MCC neurons signal behavioral shifts on subsequent trials (Kawai et al., 2015), while neural signals anticipating and signaling patch leaving or strategy switching in sequential foraging tasks are observed in phasic unitary and high-gamma activities in both primates and rodents (Karlsson et al., 2012; Quilodran et al., 2008; Rothé et al., 2011). Powell and Redish (2016) showed that on tasks where a reward criterion change was imposed, rats showed state transitions of MFC activity after they had learned about that contingency change, but before their behavior changed. In contrast, when rats were permitted to change strategies themselves (i.e., without imposed criterion changes), similar state transitions occurred before changes in behavior, therefore in a manner predictive of this un-imposed strategy switch (Powell & Redish, 2016). This implies that the MFC is signaling the need for a change in strategy, rather than passively reflecting such a change.

Across a range of studies, a picture emerges where the neural dynamics of MCC correlate with the state of uncertainty inherent to exploratory situations and the control or monitoring of information seeking behavior (Stoll, Fontanier, & Procyk, 2016; White et al., 2019). In addition, MCC activity has strong correlation with valuation processes—a multiplicity of signals reflect dynamical information accumulation in the MCC (Hunt et al., 2018; Kennerley et al., 2011). These signals are absent from OFC and LPFC activities. This integrated information is therefore used to make the decision to maintain or change the current course of action (Khamassi et al., 2015; Mansouri, Tanaka, & Buckley, 2009; Quilodran et al., 2008; Stoll et al., 2016).

Hayden, Pearson, and Platt (2011) found that neural activity during foraging scaled with the value of leaving the current exploited option and reflected an integrate-to-threshold process built across trials and leading to patch leaving (Hayden et al., 2011). Accumulation processes are observed in other frontal areas such as the FEF but they are continuous and ramp within trials, i.e., over a short timescale (Hanes & Schall, 1996). It is tempting to think that the specific neural dynamics observed during exploration or foraging provide the MCC with neural state properties that favor integration of information across time, i.e., over long timescales.

The way these correlated activity states are expressed over time is an important feature of encoding in MCC and across the frontal cortex. It may be that the capacity to shift neural state is an intrinsic feature of these cortical architectures, perhaps independent of external stimuli or constraints, which is then tuned to the needs of any ongoing task. An interesting and open question in the field is therefore how these properties are harnessed for the functional uses we ascribe to them, and a starting point is to consider how general this dynamic shifting of neural state is in MCC.

For example, we described above that information about feedback is encoded in the MCC, but this encoding occurs dynamically within the time course of a trial, starting just before outcome delivery until the next decision is made (Hunt et al., 2018; Stoll et al., 2016). The term dynamic here refers to the idea that information encoding changes over time, such that a decoder that can decode some behavioral variable from neural data at one period in time is no longer able to decode the same information later in time, while the variable is still encoded by the neural activity. In this sense the encoding is dynamic. The question of whether representations in MCC are dynamic or stable over time depends on the timescale in question, but it is increasingly clear that frontal cortical regions inherently combine elements of stable and dynamic coding. For example values in MCC are represented by mixed activity regimes composed of a stable subspace and a dynamic ensemble (Enel, Wallis, & Rich, 2020). This configuration grants the system a level of flexibility, i.e., a substrate of sustained information which can be combined with time-sensitive representations. The combination of both representations is likely crucial for the organization of behaviors in time and resembles the properties of delay activity much discussed regarding lateral prefrontal cortex (Cavanagh et al., 2018; Miller, Lundqvist, & Bastos, 2018).

It is becoming increasingly clear that there is a distinct organization in this combination of stable and dynamic coding regimes. For example, Stoll et al. (2016) used cross-temporal decoding of MCC activity to show semi-stable codes of feedback and decisions that carry over inter-trial periods during foraging tasks. This appears to be particularly the case when, as in this study, decisions rely on the history of previous choices. MCC activity in this case can be considered as a set of trajectories organized in a sequence of discrete quasi-stationary states. This organization of activities in time is referred to as metastability (Durstewitz, Vittoz, Floresco, & Seamans, 2010; La Camera, Fontanini, & Mazzucato, 2019). States can be organized around key perceptual and motor events or structured by ensemble activity regimes independently from external cues.

Metastability has been observed in other cortical areas and is proposed to form a neuronal substrate for specific internal representations of the environment (La Camera et al., 2019; Rich & Wallis, 2016). State stability is believed to reflect internal representations and the deployment of selective attention required for the current task demand (Engel et al., 2016). Transitions are more frequent when representations of internal rules are susceptible to change and are marked by an abrupt shift of dynamics in distributed neuronal ensemble (Durstewitz et al., 2010).

How might these metastable states be used to promote the functions of MCC? There are likely to be several answers to this question, again depending on the temporal scale of investigation. Value signaling in the MCC reflects temporal integration of experienced rewards (Amiez et al., 2006; Hunt et al., 2018) while also monitoring counterfactual options (Hayden, Pearson, & Platt, 2009; Kolling, Behrens, Wittmann, & Rushworth, 2016), and network transitions occur in periods when internal states are being updated and actively monitored. So, state transitions may well represent shifts in belief based on the same weighting of prior evidence and current outcomes. For example such state transitions would be signaling a shift to a more exploratory strategy of choice (Karlsson et al., 2012). Indeed, in foraging tasks, MCC trajectories describe two pseudo attractor spaces reflecting the structure of the task that alternates between phases of behavioral exploration and exploitation. Here state-space transitions are triggered by the monitoring of specific events such as feedback that reflect the need to adapt the current strategy (Enel et al., 2016; Quilodran et al., 2008).

3.3 Inhibitory control of MCC metastable states

Our most recent work provides direct evidence on how local intrinsic properties and timescales of MCC cells cause metastability, suggesting how such dynamical properties contribute to the role of MCC in adaptive behavior (Fontanier, Sarazin, Stoll, Delord, & Procyk, 2020). We studied the temporal signatures, specifically the spike auto-correlogram time constant and latency, of single neurons in monkey MCC and lateral prefrontal cortex (LPFC), recorded during a decision-making task. As suggested by Murray et al. (2014), cells demonstrated a longer intrinsic timescale in the MCC than

LPFC. We then showed that the temporal signatures were highly structured, showing anatomical organization within the MCC, and specific cell-type differences in the spike auto-correlogram features between fast and regular spiking neurons.

Importantly the timescales of certain neurons showed functional adaptation to the timescale of behaviorally relevant information in the task at hand. In MCC and for regular spiking neurons only, we observed that on average neuronal timescales lengthened when the monkeys were engaged in the task compared to when they were pausing from the task. In addition, we showed that both the neuron-type and the timescale of that neuron were determinant of the contribution of that neuron to encoding important behavioral features of the task, like reward feedback provided after choices (particularly putative interneurons for negative feedback), or value-related information used across multiple trials (Fontanier et al., 2020).

We then used these data to inform research using biophysically constrained network models to link back to the inhibitory properties discussed above. This work revealed that specific cellular conductances in the modeled cells (specifically intrinsic after-hyperpolarization (AHP) potassium and synaptic inhibitory GABA-B conductances) were critical determinants of the specificity of MCC and LPFC dynamics. These features were sufficient to provide a causal account for temporal signatures. Intriguingly, these conductances drove the cells to the recorded timescales by organizing activity into metastable states, with inhibition controlling state stability and transitions. Our model predicted therefore that the state duration in this metastable system would scale non-linearly with the timescales of different cortical regions, and we provided strong evidence to support this prediction from the monkey neurophysiological data, even reaching behavioral timescales. Our proposition on the basis of this work, therefore, is that inhibitory-controlled metastability constitutes the central dynamical process of MCC network function, locally specifying the temporal organization underlying cognitive processes operated by frontal areas.

4. Conclusion

The MCC, the region of the medial wall situated dorsal to the corpus callosum and sometimes referred to as dACC, is positioned within the densely connected core network of the primate brain, and has a rich diversity of cognitive, somatomotor and autonomic connections. The MCC shows strong local network inhibition in particular driven by GABA-B activity.

This inhibition appears to control the metastability of the region—an established feature of many cortical networks in which the neural dynamics move through a series of quasi-stationary states with jump-like modulations between them. In particular it is proposed that the strong local inhibition in MCC leads to particularly long dynamic state durations, and so less frequent transitions, with the duration of these states even approaching the timescale of behavioral phenomena. Apparently as a result of these anatomical features and synaptic and ionic determinants, the MCC cells display the longest neuronal timescales among a range of recorded cortical areas.

We have argued that the anatomical position, intrinsic properties, and local network interactions of MCC make it a uniquely positioned cortical area to perform the integration of diverse information over time that is necessary for behavioral adaptation. The functional role of the MCC is therefore defined by these properties and we consider it a single central function. The MCC will perform this integrative function on different information types that arrive, given its rich connectivity. Performing the same integrative function on different incoming types of information will lead to a diversity of behavioral correlates. While one reading of the literature might be that these diverse behavioral correlates mean that MCC has many different cognitive functions, we underline that the evidence presented above suggests that the operation performed by MCC itself is a single underlying function common to different cognitive labels.

This proposal leaves of course many important questions to be addressed. The neurobiological source of these temporal features, and the contribution of the place within the cortical hierarchy remain to be further examined. We also need to better understand the extent to which the variations in metastability and timescale are causal of the functions of MCC and other regions, or whether in fact those features result from adaptation to the context. Are these intrinsic properties that are applied to different functional uses, or are these properties tuned to fit the task in hand? Studies during task acquisition will be crucial here. Regions like the MCC do not, of course, act independently but as part of a densely connected network, for example, showing interaction with LPFC. As such there remain questions to be answered about how regions with differing timescales and differing dynamic regimes can interact in a manner that permits computations that are relevant for cognition.

The purpose of this chapter was to elaborate a synthesis of data from multiple levels of investigation, in order to seek a coherent multi-scale view of the function of the midcingulate cortex. While many pieces of information are still required from anatomical, biophysical, neurophysiological and behavioral experiments, the recent data suggest that the synaptic and cellular properties, the local network configuration as well as the embedding of MCC in specific large scale functional networks provide this region with capabilities relevant to the regulation and adaptation of behavior over long time scales.

References

Amiez, C., Joseph, J.-P., & Procyk, E. (2005). Anterior cingulate error-related activity is modulated by predicted reward. *The European Journal of Neuroscience*, *21*, 3447–3452.

Amiez, C., Joseph, J.-P., & Procyk, E. (2006). Reward encoding in the monkey anterior cingulate cortex. *Cerebral Cortex*, *16*, 1040–1055.

Amiez, C., Neveu, R., Warrot, D., Petrides, M., Knoblauch, K., & Procyk, E. (2013). The location of feedback-related activity in the midcingulate cortex is predicted by local morphology. *Journal of Neuroscience*, *33*, 2217–2228.

Amiez, C., & Petrides, M. (2014). Neuroimaging evidence of the anatomo-functional organization of the human cingulate motor areas. *Cerebral Cortex*, *24*, 563–578.

Amiez, C., & Procyk, E. (2019). Midcingulate somatomotor and autonomic functions. In B. A. Vogt (Ed.), *Handbook of clinical neurology* (1st ed., pp. 53–71). Elsevier B.V.

Amiez, C., Sallet, J., Procyk, E., & Petrides, M. (2012). Modulation of feedback related activity in the rostral anterior cingulate cortex during trial and error exploration. *NeuroImage*, *63*, 1078–1090.

Behrens, T. E. J., Woolrich, M. W., Walton, M. E., & Rushworth, M. F. S. (2007). Learning the value of information in an uncertain world. *Nature Neuroscience*, *10*, 1214–1221.

Bernacchia, A., Seo, H., Lee, D., & Wang, X.-J. (2011). A reservoir of time constants for memory traces in cortical neurons. *Nature Neuroscience*, *14*, 366–372.

Bouret, S., & Sara, S. J. (2005). Network reset: A simplified overarching theory of locus coeruleus noradrenaline function. *Trends in Neurosciences*, *28*, 574–582.

Bozkurt, A., Zilles, K., Schleicher, A., Kamper, L., Arigita, E. S., Uylings, H. B. M., et al. (2005). Distributions of transmitter receptors in the macaque cingulate cortex. *NeuroImage*, *25*, 219–229.

Bush, G., Vogt, B. A., Holmes, J., Dale, A. M., Greve, D., Jenike, M. A., et al. (2002). Dorsal anterior cingulate cortex: A role in reward-based decision making. *Proceedings of the National Academy of Sciences of the United States of America*, *99*, 523–528.

Caruana, F., Gerbella, M., Avanzini, P., Gozzo, F., Pelliccia, V., Mai, R., et al. (2018). Motor and emotional behaviours elicited by electrical stimulation of the human cingulate cortex. *Brain*, *24*. 563–17.

Cavanagh, S. E., Towers, J. P., Wallis, J. D., Hunt, L. T., & Kennerley, S. W. (2018). Reconciling persistent and dynamic hypotheses of working memory coding in prefrontal cortex. *Nature Communications*, *9*, 1–16.

Chaudhuri, R., Knoblauch, K., Gariel, M.-A., Kennedy, H., & Wang, X.-J. (2015). A large-scale circuit mechanism for hierarchical dynamical processing in the primate cortex. *Neuron*, *88*, 419–431.

Dehaene, S., Kerszberg, M., & Changeux, J. P. (1998). A neuronal model of a global workspace in effortful cognitive tasks. *Proceedings of the National Academy of Sciences of the United States of America*, *95*, 14529–14534.

Devinsky, O., Morrell, M. J., & Vogt, B. A. (1995). Contributions of anterior cingulate cortex to behaviour. *Brain*, *118*(Pt. 1), 279–306.

Dum, R. P., Levinthal, D. J., & Strick, P. L. (2016). Motor, cognitive, and affective areas of the cerebral cortex influence the adrenal medulla. *Proceedings of the National Academy of Sciences*, 201605044.

Dum, R. P., Levinthal, D. J., & Strick, P. L. (2019). The mind-body problem: Circuits that link the cerebral cortex to the adrenal medulla. *Proceedings of the National Academy of Sciences of the United States of America*, *116*, 26321–26328.

Dum, R. P., & Strick, P. L. (1991). The origin of corticospinal projections from the premotor areas in the frontal lobe. *The Journal of Neuroscience*, *11*, 667–689.

Dum, R. P., & Strick, P. L. (2002). Motor areas in the frontal lobe of the primate. *Physiology & Behavior*, *77*, 677–682.

Duncan, J. (2010). The multiple-demand (MD) system of the primate brain: Mental programs for intelligent behaviour. *Trends in Cognitive Sciences*, *14*, 172–179.

Duncan, J. (2013). The structure of cognition: Attentional episodes in mind and brain. *Neuron*, *80*, 35–50.

Durstewitz, D., Vittoz, N. M., Floresco, S. B., & Seamans, J. K. (2010). Abrupt transitions between prefrontal neural ensemble states accompany behavioral transitions during rule learning. *Neuron*, *66*, 438–448.

Enel, P., Procyk, E., Quilodran, R., & Dominey, P. F. (2016). Reservoir computing properties of neural dynamics in prefrontal cortex. O'Reilly JX, ed *PLoS Computational Biology*, *12*, e1004967.

Enel, P., Wallis, J. D., & Rich, E. L. (2020). Stable and dynamic representations of value in the prefrontal cortex. *eLife*, *9*, 1040.

Engel, T. A., Steinmetz, N. A., Gieselmann, M. A., Thiele, A., Moore, T., & Boahen, K. (2016). Selective modulation of cortical state during spatial attention. *Science*, *354*, 1140–1144.

Falkenstein, M., Hielscher, H., Dziobek, I., Schwarzenau, P., Hoormann, J., Sunderman, B., et al. (2001). Action monitoring, error detection, and the basal ganglia: An ERP study. *Neuroreport*, *12*, 157–161.

Fontanier, V., Sarazin, M., Stoll, F. M., Delord, B., & Procyk, E. (2020). Local inhibitory control of frontal network metastability underlies the temporal signature of cognitive states. *bioRxiv*, *28*, 319.

Gehring, W. J., Goss, B., Coles, M., & Meyer, D. E. (1993). A neural system for error detection and compensation. *Psychological Science*, *4*, 385–390.

Gemba, H., Sasaki, K., & Brooks, V. B. (1986). "Error" potentials in limbic cortex (anterior cingulate area 24) of monkeys during motor learning. *Neuroscience Letters*, *70*, 223–227.

Gruber, T., Zuberbühler, K., & Neumann, C. (2016). Travel fosters tool use in wild chimpanzees. *eLife*, *5*, 93.

Hahn, G., Bujan, A. F., Frégnac, Y., Aertsen, A., & Kumar, A. (2014). Communication through resonance in spiking neuronal networks. Brunel N, ed. *PLoS Computational Biology*, *10*, e1003811–e1003816.

Hanes, D. P., & Schall, J. D. (1996). Neural control of voluntary movement initiation. *Science*, *274*, 427–430.

Harlow, H. F. (1949). The formation of learning sets. *Psychological Review*, *56*, 51–65.

Hayden, B. Y., Pearson, J. M., & Platt, M. L. (2009). Fictive reward signals in the anterior cingulate cortex. *Science*, *324*, 948–950.

Hayden, B. Y., Pearson, J. M., & Platt, M. L. (2011). Neuronal basis of sequential foraging decisions in a patchy environment. *Nature Neuroscience*, *14*, 933–939.

He, S. Q., Dum, R. P., & Strick, P. L. (1995). Topographic organization of corticospinal projections from the frontal lobe: Motor areas on the medial surface of the hemisphere. *The Journal of Neuroscience*, *15*, 3284–3306.

Holroyd, C. B., & Yeung, N. (2012). Motivation of extended behaviors by anterior cingulate cortex. *Trends in Cognitive Sciences*, *16*, 122–128.

Hunt, L. T., Malalasekera, W. M. N., Berker, A. O., Miranda, B., Farmer, S. F., Behrens, T. E. J., et al. (2018). Triple dissociation of attention and decision computations across prefrontal cortex. *Nature Neuroscience*, *21*, 1–17.
Hyman, J. M., Holroyd, C. B., & Seamans, J. K. (2017). A novel neural prediction error found in anterior cingulate cortex ensembles. *Neuron*, *95*, 447–456.e3.
Janmaat, K. R. L., Byrne, R. W., & Zuberbühler, K. (2006). Primates take weather into account when searching for fruits. *Current Biology*, *16*, 1232–1237.
Jueptner, M., Stephan, K. M., Frith, C. D., Brooks, D. J., Frackowiak, R. S., & Passingham, R. E. (1997). Anatomy of motor learning. I. Frontal cortex and attention to action. *Journal of Neurophysiology*, *77*, 1313–1324.
Karlsson, M. P., Tervo, D. G. R., & Karpova, A. Y. (2012). Network resets in medial prefrontal cortex mark the onset of behavioral uncertainty. *Science*, *338*, 135–139.
Kawai, T., Yamada, H., Sato, N., Takada, M., & Matsumoto, M. (2015). Roles of the lateral Habenula and anterior cingulate cortex in negative outcome monitoring and behavioral adjustment in nonhuman primates. *Neuron*, 1–16.
Kawai, T., Yamada, H., Sato, N., Takada, M., & Matsumoto, M. (2019). Preferential representation of past outcome information and future choice behavior by putative inhibitory interneurons rather than putative pyramidal neurons in the primate dorsal anterior cingulate cortex. *Cerebral Cortex*, *29*, 2339–2352.
Kennerley, S. W., Behrens, T. E. J., & Wallis, J. D. (2011). Double dissociation of value computations in orbitofrontal and anterior cingulate neurons. *Nature Neuroscience*, *14*(12), 1581–1589.
Kennerley, S. W., Walton, M. E., Behrens, T. E. J., Buckley, M. J., & Rushworth, M. F. S. (2006). Optimal decision making and the anterior cingulate cortex. *Nature Neuroscience*, *9*, 940–947.
Khamassi, M., Quilodran, R., Enel, P., Dominey, P. F., & Procyk, E. (2015). Behavioral regulation and the modulation of information coding in the lateral prefrontal and cingulate cortex. *Cerebral Cortex*, *25*, 3197–3218.
Kolling, N., Behrens, T. E. J., Mars, R. B., & Rushworth, M. F. S. (2012). Neural mechanisms of foraging. *Science*, *336*, 95–98.
Kolling, N., Behrens, T., Wittmann, M. K., & Rushworth, M. (2016). Science direct multiple signals in anterior cingulate cortex. *Current Opinion in Neurobiology*, *37*, 36–43.
Kolling, N., Wittmann, M. K., Behrens, T. E. J., Boorman, E. D., Mars, R. B., & Rushworth, M. F. S. (2016). Value, search, persistence and model updating in anterior cingulate cortex. *Nature Neuroscience*, *19*, 1280–1285.
Kopell, N., Kramer, M. A., Malerba, P., & Whittington, M. A. (2010). Are different rhythms good for different functions? *Frontiers in Human Neuroscience*, *4*, 1–9.
Kouneiher, F., Charron, S., & Koechlin, E. (2009). Motivation and cognitive control in the human prefrontal cortex. *Nature Neuroscience*, *12*, 939–945.
La Camera, G., Fontanini, A., & Mazzucato, L. (2019). Cortical computations via metastable activity. *Current Opinion in Neurobiology*, *58*, 37–45.
Loh, K. K., Hadj-Bouziane, F., Petrides, M., Procyk, E., & Amiez, C. (2018). Rostro-caudal organization of connectivity between cingulate motor areas and lateral frontal regions. *Frontiers in Neuroscience*, *11*. 13786–17.
Loh, K. K., Procyk, E., Neveu, R., Lamberton, F., Hopkins, W. D., Petrides, M., et al. (2020). Cognitive control of orofacial motor and vocal responses in the ventrolateral and dorsomedial human frontal cortex. *Proceedings of the National Academy of Sciences of the United States of America.*, *117*, 4994–5005.
Mansouri, F. A., Tanaka, K., & Buckley, M. J. (2009). Conflict-induced behavioural adjustment: A clue to the executive functions of the prefrontal cortex. *Nature Reviews. Neuroscience*, *10*, 141–152.

Markov, N. T., Ercsey-Ravasz, M., Van Essen, D. C., Knoblauch, K., Toroczkai, Z., & Kennedy, H. (2013). Cortical high-density counterstream architectures. *Science, 342*, 1238406.

Mars, R. B., Sotiropoulos, S. N., Passingham, R. E., Sallet, J., Verhagen, L., Khrapitchev, A. A., et al. (2018). Whole brain comparative anatomy using connectivity blueprints. *eLife*, 7, 870.

Medalla, M., Gilman, J. P., Wang, J.-Y., & Luebke, J. I. (2017). Strength and diversity of inhibitory signaling differentiates primate anterior cingulate from lateral prefrontal cortex. *The Journal of Neuroscience, 37*, 4717–4734.

Meder, D., Kolling, N., Verhagen, L., Wittmann, M. K., Scholl, J., Madsen, K. H., et al. (2017). Simultaneous representation of a spectrum of dynamically changing value estimates during decision making. *Nature Communications, 8*, 1–11.

Miller, E. K., Lundqvist, M., & Bastos, A. M. (2018). Working memory 2.0. *Neuron, 100*, 463–475.

Miltner, W. H. R., Braun, C. H., & Coles, M. G. H. (2001). Event-related brain potentials following incorrect feedback in a time-estimation task: Evidence for a "generic" neural system for error detection. *Journal of Cognitive Neuroscience, 9*, 788–798. Available at: http://eutils.ncbi.nlm.nih.gov/entrez/eutils/elink.fcgi?dbfrom=pubmed&id=23964600&retmode=ref&cmd=prlinks.

Mitchell, D. J., Bell, A. H., Buckley, M. J., Mitchell, A. S., Sallet, J., & Duncan, J. (2016). A putative multiple-demand system in the macaque brain. *Journal of Neuroscience, 36*, 8574–8585.

Mongillo, G., Rumpel, S., & Loewenstein, Y. (2018). Inhibitory connectivity defines the realm of excitatory plasticity. *Nature Neuroscience, 21*(10), 1463–1470.

Monosov, I. E. (2017). Anterior cingulate is a source of valence-specific information about value and uncertainty. *Nature Communications, 8*(134), 1–12.

Morecraft, R. J., Schroeder, C. M., & Keifer, J. (1996). Organization of face representation in the cingulate cortex of the rhesus monkey. *Neuroreport*, 7, 1343–1348.

Murakami, M., Shteingart, H., Loewenstein, Y., & Mainen, Z. F. (2017). Distinct sources of deterministic and stochastic components of action timing decisions in rodent frontal cortex. *Neuron, 94*. 908–919.e7.

Murray, J. D., Bernacchia, A., Freedman, D. J., Romo, R., Wallis, J. D., Cai, X., et al. (2014). A hierarchy of intrinsic timescales across primate cortex. *Nature Neuroscience, 17*, 1661–1663.

Nácher, V., Hassani, S. A., & Womelsdorf, T. (2019). Asymmetric effective connectivity between primate anterior cingulate and lateral prefrontal cortex revealed by electrical microstimulation. *Brain Structure & Function, 224*, 779–793.

Németh, G., Hegedüs, K., & Molnár, L. (1988). Akinetic mutism associated with bicingular lesions: Clinicopathological and functional anatomical correlates. *European Archives of Psychiatry and Neurological Sciences, 237*, 218–222.

Niki, H., & Watanabe, M. (1979). Prefrontal and cingulate unit activity during timing behavior in the monkey. *Brain Research, 171*, 213–224.

Noser, R., & Byrne, R. W. (2007). Mental maps in chacma baboons (Papio ursinus): Using inter-group encounters as a natural experiment. *Animal Cognition, 10*, 331–340.

Palomero-Gallagher, N., Mohlberg, H., Zilles, K., & Vogt, B. (2008). Cytology and receptor architecture of human anterior cingulate cortex. *The Journal of Comparative Neurology, 508*, 906–926.

Palomero-Gallagher, N., Vogt, B. A., Schleicher, A., Mayberg, H. S., & Zilles, K. (2009). Receptor architecture of human cingulate cortex: Evaluation of the four-region neurobiological model. *Human Brain Mapping, 30*, 2336–2355.

Parvizi, J., Rangarajan, V., Shirer, W. R., Desai, N., & Greicius, M. D. (2013). The will to persevere induced by electrical stimulation of the human cingulate gyrus. *Neuron, 80*, 1359–1367.

Passingham, R. E. (1996). Attention to action. *Philosophical Transactions of the Royal Society of London. Series B, Biological Sciences, 351*, 1473–1479.

Passingham, R. E., & Wise, S. P. (2012). *The neurobiology of the prefrontal cortex: Anatomy, evolution, and the origin of insight.* OUP Oxford.

Pearson, J. M., Watson, K. K., & Platt, M. L. (2014). Decision making: The neuroethological turn. *Neuron, 82*, 950–965.

Petrides, M., & Pandya, D. N. (1994). Comparative architectonic analysis of the human and the macaque frontal cortex. In F. Boller, & J. Grafman (Eds.), *Handbook of neuropsychology* (pp. 17–58). Amsterdam.

Pettine, W. W., Louie, K., Murray, J. D., & Wang, X.-J. (2020). Hierarchical network model excitatory-inhibitory tone shapes alternative strategies for different degrees of uncertainty in multi-attribute decisions. *bioRxiv*. https://doi.org/10.1101/2020.01.28.923649.

Picard, N., & Strick, P. L. (1996). Motor areas of the medial wall: A review of their location and functional activation. *Cerebral Cortex, 6*, 342–353.

Powell, N. J., & Redish, A. D. (2016). Representational changes of latent strategies in rat medial prefrontal cortex precede changes in behaviour. *Nature Communications*, 7, 12830.

Premereur, E., Janssen, P., & Vanduffel, W. (2018). Functional MRI in macaque monkeys during task switching. *Journal of Neuroscience, 38*, 10619–10630.

Procyk, E., & Joseph, J.-P. (2001). Characterization of serial order encoding in the monkey anterior cingulate sulcus. *The European Journal of Neuroscience, 14*, 1041–1046.

Procyk, E., Tanaka, Y. L., & Joseph, J.-P. (2000). Anterior cingulate activity during routine and non-routine sequential behaviors in macaques. *Nature Neuroscience, 3*, 502–508.

Procyk, E., Wilson, C. R. E., Stoll, F. M., Faraut, M. C. M., Petrides, M., & Amiez, C. (2016). Midcingulate motor map and feedback detection: Converging data from humans and monkeys. *Cerebral Cortex, 26*, 467–476.

Quilodran, R., Rothé, M., & Procyk, E. (2008). Behavioral shifts and action valuation in the anterior cingulate cortex. *Neuron, 57*, 314–325.

Rich, E. L., & Wallis, J. D. (2016). Decoding subjective decisions from orbitofrontal cortex. *Nature Neuroscience, 19*, 973–980.

Rothé, M., Quilodran, R., Sallet, J., & Procyk, E. (2011). Coordination of high gamma activity in anterior cingulate and lateral prefrontal cortical areas during adaptation. *Journal of Neuroscience, 31*, 11110–11117.

Sallet, J., Mars, R. B., Quilodran, R., Procyk, E., Petrides, M., & Rushworth, M. F. (2011). Neuroanatomical basis of motivational and cognitive control: A focus on the medial and lateral prefrontal cortex. In R. B. Mars, J. Sallet, M. F. S. Rushworth, & N. Yeung (Eds.), *Neural basis of motivational and cognitive control* The MIT Press.

Sallet, J., Quilodran, R., Rothé, M., Vezoli, J., Joseph, J.-P., & Procyk, E. (2007). Expectations, gains, and losses in the anterior cingulate cortex. *Cognitive, Affective, & Behavioral Neuroscience*, 7, 327–336.

Schaeffer, D. J., Hori, Y., Gilbert, K. M., Gati, J. S., Menon, R. S., & Everling, S. (2020). Divergence of rodent and primate medial frontal cortex functional connectivity. *Proceedings of the National Academy of Sciences of the United States of America, 5*, 202003181–202021689.

Scholl, J., Kolling, N., Nelissen, N., Stagg, C. J., & Harmer, C. J. (2017). Excitation and inhibition in anterior cingulate predict use of past experiences. *Elife, 6*, e20365. https://doi.org/10.7554/eLife.20365.

Seo, H., Barraclough, D. J., & Lee, D. (2007). Dynamic signals related to choices and outcomes in the dorsolateral prefrontal cortex. *Cerebral Cortex, 17*(Suppl. 1), i110–i117.

Seo, H., & Lee, D. (2007). Temporal filtering of reward signals in the dorsal anterior cingulate cortex during a mixed-strategy game. *Journal of Neuroscience, 27*, 8366–8377.

Shen, C., Ardid, S., Kaping, D., Westendorff, S., Everling, S., & Womelsdorf, T. (2015). Anterior cingulate cortex cells identify process-specific errors of attentional control prior to transient prefrontal-cingulate inhibition. *Cerebral Cortex*, *25*, 2213–2228.

Sheth, S. A., Mian, M. K., Patel, S. R., Asaad, W. F., Williams, Z. M., Dougherty, D. D., et al. (2012). Human dorsal anterior cingulate cortex neurons mediate ongoing behavioural adaptation. *Nature*, *488*, 218–221.

Shidara, M., & Richmond, B. J. (2002). Anterior cingulate: Single neuronal signals related to degree of reward expectancy. *Science*, *296*, 1709–1711.

Shima, K., & Tanji, J. (1998). Role for cingulate motor area cells in voluntary movement selection based on reward. *Science*, *282*, 1335–1338.

Silvetti, M., Vassena, E., Abrahamse, E., & Verguts, T. (2018). Dorsal anterior cingulate-brainstem ensemble as a reinforcement meta-learner. O'Reilly J, ed. *PLoS Computational Biology*, *14*. e1006370–32.

Spitmaan, M., Seo, H., Lee, D., & Soltani, A. (2020). Multiple timescales of neural dynamics and integration of task-relevant signals across cortex. *Proceedings of the National Academy of Sciences of the United States of America*, *21*. 202005993–10.

Stoll, F. M., Fontanier, V., & Procyk, E. (2016). Specific frontal neural dynamics contribute to decisions to check. *Nature Communications*, 7, 11990.

Sul, J. H., Kim, H., Huh, N., Lee, D., & Jung, M. W. (2010). Distinct roles of rodent orbitofrontal and medial prefrontal cortex in decision making. *Neuron*, *66*, 449–460.

Talairach, J., Bancaud, J., Geier, S., Bordas-Ferrer, M., Bonis, A., Szikla, G., et al. (1973). The cingulate gyrus and human behaviour. *Electroencephalography and Clinical Neurophysiology*, *34*, 45–52.

Tervo, D., Kuleshova, E., Manakov, M., Proskurin, M., Karlsson, M., Lustig, A., et al. (2020). Anterior cingulate cortex directs exploration of alternative strategies. *bioRxiv*. https://doi.org/10.1101/2020.05.23.098822.

Tervo, D. G. R., Proskurin, M., Manakov, M., Kabra, M., Vollmer, A., Branson, K., et al. (2014). Behavioral Variabilitythrough stochastic choice and its gating by anterior cingulate cortex. *Cell*, *159*, 21–32.

Tremblay, R., Lee, S., & Rudy, B. (2016). GABAergic interneurons in the neocortex: From cellular properties to circuits. *Neuron*, *91*, 260–292.

Vogt, B. A. (2016). Midcingulate cortex: Structure, connections, homologies, functions and diseases. *Journal of Chemical Neuroanatomy*, *74*, 28–46.

Vogt, B. A., Nimchinsky, E. A., Vogt, L. J., & Hof, P. R. (1995). Human cingulate cortex: Surface features, flat maps, and cytoarchitecture. *The Journal of Comparative Neurology*, *359*, 490–506.

Vogt, B. A., & Paxinos, G. (2012). Cytoarchitecture of mouse and rat cingulate cortex with human homologies. *Brain Structure & Function*, *219*, 185–192.

Voloh, B., & Womelsdorf, T. (2018). Cell-type specific burst firing interacts with theta and beta activity in prefrontal cortex during attention states. *Cerebral Cortex*, *28*, 4348–4364.

Walsh, M. M., & Anderson, J. R. (2011). Modulation of the feedback-related negativity by instruction and experience. *Proceedings of the National Academy of Sciences of the United States of America*, *108*, 19048–19053.

Wang, X.-J. (2020). Macroscopic gradients of synaptic excitation and inhibition in the neocortex. *Nature Reviews. Neuroscience*, 1–10.

Wasmuht, D. F., Spaak, E., Buschman, T. J., Miller, E. K., & Stokes, M. G. (2018). Intrinsic neuronal dynamics predict distinct functional roles during working memory. *Nature Communications*, *9*, 1–13.

White, J. K., Bromberg-Martin, E. S., Heilbronner, S. R., Zhang, K., Pai, J., Haber, S. N., et al. (2019). A neural network for information seeking. *Nature Communications*, *10*, 1–19.

Williams, Z. M., Bush, G., Rauch, S. L., Cosgrove, G. R., & Eskandar, E. N. (2004). Human anterior cingulate neurons and the integration of monetary reward with motor responses. *Nature Neuroscience*, *7*, 1370–1375.
Wilson, C. R. E., Vezoli, J., Stoll, F. M., Faraut, M. C. M., Leviel, V., Knoblauch, K., et al. (2016). Prefrontal markers and cognitive performance are dissociated during progressive dopamine lesion. *PLoS Biology*, *14*, e1002576.
Womelsdorf, T., Valiante, T. A., Sahin, N. T., Miller, K. J., & Tiesinga, P. (2014). Dynamic circuit motifs underlying rhythmic gain control, gating and integration. *Nature Neuroscience*, *17*, 1031–1039.
Zilles, K., & Palomero-Gallagher, N. (2017). Multiple transmitter receptors in regions and layers of the human cerebral cortex. *Frontiers in Neuroanatomy*, *11*, e1000489–26.

CHAPTER FIFTEEN

Medial prefrontal cortex and the temporal control of action

Qiang Zhang, Matthew A. Weber, and Nandakumar S. Narayanan*
Department of Neurology, University of Iowa, Iowa City, IA, United States
*Corresponding author: e-mail address: nandakumar-narayanan@uiowa.edu

Contents

Abstract

Across species, the medial prefrontal cortex guides actions in time. This process can be studied using behavioral paradigms such as simple reaction-time and interval-timing tasks. Temporal control of action can be influenced by prefrontal neurotransmitters such as dopamine and acetylcholine and is highly relevant to human diseases such as Parkinson's disease, schizophrenia, and attention-deficit hyperactivity disorder (ADHD). We review evidence that across species, medial prefrontal lesions impair the temporal control of action. We then consider neurophysiological correlates in humans, primates, and rodents that might encode temporal processing and relate to cognitive-control mechanisms. These data have informed brain-stimulation studies in rodents and humans that can compensate for timing deficits. This line of work illuminates basic mechanisms of temporal control of action in the medial prefrontal cortex, which underlies a range of high-level cognitive processing and could contribute to new biomarkers and therapies for human brain diseases.

1. Deciding *when* to act

Animal behavior—or the complex symphony of coordinated action in the world—is precisely sequenced in time (Finnerty, Shadlen, Jazayeri, Nobre, & Buonomano, 2015). Although perceptual events can indicate potential

International Review of Neurobiology, Volume 158
ISSN 0074-7742
https://doi.org/10.1016/bs.irn.2020.11.004

rewards or threats, temporal control of actions over several seconds is required to maximize positive and minimize negative outcomes (Tallot & Doyère, 2020). For mammals, this is critical to foraging and successfully avoiding predators (Bateson, 2003). For humans, temporal control of action is critical for routine activities such as cooking, driving, and crossing the street, and complex activities such as math, language, poetry, and art. Perceptual decision making has been the focus of many neuroscientific studies (Gold & Shadlen, 2007). However, deciding *when* to act is as important as selecting appropriate actions; indeed, the right action at the wrong time can lead to disastrous consequences. Thus, understanding the underlying neural circuitry involved in temporal control of action has far-reaching importance (Fuster, 2001).

The timing of movements is critical to higher-level cognition. Indeed, temporal control of action underlies many higher-level executive functions such as attention, flexible behavior, reasoning, and working memory. Dual-task studies in which participants perform executive functions concurrently with timing operations reveal marked bidirectional interference; i.e., performing a time production and randomization task concurrently degrades performance of either task (Brown, 2006). Estimating a temporal interval can impair concurrent working-memory performance (Bi, Yuan, & Huang, 2013). These data imply that temporal control of action and executive functions recruit shared cognitive resources and neural circuits (Parker, Lamichhane, Caetano, & Narayanan, 2013).

We argue that the medial prefrontal cortex critically controls movements in time. We focus on medial regions of prefrontal cortex that are conserved across mammalian species. Medial prefrontal regions of frontal cortex have been defined by projections from thalamic nuclei and a lack of a prominent cortical layer IV (i.e., "agranular"; Akert, 1964; Rose & Woolsey, 1949). Lateral prefrontal human specializations in dorsal and inferior frontal regions have been implicated in specialized functions such as working-memory and language (Fuster, 2008). While these lateral regions have unclear mammalian homologies in rodents, primate medial prefrontal regions can have anatomical and functional homologies with rodents (Laubach, Amarante, Swanson, & White, 2018; Preuss, 1995; Uylings, Groenewegen, & Kolb, 2003). Although there are differences between primate and rodent nomenclature because the granular lateral prefrontal cortex does not exist in rodents, the human/primate agranular anterior cingulate cortex covering Brodmann areas 24 and 32 has anatomical homologies to rodent medial prefrontal regions including the anterior cingulate cortex and the prelimbic cortex (Fig. 1) (Laubach et al., 2018). For instance, both the human and rodent medial prefrontal cortex have a similar laminar architecture, receive similar neurotransmitter input, and have highly conserved

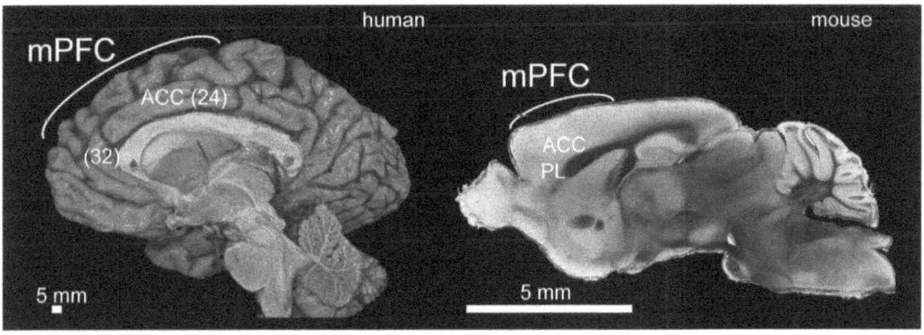

Fig. 1 Medial prefrontal cortex in humans and rodents: Sagittal brain section of the medial prefrontal cortex from humans and brightfield microscopy from the mouse. mPFC—medial prefrontal cortex; ACC—anterior cingulate; PL—prelimbic. Brodmann areas in parentheses. Human gross anatomy courtesy of Marco Hefti and the Iowa Brain Bank.

connectivity with thalamic, striatal, hypothalamic and brainstem nuclei (Gabbott, Warner, Jays, & Bacon, 2003; Gabbott, Warner, Jays, Salway, & Busby, 2005). Although medial prefrontal regions are vastly expanded in primates and humans and there are important differences between cingulate, prelimbic, and supplementary motor regions (Amiez & Procyk, 2019), in this chapter, we argue that across species, medial prefrontal regions have a significant functional homology: the temporal control of action. We start by discussing paradigms to study temporal control of action and then discuss medial prefrontal neurotransmitter systems involved in temporal control of action. We then review diseases with impairments in temporal control of action and lesion studies that disrupt the medial prefrontal cortex and impair timing of movements. Finally, we discuss both neuronal and local field potential correlates of how movements are controlled in time and then relate these signals to protocols that deliver brain stimulation that improves temporal control of action.

2. Paradigms for studying the temporal control of action

Time suffuses all paradigms and tasks in behavioral neuroscience, which generally require a specific action in a narrow temporal window. In human studies, the timing of movement is critical to inferences about cognitive processing. For instance, reaction-time linearly increases with the number of working-memory items (Narayanan et al., 2005; Sternberg, 1969) and can be harnessed in perceptual decision-making to capture trade-offs between speed and accuracy (Herz et al., 2017) or countermanding (Verbruggen et al., 2019).

However, the most elementary response in simple-reaction time tasks can be profoundly affected by temporal processing. To "control" for time in responding to a stimulus, many reaction-time tasks included a variable delay or "foreperiod" between a warning cue and the imperative stimulus (Luce, 1991). Studies with the variable delay revealed that human—and non-human participants—were more prepared to respond the longer they waited, leading to delay-dependent speeding (or "foreperiod-effect"; Fig. 2A) (Naatanen, 1971, 1972; Naatanen, Muranen, & Merisalo, 1974; Narayanan, Horst, & Laubach, 2006). Furthermore, response distributions could be influenced by the temporal probabilities of the task; that is, if a short

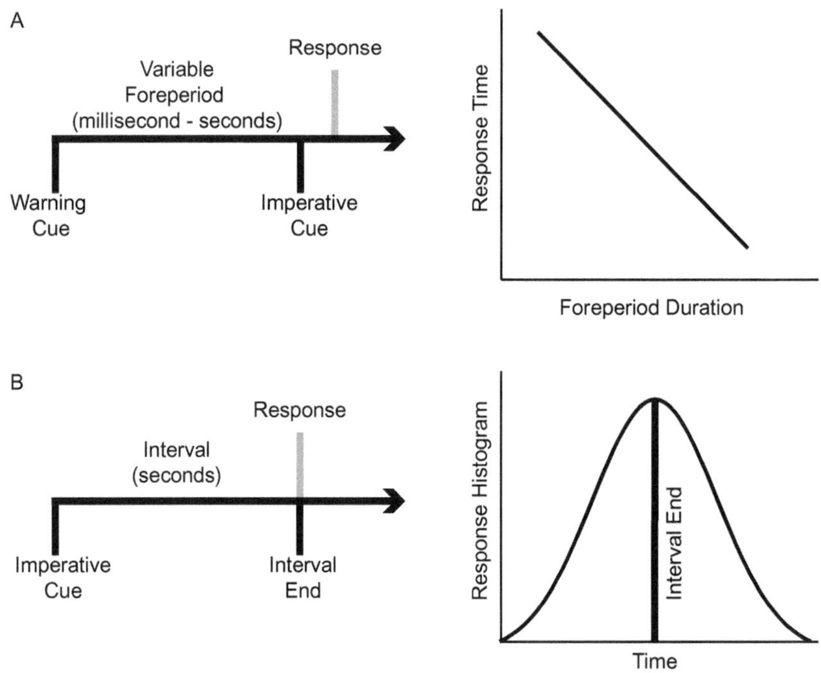

Fig. 2 Paradigms to study temporal control of action: Temporal control of action can be studied using two paradigms. (A) The simple reaction-time task requires participants to respond as fast as possible to an imperative cue. A warning cue is typically delivered at a variable interval prior to the imperative cue; the time between the warning and imperative cue is called "foreperiod" or "delay." Participants' reaction-times are higher for short foreperiods compared to long foreperiods, when they are fully prepared to respond. This phenomenon is called the "foreperiod-effect" or "delay-dependent speeding." Another paradigm is (B) interval-timing, which requires participants to estimate an interval of several seconds by making a motor response. In this task, the imperative cue indicates the start of the interval. The timing of responses is typically distributed around the end of the interval.

delay was more likely, response times would decrease (Naatanen et al., 1974). These data suggest that actions are profoundly influenced by the temporal probabilities as specified by the hazard function (Luce, 1991).

Another paradigm to explicitly study temporal control of action is interval timing (Fig. 2B) (Merchant & de Lafuente, 2014). This family of tasks requires participants to estimate an interval of several seconds with a motor response. Although interval timing was designed to capture how time is "perceived," these tasks require a motor report of temporal estimates and thus are inherently tied to action (outside of bisection tasks (Stubbs, 1976)). A beautiful aspect of these tasks is that they are conserved across many species, including mammals and birds (Buhusi & Meck, 2005) and can be performed in a broad range of conditions, from rodent operant behavior to human patients with severe brain diseases or with interoperative human neurophysiology (Carroll, O'Donnell, Shekhar, & Hetrick, 2009; Craig, Varnon, Sokolowski, Wells, & Abramson, 2014; Kelley et al., 2018). Temporal control of action may influence more complex motor sequences or language (Averbeck, Latham, & Pouget, 2006; Long et al., 2016); however, we will focus on reaction-time and interval-timing tasks as they are (1) well-suited to isolate temporal control of action from other processes, (2) they are simple enough to study in rodents and humans with cognitive impairments, and (3) complementary evidence across species facilitates convergent inferences on medial prefrontal function.

3. Neurotransmitter systems

Human and rodent prefrontal regions receive dense innervation from ascending neurotransmitter systems such as dopamine, acetylcholine, norepinephrine, and serotonin (Narayanan, Rodnitzky, & Uc, 2013). Human brain diseases and their treatments involve perturbations of these neurotransmitter systems (Arnsten, Wang, & Paspalas, 2012). Prefrontal norepinephrine and serotonin can have specific effects on goal-directed behaviors, but specific evidence for their role in temporal control of action is lacking. Conversely, dopamine and acetylcholine reliably regulate the timing of movements.

Primate prefrontal dopamine-depletion impairs behaviors across a delay (Brozoski, Brown, Rosvold, & Goldman, 1979; Goldman-Rakic, 1998). Our group has found specific effects during timing tasks. During simple reaction-time tasks, we found that rodent prefrontal dopamine-depletion and focal infusions of drugs blocking the D1-type dopamine receptors

(D1DRs) blunted delay-dependent speeding with faster responses at short intervals. However, we did not observe increased premature responding (Parker, Alberico, Miller, & Narayanan, 2013). During interval-timing tasks, we found that disrupting prefrontal dopamine by specific RNAi against tyrosine hydroxylase or dopamine depletion flattened time-response histograms (Kim et al., 2017; Narayanan, Land, Solder, Deisseroth, & DiLeone, 2012; Parker, Chen, Kingyon, Cavanagh, & Narayanan, 2015). Focal pharmacology has implicated prefrontal D1DRs but not D2DRs in temporal control of action (Narayanan et al., 2012; Parker, Chen, et al., 2015; Parker, Chen, Kingyon, Cavanagh, & Narayanan, 2014). Critically, transgenic mouse lines facilitate cell-type-specific experiments, and we have shown that optogenetic inactivation of prefrontal D1DR+ neurons in the mouse medial prefrontal cortex decreased responding at the end of a fixed-interval (Narayanan et al., 2012). These data provide convergent and consistent evidence that prefrontal D1DRs are required for the temporal control of action.

Cholinergic systems also play a clear role in timing tasks. For instance, systemic disruptions of acetylcholine with scopolamine (a muscarinic antagonist) or mecamylamine (a nicotinic antagonist) reliably impair reaction-time tasks (Jones & Higgins, 1995). Cholinergic signaling can affect prefrontal firing (Gill, Sarter, & Givens, 2000; Parikh & Sarter, 2008). These agents also impair interval timing; however, we found that scopolamine impairs stimulus-processing but not temporal processing by prefrontal neurons (Zhang, Jung, Larson, Kim, & Narayanan, 2019). These data may be consistent with a view that cholinergic systems are important for attention rather than the temporal control of action (Klinkenberg & Blokland, 2010; Sarter & Bruno, 1997). Indeed, all timing tasks require attention to the passage of time as well as imperative cues. Future work with specific manipulations of cholinergic circuits and localized manipulations in the prefrontal cortex might clarify how acetylcholine modulates prefrontal circuits.

4. Significance for human disease

The timing of movement provides a unique window into human brain diseases that affect the prefrontal cortex. Because it can be rapidly and objectively assessed, assays built on studying the timing of movement have the potential to contribute to novel biomarkers, diagnostics tools, and intraoperative assays.

Patients with Parkinson's disease have slow movements. This "bradykinesia" slows all reaction times (Evarts, Teräväinen, & Calne, 1981); nonetheless, Parkinson's disease patients have intact delay-dependent speeding (Jurkowski, Stepp, & Hackley, 2005; Wearden et al., 2008). By contrast, Parkinson's disease patients have impaired interval timing as a function of levodopa (Jones & Jahanshahi, 2014; Malapani et al., 1998; Malapani, Deweer, & Gibbon, 2002). When multiple intervals are presented, patients with Parkinson's disease are not simply slow; rather, they have increased variability by overestimating short intervals and underestimating long intervals. These deficits may reflect, in part, deficits in executive functions in Parkinson's disease patients (Narayanan, Rodnitzky, & Uc, 2013; Parker, Lamichhane, et al., 2013, p. 201).

Another disease with profound prefrontal abnormalities is schizophrenia (Abi-Dargham et al., 2002; Andreasen et al., 1997; Okubo et al., 1997). Like Parkinson's disease, schizophrenia does not consistently involve delay-dependent speeding but does involve interval-timing impairments (Ward, Kellendonk, Kandel, & Balsam, 2011). Indeed, patients with schizophrenia are more variable and less efficient in their temporal estimates but do not reliably under- or overestimate an interval compared to healthy controls (Carroll et al., 2009; Parker et al., 2017; Thoenes & Oberfeld, 2017). Interestingly, this deficit extends to first-degree relatives (Penney, Meck, Roberts, Gibbon, & Erlenmeyer-Kimling, 2005). To date, most studies include patients undergoing therapeutic interventions and little work has been done to explore the role of specific medications in patients, meaning that these interval timing deficits may be more significant than reported. This is particularly important as first-line interventions are thought to target the medial prefrontal dopamine system. While there are inconsistencies in the literature and medication status may play a role; taken together, these data suggest that medial prefrontal dysfunction drives timing deficits in schizophrenia.

By contrast, attention-deficit hyperactivity disorder (ADHD) patients have uniform deficits in timing (Barkley, Koplowitz, Anderson, & McMurray, 1997). For instance, patients with ADHD have delay-dependent deficits during reaction-time tasks (Leth-Steensen, King Elbaz, & Douglas, 2000) as well interval timing deficits (Smith, Taylor, Rogers, Newman, & Rubia, 2002). This interval timing deficit includes both consistent increases in timing variability and the overestimation of perceived time (Hove, Gravel, Spencer, & Valera, 2017; Rommelse et al., 2008). Unlike schizophrenia-related timing deficits, several studies have examined the impact of first-line interventions for ADHD. For example, medications that primarily act on the medial prefrontal dopamine system, such as methylphenidate, have been shown to normalize both timing behavior and prefrontal function in ADHD (Hart, Radua, Mataix-Cols, & Rubia, 2012; Luman, Papanikolau, & Oosterlaan, 2015). Furthermore, the medial as well as lateral prefrontal regions are dysfunctional in ADHD (Bush, 2011). These data suggest that timing dysfunction may be linked to medial prefrontal dysfunction (Bluschke, Schuster, Roessner, & Beste, 2018; Hart et al., 2012; Noreika, Falter, & Rubia, 2013; Valera et al., 2010).

A broad range of other neurological and psychiatric illness can degrade prefrontal circuits, including Alzheimer's disease and Huntington's disease (Caselli, Iaboli, & Nichelli, 2009; Hinton et al., 2007). More work needs to be done to clarify the role of medial prefrontal networks in these diseases. Nonetheless, this line of work in Parkinson's disease, schizophrenia and ADHD firmly establishes the clinical relevance of understanding how medial prefrontal regions guide actions in time. Thus, a detailed understanding of these circuits will not only have far-reaching basic-science significance but could lead to new biomarkers, diagnostic tools and neuromodulation therapies for human brain diseases.

5. Lesion studies

The critical nature of the prefrontal cortex in acting over a delay was identified by Carlyle Jacobsen, who trained macaque monkeys on delayed response and delayed-alternation tasks (Jacobsen, 1935; Jacobsen & Nissen, 1937). Animals with extensive bilateral prefrontal lesions had permanent deficits at a variety of delays, but animals with unilateral lesions or with lesions of premotor or frontal motor areas had no deficits. Although not formally measured, Jacobsen insightfully concluded that frontal lesions in humans and primates affected the "temporal patterning of movement." This finding was replicated by subsequent primate lesion studies that mapped lesions to lateral and medial regions of the baboon prefrontal cortex (Pribram, 1950; Pribram, Mishkin, Rosvold, & Kaplan, 1952) and then extended to working memory (Mishkin & Pribram, 1955; Pribram & Mishkin, 1956). Although we are unaware of medial frontal inactivation or lesions during reaction-time tasks, primate medial frontal inactivation increases timing variability during a time-production task (Wang, Narain, Hosseini, & Jazayeri, 2018).

Human lesion studies from the laboratory of Donald Stuss definitively implicate human medial prefrontal regions in the temporal control of action. These authors found that superior medial prefrontal lesions spanning Brodmann Areas 24 and 32 slowed reaction-times. By contrast, other prefrontal regions did not produce consistent effects on overall reaction-time, but right lateral prefrontal lesions blunted delay-dependent speeding and prevented speeding of responses at longer intervals (Stuss et al., 2005). Subsequent studies revealed that human prefrontal lesions consistently increased the variability of responses during time-reproduction (Picton et al., 2006; Picton, Stuss, Shallice, Alexander, & Gillingham, 2006). Taken together, these unique human data connect human medial prefrontal areas to the temporal control of action.

In rodents, medial prefrontal regions are reliably involved in guiding movements in time. Excitotoxic prelimbic/infralimbic lesions that spared cingulate regions increased premature responding during simple-reaction time performance in rats; fascinatingly, this premature responding recovered in several days (Risterucci, Terramorsi, Nieoullon, & Amalric, 2003). Work by Narayanan and Laubach that reversibly inactivated dorsal prelimbic/ventral anterior cingulate regions replicated the increased premature responding and decreased delay-dependent speeding and found that animals had faster responses at short intervals (Narayanan et al., 2006; Narayanan & Laubach, 2006). Strikingly, prefrontal inactivation also impaired the ability to optimize responses after errors and improve performance (Narayanan & Laubach, 2008).

During interval-timing, rodent medial prefrontal disruptions impaired selection of a long temporal interval during a maze-running temporal bisection task (Kim, Jung, Byun, Jo, & Jung, 2009). Prefrontal cooling of superficial areas slowed temporal estimates in a time-reproduction task (Namboodiri & Hussain Shuler, 2014; Xu, Zhang, Dan, & Poo, 2014). Our group has reliably shown that reversible prefrontal inactivation impairs fixed-interval timing by "flattening" peaks of time-response histograms, resulting in responses that are more random in time (Emmons et al., 2017; Narayanan et al., 2012; Parker et al., 2014). During peak-interval timing, which allows estimates of both the start and the stop of temporal estimates, medial prefrontal inactivation increased the variability in a dose-dependent fashion (Buhusi, Reyes, Gathers, Oprisan, & Buhusi, 2018). These data provide support for the idea that medial prefrontal lesions reliably increase the variability of how movements are controlled in time. In reaction-time tasks where response distributions are highly skewed, one form of this increased variability might include increased premature or slow responses.

6. Neuronal correlates

Recording from single neurons and networks can provide insights into how movements are guided in time (Paton & Buonomano, 2018). Although these patterns can be diverse, and representation of impending movements can be ubiquitous in frontal cortex, recordings from our group and others have identified a common theme in temporal control of action.

This pattern is characterized by "ramping"—or monotonic changes in firing rate over a temporal interval—and is a key form of temporal processing (Narayanan, 2016). In seminal recording studies by Niki and Watanabe, anticipatory patterns of neuronal activity (i.e., ramping activity) were the most common pattern (~25%) during timing and delayed-response tasks (Niki & Watanabe, 1976, 1979). These neurons were observed in the cingulate cortex and more common than sustained activity (~20%). While sustained activity may be related to working-memory (Goldman-Rakic, 1996; Niki & Watanabe, 1976), this pattern of activity may also represent "time estimation."

We have found that during timing tasks, ramping activity is the most common pattern of activity in the medial prefrontal cortex (Fig. 3A and B) (Narayanan, 2016). This can be quantified by data-driven methods such as

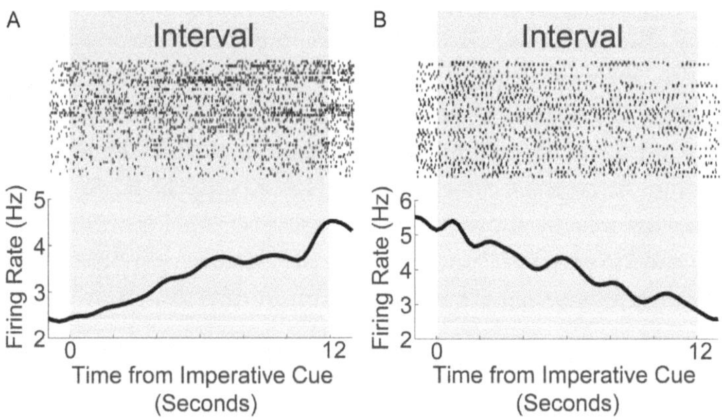

Fig. 3 Ramping neurons in medial prefrontal cortex during a 12-s fixed-interval timing task. Peri-event rasters of mouse medial prefrontal single neurons during a 12-s fixed-interval timing task. Ramping neurons can exhibit (A) monotonic increases or (B) monotonic decreases in firing over the fixed interval. *Data from recordings described in Zhang, Q., Jung, D., Larson, T., Kim, Y., Narayanan, N. S. (2019). Scopolamine and medial frontal stimulus-processing during interval timing.* Neuroscience, *414, 219–227.*

principal component analysis or by linear regression (Narayanan & Laubach, 2006, 2009). Linear patterns are a major player in higher-order representation such as neuronal manifolds (Wang et al., 2018). During interval-timing tasks, this pattern is also commonly observed (Matell, Meck, & Nicolelis, 2003; Xu et al., 2014), and our work has found that neurons anticipating the interval end are the most common pattern of activity in mice and rats, explaining ~30% of prefrontal variance (range 15%–51%) (Emmons et al., 2017; Emmons, Kennedy, Kim, & Narayanan, 2019; Kim et al., 2017; Narayanan & Laubach, 2009; Parker et al., 2014; Parker, Ruggiero, & Narayanan, 2015). In humans, scalp electrical potentials can steadily depolarize until a voluntary action and represent temporal signals (Kononowicz & van Rijn, 2014; Shibasaki & Hallett, 2006). Notably, human single-unit recordings also reveal ramping patterns of activity (Fu et al., 2019; Kamiński et al., 2017; Sheth et al., 2012).

Ramping activity is notable for several reasons. First, the temporal code is readily evident in that the firing at any given moment represents elapsed time and lends itself to quantitative models based on drift-diffusion dynamics (Luzardo, Ludvig, & Rivest, 2013; Simen, Balci, de Souza, Cohen, & Holmes, 2011). These models predict that ramping activity "scales" among prefrontal ensembles to encode short or long intervals and correlates with responses without predicting specific movements (Emmons et al., 2017; Wang et al., 2018). Second, linear models can explain high amounts of variance in ramping activity, although one study reported a logarithmic component to ramping activity during a maze-running task (Kim, Ghim, Lee, & Jung, 2013). While higher order manifolds or multivariate methods may capture additional population encoding (Paton & Buonomano, 2018; Remington, Egger, Narain, Wang, & Jazayeri, 2018; Wang et al., 2018), linear features of ramping activity are important because they are predicted by cellular and behavioral models (Durstewitz, 2003; Simen et al., 2011) and might readily be represented by downstream brain areas. In addition, the degree to which the system is linear might facilitate powerful mathematical tools to understand complex neuronal dynamics. Third, pharmacological manipulations such as dopamine-depletion and D1DR-receptor manipulations that impair temporal control of action also impair ramping patterns of activity without affecting other prefrontal patterns of activity (Parker et al., 2014; Parker, Chen, et al., 2015; Parker, Ruggiero, & Narayanan, 2015). Furthermore, inactivating medial prefrontal cortex disrupts ramping activity in other areas, such as the dorsomedial striatum (Emmons et al., 2017). Fourth, time can be robustly decoded from ramping signals, although non-ramping signals can also contribute (Emmons et al., 2017). Taken together, these data make the compelling case that ramping activity is closely linked to the temporal control of action. Notably, other patterns of neuronal activity can encode temporal features (Karmarkar & Buonomano, 2007; Laje & Buonomano, 2013; Matell & Meck, 2004), but in our hands, ramping activity explains the majority of variance and correlates with behavioral manipulations.

In our work, we argue that medial prefrontal ramping activity is related to guiding upcoming movements in time rather than controlling specific movements. Our premise is supported by three lines of evidence: (1) ramping is observed when animals are holding a lever above a force threshold and not moving (Narayanan & Laubach, 2006, 2009), (2) ramping patterns of neuronal activity are prominent prior to the initial response during interval timing tasks (Emmons et al., 2017), (3) ramping activity persists even when movement-related activity is explicitly accounted for in regression models (Emmons et al., 2017, 2019). Combined with microstimulation studies showing that medial prefrontal stimulation does not reliably evoke movements, and anatomical studies suggesting that medial prefrontal cortex does not have strong links to cortical or subcortical motor nuclei (Gabbott et al., 2005; Neafsey et al., 1986), these data suggest that medial prefrontal cortex plays a key role in planning movements in time.

At the macro-level, we have consistently observed bursts of midfrontal 4 Hz activity in "delta" or "theta" bands during timing tasks (Kim et al., 2017; Narayanan, Cavanagh, Frank, & Laubach, 2013; Parker et al., 2014; Parker, Chen, et al., 2015). These bursts originate from the human medial prefrontal cortex (Kelley et al., 2018) and are remarkable because they are highly conserved in humans and rodents and depend on dopamine in both species (Kim et al., 2017; Narayanan, Cavanagh, et al., 2013; Parker, Chen, et al., 2015). Midfrontal 4 Hz activity is a mechanism of cognitive control and is phase-locked to salient events such as novelty, errors, feedback, and to imperative stimuli during timing tasks (Cavanagh & Frank, 2014). Timing-related midfrontal 4 Hz oscillations can be dysfunctional in diseases with impaired temporal control of action such as Parkinson's disease and schizophrenia (Kim et al., 2017; Parker et al., 2017; Parker, Chen, et al., 2015). Because these oscillations can be specifically coherent with distant brain areas and with prefrontal neurons involved in temporal and error-related processing (Kim & Narayanan, 2018; Narayanan, Cavanagh, et al., 2013; Parker et al., 2014; Parker, Chen, et al., 2015), our working model is that these oscillations engage prefrontal ramping neurons instantiating temporal control of action. Accordingly, midfrontal 4 Hz power is a cognitive control signal that represents the "starting gun" that initiates temporal processing in timing models (Church, 2003).

7. Stimulation

Preceding lines of evidence from lesion, pharmacology and neurophysiology suggest that there might be opportunities to modulate the temporal control of action, particularly when it is dysfunctional. In addition to the basic-science significance of this manipulation, brain stimulation has the possibility to deliver therapies for human brain diseases (Kim, Alberico, Emmons, & Narayanan, 2015; Zhang et al., 2019). Indeed, for motor symptoms of Parkinson's disease, detailed circuit mapping and neurophysiology led to breakthroughs for clinical deep-brain stimulation, which vastly improve motor symptoms of Parkinson's disease (Deuschl et al., 2006; Limousin et al., 1998). Because temporal control of action is intimately linked to a range of cognitive functions in diseases such as Parkinson's disease, ADHD and schizophrenia, improving temporal control of action may have real-world benefits.

We found that low-frequency brain stimulation at 2 Hz—in the range of midfrontal cognitive control signals—improved interval timing in dopamine-depleted animals (Kim et al., 2017). This intervention was highly specific, targeting prefrontal D1DR + neurons in the medial prefrontal cortex of mice. Amazingly, 2 Hz brain stimulation also boosted medial prefrontal ramping activity and temporal decoding. Subsequent work from our group found that D1DR + neurons had specific coherence with ramping neurons and that stimulation early in the interval was sufficient to improve interval timing. Non-specific stimulation of all prefrontal neurons or stimulation in non-dopamine-depleted animals had no reliable effect (Kim & Narayanan, 2018). Similarly, we found that stimulation of cerebellothalamic afferents at 2 Hz could compensate for timing deficits caused by disrupting medial prefrontal D1DRs (Parker et al., 2017). Finally, we found that stimulating corticostriatal projections at 20 Hz could compensate for medial prefrontal inactivation during interval timing (Emmons et al., 2019). These data suggest that specific prefrontal stimulation has the potential to compensate for dysfunctional prefrontal circuits.

These studies also raise the possibility that brain stimulation can improve temporal control of action in human diseases. Fortuitously, clinical deep-brain stimulation targeting the subthalamic nucleus can modulate medial prefrontal networks, possibly via backfiring monosynaptic "hyperdirect" projections (Chen et al., 2020; Narayanan, Wessel, & Greenlee, 2020). Subthalamic and prefrontal neurons can have coherence at 4 Hz (Herz, Zavala, Bogacz, & Brown, 2016; Zavala et al., 2016). Low-frequency clinical deep-brain stimulation of the subthalamic nucleus is able to modulate interval-timing (Wojtecki et al., 2011). We found—to our surprise—that 4 Hz subthalamic brain stimulation increased midfrontal cognitive control signals and improved temporal control of action in Parkinson's disease patients (Kelley et al., 2018). This finding was supported by data showing that 5 Hz subthalamic nucleus stimulation could also improve conflict tasks (Scangos, Carter, Gurkoff, Zhang, & Shahlaie, 2018). This line of work is exciting because it suggests that targeted prefrontal brain stimulation has the potential to improve temporal control of action in human brain diseases, which may point the way for future neuromodulation strategies targeted at cognition.

8. Conclusion

In summary, we have reviewed the medial prefrontal neurophysiology in the temporal control of action. Much of this data comes from two paradigms: simple-reaction time tasks and interval timing, which can be markedly abnormal in human diseases such as Parkinson's disease, schizophrenia, and ADHD. Lesion evidence in humans and rodents implicates medial prefrontal regions as well as dopaminergic and cholinergic neurotransmission in temporal control of action. Medial prefrontal recordings consistently report "ramping" patterns of activity in anticipation of temporally controlled movements and $\sim$4 Hz activity in response to imperative stimuli. Low-frequency stimulation can improve temporal control in rodent models of human disease as well as in Parkinson's disease patients. Together, these studies illuminate how medial prefrontal regions orchestrate behavior in time, which could inspire novel neuromodulation therapies for human diseases.

References

Abi-Dargham, A., Mawlawi, O., Lombardo, I., Gil, R., Martinez, D., Huang, Y., et al. (2002). Prefrontal dopamine D1 receptors and working memory in schizophrenia. *The Journal of Neuroscience*, *22*, 3708–3719.

Akert, K. (1964). *The frontal granular cortex and behavior*. New York: McGraw-Hill.

Amiez, C., & Procyk, E. (2019). Chapter 4—Midcingulate somatomotor and autonomic functions. In B. A. Vogt (Ed.), *cingulate cortex. Handbook of Clinical Neurology* (pp. 53–71). Elsevier. Available at: http://www.sciencedirect.com/science/article/pii/B9780444641960000042. Accessed July 1, 2020.

Andreasen, N. C., O'Leary, D. S., Flaum, M., Nopoulos, P., Watkins, G. L., Boles Ponto, L. L., et al. (1997). Hypofrontality in schizophrenia: Distributed dysfunctional circuits in neuroleptic-naïve patients. *Lancet, 349*, 1730–1734.

Arnsten, A. F. T., Wang, M. J., & Paspalas, C. D. (2012). Neuromodulation of thought: Flexibilities and vulnerabilities in prefrontal cortical network synapses. *Neuron, 76*, 223–239.

Averbeck, B. B., Latham, P. E., & Pouget, A. (2006). Neural correlations, population coding and computation. *Nature Reviews. Neuroscience*, 7, 358–366.

Barkley, R. A., Koplowitz, S., Anderson, T., & McMurray, M. B. (1997). Sense of time in children with ADHD: Effects of duration, distraction, and stimulant medication. *Journal of the International Neuropsychological Society, 3*, 359–369.

Bateson, M. (2003). Interval timing and optimal foraging. In *Functional and neural mechanisms of interval timing* (pp. 113–141). Boca Raton, FL, US: CRC Press.

Bi, C., Yuan, X., & Huang, X. (2013). The impact of object working memory on timing. *Journal of Cognitive Psychology, 25*, 390–399.

Bluschke, A., Schuster, J., Roessner, V., & Beste, C. (2018). Neurophysiological mechanisms of interval timing dissociate inattentive and combined ADHD subtypes. *Scientific Reports, 8*, 2033.

Brown, S. W. (2006). Timing and executive function: Bidirectional interference between concurrent temporal production and randomization tasks. *Memory & Cognition, 34*, 1464–1471.

Brozoski, T. J., Brown, R. M., Rosvold, H. E., & Goldman, P. S. (1979). Cognitive deficit caused by regional depletion of dopamine in prefrontal cortex of rhesus monkey. *Science, 205*, 929–932.

Buhusi, C. V., & Meck, W. H. (2005). What makes us tick? Functional and neural mechanisms of interval timing. *Nature Reviews. Neuroscience, 6*, 755–765.

Buhusi, C. V., Reyes, M. B., Gathers, C.-A., Oprisan, S. A., & Buhusi, M. (2018). Inactivation of the medial-prefrontal cortex impairs interval timing precision, but not timing accuracy or scalar timing in a peak-interval procedure in rats. *Frontiers in Integrative Neuroscience, 12*, 20. Available at: https://www.frontiersin.org/articles/10.3389/fnint.2018.00020/full. Accessed June 28, 2020.

Bush, G. (2011). Cingulate, frontal, and parietal cortical dysfunction in attention-deficit/hyperactivity disorder. *Biological Psychiatry, 69*, 1160–1167.

Carroll, C. A., O'Donnell, B. F., Shekhar, A., & Hetrick, W. P. (2009). Timing dysfunctions in schizophrenia span from millisecond to several-second durations. *Brain and Cognition, 70*, 181–190.

Caselli, L., Iaboli, L., & Nichelli, P. (2009). Time estimation in mild Alzheimer's disease patients. *Behavioral and Brain Functions, 5*, 32.

Cavanagh, J. F., & Frank, M. J. (2014). Frontal theta as a mechanism for cognitive control. *Trends in Cognitive Sciences, 18*, 414–421.

Chen, P., Lou, S., Huang, Z.-H., Wang, Z., Shan, Q.-H., Wang, Y., et al. (2020). Prefrontal cortex corticotropin-releasing factor neurons control behavioral style selection under challenging situations. *Neuron, 106*, 301–315.e7. Available at: https://www.cell.com/neuron/abstract/S0896-6273(20)30060-X. Accessed April 5, 2020.

Church, R. (2003). A concise introduction to scalar timing theory. In *Functional and neural mechanisms of interval timing* (Meck W). CRC Press. Available at: http://www.crcnetbase.com/doi/abs/10.1201/9780203009574.sec1. Accessed July 23, 2013.

Craig, D. P. A., Varnon, C. A., Sokolowski, M. B. C., Wells, H., & Abramson, C. I. (2014). An assessment of fixed interval timing in free-flying honey bees (*Apis mellifera* ligustica): An analysis of individual performance. *PLoS One, 9*, e101262.

Deuschl, G., et al. (2006). A randomized trial of deep-brain stimulation for Parkinson's disease. *The New England Journal of Medicine, 355*, 896–908.

Durstewitz, D. (2003). Self-organizing neural integrator predicts interval times through climbing activity. *The Journal of Neuroscience, 23*, 5342–5353.

Emmons, E. B., Corte, B. J. D., Kim, Y., Parker, K. L., Matell, M. S., & Narayanan, N. S. (2017). Rodent medial frontal control of temporal processing in the dorsomedial striatum. *The Journal of Neuroscience*, *37*, 8718–8733.

Emmons, E. B., Kennedy, M., Kim, Y., & Narayanan, N. S. (2019). Corticostriatal stimulation compensates for medial frontal inactivation during interval timing. *Scientific Reports*, *9*, 1–9.

Evarts, E. V., Teräväinen, H., & Calne, D. B. (1981). Reaction time in Parkinson's disease. *Brain*, *104*, 167–186.

Finnerty, G. T., Shadlen, M. N., Jazayeri, M., Nobre, A. C., & Buonomano, D. V. (2015). Time in cortical circuits. *The Journal of Neuroscience*, *35*, 13912–13916.

Fu, Z., Wu, D.-A. J., Ross, I., Chung, J. M., Mamelak, A. N., Adolphs, R., et al. (2019). Single-neuron correlates of error monitoring and post-error adjustments in human medial frontal cortex. *Neuron*, *101*, 165–177.e5.

Fuster, J. M. (2001). The prefrontal cortex—An update: Time is of the essence. *Neuron*, *30*, 319–333.

Fuster, J. (2008). *The prefrontal cortex* (4th ed.). New York, NY: Academic Press.

Gabbott, P. L., Warner, T. A., Jays, P. R., & Bacon, S. J. (2003). Areal and synaptic interconnectivity of prelimbic (area 32), infralimbic (area 25) and insular cortices in the rat. *Brain Research*, *993*, 59–71.

Gabbott, P. L. A., Warner, T. A., Jays, P. R. L., Salway, P., & Busby, S. J. (2005). Prefrontal cortex in the rat: Projections to subcortical autonomic, motor, and limbic centers. *The Journal of Comparative Neurology*, *492*, 145–177.

Gill, T. M., Sarter, M., & Givens, B. (2000). Sustained visual attention performance-associated prefrontal neuronal activity: Evidence for cholinergic modulation. *The Journal of Neuroscience*, *20*, 4745–4757.

Gold, J. I., & Shadlen, M. N. (2007). The neural basis of decision making. *Annual Review of Neuroscience*, *30*, 535–574.

Goldman-Rakic, P. S. (1996). Regional and cellular fractionation of working memory. *Proceedings of the National Academy of Sciences of the United States of America*, *93*, 13473–13480.

Goldman-Rakic, P. S. (1998). The cortical dopamine system: Role in memory and cognition. *Advances in Pharmacology*, *42*, 707–711.

Hart, H., Radua, J., Mataix-Cols, D., & Rubia, K. (2012). Meta-analysis of fMRI studies of timing in attention-deficit hyperactivity disorder (ADHD). *Neuroscience and Biobehavioral Reviews*, *36*, 2248–2256.

Herz, D. M., Tan, H., Brittain, J.-S., Fischer, P., Cheeran, B., Green, A. L., et al. (2017). Distinct mechanisms mediate speed-accuracy adjustments in cortico-subthalamic networks. *eLife*, *6*, e21481.

Herz, D. M., Zavala, B. A., Bogacz, R., & Brown, P. (2016). Neural correlates of decision thresholds in the human subthalamic nucleus. *Current Biology*, *26*, 916–920.

Hinton, S. C., Paulsen, J. S., Hoffmann, R. G., Reynolds, N. C., Zimbelman, J. L., & Rao, S. M. (2007). Motor timing variability increases in preclinical Huntington's disease patients as estimated onset of motor symptoms approaches. *Journal of the International Neuropsychological Society*, *13*, 539–543.

Hove, M. J., Gravel, N., Spencer, R. M. C., & Valera, E. M. (2017). Finger tapping and pre-attentive sensorimotor timing in adults with ADHD. *Experimental Brain Research*, *235*, 3663–3672.

Jacobsen, C. F. (1935). Functions of frontal association area in primates. *Archives of Neurology and Psychiatry*, *33*, 558–569.

Jacobsen, C. F., & Nissen, H. W. (1937). Studies of cerebral function in primates. IV. The effects of frontal lobe lesions on the delayed alternation habit in monkeys. *Journal of Comparative Psychology*, *23*, 101.

Jones, D. N. C., & Higgins, G. A. (1995). Effect of scopolamine on visual attention in rats. *Psychopharmacology*, *120*, 142–149.
Jones, C. R. G., & Jahanshahi, M. (2014). Motor and perceptual timing in Parkinson's disease. *Advances in Experimental Medicine and Biology*, *829*, 265–290.
Jurkowski, A. J., Stepp, E., & Hackley, S. A. (2005). Variable foreperiod deficits in Parkinson's disease: Dissociation across reflexive and voluntary behaviors. *Brain and Cognition*, *58*, 49–61.
Kamiński, J., Sullivan, S., Chung, J. M., Ross, I. B., Mamelak, A. N., & Rutishauser, U. (2017). Persistently active neurons in human medial frontal and medial temporal lobe support working memory. *Nature Neuroscience*, *20*, 590–601.
Karmarkar, U. R., & Buonomano, D. V. (2007). Timing in the absence of clocks: Encoding time in neural network states. *Neuron*, *53*, 427–438.
Kelley, R., Flouty, O., Emmons, E. B., Kim, Y., Kingyon, J., Wessel, J. R., et al. (2018). A human prefrontal-subthalamic circuit for cognitive control. *Brain*, *141*, 205–216.
Kim, Y.-C., Alberico, S., Emmons, E., & Narayanan, N. (2015). New therapeutic strategies targeting D1-type dopamine receptors for neuropsychiatric disease. *Frontiers in Biology*, *10*, 230–238.
Kim, J., Ghim, J.-W., Lee, J. H., & Jung, M. W. (2013). Neural correlates of interval timing in rodent prefrontal cortex. *The Journal of Neuroscience*, *33*, 13834–13847.
Kim, Y.-C., Han, S.-W., Alberico, S. L., Ruggiero, R. N., De Corte, B., Chen, K.-H., et al. (2017). Optogenetic stimulation of frontal D1 neurons compensates for impaired temporal control of action in dopamine-depleted mice. *Current Biology*, *27*, 39–47.
Kim, J., Jung, A. H., Byun, J., Jo, S., & Jung, M. W. (2009). Inactivation of medial prefrontal cortex impairs time interval discrimination in rats. *Frontiers in Behavioral Neuroscience*, *3*, 38.
Kim, Y.-C., & Narayanan, N. S. (2018). Prefrontal D1 dopamine-receptor neurons and delta resonance in interval timing. *Cerebral Cortex*, *29*, 2051–2060.
Klinkenberg, I., & Blokland, A. (2010). The validity of scopolamine as a pharmacological model for cognitive impairment: A review of animal behavioral studies. *Neuroscience and Biobehavioral Reviews*, *34*, 1307–1350.
Kononowicz, T. W., & van Rijn, H. (2014). Decoupling interval timing and climbing neural activity: A dissociation between CNV and N1P2 amplitudes. *The Journal of Neuroscience*, *34*, 2931–2939.
Laje, R., & Buonomano, D. V. (2013). Robust timing and motor patterns by taming chaos in recurrent neural networks. *Nature Neuroscience*, *16*, 925–933.
Laubach, M., Amarante, L. M., Swanson, K., & White, S. R. (2018). What, if anything, is rodent prefrontal cortex? *eNeuro*, *5*, ENEURO.0315-18.2018.
Leth-Steensen, C., King Elbaz, Z., & Douglas, V. I. (2000). Mean response times, variability, and skew in the responding of ADHD children: A response time distributional approach. *Acta Psychologica*, *104*, 167–190.
Limousin, P., Krack, P., Pollak, P., Benazzouz, A., Ardouin, C., Hoffmann, D., et al. (1998). Electrical stimulation of the subthalamic nucleus in advanced Parkinson's disease. *The New England Journal of Medicine*, *339*, 1105–1111.
Long, M. A., Katlowitz, K. A., Svirsky, M. A., Clary, R. C., Byun, T. M., Majaj, N., et al. (2016). Functional segregation of cortical regions underlying speech timing and articulation. *Neuron*, *89*, 1187–1193.
Luce, R. D. (1991). *Response times: Their role in inferring elementary mental organization*. Oxford University Press.
Luman, M., Papanikolau, A., & Oosterlaan, J. (2015). The unique and combined effects of reinforcement and methylphenidate on temporal information processing in attention-deficit/hyperactivity disorder. *Journal of Clinical Psychopharmacology*, *35*, 414–421.
Luzardo, A., Ludvig, E. A., & Rivest, F. (2013). An adaptive drift-diffusion model of interval timing dynamics. *Behavioural Processes*, *95*, 90–99.

Malapani, C., Deweer, B., & Gibbon, J. (2002). Separating storage from retrieval dysfunction of temporal memory in Parkinson's disease. *Journal of Cognitive Neuroscience*, *14*, 311–322.

Malapani, C., Rakitin, B., Levy, R., Meck, W. H., Deweer, B., Dubois, B., et al. (1998). Coupled temporal memories in Parkinson's disease: A dopamine-related dysfunction. *Journal of Cognitive Neuroscience*, *10*, 316–331.

Matell, M. S., & Meck, W. H. (2004). Cortico-striatal circuits and interval timing: Coincidence detection of oscillatory processes. *Brain Research. Cognitive Brain Research*, *21*, 139–170.

Matell, M. S., Meck, W. H., & Nicolelis, M. A. L. (2003). Interval timing and the encoding of signal duration by ensembles of cortical and striatal neurons. *Behavioral Neuroscience*, *117*, 760–773.

Merchant, H., & de Lafuente, V. (2014). Introduction to the neurobiology of interval timing. *Advances in Experimental Medicine and Biology*, *829*, 1–13.

Mishkin, M., & Pribram, K. H. (1955). Analysis of the effects of frontal lesions in monkey. I. Variations of delayed alternation. *Journal of Comparative and Physiological Psychology*, *48*, 492–495.

Naatanen, R. (1971). Non-aging fore-periods and simple reaction time. *Acta Psychologia*, *35*, 316–327.

Naatanen, R. (1972). Time uncertainty and occurrence uncertainty of the stimulus in a simple reaction time task. *Acta Psychologia*, *36*, 492–503.

Naatanen, R., Muranen, V., & Merisalo, A. (1974). Timing of expectance peak in simple reaction time situation. *Acta Psychologica*, *38*, 461–470.

Namboodiri, V. M. K., & Hussain Shuler, M. G. (2014). Report of interval timing or action? *Proceedings of the National Academy of Sciences of the United States of America*, *111*, E2239.

Narayanan, N. S. (2016). Ramping activity is a cortical mechanism of temporal control of action. *Current Opinion in Behavioral Sciences*, *8*, 226–230.

Narayanan, N. S., Cavanagh, J. F., Frank, M. J., & Laubach, M. (2013). Common medial frontal mechanisms of adaptive control in humans and rodents. *Nature Neuroscience*, *16*, 1888–1897.

Narayanan, N. S., Horst, N. K., & Laubach, M. (2006). Reversible inactivations of rat medial prefrontal cortex impair the ability to wait for a stimulus. *Neuroscience*, *139*, 865–876.

Narayanan, N. S., Land, B. B., Solder, J. E., Deisseroth, K., & DiLeone, R. J. (2012). Prefrontal D1 dopamine signaling is required for temporal control. *Proceedings of the National Academy of Sciences of the United States of America*, *109*, 20726–20731.

Narayanan, N. S., & Laubach, M. (2006). Top-down control of motor cortex ensembles by dorsomedial prefrontal cortex. *Neuron*, *52*, 921–931.

Narayanan, N. S., & Laubach, M. (2008). Neuronal correlates of post-error slowing in the rat dorsomedial prefrontal cortex. *Journal of Neurophysiology*, *100*, 520–525.

Narayanan, N. S., & Laubach, M. (2009). Delay activity in rodent frontal cortex during a simple reaction time task. *Journal of Neurophysiology*, *101*, 2859–2871.

Narayanan, N. S., Prabhakaran, V., Bunge, S. A., Christoff, K., Fine, E. M., & Gabrieli, J. D. E. (2005). The role of the prefrontal cortex in the maintenance of verbal working memory: An event-related FMRI analysis. *Neuropsychology*, *19*, 223–232.

Narayanan, N. S., Rodnitzky, R. L., & Uc, E. Y. (2013). Prefrontal dopamine signaling and cognitive symptoms of Parkinson's disease. *Reviews in the Neurosciences*, *24*, 267–278.

Narayanan, N. S., Wessel, J. R., & Greenlee, J. D. W. (2020). The fastest way to stop: Inhibitory control and IFG-STN Hyperdirect connectivity. *Neuron*, *106*, 549–551.

Neafsey, E. J., Bold, E. L., Haas, G., Hurley-Gius, K. M., Quirk, G., Sievert, C. F., et al. (1986). The organization of the rat motor cortex: A microstimulation mapping study. *Brain Research*, *396*, 77–96.

Niki, H., & Watanabe, M. (1976). Prefrontal unit activity and delayed response: Relation to cue location versus direction of response. *Brain Research*, *105*, 79–88.

Niki, H., & Watanabe, M. (1979). Prefrontal and cingulate unit activity during timing behavior in the monkey. *Brain Research*, *171*, 213–224.

Noreika, V., Falter, C. M., & Rubia, K. (2013). Timing deficits in attention-deficit/hyperactivity disorder (ADHD): Evidence from neurocognitive and neuroimaging studies. *Neuropsychologia*, *51*, 235–266.

Okubo, Y., Suhara, T., Suzuki, K., Kobayashi, K., Inoue, O., Terasaki, O., et al. (1997). Decreased prefrontal dopamine D1 receptors in schizophrenia revealed by PET. *Nature*, *385*, 634–636.

Parikh, V., & Sarter, M. (2008). Cholinergic mediation of attention: Contributions of phasic and tonic increases in prefrontal cholinergic activity. *Annals of the New York Academy of Sciences*, *1129*, 225–235.

Parker, K. L., Alberico, S. L., Miller, A. D., & Narayanan, N. S. (2013). Prefrontal D1 dopamine signaling is necessary for temporal expectation during reaction time performance. *Neuroscience*, *255*, 246–254.

Parker, K. L., Chen, K.-H., Kingyon, J. R., Cavanagh, J. F., & Narayanan, N. S. (2014). D1-dependent 4 Hz oscillations and ramping activity in rodent medial frontal cortex during interval timing. *The Journal of Neuroscience*, *34*, 16774–16783.

Parker, K. L., Chen, K.-H., Kingyon, J. R., Cavanagh, J. F., & Narayanan, N. S. (2015). Medial frontal ~4-Hz activity in humans and rodents is attenuated in PD patients and in rodents with cortical dopamine depletion. *Journal of Neurophysiology*, *114*, 1310–1320.

Parker, K. L., Kim, Y. C., Kelley, R. M., Nessler, A. J., Chen, K.-H., Muller-Ewald, V. A., et al. (2017). Delta-frequency stimulation of cerebellar projections can compensate for schizophrenia-related medial frontal dysfunction. *Molecular Psychiatry*, *22*, 647–655.

Parker, K. L., Lamichhane, D., Caetano, M. S., & Narayanan, N. S. (2013). Executive dysfunction in Parkinson's disease and timing deficits. *Frontiers in Integrative Neuroscience*, *7*, 75.

Parker, K. L., Ruggiero, R. N., & Narayanan, N. S. (2015). Infusion of D1 dopamine receptor agonist into medial frontal cortex disrupts neural correlates of interval timing. *Frontiers in Behavioral Neuroscience*, *9*, 294.

Paton, J. J., & Buonomano, D. V. (2018). The neural basis of timing: Distributed mechanisms for diverse functions. *Neuron*, *98*, 687–705.

Penney, T. B., Meck, W. H., Roberts, S. A., Gibbon, J., & Erlenmeyer-Kimling, L. (2005). Interval-timing deficits in individuals at high risk for schizophrenia. *Brain and Cognition*, *58*, 109–118.

Picton, T. W., Stuss, D. T., Alexander, M. P., Shallice, T., Binns, M. A., & Gillingham, S. (2006). Effects of focal frontal lesions on response inhibition. *Cerebral Cortex*, *17*, 826–838. Available at: http://www.ncbi.nlm.nih.gov/entrez/query.fcgi?cmd=Retrieve&db=PubMed&dopt=Citation&list_uids=16699079.

Picton, T. W., Stuss, D. T., Shallice, T., Alexander, M. P., & Gillingham, S. (2006). Keeping time: Effects of focal frontal lesions. *Neuropsychologia*, *44*, 1195–1209.

Preuss, T. (1995). Do rats have prefrontal cortex? The Rose-Woolsey-Akert program reconsidered. *Journal of Cognitive Neuroscience*, *7*, 1–24.

Pribram, K. H. (1950). Some physical and pharmacological factors affecting delayed response performance of baboons following frontal lobotomy. *Journal of Neurophysiology*, *13*, 373–382.

Pribram, K. H., & Mishkin, M. (1956). Analysis of the effects of frontal lesions in monkey. III. Object alternation. *Journal of Comparative and Physiological Psychology*, *49*, 41–45.

Pribram, K. H., Mishkin, M., Rosvold, H. E., & Kaplan, S. J. (1952). Effects on delayed-response performance of lesions of dorsolateral and ventromedial frontal cortex of baboons. *Journal of Comparative and Physiological Psychology*, *45*, 565–575.

Remington, E. D., Egger, S. W., Narain, D., Wang, J., & Jazayeri, M. (2018). A dynamical systems perspective on flexible motor timing. *Trends in Cognitive Sciences*, *22*, 938–952.

Risterucci, C., Terramorsi, D., Nieoullon, A., & Amalric, M. (2003). Excitotoxic lesions of the prelimbic-infralimbic areas of the rodent prefrontal cortex disrupt motor preparatory processes. *The European Journal of Neuroscience*, *17*, 1498–1508.

Rommelse, N. N. J., Altink, M. E., Oosterlaan, J., Beem, L., Buschgens, C. J. M., Buitelaar, J., et al. (2008). Speed, variability, and timing of motor output in ADHD: Which measures are useful for endophenotypic research? *Behavior Genetics*, *38*, 121–132.

Rose, J. E., & Woolsey, C. N. (1949). Organization of the mammalian thalamus and its relationships to the cerebral cortex. *Electroencephalography and Clinical Neurophysiology*, *1*, 391–403 (discussion 403-4).

Sarter, M., & Bruno, J. P. (1997). Cognitive functions of cortical acetylcholine: Toward a unifying hypothesis. *Brain Research. Brain Research Reviews*, *23*, 28–46.

Scangos, K. W., Carter, C. S., Gurkoff, G., Zhang, L., & Shahlaie, K. (2018). A pilot study of subthalamic theta frequency deep brain stimulation for cognitive dysfunction in Parkinson's disease. *Brain Stimulation*, *11*, 456–458.

Sheth, S. A., Mian, M. K., Patel, S. R., Asaad, W. F., Williams, Z. M., Dougherty, D. D., et al. (2012). Human dorsal anterior cingulate cortex neurons mediate ongoing behavioural adaptation. *Nature*, *488*, 218–221.

Shibasaki, H., & Hallett, M. (2006). What is the Bereitschaftspotential? *Clinical Neurophysiology*, *117*, 2341–2356.

Simen, P., Balci, F., de Souza, L., Cohen, J. D., & Holmes, P. (2011). A model of interval timing by neural integration. *The Journal of Neuroscience*, *31*, 9238–9253.

Smith, A., Taylor, E., Rogers, J. W., Newman, S., & Rubia, K. (2002). Evidence for a pure time perception deficit in children with ADHD. *Journal of Child Psychology and Psychiatry*, *43*, 529–542.

Sternberg, S. (1969). Memory-scanning: Mental processes revealed by reaction-time experiments. *American Scientist*, *57*, 421–457.

Stubbs, D. A. (1976). Scaling of stimulus duration by Pigeons1. *Journal of the Experimental Analysis of Behavior*, *26*, 15–25.

Stuss, D. T., Alexander, M. P., Shallice, T., Picton, T. W., Binns, M. A., Macdonald, R., et al. (2005). Multiple frontal systems controlling response speed. *Neuropsychologia*, *43*, 396–417.

Tallot, L., & Doyère, V. (2020). Neural encoding of time in the animal brain. *Neuroscience and Biobehavioral Reviews*, *115*, 146–163.

Thoenes, S., & Oberfeld, D. (2017). Meta-analysis of time perception and temporal processing in schizophrenia: Differential effects on precision and accuracy. *Clinical Psychology Review*, *54*, 44–64.

Uylings, H. B. M., Groenewegen, H. J., & Kolb, B. (2003). Do rats have a prefrontal cortex? *Behavioural Brain Research*, *146*, 3–17.

Valera, E. M., Spencer, R. M. C., Zeffiro, T. A., Makris, N., Spencer, T. J., Faraone, S. V., et al. (2010). Neural substrates of impaired sensorimotor timing in adult attention-deficit/hyperactivity disorder. *Biological Psychiatry*, *68*, 359–367.

Verbruggen, F., et al. (2019). A consensus guide to capturing the ability to inhibit actions and impulsive behaviors in the stop-signal task. *eLife*, *8*, e46323.

Wang, J., Narain, D., Hosseini, E. A., & Jazayeri, M. (2018). Flexible timing by temporal scaling of cortical responses. *Nature Neuroscience*, *21*, 102.

Ward, R. D., Kellendonk, C., Kandel, E. R., & Balsam, P. D. (2011). Timing as a window on cognition in schizophrenia. *Neuropharmacology*, *62*, 1175–1181. Available at http://www.ncbi.nlm.nih.gov/pubmed/21530549. Accessed September 8, 2011.

Wearden, J. H., Smith-Spark, J. H., Cousins, R., Edelstyn, N. M. J., Cody, F. W. J., & O'Boyle, D. J. (2008). Stimulus timing by people with Parkinson's disease. *Brain and Cognition*, *67*, 264–279.

Wojtecki, L., Elben, S., Timmermann, L., Reck, C., Maarouf, M., Jorgens, S., et al. (2011). Modulation of human time processing by subthalamic deep brain stimulation. *PLoS One*, *6*, e24589. Available at: http://www.ncbi.nlm.nih.gov/pmc/articles/PMC3171456/. Accessed June 5, 2014.

Xu, M., Zhang, S., Dan, Y., & Poo, M. (2014). Representation of interval timing by temporally scalable firing patterns in rat prefrontal cortex. *Proceedings of the National Academy of Sciences of the United States of America*, *111*, 480–485.

Zavala, B., Tan, H., Ashkan, K., Foltynie, T., Limousin, P., Zrinzo, L., et al. (2016). Human subthalamic nucleus-medial frontal cortex theta phase coherence is involved in conflict and error related cortical monitoring. *NeuroImage*, *137*, 178–187.

Zhang, Q., Jung, D., Larson, T., Kim, Y., & Narayanan, N. S. (2019). Scopolamine and medial frontal stimulus-processing during interval timing. *Neuroscience*, *414*, 219–227.

CHAPTER SIXTEEN

Secondary motor cortex: Broadcasting and biasing animal's decisions through long-range circuits

Jen-Hau Yang[a] and Alex C. Kwan[a,b,*]

[a]Department of Psychiatry, Yale University School of Medicine, New Haven, CT, United States
[b]Department of Neuroscience, Yale University School of Medicine, New Haven, CT, United States
*Corresponding author: e-mail address: alex.kwan@yale.edu

Contents

International Review of Neurobiology, Volume 158
ISSN 0074-7742
https://doi.org/10.1016/bs.irn.2020.11.008

Abstract

Medial secondary motor cortex (MOs or M2) constitutes the dorsal aspect of the rodent medial frontal cortex. We previously proposed that the function of MOs is to link antecedent conditions, including sensory stimuli and prior choices, to impending actions. In this review, we focus on the long-range pathways between MOs and other cortical and subcortical regions. We highlight three circuits: (1) connections with visual and auditory cortices that are essential for predictive coding of perceptual inputs; (2) connections with motor cortex and brainstem that are responsible for top-down, context-dependent modulation of movements; (3) connections with retrosplenial cortex, orbitofrontal cortex, and basal ganglia that facilitate reward-based learning. Together, these long-range circuits allow MOs to broadcast choice signals for feedback and to bias decision-making processes.

1. Introduction

The secondary motor cortex in rodents, denoted as MOs in the Allen Mouse Brain Atlas (Wang, Ding, et al., 2020) or M2 in the Paxinos and Franklin's Atlas (Paxinos & Franklin, 2004), is a large region, likely consisting of multiple sub-regions serving different functions. In particular, the medial portion of MOs constitutes the most dorsal aspect of the medial frontal cortex (Fig. 1). It has been labeled with various names including medial agranular cortex (AGm), medial precentral cortex (PrCm), frontal orienting field (FOF), and frontal eye field MOs domain (MOs-fef). Other studies have referred to this region based on its spatial location, calling it the dorsomedial prefrontal cortex (dmPFC) or medial motor cortex (MM). The medial MOs has poorly defined boundaries with neighboring regions such as anterior cingulate cortex (ACAd) and vibrissa motor cortex (vM1). Furthermore, medial MOs contains features similar to those found in the supplementary motor cortex, premotor cortex, and frontal eye field of primates (Barthas & Kwan, 2017; Wise, 2008).

In a previous article (Barthas & Kwan, 2017), we summarized evidence from anatomical, lesion, and electrophysiological studies to propose that the main function of medial MOs is to link antecedent conditions to actions. Antecedent conditions could be sensory inputs including visual and auditory stimuli, but could also be other actions for chaining motor sequences and maintaining choice history information for decision-making. The goal of this book chapter is to build on the previous article and discuss the several dozens of new studies on MOs that have emerged over the last several years. The new studies have focused on long-range interactions between MOs and

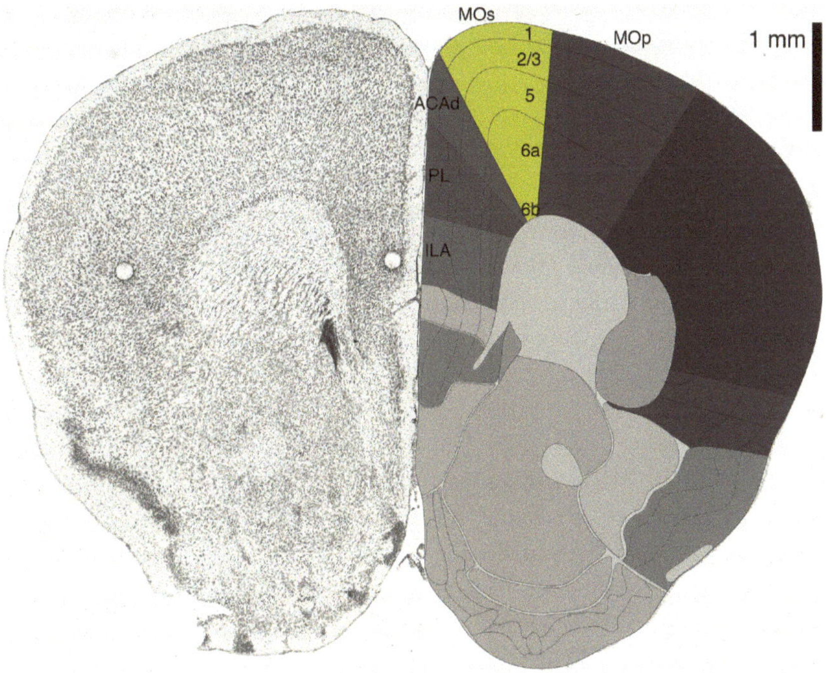

Fig. 1 The medial secondary motor cortex of the mouse. A coronal section of the frontal cortex at around +1.5 mm anterior of bregma for a 56 postnatal day old, C57BL/6J mouse, from the Allen Mouse Brain Atlas. Left half, Nissl stain. Right half, regions demarcated and named based on the Allen Mouse Common Brain Coordinate Framework (Wang, Ding, et al., 2020). MOs, secondary motor area. ACAd, anterior cingulate cortex, dorsal part. PL, prelimbic area. IL: infralimbic area.

other cortical or subcortical brain regions. These long-range connections are extensive, heterogeneous, and complex. Often, they are reciprocal and topographically organized. Overall, if MOs is indeed crucial for generating actions based on antecedents, then the novel findings on long-range circuits are beginning to illuminate how action signals are transmitted downstream to broadcast motor intent and bias the animal's decisions.

A common theme in the new studies is that the involvement of medial MOs is more pronounced in awake animals during active behavior. For example, as we will discuss below, there is evidence that MOs conveys locomotor information for predictive sensory processing and injects biases into action selection for decision-making. Its contributions therefore are less about explicit motor actions, but more about motivated behaviors. As such, instead of thinking of the region as the 'secondary motor cortex' with the word *motor* immediately emphasizing movement generation, thinking of medial MOs as the 'secondary motivation cortex' may be a more apt name that encompasses its broad range of behavioral functions.

2. Organization of long-range circuits

The foundations for our current understanding of connections to and from the medial frontal cortex, including medial MOs, ACAd, prelimbic area (PL), and infralimbic area (ILA), were built on tried-and-true anterograde tract-tracing (Sesack, Deutch, Roth, & Bunney, 1989) and retrograde labeling studies (Gabbott, Warner, Jays, Salway, & Busby, 2005). Scaling up the methods and systematic analyses of big data produced extensive databases for cortico-cortical (Oh et al., 2014; Zingg et al., 2014), cortico-striatal (Hintiryan et al., 2016), and cortico-thalamic (Hunnicutt et al., 2014) connectivity. Therefore, in bulk where many axons are mapped at once, detailed information is available for medial MOs. Based on early neuroanatomical tracing studies and large databases, we noted differences in long-range connectivity between MOs and the neighboring ACAd and primary motor cortex (MOp) (Barthas & Kwan, 2017). It should also be mentioned that there is no detectable sex difference in the connectivity pattern of MOs afferents (Billeh et al., 2016).

Recent efforts have focused on tracing a single neuron's axon including all of the collaterals. Although this approach was not new even for medial MOs (Kita & Kita, 2012), large-scale studies based on automated methods can map axonal collaterals of thousands of single neurons, revealing striking diversity in the projection patterns (Winnubst et al., 2019). Specifically, a single cortical neuron is likely to send axonal collaterals to numerous brain regions, and few neurons share the same exact set of efferent targets. In this regard, MOs is no exception to the principle. In one study, efferents from 80 pyramidal cells including 36 intratelencephalic (IT) neurons in the mouse MOs were mapped in a brain-wide manner (Lin et al., 2018). Individual MOs IT neurons had extensive axonal collaterals, with axonal length of single neurons reaching as much as 300 mm. Given the remarkable axonal length, unsurprisingly each neuron had terminal fields in a variety of brain regions and the axonal trajectory was distinct between neurons.

Characterization of inputs can be performed at single-cell resolution and with monosynaptic specificity using rabies viruses (Wickersham et al., 2007) [although see caveats, (Svoboda, 2019)]. The trans-synaptic tracing approach gave clues into the connectivity principles for distinct inputs impinging on

MOs and ACAd (Zhang et al., 2016). Luo and colleagues used a similar approach to map long-range inputs into glutamatergic and GABAergic neurons in MOs and MOp (Luo et al., 2019). Consistent with bulk tracing (Zingg et al., 2014), MOs was found to receive inputs from neurons in the medial subnetwork such as orbitofrontal cortex (ORB), retrosplenial cortex (RSP), visual cortex (VIS), and auditory cortex (AUD) (Fig. 2). By contrast, MOp received inputs primarily from somatosensory regions. Glutamatergic and GABAergic neurons in MOs had similar proportions of inputs from the various regions.

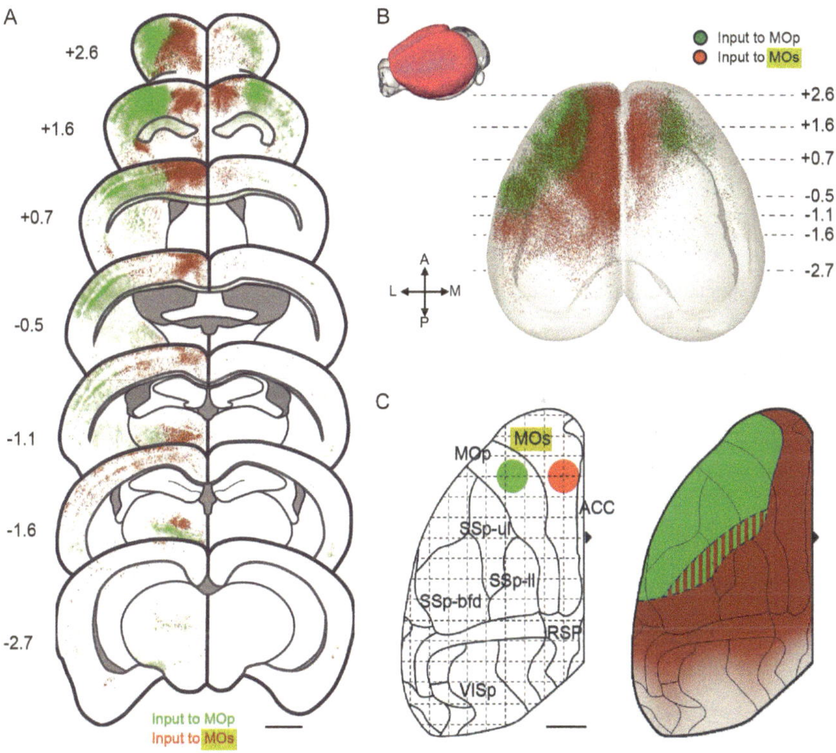

Fig. 2 Long-range cortical connections with the secondary motor cortex. (A) Coronal sections showing cortical neurons that send long-range connections to the primary motor cortex (MOp, in green) and secondary motor cortex (MOs, red). Each dot is a neuron. Scale bar = 1 mm. (B) Three-dimensional visualization. (C) Cartoon illustrating the input cortical regions areas projecting to MOp (green), MOs (red), or both (green and red stripes). The green and red circles indicate the injection sites. Scale bar = 1 mm. *Adapted from Luo, P., Li, A., Zheng, Y., Han, Y., Tian, J., Xu, Z., et al. (2019). Whole brain mapping of long-range direct input to glutamatergic and gabaergic neurons in motor cortex.* Frontiers in Neuroanatomy, 13, *44.*

Among neurons that send long-range projections, less is known about their microcircuit organization within MOs. In MOp, a series of elegant studies by Shepherd and colleagues have delineated directional preferences for connectivity between subpopulations of cortico-striatal, cortico-spinal, and cortico-thalamic pyramidal neurons (Anderson, Sheets, Kiritani, & Shepherd, 2010; Kiritani, Wickersham, Seung, & Shepherd, 2012; Yamawaki & Shepherd, 2015). Similar microcircuit motifs may be present for projection neurons in MOs. Although the organization is not fully tested, findings from current studies are at least consistent with this view. For example, Kawaguchi and colleagues found that crossed-corticostriatal cells innervate corticopontine cells unidirectionally with depressing synapses in MOs (Kawaguchi, 2017). Moreover, corticopontine cells frequently had reciprocal connections with short-term facilitating synapses in MOs.

Genetics likely play a role in shaping long-range and local connectivity, however neural activity could also be a powerful factor in sculpting neuronal networks. Ren and colleagues examined columnar microcircuits of excitatory neuron clones that are sister cells in primary somatosensory cortex (SSp) and primary visual cortex (VISp), and characterized their presynaptic partners in the medial frontal cortex (Ren, Li, Lin, Bergami, & Shi, 2019). Interestingly, like the sister cells, the presynaptic partners were organized vertically and preferentially synapse on each other. The authors went on to show that formation of these microcircuits in the frontal cortex depended on the clonal relationship between the sister cells and required synaptic communication from the sensory cortex. The interpretation is that the assembly of specific microcircuits in frontal cortex relies on the reciprocal microcircuit-to-microcircuit communication between frontal and sensory cortices.

3. Circuits for perceptual behavior

3.1 Behavioral engagement during perceptual tasks

During perceptual tasks, MOs is often seen to respond after initial activity in sensory cortices. For example, in a whisker-based, delayed-response task, Gilad and colleagues reported stimulus-driven activation in a subdivision of posterior parietal cortex (PTLp), which then transitioned into activity in the frontomedial area (Gilad, Gallero-Salas, Groos, & Helmchen, 2018). The response in medial frontal cortex was specific to trials in which animals made limb movements, body stretching, or vigorous whisking during stimulus presentation, but absent for trials when animals were still.

This could be because animals are using an active versus passive strategy to solve the discrimination problem, although alternatively active trials might reflect heightened arousal. The involvement of MOs was not unique to tactile stimuli, because neural responses in the medial frontal cortex was present in an auditory version of the task (Gallero-Salas et al., 2020). Locations of frontal responses overlapped for the two tasks, with auditory and tactile responses observed in anterior and posterior parts of medial frontal cortex respectively.

Likewise, visually evoked cortical activity was observed to emerge in VIS, followed by response in medial MOs (Musall, Kaufman, Juavinett, Gluf, & Churchland, 2019; Salkoff, Zagha, McCarthy, & McCormick, 2020; Zatka-Haas, Steinmetz, Carandini, & Harris, 2020). The frontal cortical response is dependent on behavioral engagement. Take for instance the study by Salkoff and colleagues. In a go/no-go task, the authors detected calcium transients in MOs during hit trials, but not in miss trials when the same visual stimuli were presented (Salkoff et al., 2020). Causal perturbations corroborate with the imaging results. Muscimol-mediated inactivation of MOs increased response time and partially weakened discriminability (Salkoff et al., 2020). Optogenetic inactivation of MOs reduced the probability of choosing contralateral option, but had no effect during no-go trial (Zatka-Haas et al., 2020). Altogether, these studies point to the involvement of MOs in perceptual behavior selectively during situations when animals are engaged in the task.

3.2 Processing of sensory stimuli into choices

What is the function of MOs in perceptual tasks? In a previous article (Barthas & Kwan, 2017), we posited that a major function of MOs is to flexibly link antecedent conditions, including sensory inputs, to actions. MOs is active when the animals are engaged, presumably more so for complex and demanding tasks. This precondition is illustrated by a study by Siniscalchi and colleagues, who developed a task that switches between multiple auditory-motor mapping rules (Siniscalchi, Phoumthipphavong, Ali, Lozano, & Kwan, 2016). That is, depending on the rule, for the same auditory stimulus, the animals had to respond differently to gain reward. Bilateral inactivation of MOs led to preservation and impaired flexibility, specifically for the more difficult switches from non-conditional responding (e.g., always choose left) to conditional actions (e.g. depending on stimuli, choose left or right), while sparing the simpler rule switches. Further support for a role of MOs in adaptive sensory-motor selection came from a new study (Wang, Liu, & Yao, 2020). Using a task where animals had to categorize visual stimulus, but with the categorization boundary changing in blocks of trials within a session, Wang and colleagues found that decisions are influenced by the history of sensory stimuli, which was diminished following bilateral inactivation of MOs. Together, the results highlight the role of MOs in an uncertain sensory environment where the animals need to adjust cue-action associations.

The mechanisms for how sensory cues are associated with choice signals remain poorly understood. Local GABAergic neurons may contribute to this process. In an auditory go/no-go task, during delay period between the stimulus and choice, optogenetic activation of somatostatin- (SST) or parvalbumin-expressing (PV) interneurons would increase the number of false alarms (Kamigaki & Dan, 2017). Intriguingly, activation of vasoactive intestinal peptide-expressing (VIP) interneurons had the opposite effect of reducing false alarms to improve performance. The impact of these GABAergic signals onto pyramidal neurons, which presumably are responsible for the sensorimotor coupling, is not fully known.

3.3 Circuit with visual cortex

Medial MOs has reciprocal connections with visual cortical regions including VISp and higher visual areas (Itokazu et al., 2018; Zhang et al., 2016; Zingg et al., 2014). In terms of laminar specificity, MOs neurons in layer 5 mainly target pyramidal neurons in layer 2/3 (through axons terminating in superficial layer 1) and layer 6 (through axons terminating in layer 6) in VISp, as well as GABAergic interneurons including all of the PV, SST, and VIP subtypes in VISp (Leinweber, Ward, Sobczak, Attinger, & Keller, 2017). There is topographic organization, at least in the macroscale, with posterior MOs and ACAd projecting to medial VISp, versus anterior MOs and ACAd projecting to anterior VISp. Projections from MOs to higher visual areas are localized to rostrolateral portions of VIS (Itokazu et al., 2018).

Numerous recent studies provided solid evidence that the MOs-to-VISp connection is crucial for providing feedback to predict optic flow. As an animal moves, the visual scene shifts and the movement should be proactively subtracted for accurate perception (e.g., for a predator to estimate the speed of a prey during chase). Leinweber and colleagues imaged neural activity in MOs and ACAd axons in VISp in a head-fixed mouse as it navigates a 2D virtual environment (Leinweber et al., 2017). Calcium transients in the frontal cortical axons contained motor-related signals that could be used for estimating optic flow. Moreover, optogenetic stimulation of MOs axons over VISp caused turning behavior—consistent with disrupting optic flow. The corollary discharge available for predictive coding in the sensory cortex is not limited to locomotion, because other motions such as eye movements also involves feedback from MOs (Itokazu et al., 2018). The specifics for how the predicting coding occur and how feedback shapes activity dynamics in VIS is less known. Conceivably, feedback axons could sculpt excitatory dynamics through their actions on GABAergic neurons (Zhang et al., 2014), which in turn produce gain modulation in the sensory cortex to adapt for the environment (Ferguson & Cardin, 2020).

One caveat with the aforementioned experiments is that they were performed on head-fixed mice. To overcome this limitation, Guitchounts and colleagues recorded from neurons in layer 2/3 of VISp in freely moving rats equipped with a head-mounted inertial measurement unit (Guitchounts, Masis, Wolff, & Cox, 2020). The VISp neurons encoded the direction of head orientation in light, and more importantly, also in dark conditions. The head orientation-related signal in VISp rose before movement onset, suggesting that it was predictive in nature, and was substantially reduced following lesion of MOs with ibotenic acid. Therefore, extending the idea that MOs is transmitting signals for optic flow prediction, these results further suggest that feedback signals broadcasted from MOs to VIS contain a rich multitude of movement and postural information. As there is increasing recognition that vision should be studied in context and as part of a sensorimotor loop (Froudarakis et al., 2019; Gomez-Marin & Ghazanfar, 2019), this connection between MOs and VIS should be an important pathway towards understanding naturalistic visual processing.

3.4 Circuit with auditory cortex

Auditory cortical regions including primary auditory area (AUDp) and anterior auditory field innervate the medial part of MOs (Nakata, Takemoto, & Song, 2020; Zhang et al., 2016). The connection is reciprocal, because MOs axons extend to auditory cortex, projecting ipsilaterally to superficial and deep cortical layers (Nelson et al., 2013).

Activation of MOs by auditory stimuli depends on the behavioral state of the animal. In one clear demonstration, voltage-sensitive dye imaging revealed robust activation of medial frontal cortex by an auditory stimulus in awake mice (Mohajerani et al., 2013). This response was absent in anesthetized animals. Single-unit recordings showed spiking responses to auditory stimulus in the rat medial agranular cortex (Handa, Takekawa, Harukuni, Isomura, & Fukai, 2017). Additionally, even when frontal cortical neurons do not overtly change firing rate, they may still encode information about the auditory stimulus via other means such as spiking timing (Insanally et al., 2019).

The primary effect of feedback from MOs on auditory cortical areas is suppression. Aside from excitatory synapses onto pyramidal neurons, MOs axons innervate PV interneurons in AUDp which mediate powerful inhibition on the spiking activity of pyramidal neurons (Nelson et al., 2013). Indeed, optogenetic activation of MOs axonal terminals in AUD reduced spontaneous and stimulus-evoked activity in auditory cortical neurons in mice (Schneider, Nelson, & Mooney, 2014). In a follow-up study, Schneider and colleagues tested whether this suppression can be specific to certain auditory stimuli. The rationale is that, in order to detect sounds from the external environment, it is advantageous for animals to suppress sounds generated from their own movements (Schneider, Sundararajan, & Mooney, 2018). Using a clever acoustic virtual reality system, they found that MOs connected preferentially to auditory cortical interneurons that respond to the frequency of the movement-related sound, such that interneurons can be selectively recruited during locomotion to mediate the filtering. Thus, not only can MOs suppress neural activity in sensory cortex, but the strength and specificity of the predictive coding scheme is shaped by experience.

4. Circuits for adaptive movements

4.1 Higher order control of movements

One of the more consistent findings for medial MOs is that inactivating the region has minimal impact on simple motor movements. For example, bilateral inactivation of MOs with muscimol led to no detectable change in lick response time (Siniscalchi et al., 2016). Some studies reported subtle effects. In a delay response task, perturbation of MOs during the delay period prompted the mouse to take a more passive stance, but overall had little effect on discrimination performance (Gilad et al., 2018). In cases when movements were altered, the deficits were mild. Kawai and colleagues lesioned the entire motor cortex including both MOp and MOs in rats (Kawai et al., 2015). In a skilled task that requires pressing a lever in quick succession within a time gap, lesion had no effect on well-trained rats. There were only transient behavioral impairments immediately after the lesion.

Whereas execution is intact, MOs appears to be important for learning new movements, chaining motor sequence, and timing self-paced movements. Naïve rats with motor cortex lesion failed to master the lever-pressing task with temporal gap (Kawai et al., 2015). Moreover, for rats that learned the timing, subsequent lesion rendered them unable to adjust to learn a new timing. These findings echo an earlier demonstration that animals with dorsomedial frontal lesions had difficulty figuring out motor sequence and could not reverse learned sequences (Ostlund, Winterbauer, & Balleine, 2009). The defects in motor learning and sequencing may be explained by a role for MOs in controlling the stochastic variability of action timing (Makino et al., 2017; Murakami, Shteingart, Loewenstein, & Mainen, 2017). Variability should be controlled for learning, because animals need to initially stray from the previously learned timing, and then subsequently hone in on the new timing. By contrast, mismanaging variability is less harmful to execution, as animals can still perform the action on average.

Results of causal perturbations emphasized that MOs can orchestrate the activity of other cortical regions (Allen et al., 2017; Kondo & Matsuzaki, 2020; Makino et al., 2017). In one such study, Makino and colleagues studied mice as they learned a lever pressing task (Makino et al., 2017). Using a suite of modern imaging methods and sophisticated analytical methods, they showed that learning was associated with a temporal compression of sequential activation of cortical regions. Granger analysis highlighted causality from MOs; there was an activity stream emanating from MOs that predicted the activity of other regions in expert mice. Importantly, muscimol inactivation of MOs disrupted the learned, compressed temporal sequence of activity. To summarize, multiple lines of evidence indicate that MOs exerts top-down modulation on MOp and other cortical regions for the learning of new movements and for the generation of timed or sequential actions.

4.2 Neural representation of actions and postures

Action—whether the animal made a specific movement—is robustly encoded by neurons in MOs, although the signal is not unique to MOs and widely represented in most of the brain (Allen et al., 2017; Musall et al., 2019; Steinmetz, Zatka-Haas, Carandini, & Harris, 2019). Based on an elegant set of analyses, Musall and colleagues captured a wide array of movements, and dissociated influence of instructed movements, which is licking for their task, versus uninstructed movements, which include whisking, facial movements, pupil diameter, hindpaws, and other motions. They found, surprisingly, that uninstructed movements correlate strongly with activity in many cortical regions. In fact, incorporating uninstructed movements into their model was more effective than using task variables in terms of capturing the variance of single-cell activity in MOs. The cortex-wide representation of action could be observed at the level of single neurons. Using a novel method for cellular-resolution imaging over large area of the dorsal cortex, Kauvar and colleagues identified representation of actions in numerous regions in a three-option lick-to-target odor task (Kauvar et al., 2020). Despite the widespread representation, it is worth noting that the action- and choice-related signals in MOs are among the earliest to emerge relative to other brain regions (Musall et al., 2019; Sul, Jo, Lee, & Jung, 2011).

Not all movements are the same, and there is evidence that MOs may be particularly sensitive to different types of postures and naturalistic behaviors. In a study that tracked freely foraging rats in three dimensions, neurons in MOs and PTLp exhibited tuning curves for postures of the animal (Mimica, Dunn, Tombaz, Bojja, & Whitlock, 2018). Spikes in MOs neurons were related to the pitch, azimuth, and roll of the head, as well as neck, back, or combinations of these features. The most prominently encoded feature in medial MOs was head-related movements, consistent with the earlier discussion on the role of feedback for predictive coding in sensory cortex (Guitchounts et al., 2020). The representation of a broad array of postural information may contribute to the distinct neural activity patterns observed in MOs during naturalistic behaviors such as grasping, eating, grooming, rearing, and turning (Tombaz et al., 2020).

4.3 Circuit with primary motor cortex

In a series of studies, Kawaguchi and colleagues mapped the recurrent connections between MOs and MOp (Kawaguchi, 2017; Ueta, Hirai, Otsuka, & Kawaguchi, 2013; Ueta, Otsuka, Morishima, Ushimaru, & Kawaguchi, 2014). In MOs, the MOp-projecting pyramidal cells are distributed mainly in lower layer 2/3 and upper layer 5, including commissural neurons and cortico-thalamic projection neurons, but not corticospinal neurons that lay in deep layer 5. These afferents to MOp preferentially terminate in upper layer 1 of MOp. There is some topography in the projections: lateral-medial axis in MOs aligns with the rostral-caudal axis of MOp (Ueta et al., 2014).

How do frontal cortical signals influence MOp? In rats, recording from neurons in the dorsomedial prefrontal cortex and MOp uncovered a coupling between the two regions during the delay period in a delayed lever-pressing task (Narayanan & Laubach, 2006). The inter-areal coordination is consistent with a top-down control on the neural ensembles in the motor cortex to inhibit inappropriate responding. In another study, recording from mice running different types of ladder wheels and carefully visualizing the kinematics of joints and muscles, Omlor and colleagues found that MOs is responsible for context-dependent modulation of neural representation of joint movements in MOp (Omlor et al., 2019). Together, these studies illustrate a modulatory role for MOs on MOp, which is to provide contextual information for adjusting movements.

4.4 Circuit with brainstem

Anterograde tracing of the axon emanating from MOs have shown that MOs and the neighboring vM1 project to various subcortical targets (Fig. 3), including nuclei in the brainstem (Grinevich, Brecht, & Osten, 2005; Kita & Kita, 2012). How different frontal cortical locations maps onto motor output and their relations to brainstem connections were explored recently in a rigorous study by Mercer Lindsay and colleagues (Mercer Lindsay et al., 2019). They systematically tested focal activation of MOs via optogenetics and measured motor outputs, including forelimb, jaw, nose, and vibrissae movements, as well as more ethologically relevant actions, such as bringing forelimb to mouth, using electromyography and high-speed videography. Tracing backwards from muscle-innervating motor neurons, they identified spinal trigeminal pars oralis (SpVO) and spinal trigeminal interpolaris rostralis (SpVIr) as trigeminal premotor nuclei in the medulla portion of the brainstem, which received axonal collaterals from MOs. The implication of the results is that neighboring neurons in MOs can send parallel pathways to pools of neurons in premotor brainstem nuclei to initiate complex movements.

5. Circuits for decision-making

5.1 Choice, outcomes, and choice history

In the previous section, we discussed a potential role for MOs in learning new movements. However, learning for animals in the laboratory is in essence a process of optimizing action plans based on the outcomes of prior actions. Namely, rewarded choices should be repeated, whereas mistakes should be avoided. Accordingly, numerous studies have shown that neurons

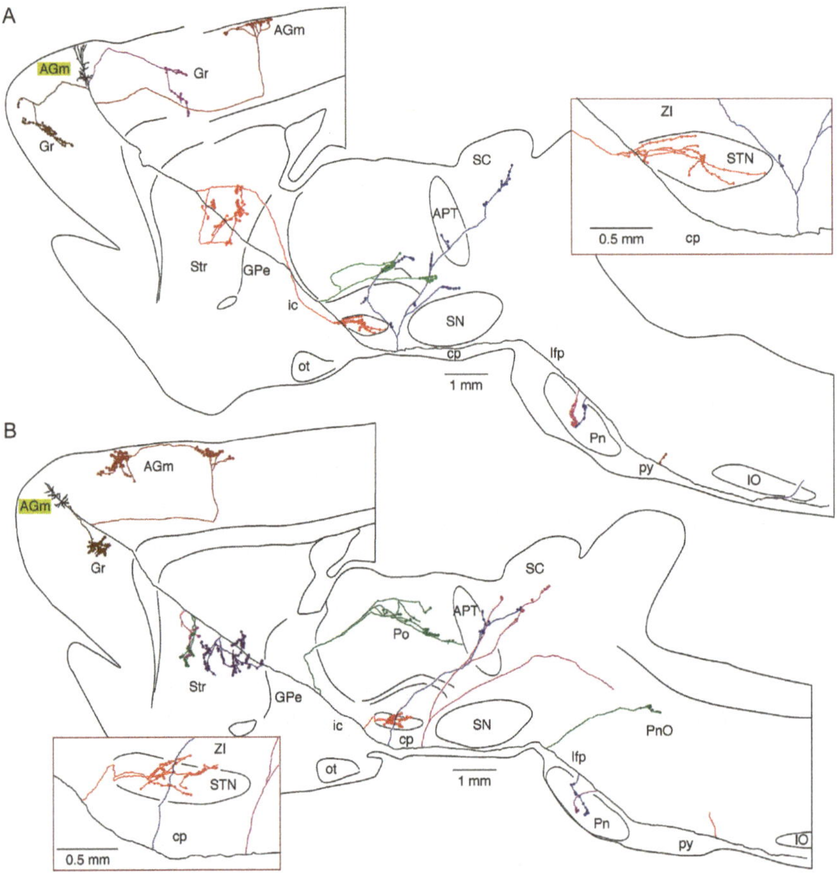

Fig. 3 Long-range subcortical projections from the secondary motor cortex. The axonal field of two pyramidal tract neurons in the rat medial agranular cortex (AGm). The axons emit multiple collaterals to various subcortical sites. Note that both neurons had collaterals innervating the AGm, granular cortex (Gr), and striatum (Str). Also note the differences, for example, in the trajectory of the thalamic collateral. GPe, globus pallidus external segment; ic, internal capsule; ZI, zona incerta; STN, subthalamic nucleus; cp, cerebral peduncle; ot, optic tract; Po, posterior thalamic nucleus; APT, anterior pretectal nucleus; SC, superior colliculus; SN, substantia nigra; lfp, longitudinal fasciculus of the pons; Pn, pontine nucleus; py, medullary pyramid; IO, inferior olive. Adapted from (Kita & Kita, 2012).

in MOs carry signals related to choices (e.g. left versus right in a two-choice task) (Erlich, Bialek, & Brody, 2011; Siniscalchi et al., 2016; Steinmetz et al., 2019; Sul et al., 2011). Moreover, the neural representation for choice in MOs arises early and thus can participate in the decision process ((Sul et al., 2011), although see (Chen, Li, Daie, & Svoboda, 2017)). The choice-related signals are then maintained until at least the next trial

(Jiang, Liu, Zhang, Xie, & Yao, 2019; Scott et al., 2017; Siniscalchi et al., 2016; Sul et al., 2011). These persistent neural signals for chosen actions are modulated by error (Narayanan, Cavanagh, Frank, & Laubach, 2013) and reward (Siniscalchi, Wang, & Kwan, 2019). The enhanced and more persistent representation of rewarded choice in MOs provides a plausible substrate for biasing the animal to repeat that choice in the near future, which could promote reward-guided learning (Siniscalchi et al., 2019).

When MOs is silenced prior to decision during a two-choice perceptual task, the effect is a bias towards the ipsilateral choice at the expense of the contralateral choice (Erlich, Brunton, Duan, Hanks, & Brody, 2015; Guo et al., 2014; Zatka-Haas et al., 2020). However, acute inactivation could have off-target effects. Whether permanent lesions have similar impact on choices and whether MOs participates in higher-order decision-making processes is less understood. Employing a multisensory task involving visual and/or auditory stimuli, Pisupati and colleagues used computational models to fit animal's behavior to make an astute observation that even on easy trials, animals make errors, which could be attributed to exploration (Pisupati, Chartarifsky-Lynn, Khanal, & Churchland, 2019). Unilateral inactivation of MOs alters the lapse rate on one side, suggesting a role in action value encoding. More specifically, the results indicate a selective devaluation of contralateral actions following the loss of MOs.

We do not know much about the neural computations that occur within MOs during decision-making. A challenge is that current decision-making tasks for rodents tend to be simplistic. Consequently, learning-related shaping of neural dynamics can be highly individualized, because an almost countless number of circuit motifs might be sufficient to solve simple tasks (Atilgan & Kwan, 2018). Indeed, empirical measurements of the neural ensemble dynamics in the medial frontal cortex showed diversity across individuals (Kurikawa, Haga, Handa, Harukuni, & Fukai, 2018). It is probable that MOs engages other brain regions for reward-guided learning. In a classical conditioning task where cues are associated probabilistically with rewards, Kondo and colleagues identified the dorsomedial frontal cortex as a hub with causal influence on other cortical regions (Kondo & Matsuzaki, 2020).

5.2 Circuit with retrosplenial cortex

The projection from RSP to MOs is dense and has been characterized in detail (Li, Yamawaki, Barrett, Kording, & Shepherd, 2018; Yamawaki, Radulovic, & Shepherd, 2016). To gain insight into the function of this long-range connection, Olson and colleagues made unit recordings in MOs of rats navigating a maze with turns and tracks (Olson, Li, Montgomery, & Nitz, 2020). They found neural activity related to turn direction at all of the turn locations in the maze. These choice-related signals were detected just before each turn and then persisted, suggesting contributions to action planning and execution. Moreover, contextual variables, such as progress and route, modulate the choice-related signals. Overall, these observations during navigation agree with earlier reports of early rise and context dependence of choice-related signals in two-choice tasks (Erlich et al., 2011; Siniscalchi et al., 2016; Sul et al., 2011). However, Olson and colleagues noted also that action encoding is still present when a decision is not needed (e.g. a corner to turn for the rat when that turn is the only option). This result appears to suggest a role for MOs in orientation, navigation, and generation of feedback for sensory cortex, rather than a decision-making role—although these functions are not mutually exclusive and can be supported by MOs in parallel. Based on comparing these results to prior studies of RSP in related tasks (e.g., (Alexander & Nitz, 2015)), it was suggested that navigation-related signals may originate from RSP and/or PTLp and propagate to MOs.

5.3 Circuit with orbitofrontal cortex

ORB is thought to be involved in the formation of stimulus-outcome associations for decision-making, although more recently also suggested to signal expected outcomes and maintain cognitive maps (Schoenbaum, Roesch, Stalnaker, & Takahashi, 2009). In a lever-pressing paradigm, chemogenetic attenuation of ORB projection to MOs biased the exploratory use of a novel lever (Schreiner & Gremel, 2018). The idea is that experience-dependent strengthening of the ORB-to-MOs pathway may promote the exploitation of learned associations, and therefore inactivation leads to exploration. Modification of the ORB projection in MOs over longer time scale can also influence decision behaviors. In a rule learning task, mice were trained to discriminate between odor options (Johnson, Peckler, Tai, & Wilbrecht, 2016). Longitudinally across learning, the ORB boutons in MOs undergo structural plasticity such that the gain of boutons was correlated to exploiting a learned rule, whereas the loss of bouton loss was related to exploratory behavior. Therefore, both in the short and long timescales, the pathway from ORB to MOs may be important for regulating the balance between exploration and exploitation during decision-making.

5.4 Circuit with basal ganglia

The basal ganglia circuitry is implicated in reward-guided decision-making, particularly as a substrate for action value encoding and updating (Burton, Nakamura, & Roesch, 2015). MOs has multiple points of entry into the basal ganglia. First, medial MOs projects to dorsomedial striatum (Hintiryan et al., 2016). Ipsilaterally, MOs neurons project to both the patch and matrix compartments of the striatum, with a slight preference for matrix (Smith et al., 2016). Second, there are cortico-pallidal innervations (Abecassis et al., 2020; Karube, Takahashi, Kobayashi, & Fujiyama, 2019). Cortical axons from MOs are found exclusively in the ipsilateral globus pallidus (GP). Intriguingly, these cortical axons preferentially innervate GP neurons that project to striatum, more so than GP neurons that project to the subthalamic nucleus (STN) (Karube et al., 2019).

The third point of entry is a direct projection to STN, which originates from layer 5 corticospinal neurons in MOs (Kita & Kita, 2012). In a new study, Li and colleagues reported that STN-projecting neurons in the dorsomedial frontal cortex are largely distinct from other subpopulations that project to VISp and lateral hypothalamic area (LHA) (Li, Nguyen, Ma, & Dan, 2020). In a go/no-go task, the STN-projecting population has elevated activity when animal inhibited responding correctly. Furthermore, optogenetic stimulation of the MOs-to-STN projection reduced false alarms. The results indicate that the MOs-to-STN projection provides a stopping signal to halt inappropriate response for impulse control, versus the MOs-LHA pathway which acts in the opposite direction to promote responding. This conclusion aligns with the observation that stimulation of MOs terminals in STN led to premature stopping in a virtual stop-and-go task (Adam, Johns, & Sur, 2020).

It should be noted that MOs sends some projections to substantia nigra pars compacta (SNc), which is a site with dopaminergic neurons (Watabe-Uchida, Zhu, Ogawa, Vamanrao, & Uchida, 2012). The connection to SNc does not mean that MOs targets other neuromodulatory centers, because for example there is negligible projection to the ventral tegmental area (VTA) (Watabe-Uchida et al., 2012) or the cholinergic centers in the basal forebrain (Do et al., 2016). The projection to SNc is another potential mechanism for MOs to exert influence on choice behavior and additionally effort-related decision-making, through modulating the dopaminergic tone (Salamone et al., 2018).

6. Other projections

There are a few other prominent projections where the functions are largely unknown. Medial MOs sends extensive callosal axons to MOs in the opposite hemisphere. If MOs in one hemisphere is repeatedly perturbed, the contralateral MOs will adapt, a situation that was observed when mice needed to control eye movements (Sato et al., 2019). This result agrees with findings in the anterior lateral motor cortex (ALM), where interhemispheric connections facilitate recovery from transient perturbations, suggesting that regions in the two hemispheres may carry redundant information for robustness (Li, Daie, Svoboda, & Druckmann, 2016). Indeed, in a more physiological context, the halves of MOs were suggested to work mostly independently with only weak coupling (Scott et al., 2017).

Parietal cortex in rodents is implicated in decision-making and multisensory integration (Choi, Lee, & Lee, 2018), suggesting that recurrent connections between MOs and PTLp could carry both choice- and perceptual-related signals. The dual role might reflect subdivisions within PTLp, which are all connected with MOs: Medial PTLp connects with caudal and rostral MOs, whereas in between, the intermediate rostrocaudal portion of MOs connects with lateral and caudolateral subdivisions of PTLp (Olsen et al., 2019). In one report, the acute inactivation of PTLp-to-MOs neurons altered movement kinematics by reducing trial-to-trial reliability and decreasing peak velocity of a lever press action (Hwang et al., 2019).

There have been progress in mapping of the connection between MOs and various other brain regions including the ACAd (Fillinger, Yalcin, Barrot, & Veinante, 2017, 2018), thalamus (Hunnicutt et al., 2014), superior colliculus (Zhang et al., 2016), and perirhinal cortex (Ueta et al., 2013; Ueta et al., 2019). Other notable connections are between MOs and the ventral portion of the medial frontal cortex, including ILA and PL, as well as the claustrum (Zingg et al., 2014). The functional significance of these connections in perceptual, motor, and decision behaviors remains to be determined.

7. Relevance for mental disorders

7.1 Motor dysfunctions: Obsessive-compulsive disorder and Parkinson's disease

Alterations to corticostriatal circuits have long been implicated in diseases with motor dysfunctions. A few studies have begun to illuminate the role of MOs in mental disorders. Characterizing a *Sapap3* knockout mouse model for compulsive behaviors, Corbit and colleagues found hyperactive neurons in central striatum (Corbit, Manning, Gittis, & Ahmari, 2019). There was no difference in intrinsic excitability, yet hyperactivity arises because of increased excitatory drive to striatal projection neurons and fast-spiking interneurons. The culprit was a striking 6-fold elevation in the postsynaptic responses when evoking afferents from MOs, due in part to exuberant expression of postsynaptic glutamate receptors. This aberrant increase was selective to MOs and absent for inputs from lateral ORB, suggesting a pathological shift in the balance of MOs versus ORB inputs into the striatum in the mouse model.

In a different study, dopamine-depleted male mice were characterized as a model for Parkinson's disease (Magno et al., 2019). Optogenetic activation of glutamatergic neurons in MOs alleviated some of the motor disturbances, as evident by increased mobility in terms of rotations, distance traveled, and speed. There may even be a restoration of dopaminergic tone—presumably through the SNc pathway (Watabe-Uchida et al., 2012)—because the photostimulation protocol induced an increase of dopamine in basal ganglia.

Pre-pulse inhibition (PPI) is an assay of sensorimotor gating and is thought to a translational assay. Although PPI primarily involves thalamic circuits, top-down modulation from the frontal cortex can influence the measure. For example, fear conditioning with the prepulse can enhance subsequent PPI, especially if the prepulse is masked by a background noise. Inactivation of MOs in rats eliminated this attentional enhancement of PPI (Meng, Ding, Chen, & Li, 2020). Furthermore, in mice with a *Shank3* mutation that exhibited frontal cortical defects, animals had deficient PPI (Ali, Shao, et al., 2020). By manipulating GABAergic neurons in MOs to return synaptic excitability to normal levels, PPI performance was restored in mutants to be comparable to control animals. Taken all together, although MOs is traditionally not thought of as a disease-related region, these studies are demonstrating that it underpins motor and sensorimotor dysfunctions.

7.2 Reward processing dysfunctions: Stress-related disorders and depression

Based on the role of MOs in reward-guided decision-making, pathological disruption of the brain region is expected to compromise reward processing, which is a hallmark of stress-related disorders including depression. Indeed, brain-wide mapping of cortical activity alterations in the learned helplessness model for depression identified MOs as one of several cortical regions targeted by stress (Kim et al., 2016). In one recent study, two-photon calcium imaging was used to characterize longitudinally the impact of social defeat stress on the activity of neurons in MOs and ACAd (Barthas et al., 2020). Acute and chronic stress had significant impact on the spontaneous neural activity. Notably, stress-induced activity changes were distinct between mice that were susceptible and resilient to the stress. The results suggest that reorganization of the neural activity pattern in the medial frontal cortex may contribute to stress resilience.

In contrast to stress, antidepressants act to relieve the symptoms of depression. Subanesthetic ketamine exerts antidepressant effects with a rapid onset and relatively sustained time course. Brain-wide mapping showed that ketamine increases metabolic activity in numerous brain regions including all of the sub-regions of medial frontal cortex: ILA, PL, ACAd, and medial MOs (Duncan, Miyamoto, Leipzig, & Lieberman, 1999; Miyamoto, Mailman, Lieberman, & Duncan, 2001). Given the convergence of stress effects, ketamine action, and role in reward processing in the same region, the medial frontal cortex is a site for investigating the neural mechanisms underlying antidepressant actions. Our lab has characterized the actions of ketamine in MOs and ACAd to show that a single dose of subanesthetic ketamine increases the density of dendritic spines by promoting new spine formation (Phoumthipphavong, Barthas, Hassett, & Kwan, 2016) and elevates dendritic calcium signals that could be the precursor to long-term synaptic plasticity (Ali, Gerhard, et al., 2020).

8. Conclusion: Many circuits, many functions

We reviewed recent studies to highlight long-range connections with MOs that subserve perceptual, motor, and decision-related functions. We divided the functions into three broad categories, although undoubtedly decisions could involve sensory stimuli and would require a motor response, therefore the long-range interactions are expected to occur concurrently in a brain-wide fashion.

One possibility is that distinct subregions within medial MOs subserve the different functions. A challenge to test this possibility is the lack of consensus on the classification and borders of frontal cortical regions (Amarante, Caetano, & Laubach, 2017; Brecht, 2011; Ebbesen et al., 2018; Svoboda & Li, 2018). Instead of arguing about how to divide and name subregions, a concrete path to move forward is to exhaustively map the neural activity in the medial frontal cortex, with exact anatomical coordinates and over a large swath of area. Although several studies have recorded from multiple regions (e.g., (Siniscalchi et al., 2016; Sul et al., 2011)), arguably the only dense mapping of frontal cortical responses during behavior to date was performed by Chen and colleagues (Chen et al., 2017). By carefully aligning fields of view and densely sampling with two-photon calcium imaging, they measured stimulus, choice, and outcome-related activity from 15,431 neurons in a roughly 1 x 1 mm area spanning the anterior lateral and medial portions of MOs. In a whisker-based tactile task, there was early stimulus-related activity in medial MOs, whereas choice in terms of lick direction arose earliest in deep layers of ALM, and then subsequently spread to medial MOs. The result suggests that, at least for tactile stimuli from whiskers, the genesis of the choice signal may occur in ALM rather than medial MOs. Looking forward, dense recording of neural activity using Neuropixels followed by histological reconstruction of the probe location will be a promising approach (Steinmetz et al., 2019). In a landmark proof-of-principle study, Steinmetz and colleagues searched for the emergence of choice signals in the entire mouse brain and identified set of regions including the medial frontal cortex, as well as midbrain areas including midbrain reticular nucleus, superior colliculus, zona incerta, and parts of basal ganglia.

An alternative possibility is that the multiple functions may be supported by the same neurons in medial MOs. Frontal cortical neurons can exhibit activity with remarkable task dependence, as seen in the differential neural responses when animals have to perform the same tasks using different effectors, such as lever press versus nose-poke (Murakami, Vicente, Costa, & Mainen, 2014), or rotarod versus wheel running (Cao et al., 2015). A strong case for task dependence came from a new study by Pinto and colleagues, in which they tested the involvement of dorsal cortical sites including medial MOs in three tasks with different demands (Pinto et al., 2019). Inactivation of MOs led to significant decrease in performance in memory-guided or accumulating-towers task, but not in visual-guided task, and modeling suggested the region contributes to these tasks in different ways to meet the behavioral demands. The results imply that the function of MOs in behavior may be highly task dependent.

To summarize, we will start by reciting a classic critique by Allen Newell, who declared that 'You can't play 20 questions with nature and win" (Newell, 1973). The idea is that for simple tasks, one may construct a toy computational model to reproduce the observed behavior for each task. However, the models generalize poorly and would fail to solve other tasks, highlighting the difficulty in integrating simple models into a cohesive framework that can broadly predict and mimic behavior. Here, in reviewing the recent surge of studies of the rodent medial MOs, there is an appreciation for copious amount of new information on the long-range pathways. But given that brain regions are densely interconnected, axonal projections are diverse and heterogeneous, and cortical function can be task-dependent, we arrive at a similar conceptual crossroads as Newell in terms of consolidating these findings to find common grounds. Indeed, many of the long-range pathways may be involved simultaneously and be interacting during more complex behaviors. To overcome the conundrum, future studies that employ naturalistic behaviors or apply rigorous computational models that can dissect the animal's intent will be the most informative and are urgently needed towards uncovering the neural computations in MOs during motivated behaviors.

Acknowledgments

We thank Michael Siniscalchi for early discussions of presented ideas; Marina Picciotto and Jane Taylor for coming up with the term "Secondary Motivation Cortex" in a Friday joint-lab meeting. This work was supported by NIH/NIMH grants R01MH112750 (A.C.K.), R01MH121848 (A.C.K.), R21MH118596 (A.C.K.).

References

Abecassis, Z. A., Berceau, B. L., Win, P. H., Garcia, D., Xenias, H. S., Cui, Q., et al. (2020). Npas1(+)-Nkx2.1(+) neurons are an integral part of the cortico-pallido-cortical loop. *The Journal of Neuroscience*, *40*, 743–768.

Adam, E. M., Johns, T., & Sur, M. (2020). Cortico-subthalamic projections send brief stop signals to halt visually-guided locomotion. *bioRxiv*. https://doi.org/10.1101/2020.02.05.936443.

Alexander, A. S., & Nitz, D. A. (2015). Retrosplenial cortex maps the conjunction of internal and external spaces. *Nature Neuroscience*, *18*, 1143–1151.

Ali, F., Gerhard, D. M., Sweasy, K., Pothula, S., Pittenger, C., Duman, R. S., et al. (2020). Ketamine disinhibits dendrites and enhances calcium signals in prefrontal dendritic spines. *Nature Communications*, *11*, 72.

Ali, F., Shao, L. X., Gerhard, D. M., Sweasy, K., Pothula, S., Pittenger, C., et al. (2020). Inhibitory regulation of calcium transients in prefrontal dendritic spines is compromised by a nonsense Shank3 mutation. *Molecular Psychiatry*. https://doi.org/10.1038/s41380-020-0708-6.

Allen, W. E., Kauvar, I. V., Chen, M. Z., Richman, E. B., Yang, S. J., Chan, K., et al. (2017). Global representations of goal-directed behavior in distinct cell types of mouse neocortex. *Neuron*, *94*, 891–907. e896.

Amarante, L. M., Caetano, M. S., & Laubach, M. (2017). Medial frontal theta is entrained to rewarded actions. *The Journal of Neuroscience*, *37*, 10757–10769.

Anderson, C. T., Sheets, P. L., Kiritani, T., & Shepherd, G. M. (2010). Sublayer-specific microcircuits of corticospinal and corticostriatal neurons in motor cortex. *Nature Neuroscience*, *13*, 739–744.

Atilgan, H., & Kwan, A. C. (2018). Same lesson, varied choices by frontal cortex. *Nature Neuroscience*, *21*, 1648–1650.

Barthas, F., Hu, M. Y., Siniscalchi, M. J., Ali, F., Mineur, Y. S., Picciotto, M. R., et al. (2020). Cumulative effects of social stress on reward-guided actions and prefrontal cortical activity. *Biological Psychiatry, 88*, 541–553.

Barthas, F., & Kwan, A. C. (2017). Secondary motor cortex: Where 'sensory' meets 'motor' in the rodent frontal cortex. *Trends in Neurosciences, 40*, 181–193.

Billeh, Y. N., Rodriguez, A. V., Bellesi, M., Bernard, A., de Vivo, L., Funk, C. M., et al. (2016). Effects of chronic sleep restriction during early adolescence on the adult pattern of connectivity of mouse secondary motor cortex. *eNeuro, 3*. ENEURO.0053-0016.2016.

Brecht, M. (2011). Movement, confusion, and orienting in frontal cortices. *Neuron, 72*, 193–196.

Burton, A. C., Nakamura, K., & Roesch, M. R. (2015). From ventral-medial to dorsal-lateral striatum: Neural correlates of reward-guided decision-making. *Neurobiology of Learning and Memory, 117*, 51–59.

Cao, V. Y., Ye, Y., Mastwal, S., Ren, M., Coon, M., Liu, Q., et al. (2015). Motor learning consolidates arc-expressing neuronal ensembles in secondary motor cortex. *Neuron, 86*, 1385–1392.

Chen, T. W., Li, N., Daie, K., & Svoboda, K. (2017). A map of anticipatory activity in mouse motor cortex. *Neuron, 94*, 866–879. e864.

Choi, I., Lee, J. Y., & Lee, S. H. (2018). Bottom-up and top-down modulation of multisensory integration. *Current Opinion in Neurobiology, 52*, 115–122.

Corbit, V. L., Manning, E. E., Gittis, A. H., & Ahmari, S. E. (2019). Strengthened inputs from secondary motor cortex to striatum in a mouse model of compulsive behavior. *The Journal of Neuroscience, 39*, 2965–2975.

Do, J. P., Xu, M., Lee, S. H., Chang, W. C., Zhang, S., Chung, S., et al. (2016). Cell type-specific long-range connections of basal forebrain circuit. *eLife, 5*, e13214.

Duncan, G. E., Miyamoto, S., Leipzig, J. N., & Lieberman, J. A. (1999). Comparison of brain metabolic activity patterns induced by ketamine, MK-801 and amphetamine in rats: Support for NMDA receptor involvement in responses to subanesthetic dose of ketamine. *Brain Research, 843*, 171–183.

Ebbesen, C. L., Insanally, M. N., Kopec, C. D., Murakami, M., Saiki, A., & Erlich, J. C. (2018). More than just a "motor": Recent surprises from the frontal cortex. *The Journal of Neuroscience, 38*, 9402–9413.

Erlich, J. C., Bialek, M., & Brody, C. D. (2011). A cortical substrate for memory-guided orienting in the rat. *Neuron, 72*, 330–343.

Erlich, J. C., Brunton, B. W., Duan, C. A., Hanks, T. D., & Brody, C. D. (2015). Distinct effects of prefrontal and parietal cortex inactivations on an accumulation of evidence task in the rat. *eLife, 4*, e05457.

Ferguson, K. A., & Cardin, J. A. (2020). Mechanisms underlying gain modulation in the cortex. *Nature Reviews. Neuroscience, 21*, 80–92.

Fillinger, C., Yalcin, I., Barrot, M., & Veinante, P. (2017). Afferents to anterior cingulate areas 24a and 24b and midcingulate areas 24a' and 24b' in the mouse. *Brain Structure & Function, 222*, 1509–1532.

Fillinger, C., Yalcin, I., Barrot, M., & Veinante, P. (2018). Efferents of anterior cingulate areas 24a and 24b and midcingulate areas 24a' and 24b' in the mouse. *Brain Structure & Function, 223*, 1747–1778.

Froudarakis, E., Fahey, P. G., Reimer, J., Smirnakis, S. M., Tehovnik, E. J., & Tolias, A. S. (2019). The visual cortex in context. *Annual Review of Vision Science, 5*, 317–339.

Gabbott, P. L., Warner, T. A., Jays, P. R., Salway, P., & Busby, S. J. (2005). Prefrontal cortex in the rat: Projections to subcortical autonomic, motor, and limbic centers. *The Journal of Comparative Neurology, 492*, 145–177.

Gallero-Salas, Y., Han, S., Sych, Y., Voigt, F. F., Laurenczy, B., Gilad, A., et al. (2020). Sensory and behavioral components of neocortical signal flow in discrimination tasks with short-term memory. *Neuron*. https://doi.org/10.1016/j.neuron.2020.10.017.

Gilad, A., Gallero-Salas, Y., Groos, D., & Helmchen, F. (2018). Behavioral strategy determines frontal or posterior location of short-term memory in neocortex. *Neuron*, *99*, 814–828. e817.

Gomez-Marin, A., & Ghazanfar, A. A. (2019). The life of behavior. *Neuron*, *104*, 25–36.

Grinevich, V., Brecht, M., & Osten, P. (2005). Monosynaptic pathway from rat vibrissa motor cortex to facial motor neurons revealed by lentivirus-based axonal tracing. *The Journal of Neuroscience*, *25*, 8250–8258.

Guitchounts, G., Masis, J., Wolff, S. B. E., & Cox, D. (2020). Encoding of 3d head orienting movements in the primary visual cortex. *Neuron*, *108*, 512–525.

Guo, Z. V., Li, N., Huber, D., Ophir, E., Gutnisky, D., Ting, J. T., et al. (2014). Flow of cortical activity underlying a tactile decision in mice. *Neuron*, *81*, 179–194.

Handa, T., Takekawa, T., Harukuni, R., Isomura, Y., & Fukai, T. (2017). Medial frontal circuit dynamics represents probabilistic choices for unfamiliar sensory experience. *Cerebral Cortex*, *27*, 3818–3831.

Hintiryan, H., Foster, N. N., Bowman, I., Bay, M., Song, M. Y., Gou, L., et al. (2016). The mouse cortico-striatal projectome. *Nature Neuroscience*, *19*, 1100–1114.

Hunnicutt, B. J., Long, B. R., Kusefoglu, D., Gertz, K. J., Zhong, H., & Mao, T. (2014). A comprehensive thalamocortical projection map at the mesoscopic level. *Nature Neuroscience*, *17*, 1276–1285.

Hwang, E. J., Link, T. D., Hu, Y. Y., Lu, S., Wang, E. H., Lilascharoen, V., et al. (2019). Corticostriatal Flow of Action Selection Bias. *Neuron*, *104*, 1126–1140. e1126.

Insanally, M. N., Carcea, I., Field, R. E., Rodgers, C. C., DePasquale, B., Rajan, K., et al. (2019). Spike-timing-dependent ensemble encoding by non-classically responsive cortical neurons. *eLife*, *8*, e42409.

Itokazu, T., Hasegawa, M., Kimura, R., Osaki, H., Albrecht, U. R., Sohya, K., et al. (2018). Streamlined sensory motor communication through cortical reciprocal connectivity in a visually guided eye movement task. *Nature Communications*, *9*, 338.

Jiang, W., Liu, J., Zhang, D., Xie, T., & Yao, H. (2019). Short-term influence of recent trial history on perceptual choice changes with stimulus strength. *Neuroscience*, *409*, 1–15.

Johnson, C. M., Peckler, H., Tai, L. H., & Wilbrecht, L. (2016). Rule learning enhances structural plasticity of long-range axons in frontal cortex. *Nature Communications*, *7*, 10785.

Kamigaki, T., & Dan, Y. (2017). Delay activity of specific prefrontal interneuron subtypes modulates memory-guided behavior. *Nature Neuroscience*, *20*, 854–863.

Karube, F., Takahashi, S., Kobayashi, K., & Fujiyama, F. (2019). Motor cortex can directly drive the globus pallidus neurons in a projection neuron type-dependent manner in the rat. *eLife*, *8*, e49511.

Kauvar, I. V., Machado, T. A., Yuen, E., Kochalka, J., Choi, M., Allen, W. E., et al. (2020). Cortical observation by synchronous multifocal optical sampling reveals widespread population encoding of actions. *Neuron*, *107*, 351–367. e319.

Kawaguchi, Y. (2017). Pyramidal cell subtypes and their synaptic connections in layer 5 of rat frontal cortex. *Cerebral Cortex*, *27*, 5755–5771.

Kawai, R., Markman, T., Poddar, R., Ko, R., Fantana, A. L., Dhawale, A. K., et al. (2015). Motor cortex is required for learning but not for executing a motor skill. *Neuron*, *86*, 800–812.

Kim, Y., Perova, Z., Mirrione, M. M., Pradhan, K., Henn, F. A., Shea, S., et al. (2016). Whole-brain mapping of neuronal activity in the learned helplessness model of depression. *Frontiers in Neural Circuits*, *10*, 3.

Kiritani, T., Wickersham, I. R., Seung, H. S., & Shepherd, G. M. (2012). Hierarchical connectivity and connection-specific dynamics in the corticospinal-corticostriatal microcircuit in mouse motor cortex. *The Journal of Neuroscience*, *32*, 4992–5001.

Kita, T., & Kita, H. (2012). The subthalamic nucleus is one of multiple innervation sites for long-range corticofugal axons: A single-axon tracing study in the rat. *The Journal of Neuroscience*, *32*, 5990–5999.

Kondo, M., & Matsuzaki, M. (2020). Independent representations of reward-predicting cues and reward history in frontal cortical neurons. *bioRxiv*. https://doi.org/10.1101/2020.01.23.916460.

Kurikawa, T., Haga, T., Handa, T., Harukuni, R., & Fukai, T. (2018). Neuronal stability in medial frontal cortex sets individual variability in decision-making. *Nature Neuroscience*, *21*, 1764–1773.

Leinweber, M., Ward, D. R., Sobczak, J. M., Attinger, A., & Keller, G. B. (2017). A sensorimotor circuit in mouse cortex for visual flow predictions. *Neuron*, *95*, 1420–1432. e1425.

Li, N., Daie, K., Svoboda, K., & Druckmann, S. (2016). Robust neuronal dynamics in premotor cortex during motor planning. *Nature*, *532*, 459–464.

Li, B., Nguyen, T. P., Ma, C., & Dan, Y. (2020). Inhibition of impulsive action by projection-defined prefrontal pyramidal neurons. *Proceedings of the National Academy of Sciences of the United States of America*, *117*, 17278–17287.

Li, X., Yamawaki, N., Barrett, J. M., Kording, K. P., & Shepherd, G. M. G. (2018). Scaling of optogenetically evoked signaling in a higher-order corticocortical pathway in the anesthetized mouse. *Frontiers in Systems Neuroscience*, *12*, 16.

Lin, H. M., Kuang, J. X., Sun, P., Li, N., Lv, X., & Zhang, Y. H. (2018). Reconstruction of intratelencephalic neurons in the mouse secondary motor cortex reveals the diverse projection patterns of single neurons. *Frontiers in Neuroanatomy*, *12*, 86.

Luo, P., Li, A., Zheng, Y., Han, Y., Tian, J., Xu, Z., et al. (2019). Whole brain mapping of long-range direct input to glutamatergic and gabaergic neurons in motor cortex. *Frontiers in Neuroanatomy*, *13*, 44.

Magno, L. A. V., Tenza-Ferrer, H., Collodetti, M., Aguiar, M. F. G., Rodrigues, A. P. C., da Silva, R. S., et al. (2019). Optogenetic stimulation of the M2 cortex reverts motor dysfunction in a mouse model of parkinson's disease. *The Journal of Neuroscience*, *39*, 3234–3248.

Makino, H., Ren, C., Liu, H., Kim, A. N., Kondapaneni, N., Liu, X., et al. (2017). Transformation of cortex-wide emergent properties during motor learning. *Neuron*, *94*, 880–890. e888.

Meng, Q., Ding, Y., Chen, L., & Li, L. (2020). The medial agranular cortex mediates attentional enhancement of prepulse inhibition of the startle reflex. *Behavioural Brain Research*, *383*, 112511.

Mercer Lindsay, N., Knutsen, P. M., Lozada, A. F., Gibbs, D., Karten, H. J., & Kleinfeld, D. (2019). Orofacial movements involve parallel corticobulbar projections from motor cortex to trigeminal premotor nuclei. *Neuron*, *104*, 765–780. e763.

Mimica, B., Dunn, B. A., Tombaz, T., Bojja, V., & Whitlock, J. R. (2018). Efficient cortical coding of 3D posture in freely behaving rats. *Science*, *362*, 584–589.

Miyamoto, S., Mailman, R. B., Lieberman, J. A., & Duncan, G. E. (2001). Blunted brain metabolic response to ketamine in mice lacking D(1A) dopamine receptors. *Brain Research*, *894*, 167–180.

Mohajerani, M. H., Chan, A. W., Mohsenvand, M., LeDue, J., Liu, R., McVea, D. A., et al. (2013). Spontaneous cortical activity alternates between motifs defined by regional axonal projections. *Nature Neuroscience*, *16*, 1426–1435.

Murakami, M., Shteingart, H., Loewenstein, Y., & Mainen, Z. F. (2017). Distinct sources of deterministic and stochastic components of action timing decisions in rodent frontal cortex. *Neuron*, *94*, 908–919. e907.

Murakami, M., Vicente, M. I., Costa, G. M., & Mainen, Z. F. (2014). Neural antecedents of self-initiated actions in secondary motor cortex. *Nature Neuroscience*, *17*, 1574–1582.

Musall, S., Kaufman, M. T., Juavinett, A. L., Gluf, S., & Churchland, A. K. (2019). Single-trial neural dynamics are dominated by richly varied movements. *Nature Neuroscience*, *22*, 1677–1686.

Nakata, S., Takemoto, M., & Song, W. J. (2020). Differential cortical and subcortical projection targets of subfields in the core region of mouse auditory cortex. *Hearing Research*, *386*, 107876.

Narayanan, N. S., Cavanagh, J. F., Frank, M. J., & Laubach, M. (2013). Common medial frontal mechanisms of adaptive control in humans and rodents. *Nature Neuroscience*, *16*, 1888–1895.

Narayanan, N. S., & Laubach, M. (2006). Top-down control of motor cortex ensembles by dorsomedial prefrontal cortex. *Neuron*, *52*, 921–931.

Nelson, A., Schneider, D. M., Takatoh, J., Sakurai, K., Wang, F., & Mooney, R. (2013). A circuit for motor cortical modulation of auditory cortical activity. *The Journal of Neuroscience*, *33*, 14342–14353.

Newell, A. (1973). You can't play 20 questions with nature and win: Projective comments on the papers of this symposium. In W. G. Chase (Ed.), *Visual information processing*. New York: Academic Press.

Oh, S. W., Harris, J. A., Ng, L., Winslow, B., Cain, N., Mihalas, S., et al. (2014). A mesoscale connectome of the mouse brain. *Nature*, *508*, 207–214.

Olsen, G. M., Hovde, K., Kondo, H., Sakshaug, T., Somme, H. H., Whitlock, J. R., et al. (2019). Organization of posterior parietal-frontal connections in the rat. *Frontiers in Systems Neuroscience*, *13*, 38.

Olson, J. M., Li, J. K., Montgomery, S. E., & Nitz, D. A. (2020). Secondary motor cortex transforms spatial information into planned action during navigation. *Current Biology*, *30*, 1845–1854. e1844.

Omlor, W., Wahl, A. S., Sipila, P., Lutcke, H., Laurenczy, B., Chen, I. W., et al. (2019). Context-dependent limb movement encoding in neuronal populations of motor cortex. *Nature Communications*, *10*, 4812.

Ostlund, S. B., Winterbauer, N. E., & Balleine, B. W. (2009). Evidence of action sequence chunking in goal-directed instrumental conditioning and its dependence on the dorsomedial prefrontal cortex. *The Journal of Neuroscience*, *29*, 8280–8287.

Paxinos, G., & Franklin, K. B. J. (2004). *The mouse brain in stereotaxic coordinates*. Elsevier Academic Press.

Phoumthipphavong, V., Barthas, F., Hassett, S., & Kwan, A. C. (2016). Longitudinal effects of ketamine on dendritic architecture in vivo in the mouse medial frontal cortex. *eNeuro*, *3*. ENEURO.0133-0115.2016.

Pinto, L., Rajan, K., DePasquale, B., Thiberge, S. Y., Tank, D. W., & Brody, C. D. (2019). Task-dependent changes in the large-scale dynamics and necessity of cortical regions. *Neuron*, *104*, 810–824. e819.

Pisupati, S., Chartarifsky-Lynn, L., Khanal, A., & Churchland, A. K. (2019). Lapses in perceptual judgments reflect exploration. *bioRxiv*. https://doi.org/10.1101/613828.

Ren, S. Q., Li, Z., Lin, S., Bergami, M., & Shi, S. H. (2019). Precise long-range microcircuit-to-microcircuit communication connects the frontal and sensory cortices in the mammalian brain. *Neuron*, *104*, 385–401. e383.

Salamone, J. D., Correa, M., Ferrigno, S., Yang, J. H., Rotolo, R. A., & Presby, R. E. (2018). the psychopharmacology of effort-related decision making: Dopamine, adenosine, and insights into the neurochemistry of motivation. *Pharmacological Reviews*, *70*, 747–762.

Salkoff, D. B., Zagha, E., McCarthy, E., & McCormick, D. A. (2020). Movement and performance explain widespread cortical activity in a visual detection task. *Cerebral Cortex*, *30*, 421–437.

Sato, T. R., Itokazu, T., Osaki, H., Ohtake, M., Yamamoto, T., Sohya, K., et al. (2019). Interhemispherically dynamic representation of an eye movement-related activity in mouse frontal cortex. *eLife*, *8*, e50855.

Schneider, D. M., Nelson, A., & Mooney, R. (2014). A synaptic and circuit basis for corollary discharge in the auditory cortex. *Nature*, *513*, 189–194.

Schneider, D. M., Sundararajan, J., & Mooney, R. (2018). A cortical filter that learns to suppress the acoustic consequences of movement. *Nature*, *561*, 391–395.

Schoenbaum, G., Roesch, M. R., Stalnaker, T. A., & Takahashi, Y. K. (2009). A new perspective on the role of the orbitofrontal cortex in adaptive behaviour. *Nature Reviews. Neuroscience*, *10*, 885–892.

Schreiner, D. C., & Gremel, C. M. (2018). Orbital frontal cortex projections to secondary motor cortex mediate exploitation of learned rules. *Scientific Reports*, *8*, 10979.

Scott, B. B., Constantinople, C. M., Akrami, A., Hanks, T. D., Brody, C. D., & Tank, D. W. (2017). Fronto-parietal cortical circuits encode accumulated evidence with a diversity of timescales. *Neuron*, *95*, 385–398. e385.

Sesack, S. R., Deutch, A. Y., Roth, R. H., & Bunney, B. S. (1989). Topographical organization of the efferent projections of the medial prefrontal cortex in the rat: An anterograde tract-tracing study with Phaseolus vulgaris leucoagglutinin. *The Journal of Comparative Neurology*, *290*, 213–242.

Siniscalchi, M. J., Phoumthipphavong, V., Ali, F., Lozano, M., & Kwan, A. C. (2016). Fast and slow transitions in frontal ensemble activity during flexible sensorimotor behavior. *Nature Neuroscience*, *19*, 1234–1242.

Siniscalchi, M. J., Wang, H., & Kwan, A. C. (2019). Enhanced population coding for rewarded choices in the medial frontal cortex of the mouse. *Cerebral Cortex*, *29*, 4090–4106.

Smith, J. B., Klug, J. R., Ross, D. L., Howard, C. D., Hollon, N. G., Ko, V. I., et al. (2016). Genetic-based dissection unveils the inputs and outputs of striatal patch and matrix compartments. *Neuron*, *91*, 1069–1084.

Steinmetz, N. A., Zatka-Haas, P., Carandini, M., & Harris, K. D. (2019). Distributed coding of choice, action and engagement across the mouse brain. *Nature*, *576*, 266–273.

Sul, J. H., Jo, S., Lee, D., & Jung, M. W. (2011). Role of rodent secondary motor cortex in value-based action selection. *Nature Neuroscience*, *14*, 1202–1208.

Svoboda, K. (2019). Using rabies virus for tracing neural connections: Caveats and limitations exposed by studies of barrel cortex circuits. In *Spikesphotons (Wordpress.com)*.

Svoboda, K., & Li, N. (2018). Neural mechanisms of movement planning: Motor cortex and beyond. *Current Opinion in Neurobiology*, *49*, 33–41.

Tombaz, T., Dunn, B. A., Hovde, K., Cubero, R. J., Mimica, B., Mamidanna, P., et al. (2020). Action representation in the mouse parieto-frontal network. *Scientific Reports*, *10*, 5559.

Ueta, Y., Hirai, Y., Otsuka, T., & Kawaguchi, Y. (2013). Direction- and distance-dependent interareal connectivity of pyramidal cell subpopulations in the rat frontal cortex. *Frontiers in Neural Circuits*, 7, 164.

Ueta, Y., Otsuka, T., Morishima, M., Ushimaru, M., & Kawaguchi, Y. (2014). Multiple layer 5 pyramidal cell subtypes relay cortical feedback from secondary to primary motor areas in rats. *Cerebral Cortex*, *24*, 2362–2376.

Ueta, Y., Sohn, J., Agahari, F. A., Im, S., Hirai, Y., Miyata, M., et al. (2019). Ipsi- and contralateral corticocortical projection-dependent subcircuits in layer 2 of the rat frontal cortex. *Journal of Neurophysiology*, *122*, 1461–1472.

Wang, Q., Ding, S. L., Li, Y., Royall, J., Feng, D., Lesnar, P., et al. (2020). The allen mouse brain common coordinate framework: A 3D reference atlas. *Cell*, *181*, 936–953. e920.

Wang, T. Y., Liu, J., & Yao, H. (2020). Control of adaptive action selection by secondary motor cortex during flexible visual categorization. *eLife*, *9*, e54475.

Watabe-Uchida, M., Zhu, L., Ogawa, S. K., Vamanrao, A., & Uchida, N. (2012). Whole-brain mapping of direct inputs to midbrain dopamine neurons. *Neuron, 74*, 858–873.

Wickersham, I. R., Lyon, D. C., Barnard, R. J., Mori, T., Finke, S., Conzelmann, K. K., et al. (2007). Monosynaptic restriction of transsynaptic tracing from single, genetically targeted neurons. *Neuron, 53*, 639–647.

Winnubst, J., Bas, E., Ferreira, T. A., Wu, Z., Economo, M. N., Edson, P., et al. (2019). Reconstruction of 1,000 projection neurons reveals new cell types and organization of long-range connectivity in the mouse brain. *Cell, 179*, 268–281. e213.

Wise, S. P. (2008). Forward frontal fields: Phylogeny and fundamental function. *Trends in Neurosciences, 31*, 599–608.

Yamawaki, N., Radulovic, J., & Shepherd, G. M. (2016). A corticocortical circuit directly links retrosplenial cortex to M2 in the mouse. *The Journal of Neuroscience, 36*, 9365–9374.

Yamawaki, N., & Shepherd, G. M. (2015). Synaptic circuit organization of motor corticothalamic neurons. *The Journal of Neuroscience, 35*, 2293–2307.

Zatka-Haas, P., Steinmetz, N. A., Carandini, M., & Harris, K. D. (2020). A perceptual decision requires sensory but not action coding in mouse cortex. *bioRxiv*. https://doi.org/10.1101/501627.

Zhang, S., Xu, M., Chang, W. C., Ma, C., Hoang Do, J. P., Jeong, D., et al. (2016). Organization of long-range inputs and outputs of frontal cortex for top-down control. *Nature Neuroscience, 19*, 1733–1742.

Zhang, S., Xu, M., Kamigaki, T., Hoang Do, J. P., Chang, W. C., Jenvay, S., et al. (2014). Selective attention. Long-range and local circuits for top-down modulation of visual cortex processing. *Science, 345*, 660–665.

Zingg, B., Hintiryan, H., Gou, L., Song, M. Y., Bay, M., Bienkowski, M. S., et al. (2014). Neural networks of the mouse neocortex. *Cell, 156*, 1096–1111.

FSC
www.fsc.org
MIX
Papier | Fördert
gute Waldnutzung
FSC® C083411

Kolibri 360 GmbH
Ferdinand-Jühlke-Straße 7
99095 Erfurt, Deutschland
produktsicherheit@kolibri360.de